T0143365

5

Primates of Colombia

Thomas Richard Defler

Illustrated by:
Stephen D. Nash
César Landazábal Mendoza
Margarita Nieto Díaz

Editors:
José Vicente Rodríguez-Mahecha
Anthony Rylands
Russell A. Mittermeier

CONSERVATION
INTERNATIONAL

BOGOTÁ, D.C. - COLOMBIA

2004

Enquiries to the publisher should be directed to
the following address:

Thomas R. Defler
Instituto Imani
Universidad Nacional de Colombia
Leticia, Amazonas
thomasdefler@hotmail.com
caparu@tutopia.com

Editors of this number

José Vicente Rodríguez- Mahecha
Biodiversity Science Unit & Analysis
CBC de los Andes
Conservation International
jvrodriguez@conservation.org

Russell A. Mittermeier
ChairIUCN/SSC Primates Specialist Group
President of conservation International
r.mittermeier@conservation.org

Editor of the Series
Russell A. Mittermeier
Anthony Rylands
Conservation International
Tropical Field Guide Series
1919 M Street, NW, Suite 600
Washington, D.C. 20036 USA

Design:
Luis Felipe Sossa R.
Símbolo Ltda.

ISBN 1-881173-83-6
Printed in Colombia
By Panamericana Formas e Impresos S.A.

To my father and mother
for giving me life and love of nature

THOMAS

In memoriam

Jorge Ignacio Hernández-Camacho.

It was his gentleness and his respect for nature and for people which impressed so many people, but in my case, his genius together with his human qualities are what I shall miss most of all. I shall never stop missing him and his intellectual company, and I shall always express my gratitude to him for his friendship, his many stimulating conversations, and his support for my own efforts.

THOMAS R. DEFLER

Alouatta seniculus by IAN STEPHENS.

Primates of Colombia

Prologue

It is with great pride that I have the privilege of editing the work of my colleague and dear friend Thomas Richard Defler in producing this field guide. I am absolutely convinced that its publication will be a basic tool that will permit us to leave the scenario of generalized ignorance about one of the most spectacular groups of Colombia´s national biota, the primates. But this would not have been possible without a lifetime of dedication, effort and sacrifice of the author.

Honestly I don't remember when I first met Thomas, it seems like we have been friends all of my life. I have followed his steps with interest and with recognition for his battle-hardened decision to live in the Colombian forests and his impassioned capacity for concentration on the Colombian primates. His focus has led him to travel large sectors of the Colombian national territory, assuming the natural risks of ambling through primitive landscapes saturated with geographic and climatic complexities and, why not say it, also artificial landscapes which, since his arrival in Colombia more than 27 years ago, unfortunately began to propagate the evils of violence which have spread out over so much over Colombian territory and which resulted in the expulsion of the author from various parts of the country where he had put down roots and consolidated field projects.

I have known him in the forest in expeditions we organized together, and I have been witness to his work, his passions and caprices, frustrations and successes. I have been witness to his naivety and his good will towards his assistants and students, as well as his total lack of interest in anything material except for his books. The effort to adapt his book on the primate fauna of Colombia as a field guide constitutes recognition of his work and bravery as well as a profound gratitude for his dedication to the knowledge of the Colombian fauna and for his belief in this country and in its institutions, which have always opened him their doors.

The El Tuparro National Park (Vichada) was his first home after arriving in Colombia as a member of the American Peace Corps; from there he dedicated himself to an exploration of the Orinoco region, using all methods of transportation, attempting to know the secrets of nature as they related to fauna and especially to primates. These wonderful experiences he shared and analyzed with Jorge Hernández Camacho our teacher, colleague and friend. It was Jorge´s influence which forged Thomas´ personality as a primatologist and which oriented many of his research activities. Sharing together the enthusiasm of resolving some historic problems in the history of primatology led them to describe a neotype for *Cebus albifrons*, a species previously described as *Simia albifrons* by Alexander von Humboldt in 1812, from an imprecise locality, using a confusing diagnosis, and without the preservation of a holotype to record the species´ existence, according to the rules of the International Code of Zoological Nomenclature. They also decided to clarify the taxonomy of *Aotus hershkovitzi*, which had remained without an adequate description after the tragic death in the Armero disaster of the taxon´s discoverer Jairo Ramírez-Cerquera, before he was able to finish his own manuscript. In the scenario of Colombian primatology and conservation Thomas has contributed arduously with his more than 60 publications on diverse aspects of ecology, taxonomy and natural history of this group and others.

After many years in the Orinoco and having studied many aspects of the primates of this region, Thomas moved to the Amazon on the lower Apaporis river, a tributary of the Caquetá on the border with Brazil, where he founded the Caparú Biological Station along with the biologist-artist Sara Bennett. There he commenced to accumulate information on the resident primates of the zone and together with Sara on the climate, phenology and flora in an effort to understand the dynamics of our Amazonian forests and their capacity to maintain healthy populations of primates. Nevertheless, groups of insurgents obligated him to abruptly and brutally leave his beloved research station, and with tenacity and desire to survive he bravely liberated himself from his captors and with the forest´s help he was able to arrive in a safe harbor. These dark moments did not stop him and soon he returned to the Amazon to stablish the Ecological Station Omé, now as a professor of the Instituto Amazónico de Investigaciones of the National University of Colombia in Leticia.

This life of sacrifice has motivated us to promote all of this information compiled by Thomas to be made public, since we are absolutely certain that in a country where we have so much to know and appreciate, we should facilitate the spread of knowledge, since knowledge is adquired not only by seeing and observing, but also by understanding and appreciating. This field guide will be the best tool to manage this, providing the reader the opportunity to get first hand information which will help resolve questions which are listed here; all of this as a first step towards the necessary conservation of this beautiful group of animals.

José Vicente Rodríguez-Mahecha

Short Review of Author and Illustrators

Thomas Richard Defler, Ph.D.

Associate professor at the National University of Colombia and adjunct professor in the Archaeology Department of the University of Calgary (Canadá), Thomas Defler is a well-known primatologist who has worked for more than 27 years in the Orinoco and Amazonian regions of Colombia, orienting his studies towards the ecology and conservation of primate species in these two regions of the country. His Colombian primate research began in cooperation with various national agencies of conservation and faunal management, especially with the now disappeared INDERENA of the Ministry of Agriculture. During the last 19 years he has been profoundly interested and connected to the Amazonian region, with which he frankly confesses to be in love. There he has managed to establish two research stations, one on the Apaporis river, called "Caparú", which he founded in company with the biologist-artist Sara Bennett and who for many years accompanied Thomas in his research. Recently he divides his time between his university duties in Leticia and a new research station, founded five years ago and baptized with the name "Omé". His most recent focus has been on diversity and densities of communities of primates as well on some taxonomic and systematic problems of *Aotus* and *Cebus*.

Stephen Nash

A native of Great Britain, Stephen Nash has been the illustrator of a great part of the books, posters and other visuals used en diverse campaigns by Conservation International since he joined in 1989. He illustrated the first book in the field book series, "*The Lemurs of*

Madagascar". After finishing his academic career in the Department of Natural History Illustration of the Royal College of Art in London he initiated his professional career as an artist in the primate program of the World Wildlife Fund. After this he joined Conservation International and since that time he has worked in his studio at the State University of New York at Stony Brook where he is Associate Researcher in the Department of Anatomic Sciences and Adjunct Associate Professor in the Department of Art.

Cesar Landazábal

A graduate in plastic arts of the Superior Academy of Fine Arts in Bogotá, César Landazábal has been always related to nature and has developed innumerable works which have earned him prestige and recognition several times. He has participated in 15 artistic exhibitions, including the Museum of Contemporary Art in Bogotá (1977) and the Foundation Joan Miró of Spain (1976). His greatest creativity was in the old INDERENA, where he illustrated an important number of the publications of that public agency and which now constitutes the country's memory about natural resources. As special works he has designed and illustrated more than 60 stamps for the Colombian national postal service and was the illustrator of the educational series *Nuestra Fauna* and of the following books: *Aves de Colombia*, *Parque Nacional Isla de Salamanca*, *Literatura Oral Sikuani* and *Selva y Futuro*.

Margarita Nieto

Born in the Colombian capital, Bogotá, Margarita Nieto studied plastic arts in the School of Art in Bogotá. She is expert in water color and pen and ink drawings. She began her carrer as an illustrator working in the Public Education Department of INDERENA, where she illustrated various themes published by this agency. Later she began illustrating wildlife, beginning with the *Catalogue of Ornamental Fishes of Colombia*. As special projects she designed the stamp "Amazonia" for the postal department and other stamps for the United Nations, all with the theme of colombian fauna and flora. She has been illustrator of books and journals such as *Aves del Parque Nacional Natural Los Katíos*, *Colombia Fauna en Peligro*, *Nuevos Parques Nacionales*, *Aves del Bajo Magdalena*, *Desiertos y Zonas Áridas y Semiáridas de Colombia* and the Scientific journal *TRIANEA*. She has also participated in the production of albums such as *Nuestra Fauna* and other works that have contributed to the knowledge of our biodiversity, such as the *Calendario de Aves de la Guajira*. Presently she works as an independent consultant for various institutions.

Cebus apella using rock as hammer by PETE OXFORD.

Table of contents

Table of contents

1. Introduction and acknowledgements

Colombia is one of the most biologically diverse countries in the world: it is a "megadiversity country" due to its amazingly rich and varied flora and fauna (MITTERMEIER, 1988; MAST et al., 1997). Although its 1,141,748 km² represents only 0.8% of the earth's surface, Colombia harbours about 10% of our planet´s terrestrial plants and animals (McNEELY et al., 1990), underlining its importance in international efforts to conserve the world's natural resources for the future. This biotic richness is undoubtedly due to the equatorial location and topographic complexity of the country, although there may be other contributing factors. Although our knowledge of this diversity is limited, it is obvious that Colombia has more species of birds, crocodiles, tapirs, orchids and frogs than any other country in the world, and it is one of the top countries for snakes, lizards, fishes and angiosperm plants. The country is at least fourth in the world in number of mammals; 471 species have been confirmed (ALBERICO et al., 2000).

This book is about one group of animals, the Primates, which are particularly well represented in Colombia. In terms of the diversity of species, Colombia is also among the top primate countries in the world; superceded only by Brazil, and Peru in South America. This field guide illustrates and describes 28 species comprising 43 different taxa, 15 of these taxa being endemic to Colombia. The aim is to help Colombians and other people interested in primates to strengthen their interest with appropriate information and to provide an introduction for those who would like to learn something about these fascinating animals.

The initial chapters provide a general introduction to the history, zoogeography and conservation of Colombian primates, providing a context for the treatment of the individual species. With this information the reader will better be able to appreciate the importance of Colombia's contribution to our understanding of the fossil history of these animals, as well as to elucidate some of the details of present primate distributions and important aspects of their conservation.

Basically the book was written with primate conservation in mind, since one of the principle challenges of the century is to safeguard our planet's biodiversity, which includes all plants, animals, microorganisms, ecosystems, and the ecological processes in which they are involved. Originally the book was to be published only in Spanish and was particularly meant for Colombians and other Spanish speakers, who have very little literature about primates and biodiversity, handicapping many non-English readers who would like to add to their knowledge of the natural world. In its writing, however, it soon became evident that the information would be of interest to a wider audience. Now that the Spanish edition has been published the author wishes to offer the present edition in English, somewhat changed in format and slightly expanded.

The heart of the book is the section presenting individual species accounts. Here an effort has been made to present fieldwork and, when appropriate, some captive studies, in order to describe the natural history of the Colombian species. The references to field work are intended to be exhaustive, and, indeed, the listed references should provide a fair picture of the work done on each species. Hopefully the basic information and references will provide a strong basis for students of primates who wish to study these animals. It is inevitable that some references to field work will be inadvertently overlooked, but a great effort has been made to include especially all those carried out in Colombia and as many as possible from other neotropical countries.

It is of course not possible to summarize all of the research and available knowledge in each species account, though I make reference to much of this knowledge. Each account is accompanied by a detailed distribution map of that species in Colombia and by various illustrative materials, both drawings and photographs. The distribution maps compile information from at least seven museum collections of Colombian material as well as selected observations, mostly by the author. The geo-referenced points which were the basis of these maps will be listed in the near future in a separate publication. These are the most up-to-date distribution maps for the species that exist for Colombia, although I happily admit to a tremendous debt to Philip Hershkovitz and Jorge Hernández-Camacho. who had traveled this road before. Nevertheless, there are changes and updates to Hershkovitz' important works, as a result of data which was not available to him and due to my own fieldwork. A gazetteer of localities where the primates have been collected and observed by the author will also be included in a future publication, to avoid making this book too lengthy.

The excellent drawings and illustrations by Stephen D. Nash, César Landozabal and Margarita Nieto help to give an idea of some aspects of the natural history of these primate species. The photographs are intended as an aid in picturing the morphological characteristics of the animal. I have added a glossary, since it is inevitable that some terms used will be unfamiliar to readers.

The book is the culmination of my twenty seven years of primate work in Colombia, and as such I would like to thank the many individuals and organizations who have helped me over the years. In particular I want to thank the person to whom this book is dedicated, the late Professor Jorge Hernández-Camacho, who dedicated his life to the study of the Colombian flora and fauna and so generously gave his time to educate me in many aspects of Colombian primatology. Jorge´s eloquent capacity for recording the details of the natural history and systematics of Colombian fauna made him a very agreeable friend to spend time with, since a conversation with him was a profoundly enriching experience. Since the beginnings of my association with him, he provided the pillars which strengthened my knowledge of Colombian primates, and I shall always be grateful that I knew him and that he counted me among his friends. Thank you, Jorge. It was partly the influence of a seminal article published by him and Robert Cooper (1976) "The non-human primates of Colombia" that inspired me to write this book and to update their important synthesis, providing also the now-published Spanish version of this book.

Also, I particularly want to thank José Vicente Rodríguez-Mahecha of Conservation International for his dedication to the study of Colombian fauna, and his tireless efforts for its protection, as well as for his constant support of my own work, including as one of the editor's of this book. Although Vicente´s dedication to Colombian fauna has cost him personally very dearly, he continues to dedicate himself to what he loves, always exhibiting his personal faith that our efforts will not be in vain and that at the end, will be all worth the work and sacrifice. Other friends who have given strong and important personal and practical inspiration over the years are Professor Carlos Moreno and his wife Professor Amanda Rodríguez, as well as Sr. Pedro Nel Pinzón and his wife Sra. Consuelo Acosta; all of these friends I would like to thank with all of my heart.

Particular thanks also to Russell A. Mittermeier, who besides his own dedication to primate research and conservation, has so often supported my activities and unselfishly helped so many other primatologists and conservationists. My heartfelt thanks also go out to Anthony Rylands, who has constantly been available to my primatological needs, has been a source of much-needed information and has steadfastly edited my bad English. Many times Anthony has been my only source of much-desired technical conversations about the ins and outs of neotropical primates, and I should like to recognize his dedication to primates and their conservation and his kind support of my own efforts. I thank both Rylands and Pablo Stevenson for agreeing to critically read my manuscript and in that way improve it over the Spanish edition and along the way to correct the many errors which have cropped up in the first edition. Any errors remaining are of course my own responsibility.

Many others have also made important contributions to this book. Among those I would like to thank particularly: Juan Manuel Renjifo, Noel Rowe, Germán Andrade, Javier Ardila, Amanda Barrera de Jorgenson, Ernesto Barriga, Marta Lucia Bueno, Archie Carr III, John Cassidy, Alice Defler, Raymond Defler, Louise Emmons, Natalia Florez, Marcelle Gianellonni, Andrés González Hernández, Cristina Habibe, Kosei Izawa, Angela Maldonado, Rod Mast, Mike Melampy, Hernando Orozco, Bernardo Ortiz, Erwin Palacios, John Robinson, Adriana Rodríguez, Marc G. M. van Roosmalen, Juan Pablo Ruíz, Paul Salaman, Nancy Vargas, and Brent White. The late Philip Hershkovitz provided invaluable contributions to our understanding of the taxonomy and distributions of the Colombian primates and he was always available to any technical discussion that I might throw his way. He also gave me a wonderful tour of the architectural wonders of downtown Chicago!

Over the years Thomas Urbanek, Olga Forrero, Celestino Yukuna, Angel Yukuna, Marco Cruz and Estéban Meléndez have been especially helpful and important assistants to me in the field, and I am most grateful to them, recognizing the enormous value of willing, capable, and knowledgeable assistants during my work. Thanks also to Mario Salazar and Jorge Eduardo Pinzón who helped move my expeditions along, especially in Vichada, and helped keep me safe and comfortable in the air and on the ground. Thanks, too, to the artist-biologist Sara E. Bennett, who was a long-time companion during our 17 year stay at Caparú and who proved an able monkey-mother to the many orphans who came our way.

Besides the university Institute (Imani) which has been my academic home since 1997, several other organizations have provided key support. I would not have been able to work in Colombia without the active backing of INDERENA of the Ministry of Agriculture (now superceded by the Ministry of the Environment), which provided permission to do research over the years and many indispensable services. Later many employees of the Corporación Autónoma del sur de Amazonas became friends and extended many services as well as friends and services of the National Park Service. I could not begin to list all of the employees of those organizations who helped me in one way or another, and who became good friends through the years. For important collection data I should like to thank the following institutions: Unidad Investigativa Federico Medem (INDERENA, Bogotá), the Instituto de Ciencias Naturales (Bogotá), the United States National Museum (Smithsonian Institution, Washington, D.C.), the American Museum of Natural History (New York), the Field Museum of Natural History (Chicago), and the British Museum of Natural History (London).

I would like also to mention Conservation International (USA), which has helped to support my research and conservation work, as well as Conservación Internacional Colombia, which published the Spanish version of this book, COLCIENCIAS (Colombia), which supported the writing of this book and my research, and the Fundación Natura (Colombia), which sponsored my research and conservation activities for many years and provided a congenial group of like-minded Colombian conservationists in Bogotá. Additionally I would like to thank the following for financial support of various research projects: The Wildlife Conservation Society, The Center for Field Research, INDERENA, The National Geographic Society, the U.S. Peace Corps, The Woolly Monkey Foundation, WWF-U.S., the International Primate Protection League, my many Ohio friends, including the local Cleveland Sierra Club Chapter, the Cleveland Natural History Museum, and the Louisville Zoological Garden. I sincerely thank them all for their support.

2. Definitions, classification and fossil history

Colombia's Primate Biodiversity

We can list Colombia as one of the most diverse in the world in number of primates, along with Brazil (69 spp.), Zaire (32 spp.), Indonesia (35 spp.), Madagascar (32 spp.). and Perú (30 spp.) (MITTERMEIER & OATES, 1985; RYLANDS et al., 1995; MAST et al., 1993) although recent taxonomic considerations split taxa into even more species, a complete discussion of which is outside the scope of this book (RYLANDS et al., 2000; GROVES, 2001). It is not possible to state with complete security the number of species of Colombian primates, the total number lies between 26-32 or more. There are difficult taxonomic questions which have not been resolved and which produce a different count, according to the viewpoint of the person counting. These taxonomic problems will be discussed in this book as each species is described. There is a high probability that after additional field and karyotypic work, still more species will be identified for the country, and that future revisions of some taxa will change the currently confirmed number of species. Primatology in Colombia, as elsewhere, is a dynamic and changing science in all of its aspects, so that we can look forward to many revisions in the next few years.

Why study primates?

During the past twenty years the study of primates has become increasingly popular as a research topic, and field studies of these interesting animals have increased concomitantly. But why spend so much time and energy studying these mammals? What is it about this increasingly popular field of human endeavor that continues to attract biologists, anthropologists, psychologists, sociologists, psychiatrists, medical researchers, veterinarians, conservationists and other animal fans?

There are many characteristics of these animals which are attractive to humans and which cause us to want to know more about them. On the most accessible level to most of us, primates are appealing in their multi-faceted and complex behaviors. They make us laugh and smile. They remind us of ourselves either comically or in a serious vein or in a manner which is for some of us profoundly disturbing. This is mainly because primates are our closet relatives in the animal kingdom. Because we are also primates we often feel a close relationship and sympathy towards other primates. When we say that a primate "acts like a human", we sometimes realize more correctly and on a deeper level that the primate is acting like many primates do, including human beings. This is often a forceful reminder of our evolutionary links with the animal kingdom.

Since we realize that these animals are our closest animal relatives, many researchers study them with the object of learning something about human beings. Through other primates we clarify our understanding of evolution, social behaviors, learning behaviors and pre-cultural phenomena that may be common to many species besides humans. Medical researchers find in primates models of human illness, and they use primates in experimental ways that would be impossible using human beings as subjects, although there are increasingly vocal opponents of this type of research.

Ecologists study primates because they feel that these animals should be able to supply us with the ecological data that will enable us to better understand the workings of the very complex ecosystems where these creatures are naturally found. Most primates are at home in tropical forests and are visible enough to be able to teach us something about the mechanisms of such phenomena as predation, foraging, behavioral ecology, social evolution and demographic adaptations. Since there is now a large and growing data base on many species of primates, this allows us to compare and contrast species and different habitats, with the object of searching for some patterns which hold true in complex tropical ecosystems, since these ecosystems represent our world at its richest and most diverse.

Understanding tropical ecosystems allows conservationists to offer usable management models for the future. Primates provide likeable and visible flagship species that can very effectively interest others in the conservation of tropical ecosystems, since a monkey can serve as an attractive focus to many who would not be immediately interested in the mechanisms of the conservation aspects of that monkey. Seeing and learning about an interesting animal has often been the first step for an individual's developing interest in helping to guarantee that animal's future. After understanding the need to conserve one species of animal, it is only a short step to the realization of the need for conservation of representative samples of all of the ecosystems represented in our worldwide heritage.

Finally these animals are studied because they are admired for their own sake. All of the complex characteristics of primates, their intelligence, personalities, their general need for great tracts of forest, their beauty and their role as our closest relatives makes many people wish to protect and to preserve them and their habitats because of a fundamental belief that primates (and other animals) have an inherent right to exist in this world alongside of us. Primates may represent a conceptual door for some of us to an appreciation of all of that vast natural world that we had better learn very quickly to appreciate and to value, since our opportunity to insure the survival of nature into the future as part of humanity's rich, natural heritage is fast becoming limited.

What is a primate?

But what is a primate? How can we define this term? This group of warm-blooded or homoeothermic mammals is not easily defined, since the group shares a complex of generalized characters which are found in different mixes in other broad groups of mammals. Perhaps the most complete definition available, based on a consideration of living and extinct representatives of the order, has been offered by the respected North American mammalogist P. HERSHKOVITZ (1977) in his classic work on the Callitrichidae as follows (see glossary for terms):

"Hands and feet, with few living exceptions, pentadactyl, usually palmigrade but always plantigrade; hallux opposable with a short, blunt, degenerate claw, nail, or inunguiculate; manual digits capable of divergence, convergence, and flexions, scaphoid, lunate, triquetrum, centrale, and pisiform bones always present and discrete; tail primitively present and well developed, but secondarily reduced or absent in some species; penis bone present but secondarily lost in a few species; auditory bulla complete or nearly so, the main portions formed by extension of petrosal bone; entotympanic bone absent or rudimentary when present; perpendicular plate of ethmoid (ossified mesethmoid) present; supraorbital region broad with edges ridged or overhanging and more or less divergent; orbits large, postorbital process well-developed and in all recognized forms continuous with comparable developed orbital process of malar to form a complete ring; pterygoid process of

sphenoid bifurcate, the well-developed, often greatly enlarged lateral plate arising anteriorily at the sphenopalatine suture, the medial plate smaller than the lateral, sometimes nearly obsolete molars euthemorphic, occlusal surface bunodont to bilophodont".

For the specialist the above may be complete, but as a simple definition it may seem impenetrable. Nevertheless, it remains true that there is no one characteristic which defines a primate.

LE GROS CLARK (1959) attempted to define the Primate Order with a series of evolutionary trends (in NAPIER & NAPIER, 1967), which are critically commented upon by HERSHKOVITZ (1977) as well. Clark's trends are as follow:

1. Preservation of a generalized structure of the limbs with primitive pentadactyly and the retention of certain elements of the limb skeleton (such as the clavicle) which tend to be reduced or to disappear in some groups of mammals.

2. An enhancement of the free mobility of the digits especially of the thumb and big toe (which are used for grasping purposes).

3. The replacement of sharp, compressed claws by flattened nails associated with the development of highly sensitive tactile pads on the digits.

4. The progressive abbreviation of the snout or muzzle.

5. The elaboration and perfection of visual apparatus with development of varying degrees of binocular vision.

6. Reduction of the apparatus of smell.

7. The loss of certain elements of the primitive mammalian dentition and the preservation of a simple cusp pattern of the molar teeth.

8. Progressive elaboration of the brain affecting predominantly the cerebral cortex and its dependencies.

9. Progressive and increasingly efficient development of those gestational processes concerned with the nourishment of the fetus before birth.

NAPIER & NAPIER (1967) add to the above list:

10. Progressive development of truncal uprightness leading to facultative bipedalism.

11. Prolongation of post-natal life periods.

Although some of these "trends" might be called into question, such as, for example, the view of compressed claws as necessarily primitive as in *Saguinus* and *Callithrix,* now generally thought to be derived characteristics, these trends contribute to a general picture of what is a primate.

How primates are classified in the animal kingdom

It is generally accepted that the Order Primates may be divided into two separate Suborders: the Strepsirhini, comprising lemurs (lemurids, cheirogaleids, doubentonioids, lepilemurids, and indriids) and lorisoids (lorisids and galagids) and the Haplorhini, comprising the tarsioids, catarrhines (Old World monkeys and apes) and platyrrhines (New World monkeys). This division of the suborder is based on differences in the rhinarium or glandular skin surrounding the nostrils, the type of external ear, pedal digit, mammae, facial portion of the skull, the orbits, lacrimal bones, auditory bullae, symphysis of the mandible, dental formula, form of incisors and canines, sublingua and type of placenta (HERSHKOVITZ, 1977). Another scheme of division into Suborders places all of the Stresirhini as well as the tarsiers into the Prosimii, leaving the monkeys, apes and humans in the Suborder Anthropoidea, but this way of classifying them is now generally falling into disuse, due to our appreciation of the phylogenetic place of the tarsiers, near the root of modern primates.

The Strepsirhini are generally considered to exhibit the more primitive of the above characteristics, but the reader should understand that the use of the term "primitive" is used to mean characteristics closer to the common origin of the two Suborders. It ought to be underlined here that the terms "primitive" and "advanced" are often misunderstood in biology and evolution and should be understood only in a narrow comparative sense to the stock which may have given rise to the groups compared.

All Colombian primates are classified along with all neotropical primates in the Infraorder Platyrrhini (platyrrhines), referring to the placement of the nares or nostrils directed to each side, rather than forward as in the Catarrhini (catarrhines) of the Old World (Figure 1). Of course these two categories reflect a host of evolutionary differences shown in the fact that the two groups have evolved separately from at least the Oligocene of 35 million years ago, when the founding Platyrrhines were isolated in the then island

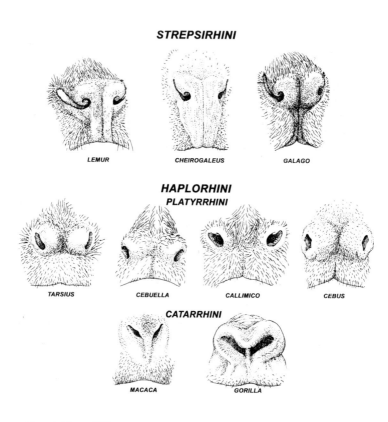

Figure 1. Noses of different species of primates showing the differences in the rhinarium or glandular skin which surrounds the nares in the two suborders. (Adapted from HERSHKOVITZ, 1977).

continent of South America and these differencies include dental formulae including three premolars (one of which has been lost in the Catyhrrines), the tympanic ring fused to the side of the auditory bulla and not extending laterally as a bony tube, the parietal and zygomatic bones joined and separating the frontal bone above from the sphenoid bone below and with cranial sutures which fuse rather late. The earliest known higher primates, the early Oligocene parapithecids from Africa are known to have some of these very characteristics reflected in the Platyrrhines and apparently lost in the Catarrhines (FLEAGLE, 1988, 1999). There are no known South American primate fossils before the above date.

Evidence is very strong that 35 million years ago the African and South American continents were much closer together and were in the process of moving apart to the positions which they occupy today (CIOCHON & CHIARELLI, 1980). The most evident fact is that this 35 million years or more of isolation has produced an extraordinary primate fauna that is unique to the New World and comprises about 98 species, according to a recent species list (RYLANDS *et al.* 1995) or according to RYLANDS *et al.* (2000) 110 or following VAN ROOSMALEN & MITTERMEIER (2003) on raising all subspecies of *Callicebus* to species level, 115 species. The upper and more recently published sum represents recent discoveries of new primate species and revisions which, based on more ample knowledge or a different philosophy of what comprises a species of these primates, have tended to split older taxa into several species. This trend is likely to continue as the systematics of these primates develops, and the uncertainty reflects the problem of stating exactly the number of species found in Colombia.

Neotropical Fossil Primates

The fossil record of the antecedents of Colombian and other neotropical primates is very incomplete and inconclusive for many decisions to be made with respect to origins or phylogenetic relationships, but some early fossils should be noted in the continuing effort to understand the evolutionary history of these mammals. ROSE & FLEAGLE (1981) and FLEAGLE (1988, 1999) provide an interesting review of the little more than a dozen neotropical primate fossils found up to that date, although new discoveries have continued to be made since these publications so that more than twenty fossil primate species can be listed today. MacFadden (1990) chronicles the known Cenozoic primate localities in South America.

The primate radiation which gave rise to all living forms of "higher primates" (the Anthropoidea) is but one of three primate radiations, including as well the first radiation of the plesiadapiform primates of North America and Europe (though many do not accept them as primates) and the second radiation of the prosimians, including the lemurs and the tarsiers.

The living and extinct members of the higher primates have in each mandibular quadrant two incisors, canines, at least two premolars and molar teeth present with a distinctive tooth morphology. These primates (unlike the two earlier radiations) have the two halves of the lower jaw fused in a bony symphysis and their eye socket is enclosed in a bony plate. Higher primates tend to have complex brains and reduced snouts and frontally directed orbits. The New World primates can be distinguished from the Old World primates by the presence of three premolars and an auditory bulla where the tympanic rings forms the external boundary, not extended to form the auditory meatus (ROSE & FLEAGLE, 1981).

The direct evidence of Ceboid (neotropical primate) evolution is scant and evolutionary affinities of this group with extinct North American primates or some Afro-Asian ancestor has been until lately scarce and equivocal, since fossil evidence in South America begins abruptly in the Oligocene with no bridging material to earlier animals. Some support a connection of the neotropical primates with as yet unknown anthropoids of the Eocene and Oligocene of North America, Europe and Asia (GINGERICH, 1980), although this requires a dispersal route for the present-day primate ancestors from the North American continent to South America against the probable ocean currents of the time. Since there is no fossil evidence of advanced primates in North America, this hypothesis requires that the Platyrrhines would have evolved in South America. Nevertheless, neotropical prosimians have not been discovered in South America. The most ancient primate fossils so far known in South America are from the Oligocene, and they already had many characteristics in common with modern-day neotropical primates (FLEAGLE, 1988, 1999).

The other major hypothesis is that the neotropical primates had their origin on the continent of Africa (HOFFSTETTER, 1972, 1974, 1980; HOFFSTETTER et al., 1981; LAVOCAT, 1969). Increasing support for this hypothesis comes from Oligocene primate fossils of the El Fayum fauna of Egypt, which have several characteristics in common with neotropical primates from the same age. For example, the parapithecid primates of the El Fayum formation of the early Oligocene have a dental formula that is the same as the Platyrrhines (2/2, 1/1, 3/3, 3/3) and they possess canine characteristics (in *Apidium*), limb structure, position of cranial bones, and morphology of otic bones that are quite similar to the neotropical primates (FLEAGLE, 1988, 1999) (see CIOCHON & CHIARELI, 1980 for several viewpoints).

Besides the above strong evidence, the South American caviomorph rodents also have some characteristics in common with African animals of the Oligocene and Miocene. This strengthens the proposal of an African origin for at least a part of the South American mammal fauna (FLEAGLE, 1988, 1999) This hypothesis, however, also has the problem of the dispersal route, which would have had to be over the Atlantic Ocean.

Despite such a daunting route, new data on the tectonics of the continental plates suggests that the South American continent and the African continent were closer together and that paleo-currents probably moved from Africa towards South America, just as they probably moved from South America towards North America. Additionally, apparently in the south Atlantic the water was not very deep, and there were probably numerous volcanic islands, which would have supported the elements of fauna, which passed from one continent to the other. It seems likely that the parapithecid primates of the El Fayum fauna probably gave rise to the catarrhine primates as well as to platyrhine primates, giving these two modern primate groups a common Oligocene origin (FLEAGLE, 1988, 1999).

The Fossil Evidence

Until recently only a few neotropical fossils had been known, and of these the phylogenetic position is clear for only the youngest. Before discussing each fossil it is worthwhile underlining two interesting and important points about them. One is the importance of Colombia and more specifically the importance of the Miocene La Venta formation in the upper Magdalena River valley just north of Neiva. La Venta has produced a great diversity of neotropical primates (GEBO *et al.*, 1990; FLEAGLE *et al.*, 1997) that is equivalent to the fauna of any modern lowland tropical rainforest. However, possible ancestors of the widespread *Cebus, Callicebus* and *Lagothrix* have not yet been discovered there.

The second point is the evidence for a recent primate fauna from the Caribbean of at least two and possible six or more species (FORD, 1990a). This fauna was probably driven extinct by the first human settlers of these Caribbean Islands (FORD, 1986, 1990a; FORD & MORGAN, 1986). Island extinctions show patterns which are well-known for many islands such as Madagascar (MITTERMEIER *et al.*, 1994), New Zealand and Hawaii (DIAMOND, 1992), all of which possessed extraordinary faunas which disappeared to one degree or another when the first human beings arrived, apparently due to human activities such as hunting and habitat destruction and introduction of exotic species. We are just becoming aware that this common, human-driven extinction spasm also occurred on the Caribbean islands as well. Understanding the details of these extinctions will not only increase our appreciation of the biodiversity that evolved in these places, but it will also allow us to appreciate more fully the destructive influence of human beings upon this diversity. It is worthwhile emphasizing that much of this human-driven extinction happened even before the technological era of today.

Our knowledge of Colombian fossils and of other neotropical primates is very incomplete and inconclusive and does not permit many decisions about phylogenetic relationships. Nevertheless, this account would not be complete without a brief description of some of the early fossils to lay the basis for future efforts at understanding the evolutionary history of these animals. All known neotropical primate fossils come from only six groups representing the last 26 million years. The groups are (1) late Oligocene-early Miocene Salla beds in Bolivia; (2) early Miocene of the Chilean Andes; (3) early to middle Miocene rocks in Argentina; (4) middle Miocene Honda group rocks in Colombia; (5) late Miocene rocks from Río Acre, Brazil; (6) Pleistocene-Recent cave deposits in Brazil and Caribbean (FLEAGLE *et al.*, 1997).

In the last few years several interesting new fossils have been found in the La Venta formation of Colombia, totaling perhaps half of the fossils that are known.

Late Oligocene/early Miocene of Bolivia (25-26 million years) (MACFADDEN, 1990):

(1) The oldest neotropical fossil known, **Branisella boliviana** (HOFFSTETTER, 1968, 1969) is from the Lower Oligocene of Bolivia and consists of a maxillary fragment preserving P4-M2, the roots of P3 and P2 and the edge of the alveolus that indicates the presence of M3. The teeth most closely resemble *Saimiri*, although HERSHKOVITZ (1977) disagreed, sustaining that *Branisella* was not ancestral to the platyrrhines. However, new material found since

HERSHKOVITZ (1977) published his above opinion argue that this primate is a true platyrrhine; the new material contains a third molar, considered diagnostic for this group of monkeys (WOLF, 1984; HOFFSTETTER, 1969, 1974; ROSENBERGER, 1981, 1984; TAKAI & ANAYA, 1994). MASANURU & ANAYA (1996) recently suggested that *Branisella* has a close affinity with the callitrichines due to the shared-derived character of an upper molar with a small hypocone rather than a moderate to large hypocone with well developed talonid, which would connect it to the ancestral upper molar morphotype for platyrrhines. The living animal probably weighed around 1 kg. and may have lived in a non-rainforest habitat, suggesting the possibility that the use of rainforest habitat by platyrhines may be a secondary adaptation (MACFADDEN, 1990).

(2) ***Szalatavus attricuspis*** (ROSENBERGER, HARTWIG & WOLFF, 1991c) was probably about 1000 g, but any connection with a particular modern group is difficult to say with this material. However, new material collected of *Branisella boliviana* suggest that *B. boliviana* is actually a senior synonym for *Szalatavus attricuspis* (MASANURU & ANAYA, 1996).

Early miocene of Chilean Andes

(3) ***Chilecebus carrascoensis*** (FLYNN, WYSS, CHARRIER & SWISHER, 1995) was a complete platyrrhine skull found in volcaniclastic deposits of the early Miocene (argon dated at 20.9 +/- 0.27 Myr) and representing the first extra-Patagonian fossils of this age. This skull's characteristics and its complete state of preservation seem to unambiguously connect the origin of platyrrhine primates to Africa making this one of the most important platyrrhine fossils ever found. The intact skull has a dental formula of 2/2, 1/1, 3/3, 3/3 the size of the adult animal is estimated around 1000-1200 g and the skull is unusual for its low vaulting cranium, small size and small canines and its various characteristics show discordant derived resemblances to a variety of fossil and living platyrrhine monkeys in general as well as poor resolution of relationships.

Miocene of Argentina (16 - 23 million years):

(4) ***Dolichocebus gaimensis*** (a nearly complete skull with some teeth and a talus) is from the Upper Oligocene of Argentina (KRAGLIEVICH, 1951) and is often said to be related to *Homunculus*. ROSENBERGER (1979, 1982)

suggests that this genus is ancestral to *Saimiri*, because interorbital foramen link the two orbits and molar morphology is similar to *Aotus* or *Callicebus*. HERSHKOVITZ (1982a) vigorously disagreed with this assessment and preferred to place this species in a phylogenetic position that is uncertain at this moment. Other primitive anthropoid characteristics (such as wide molars and a paraconule) are shared with the Oligocene parapithecids of Egypt. This species may have weighed as much as 3 kg while alive and was probably diurnal (KAY & CHRISTOPHER, 2000). Talar morphology of this genus is similar to *Saimiri, Callicebus, Cebus* and *Aotus* (GEBO & SIMONS, 1987). New fossil teeth ascribed to this species and to *Tremacebus harringtoni* from late Oligocene (Colhuehuapian) localities of Patagonian Argentina are more primitive than any previously known platyrrhine teeth and conform well with hypothetical ancestral morphotype for New World monkeys and are very similar to teeth of Oligocene catarrhines from Egypt (FLEAGLE & BROWN, 1983).

(5) A later fossil skull and lower mandible with broken teeth **Tremacebus harringtoni** (HERSHKOVITZ, 1974) from the Upper Oligocene of Argentina and originaly described by RUSCONI, (1933, 1935) may be ancestral to the callitrichids, according to HERSHKOVITZ (1974) who placed it in this family. ROSENBERGER (1984) and FLEAGLE (1988) suggest this species was ancestral to *Callicebus* or to *Aotus*. The fossil additionally has similarities to *Branisella*, sharing rectangular molars, low bulbous cusps, a large lingually placed hypocone and a slight internal cingulum with a paracone on M1. The orbitals of this animal were much larger than other platyrrhines, suggesting a possible nocturnal or crepuscular adaptation. The animal probably weighed 1-2 kg.

(6) **Soriacebus ameghinorum** (FLEAGLE, 1990) and the closely related **Soriacebus adrianae** (FLEAGLE *et al.*, 1987; FLEAGLE, 1990) of the Lower Miocene from Argentina also has a dental formula of 2/2, 1/1, 3/3, 3/3 and was the size of *Pithecia* (2 kg). The molars were similar to *Saguinus* with large procumbent incisors and large premolars similar to *Cebupithecia* and probably adaptations to hard-shelled fruits and seed eating (KINZEY, 1992). It is difficult to assign this genus to any particular family (FLEAGLE *et al.*, 1987) and may represent an early offshoot of living platyrrhines (KAY, 1990), although ROSENBERGER *et al.* (1990) recently assigned it to the pithecines. This and the species following date from 17.5-16.5 Million years or Middle-Miocene (MACFADDEN, 1990).

(7) **Carlocebus carmensis** (dental formula 2.1.3.3) and **Carlocebus intermedius** were described by FLEAGLE (1990) from the early-middle Miocene Pinturas formation of Argentina. The fossils consisted of various parts of mandibles. At first FLEAGLE *et al.* (1987) referred this material to *Homunculus* sp. for the similarities to that species. Presently FLEAGLE (1990) believes that the genus shows greatest similarities to *Callicebus*.

(8) **Homunculus** of several species have been recovered from early Miocene sediments in Argentina (AMEGHINO, 1891) of about 16 million years BP. **Homunculus patagonicus** probably weighed in at about 3 kg. or the size of a *Cebus*. The dental formula is 2/2,1/1,3/3,3/4. It has been related at various times to *Aotus* or to *Callicebus* (BLUNTSCHLI, 1931; SCOTT, 1937), although STIRTON (1951) pointed out molar similarities to *Alouatta*. CIOCHON & CORRUCCINI (1975) indicated that some limb elements were similar to *Callicebus* and that this primate was probably saltorial in locomotion. HERSHKOVITZ (1970) preferred to place *Homunculus* into its own family Homunculidae. The muzzle was short and the orbits moderate in size as in diurnal primates (AMEGHINO, 1891). It was probably a frugivore-folivore.

Colombian miocene primates La Venta (11.8 – 13.5 million years)

(9) **Neosaimiri fieldsi** (=**Saimiri fieldsi**) (STIRTON, 1951) from the Upper Miocene of the La Venta Colombian fauna is generally agreed to be closely related to *Saimiri*. This fossil lower mandible presents perhaps the least controversy of any of the others (ROSENBERGER *et al.* 1991a). A recent morphometric study of the skull concluded that *Neosaimiri* is probably synonymous with *Saimiri* (SZALAY & DELSON, 1979; HARTWIG *et al.*, 1990; ROSENBERGER, HARTWIG *et al.* 1991b). TAKAI (1994) reported on the collection of more than 200 new specimens of teeth of the species, allowing a close match to *Saimiri* but pointing out the differences in the proportions of tooth series with smaller incisors and larger molars, the structure of the P-4 hypocone and the strong polymorphic morphology of the M-1-2, arguing for a separate genus (*Neosaimiri*) from *Saimiri*. The discovery of postcranial specimens suggests that the species was more heavily built than *Saimiri* but that it was an arboreal quadruped (NAKATSUKASA *et al.*, 1997; TAKAI, 1994).

(10) **Laventiana annectens** (ROSENBERGER, SETOGUCHI & HARTWIG 1991) group this fossil with *Neosaimiri* and *Saimiri* and intermediate between *Saimiri* and the callitrichines and TAKAI (1994) agrees with this. Nevertheless, KAY

& MELDRUM (1997) believe that its lower dentition and other molar differences make it distinct enough to be in a separate genus, though generally the species currently is thought to be a lineal ancestor of modern squirrel monkeys.

(11) **Patasola magdalenae** (KAY, 1989, 1994) was a small platyrhine of the mid-Miocene about the size of *Leontopithecus* or about 400-600 g and related to the Callitrichinae, perhaps mid-way between callitrichids and *Callimico*. The fossil is found in both the La Victoria and Villavieja formations. The animal shows many derived characters from *Neosaimiri* and *Saimiri* and with living callitrichids and *Callimico*, which suggests that these groups are closely related (FLEAGLE *et al.*, 1997).

(12) **Lagonimico conclucatus** is a large callitrichid skull found in the La Victoria formation (13.5-12.9 million years BP) at La Venta and having elongate, compressed lower incisors that are procumbent and slightly elongated, a reduced third molar, absence of upper molar hypocones. The skull has small orbits, inflated lower crown (bunodont) teeth are shortened and have rounded shearing crests (KAY, 1994). This animal is probably not a direct ancestor of the modern callitrichids, according to KAY (1994). It was probably the size of a *Callicebus*.

(13) **Micodon kiotensis** (SETOGUCHI & ROSENBERGER, 1985) from the La Venta of Colombia (Miocene) may belong to the callitrichids, based on its small size of the type specimen, which is a single upper molar, which does not have characteristics of living callitrichids. Contrarywise it may belong to the pithecines, based on occlusal morphology of its upper molar (FLEAGLE, 1988, 1999). More material is needed to determine its exact phylogenetic relationship.

(14) **Cebupithecia sarmientoi** is another Upper Miocene fossil from the La Venta fauna of Colombia. It was about the size of *Pithecia* at 2-3 kg. STIRTON (1951) felt that this species was definitely allied with *Pithecia* because of the large and procumbent upper incisors, indicated by the alveoli and root of a central incisor and on the size and splayed orientation of the upper and lower canines. KAY (1990) and FLEAGLE *et al.* (1997) agree with this assessment. Skeletal fragments suggest a leaping *Pithecia*-like animal as well, since it possessed thoracic vertebrae, a long, non-prehensile tail and other features associated with adaptations of clinging and leaping as in *Pithecia* (MELDRUM & LEMELIN, 1991; MELDRUM *et al.*, 1990). Comparisons

of skeletal morphology to that of *Pithecia, Cacajao,* & *Chiropotes* show as well that this fossil primate is closest to *Pithecia* (FLEAGLE & MELDRUM, 1988). HERSHKOVITZ (1970) believed that a misinterpretation and certain absent features argue for a separate subfamily within the Cebidae. Others disagree with the Hershkovitz assessment (ORLOSKY, 1973; OROSKY & SWINDLER, 1975).

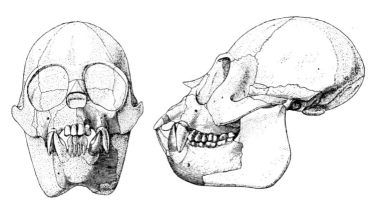

Figure 2. Skull of *Cebupithecia sarmientoi.*

(15) ***Stirtonia tatacoensis***, (2.1.3.3. lower jaw) collected in the La Venta fauna of Colombia is one of the largest fossil ceboids known, probably weighing about 6 kg. (HERSHKOVITZ, 1970). STIRTON (1951) places this fossil in the same subfamily as the howling monkey, although HERSHKOVITZ (1970), noting certain unique traits such as a reduced third molar and the divergent shape of the mandible, argues for placing it in its own subfamily, Stirtoninae. A slightly older fossil, *S. victoriae* has also been found at La Venta (KAY *et al.,* 1987). FLEAGLE *et al.* (1997) prefer to ally this fossil with the howling monkeys as well. Fossil teeth collected from several sites on the Río Acre on the border between Brazil and Peru best fit *Stirtonia* sp. Another very large tooth is assigned to the Cebidae, but cannot further be identified (KAY & FRAILEY, 1993).

(16) **Mohanamico hershkovitzi** was a small primate (1 kg) from the Miocene, discovered in the La Venta of Colombia. Because of its large lateral incisors and the structure of its canine and anterior premolar, this primate has been assigned to the base of the pithecine radiation (LUCHTERHAND *et al.,* 1986). However, KAY (1990) believes that this species is synonymous with *Aotus dindensis* and ROSENBERGER *et al.* (1990) believed *Mohanamico* shares some characters with *Callimico*.

(17) **Nuciruptor rubricae** (MELDRUM & KAY, 1997) is a middle Miocene fossil from La Venta with obvious connections to present-day pithecines as well as to *Cebupithecia sarmientoi. Nuciruptor* has procumbent, moderately elongated incisors and low-crowned molars, suggesting it was a seed predator as living pitheciines of about 2 kg. The dental morphology makes it less likely that *Mohanamico hershkovitzi* of the m. Miocene, Colombia is a pithecine.

(18) **Aotus dindensis** (SERTOGUCHI & ROSENBERGER, 1987) of the middle Miocene of La Venta was closely related to the modern *Aotus,* although it has much narrower lower incisors. This species had large orbits and was probably nocturnal, according to SETOGUCHI & ROSENBERGER (1987), but KAY (1990) disagrees with this assessment and believes this species to be synonymous with *Mohanamico hershkovitzi.*

Late Miocene (6-9 million years) río Acre, Brazil

(19) A tooth from the río Acre in Brazil is tentatively placed in **Stirtonia**, extending the time frame for this genus closer to *Alouatta.* Another, unamed upper molar appears to be an acestor of *Cebus,* but the animal was twice as large as modern *Cebus* (KAY & FRAILEY, 1993).

Pleistocene/Recent Primate Fossils from Several Sites

(20) **Protopithecus brasiliensis** was first described by LUND (1837) from subfossil limb bone fragments (left proximal femur and right distal humerus) found in Pleistocene cave deposits of Lagoa Santa, Brazil and seem to be from a very large ateline. The linear dimensions of these bones suggest that they came from an animal of 21 kg, 40% larger than present day *Brachyeles* body weights (HARTWIG, 1995; HARTWIG *et al.,* 1996). The species was probably sympatric with the other large extinct atelid *Caipora bambuiorum* and was not necessarily totally arboreal, since for its size and the probable vegetation type at the time, it might have been at least partially terrestrial (HEYMANN, 1998).

(21) **Caipora bambuiorum** (HARTWIG & CARTELLE, 1996; CARTELLE & HARTWIG, 1996) was discovered from Pleistocene deposits in Bahia State, Brazil. It was similar to *Ateles* in body proportions although twice as large at around 20 kg. Although this species had an intermembral index and other proportions similar to other brachiators, its size and the type of vegetation that was probably present in the area where it was found suggest that it could have been at least semi-terrestrial (HEYMANN, 1998). This primate was apparently sympatric with the other large, extinct atelid *Protopithecus brasiliensis*.

(22) **Xenothrix mcgregori** was discovered in Jamaica from a Pleistocene or Recent deposit and is placed in a separate family by HERSHKOVITZ (1970, 1977). However, an analysis by ROSENBERGER (1977) placed the species within the Cebidae and phylogenetically closest to *Aotus* or *Callicebus*, although FORD (1986, 1988) and FORD & MORGAN 1986) suggest the fossil should be investigated in light of a implying cladistic analysis of a primate distal right tibia found on the island of Hispaniola from probable Recent deposits. The tibia was the size of *Cebus*, but seems most closely related to *Saguinus*. This suggests the possibility of a large *Saguinus*-like primate from the Caribbean islands, suggesting the development of giganticism, a common trend of island fauna. Actually there may have been a minimum of at least three species of primates on Jamaica until humans appeared to wipe them out, according to post-cranial fossils recently collected (FORD, 1990a). However, a more recent cladistic analysis of HOROVITZ & MACPHEE (1999) and HOROVITZ (1999) suggest that a clade including *Antillothrix* and *Paralouatta* was most closely related to *Xenothrix* which seems to have an affinity to *Callicebus*; all of this suggesting a single colonization event from South America for the known species of Antillean primate fauna.

(23) **Antillothrix bernensis** (RIMOLI, 1977 = *Saimiri bernensis*) was described from deposits about 3800 years old in the Dominican Republic from a jaw fragment that was about twice as large as modern *Saimiri* from the mainland. It is apparent from this and from *Xenothrix* that a Caribbean primate fauna did exist and was probably extinguished by the arrival of human beings. The size may illustrate the common development of giganticism on many islands. Fossil remains already collected on Hispaniola may prove to represent more than one species of primate according to FORD (1990a). HOROVITZ (1999), HOROVITZ & MACPHEE (1999) and MACPHEE & HOROVITZ (2002) have shown that this taxon was actually quite distinct from *Saimiri* in many ways and actually

belongs in a clade with *Paralouatta varonai* of Cuba, which is a clade sister to the group containing the atelids (including *Stirtonia*).

(24) **Paralouatta varonai**, RIVERO & ARREDONDO (1991) is identified with a well-preserved howler monkey-like skull that was recently found in Cuba and that is much larger than living howler monkey skulls (RIVERO DE LA CALLE, 1988). Analysis of new dental and mandibular remains from Cuba suggest that the species was actually very different from *Alouatta* and was actually more closely related to the Hispanionalan primate *Antillothrix bernensis* and more distantly related to *Xenothrix* and *Callicebus*, implying that there was only one primate colonization from the South American mainland (HOROVITZ & MACPHEE, 1999).

(25) **Ateles anthropomorpha** (AMEGHINO, 1910) were teeth found in association with human remains in Cuba and originally believed to be a new genus (*Montaneia*). MILLER (1916) identified the teeth as *Ateles* and most researchers assumed that the teeth came from an animal brought over from the mainland. However, ARREDONDO & VARONA (1983) suggested the teeth are substantially different from living *Ateles*, implying that this species was a Cuban endemic. However, MACPHEE & RIVERO DE LA CALLE (1996) have dated the specimens as being less than three centuries old, making it probable that the teeth belong to *Ateles geoffroyi*, probably imported from Central America.

Primate Research in Colombia

The earliest work with non-human primates in Colombia was accomplished via the observations of various species by the first Spaniards now found in early chronicles. Early accounts of von Jacquin and others of *Saguinus oedipus, Cebus capucinus, Alouatta seniculus* were used by Linnaeus for his discriptions of these species.

Some of the earliest and still very important work done in Colombia was by Alexander von Humboldt, who first arrived on Colombian soil on the left bank of the Orinoco River in April, 1801. Von Humoldt and the French botanist Bonpland's explorations of the Orinoco resulted in the first descriptions of *Cebus albifrons* and *Lagothrix lagothricha* from the Colombian side of the river, as well as descriptions of *Callicebus torquatus* from the Venezuelan side and *Aotus trivirgatus* from either Colombian or Venezuelan territory. At that time both sides of the river were considered Gran Colombia. Later, von Humboldt's description of *Cebus capucinus* seen near the mouth of the Río Sinú was used by Goldman to establish the type locality of the species.

Modern collecting began in the late 1800's via British and North American expeditions to the country. Using Colombian material FOODEN (1963) described *Lagothrix lagothricha lugens*, a subspecies restricted perhaps entirely to Colombia, and whose type locality was established as the "Tolima Mountains, located by coordinates at 2°20'N on the Magdalena River at 5,000 and 7,000 ft." by an unnamed collector for the British Museum.

During modern times the naturalist and Christian Brother Nicéforo María made extensive collections of Colombian fauna (including primates) which helped define the geographic ranges within the country. Later the North American Philip Hershkovitz gathered what is the most extensive collection of primates (and other mammals) of the country. Based on these collections he described *Cebus albifrons cesarae* and *C. a. pleei* (Hershkovitz, 1949). Other collections were made by various workers with the material usually resulted in the material being deposited in museums of North America and Great Britain. However, in the past 30 years increasingly extensive collections and observations have been made by Colombian biologists, such as Carlos Velásquez, Jorge Hernández Camacho, Ernesto Barriga, Alberto Cadena and various collectors for INDERENA, a government agency in charge of natural resources from 1971-1995 (especially Jorge Morales and José Vicente Rodríguez). This material forms the basis of the important and best knwn Colombian collections of primate material found in the museum of the Instituto von Humboldt and at the Instituto de Ciencias Naturales of the Colombian National University.

Field work not involving collection was probably initiated by MARTIN MOYNIHAN (1976a, 1976b), who extensively observed primates in the Amazonian piedmont during the 1960's. A group led by Michael Kavanaugh made important observations in the Serranía de San Lucas, a part of Colombia that, because of the presence of political insurgents, has been isolated from Colombian primatology since that time. The group studied the northernmost known population of *Lagothrix lagothricha* for a few weeks (KAVANAUGH & DRESDALE, 1975) . Later K. GREEN (1978), then of the U.S. Peace Corps, censused populations of *Saguinus leucopus, Aotus lemurinus griseimembra, Ateles hybridus brunneus,* and *Lagothrix lagothricha lugens* in the same zone (Serranía de San Lucas, Bolívar).

THORINGTON (1967, 1968) studied squirrel monkeys for 500 hours east of San Martín, Meta at Hacienda Barbascal. A few years before, Mason had studied the behavior of *Callicebus cupreus ornatus* (called in his publications *Callicebus moloch ornatus* before HERSHKOVITZ' 1990 revision) at the same Hacienda. Because the forest where the animals were studied was a remnant forest and many of the animals in it were apparently refugees from other forests, the calculated density for this forest was very high and probably did not represent a density in a more natural situation.

Robinson collected data for his Ph.D. on *Callicebus cupreus ornatus* in two localities of low gallery forests to the south of Hacienda Barbascal and west of San Juan de Arama (ROBINSON, 1977, 1979a, 1979b, 1981, 1982a, 1982b). Robinson found densities that seemed more in accordance with a basically undisturbed vegetation, although cattle did roam through his study forests. In northern Colombia (Sucre) Pat Neyman completed her Ph.D. thesis research on the endangered *Saguinus oedipus* at a farm in Sucre, near the Caribbean coast. Even then her data suggested the extreme danger of the species' possible extinction, because of habitat loss (NEYMAN, 1977, 1978).

A team of several North Americans and Colombians performed primate censuses in northern Colombia for the Pan American Health Organization. They confirmed that, not only were there few *Saguinus oedipus*, but that *Aotus lemurinus griseimembra* were also extremely endangered (SCOTT *et al.* 1976; STRUHSAKER *et al.*, 1975; HELTNE & MEJÍA, 1978). One of the results of the primate censuses of PAHO was the establishment of a primate center in northern Colombia near Colosó, Sucre, which was to serve as holding facilities for some of the endangered taxa and as a center for research and conservation. Nevertheless, only a few studies have come from this facility (BARBOSA, 1988), although lately it served as a base for Anne Savage who with her Colombian collaborators collected data for her own Ph.D. thesis and designed a conservation program for the area (SAVAGE, 1980, 1990a; Savage *et al.,* 1986, 1987, 1988, 1990a, 1990b, 1990c, 1993, 1995a, 1995b, 1995c, 1996a, 1996b, 1997, 2002; SAVAGE & GIRALDO, 1990; SAVAGE & BAKER, 1996).

During 1975-1976 Steve Gaulin with his wife Cynthia studied *Alouatta seniculus* in a small and isolated cloud forest at Finca Merenberg (Huila) and this study became the basis for a Ph.D. thesis (GAULIN, 1977; GAULIN & GAULIN, 1982). Another interesting and more recent study of *Alouatta seniculus* took place in cloud forest of the Ucumarí Regional Park of the department of Risaralda at an altitude of 1900-2450 m (CABRERA, 1994).

Since the description of the first La Venta fossils by STIRTON (1951) other material continues to come to light, providing the first descriptions of the genera *Mohanamico, Micodon, Stirton, Neosaimiri, Cebupithecia*, and an earlier species of *Aotus* (SZALAY & DELSON, 1979; SETOGUCHI & ROSENBERGER, 1985, 1987; LUCHTERHAND *et al.*, 1986; KAY *et al.*, 1987). These fossils are considered to be from the middle Miocene, usually considered of Friasian and perhaps Santacrusian ages (15-16 million years B.P.). Following is a list of major publications resulting from studies of La Venta primate fossils: *Lagonimico conclucatus* KAY (1994); *Mohanamico* LUCHTERHAND *et al.*, (1986); KAY (1990); ROSENBERGER *et al.* (1990); SERTOGUCHI & ROSENBERGER (1987); *Micodon*, STIRTON (1951); *Stirtonia*, STIRTON (1951); HERSHKOVITZ, (1970, 1977); SETOGUCHI, WATANABE & MOURI (1981); KAY, (1980, 1989, 1990); *Neosaimiri* (=*Laventiana*; =*Saimiri* ?), TAKAI (1994); TAKAI *et al.*, (1992a); TAKEMURA *et al.* (1992) ; STIRTON, (1951); HERSHKOVITZ, (1970, 1977); SZALAY & DELSON, (1979); ROSENBERGER *et al.*, (1990); ROSE & FLEAGLE, (1981); FLEAGLE, (1988, 1999); *Cebupithecia,* STIRTON & SAVAGE (1951); STIRTON (1951); DAVIS (1987, 1988); MELDRUM & FLEAGLE (1988); FLEAGLE & MELDRUM (1988); FORD (1990a, 1990b); KAY (1990); HERSHKOVITZ (1970, 1977); FLEAGLE (1988, 1999); DELSON & ROSENBERGER (1984); ORLOSKI & SWINDLER (1975); *Aotus*, GEBO *et al.* (1990); GEBO (1988, 1989); SETOGUCHI & ROSENBERGER (1987); FLEAGLE (1988, 1999); La Venta community, KAY & MADDEN (1997).

Aside from various collections, the earliest primatological observations in the Colombian Amazon were accomplished by K. IZAWA (1975, 1976, 1978a) NISHIMURA (1987), and NISHIMURA & IZAWA (1975) from study sites on the Peneya River, a left bank affluent of the Caquetá River in southern Caquetá . Later, Izawa's group signed an agreement with the University of Los Andes to develop a research site on the Duda River, immediately west of the Serranía de la Macarena, where the joint Colombian-Japanese group continued to study primates in Tinigua National Park at the Primate Research Center (later named the Ecological Research Station) of southern Meta. The following lists most of the numerous publications which have appeared, mostly in the Japanese "serial Field Studies of New World Monkeys, La Macarena, Colombia": AHUMADA (1989, 1990); CALLE (1992a, 1992b);

CEBALLOS (1989); ESCOBAR-PARAMO (1989a, 1989b, 1990); FIGUEROA (1989); HIRABUKI & IZAWA (1990); IZAWA (1978a, 1978b, 1979a, 1979b, 1980, 1988a, 1988b, 1989a,1989b, 1990a, 1990b, 1990c, 1990f, 1992, 1993, 1994a,1994b, 1997a, 1987b, 1997c, 2002); IZAWA et al. (1979, 1990a, 1990b); IZAWA & LOZANO (1989, 1990a, 1990b, 1991, 1992, 1994); IZAWA & MISUNO (1977, 1990a, 1990b, 1990c); IZAWA & NISHIMURA (1988); IZAWA & TOKUDA (1988); KIMUARA (1992); KOBAYASHI & IZAWA (1992); NISHIMURA (1990a, 1990b, 1990c, 1994, 1996, 1997, 1998a, 1998b, 1998c, 1999a, 1999b, 1999c, 2001, 2002a, 2002b, 2003); NISHIMURA et al. (1992k, 1995, 1990c); POLANCO (1992); POLANCO & CADENA (1993); POLANCO et al. (1994); SOLANO (1995); STEVENSON (1992); STEVENSON et al. (1992, 1993, 1994); TOKUDA (1988); VALENZUELA (1992, 1993); VARGAS (1994b); YONEDA (1988, 1990); YOSHIHIKO & IZAWA (1990).

Undoubtedly the most active researchers in this group have been Kosei Izawa, T. Nishimura and Pablo Stevenson, having studied several species of primate each, but with particular emphasis on *Cebus apella* (Izawa) and *Lagothrix lagothricha* (Nishimura and Stevenson). Despite many problems of personal security, Stevenson has continued work in the Río Duda, attempting to support students and to advance his own research agenda (STEVENSON, 1992, 1997a, 1997b, 1998a, 1998b, 2000, 2001, 2002a, 2002b, 2003; STEVENSON & AHUMADA, 1994; STEVENSON & CASTELLANOS, 2000, 2001; STEVENSON & QUIÑONES, 1993; STEVENSON & DEL PILAR MEDINA, 2003; STEVENSON et al. 1992, 1994, 1998, 1999, 2000, 2002; TAKEHARA & STEVENSON, 1997). These activities have recently resulted in Stevenson´s Ph.D., the first awarded to a Colombian primatologist. Sadly the Japanese group has been forced to suspend their research in Colombia because of problems of security.

During 1977-1982 T. Defler censused and studied the primates *Cebus albifrons, Alouatta seniculus, Cebus apella,* and *Callicebus torquatus* in the El Tuparro National Park in eastern Vichada (DEFLER, 1979a, 1979b, 1980, 1981, 1982, 1983a, 1983b, 1983c, 1985a, 1985b). In 1983, Defler moved to the lower Apaporis River in Vaupés where he founded the Estación Biológica Caparú, at first emphasizing *Lagothrix lagothricha* and later adding *Cacajao melanocephalus* and *Saguinus inustus* (DEFLER, 1987, 1989a, 1989b, 1989c, 1990a, 1990b, 1991, 1994a, 1994b, 1994c, 1995, 1996a, 1996b, 1996c, 1999a, 1999b, 2001, 2003a, 2003b; DEFLER & PINTO (1985); DEFLER & DEFLER, 1996). One of Caparú's main tasks has been to support primate and other ecological and conservation-based research of young Colombian biologists as follows: *Callicebus torquatus* (FORRERO, 1987); (PALACIOS & RODRÍGUEZ, 1995); (PALACIOS *et al.*, 1997); HERNÁNDEZ-B. & CASTILLO-A. (2002); *Lagothrix lagothricha* (MUÑOZ, 1991), *Lagothrix lagothricha* (ARDILA & FLOREZ, 1994) and *Cebus apella* a M.A. thesis by GÓMEZ (2003) and bachelor´s thesis by CASTILLO (2002). Field research of Palacios (*Alouatta seniculus*) have continued at Caparú and other sites in the lower Caquetá and Putumayo rivers (PALACIOS, 1997, 2000, 2003; PALACIOS & RODRÍGUEZ, 2001, in review; PALACIOS & PERES, in press).

Other scattered primate research projects have included work on *Saguinus leucopus* (CALLE, 1992a, 1992b, 1992c; VARGAS & SOLANO,1996a, 1996b; POVEDA, 2000), *Saguinus nigricollis hernandezi* (VARGAS, 1992, 1994a, 1994b). Two projects on *Alouatta seniculus* included work close to Colosó, Sucre and the local primate center there (BARBOSA, 1988) and a research project of CABRERA (1994) in cloud forest of the Cordillera Central and POLANCO (1992) with *Callicebus cupreus ornatus* and more which are not listed here. East of the Serranía La Macarena during the 1960's L. Klein, assisted by his wife Dorothy, studied *Ateles belzebuth* and other sympatric primates near the Río Guayabero (KLEIN, 1971, 1972, 1974: KLEIN & KLEIN, 1971a, 1971b, 1973a, 1973b, 1975, 1976, 1977).

Research on *Saguinus oedipus* has included censuses by BARBOSA *et al.* (1988) and the research around Colosó by Savage cited above as well as the earlier work by NEYMAN (1977) and others such as J. V. Rodríguez and Hernández-C. as well as a position paper by DEFLER & RODRÍGUEZ (2003) delivered to the IUCN. Recently RAMÍREZ (1998) finished her work on *Alouatta palliata* on the Pacific coast and SÁNCHEZ (1998) finished a project on *Callicebus cupreus ornatus* SE of Villavicencio in Meta department. Other theses have been finished in intervening years.

Finally some censuses were conducted on the "Isla de los Micos"on the Amazon River, west of Leticia by a group of North Americans and a Colombian in order to evaluate the populations of *Saimiri sciureus* kept by the animal dealer Mike Tsalickis prior to shipment to Florida (BAILEY *et al.*, 1974). Other research projects continue proliferating.

Primate research by Colombians has continued to expand in the last few years, and the nucleus of a committed group of primatologists forms the recently organized Colombian Primatological Association led by the veterinarian Victoria Periera as the Association's first president (2003) with her associate Fernando Nassar, themselves the organizers of Fundación Araguatos, which has begun an active program of publication and courses in primatology. It is the hope of this author that primatology might continue to develop apace in Colombia so that our knowledge of these fascinating animals may do justice to the richness to be found in the Colombian primate fauna.

Callicebus cupreus ornatus

3. Zoogeography of colombian primates

Introduction

Geographically Colombia is a topographically complex country located directly on the equator and possessing very high mountain ranges, numerous wide rivers, extreme temperature differences, and varied precipitation. These characteristics affect to a high degree the distribution of the many plants and animals found within the country, so it is important to understand the country´s basic geography and especially the aspects which help determine primate distribution.

The fact that anthropogenic activities now affect so much of Colombian flora and fauna must also be taken into consideration, since we have entered an era of worldwide extreme risk for many species of plants and animals. It is vital that we understand the effects of human population pressures upon the distribution of Colombian organisms such as primates, in order to try to counter them. These aspects of Colombian geography are discussed in more detail in Chapter 4.

Colombian Topography (See Figure 3)

Colombia is divided down the middle by the Cordillera de los Andes, from the northern Caribbean coastal regions where the Andes terminate, to the frontier with Ecuador in the south where they continue southward through Ecuador, Peru and Chile. In the southern part of the country this great chain of mountains is divided into three separate cordilleras: the Cordillera Occidental (in the west), the Cordillera Central, and the Cordillera Oriental (in the east). The Cordillera Occidental and Oriental originate in the department of Nariño and the Central Cordillera, which is the most ancient and the highest of the three, and continues as a prolongation of the Cordillera Oriental of Ecuador. These mountain ranges have been partial and complete biogeographic barriers for a great variety of Colombian flora and fauna. They have isolated the Pacific coast from the eastern part of the country, restricting the expansion of many species, permitting the classification of Colombian species into two groups: the cis-Andean and the trans-Andean species, according to whether or not a particular species is found east or west of the Cordillera.

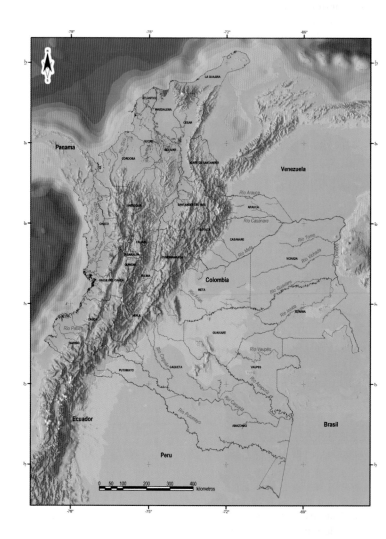

Figure 3. Topography of Colombia, including major rivers and altitudes.

In western Colombia a rather wide strip of lowland extends from the north to the south, allowing the dispersal of primates and other fauna and flora along the Pacific coast in a north to south direction. In the north of this region some low serranías are located, such as the Serranía del Baudó (which reaches 1,400 m of altitude in the Alto de Buey) and the Serranía del Darién, which forms the Colombo-Panamanian border. Additionally in those northern parts are the Serranía de los Saltos and the Serranías del Limón and Quía . Some of those low chains of hills may have influenced the southward penetration of faunistic elements such as the primate *Ateles geoffroyi grisescens*, which may be restricted in Colombia to the Serranía de Baudó.

The principal rivers of the region, originating in the Cordillera Occidental are (from south to north) the Mataje, the Mira, the Patía, Sanguianga, Tapaje, Iscuandé, Guapi, Timbiquí, Buey, the San Juan de Micay, the Naya, Yurumanguí, Cajambre Raposo, Anchicaya, Dagua and in the north-central zone of the region the San Juan River originating in Cauca, and the Atrato River which originates in the Chocó and flows into the Caribbean in the Gulf of Urabá.

The Cordillera Occidental is the lowest of the three cordilleras, reaching only 3,000 m altitude in a few places. This Cordillera terminates in the north with the Paramillo del Sinú, which ramifies into the Serranía de Abibe, the Serranía de San Jerónimo and the Serranía de Ayapel in southern Córdoba and northern Antioquia Departments; here the Sinú and San Jorge rivers have their headwaters, flowing north to empty into the Caribbean. A very deep river valley between the Cordillera Occidental and the Cordillera Central contains the Cauca river. Its headwaters have their origin south of the Pan de Azúcar glacier in Puracé National Park, and this river eventually empties into the Magdalena River.

The high Cordillera Central contains several lofty volcanic massifs, from the Pan de Azúcar (Puracé National Park) in the south (which reaches an altitude of 5,500 m), the Nevado del Huila (also reaching 5,500 m in the Nevada del Huila National Park), and los Nevados (which reaches 5,400 m in the National Park Los Nevados). The Cordillera Central ends in the Serranía de San Lucas in the Department of Bolívar.

The easternmost chain of the Andes, the Cordillera Oriental, is the widest of the three cordilleras and geologically it is the more recent. Heights of this

Cordillera vary from the 1,600 m of La Uribe, a pass between the Departments of Huila and Meta, to the 5,300 m of the Serranía del Cocuy (El Cocuy National Park). The maximum width of the Cordillera Oriental is about 250 km around the region of Bogotá. To the north, the Cordillera Oriental divides and forms the Serranía de los Motilones or Perijá (Catatumbo-Barí National Park) and the Massif of Tamá or Táchira. Between the Cordillera Central and the Cordillera Oriental flows one of the principal rivers of the country, the Magdalena, which originates on the volcano Puracé in the south of the department of Huila, later flowing to the north and receiving the waters of the Cauca river before flowing into the Caribbean.

To the north of the three Cordilleras is an extensive zone of lowlands, the Llanura Costera del Caribe or the Caribbean lowlands. In this part of Colombia several small ranges of hills (serranías) are found such as the Serranía de San Jacinto or Montes de María in Sucre (important as a reserve for the endangered primate *Saguinus oedipus*). Perhaps the most important characteristic of the region is the enormous volcanic massif the Sierra Nevada de Santa Marta (Sierra Nevada de Santa Marta National Park), with the highest point found in the country (Colón Peak, 5,770 m). To the east and to the north extends the xerophytic Guajira peninsula, a basically flat land which also possesses several isolated ranges of hills, particularly important in biogeographic terms being the Serranía de Macuira (Macuira National Park), which reaches 865 m elevation.

The most important rivers of the Caribbean lowlands (from west to east) are the Atrato, Mulatos, San Juan, Córdoba, Canalete, Sinú, Pechelín, el canal del Dique, Magdalena, Fundación, Aracataca, Tucurinca, Sevilla, Frío, Córdoba, Manzanares, Piedras, Guachaca, Mendiguaca, Don Diego, Palomino, Ancho Dibulla, Eneal and the Ranchería, as well as the Catatumbo, which empties into Maracaibo Lake in Venezuela. The Magdalena river has been counted as the principle river of the country, and its watershed forms the two inter-Andean valleys, since its major tributary, the Cauca, flows between the western and central cordilleras, with the principle branch of the middle and upper Magdalena between the eastern and central cordilleras.

Finally to the east of the Andes and comprising about half of Colombia, there is a vast region of lowlands which are predominantly flat and which form the Llanos Orientales in the north and the Amazonian forest ("selva amazónica") in the south (generally between 100-650 m altitude). The flat character of this great region is interrupted here and there by a few ranges of low mountains and hills as well as some inselbergs and mesas. Examples of these are the geologically ancient Serranías of Naquén and Taraira, as well as geologically younger Chiribiquete. Most noteworthy perhaps is the range of mountains most known in this eastern region, the Serranía de la Macarena in Meta department, which reaches 2,000 m altitude and is 120 km long running north to south (Sierra de la Macarena National Park). Numerous rivers originate in the Cordillera Oriental and flow to the east throughout these lowlands, and these form part of the watersheds of both the Orinoco and the Amazon Rivers. The major affluents include the Arauca, Casanare, Meta, Vichada, Guaviare and Inírida Rivers which flow into the Orinoco River and the Guanía, Vaupés, Apaporis, Caquetá , and the Putumayo Rivers which in turn flow into the Amazon. Finally, bordering the Colombian trapezium in the south is a small stretch of the Amazon River.

This complex Colombian topography has a range of altitudes from 0-5,770 m. Besides being located over the equator, Colombian mainland is located between the latitude 12°30'40"N in the Guajira and 4°13'30"S in Amazonas Department and between longitude 66°50'54"W in the Río Negro and 79°01'23"W in the Department of Nariño. The national territory includes as well the archipelagos of San Andrés and Providencia, Rosario and San Bernardo and the islands Fuerte and Tortuguilla in the Caribbean and in the Pacific the islands of Malpelo, Gorgona and Gorgonilla. Of these archipelagos and islands only Gorgona Island provides habitat for primates.

This vast region of 1,141,748 km² provides an enormous quantity of habitats, which are principally influenced by two factors which vary and interact with topography: temperature and precipitation. The variations in these two factors are reflected by the type of vegetation cover, and these exercise primary effects over the type of fauna that is found in any particular zone. The added effects of soils and historical processes also likely contribute to the distribution of primates, although a precise understanding of these elements is still to be clarified.

HERNÁNDEZ C. et al.(1992a, 1992b) identified biogeographic divisions and terrestrial biomas for Colombia. They divided the country into nine Biogeographic Provinces: Carribean Insular Territories, Pacific Insular Territories, the Carribean Dry Belt, the Sierra Nevada of Santa Marta, the Chocó-Magdalena, the Northern Andes, Orinoquia, Guyana and Amazonia, which were then divided up into 99 Districts, the majority of which contain non-human primates.

Ecological Factors

Temperature (**Table 1. Isothermic values for Colombia**)

Figure 2 shows the isothermic values for Colombia. The variations in the altitudinal thermic levels include thermic level, annual isotherms, total area in square kilometers and synonyms for the name of the thermic level. These altitudinal thermic levels are as follow: (1) **Hot Thermic Level** (24°C, 932,866 km² or 80.9% of the country, also called *tierra caliente*; (2) **Temperate Thermic Level** (18-24°C, 110,431 km² or 9.7% of the country, also called *tierra templada*; (**Subtropical Zone** and **Premontane Zone** in the Holdridge system); (3) **Cold Thermic Level**, 12-18°C, 77,747 km² or 6.8% of the country, *tierra fría*, (**Temperate Zone** and **Lower Montane Zone** in the Holdridge system); (4) **Subparamo Thermic Level**, 6-12°C, 27,559 km² of the country, (**Subalpine Zone** or **Montane zone** in the Holdrige system); (5) **Paramo Thermic Level**, 1.5-6°C, 2,027 km² or 0.18% of the country, (**Alpine Zone** or **Subalpine and Alpine Zone)**; and (6) **Perpetual Snow Thermic Level**, 1.5°C or below with 74 km² (HERNÁNDEZ-C. AND DEFLER, 1989).

The Holdridge humidity provinces vary from perarid desertic regions of the Guajira peninusla to the superhumid regions, the perhumid and humid provinces being the most important humidity province for Colombian primates, since the following numbers of primate species are to be found per province: (1) Superhumid (16 species) (2) Perhumid (29 species), (3) Humid (29 species, (4) Subhumid (11 species), (5) Semiarid (3 species), (6) Arid and (7) Perarid (0 species) (HERNÁNDEZ-C. & DEFLER, 1989). Especially in the lower altitudes there is little temperature fluctuation throughout the year, due to the location in the equatorial zone. It is typical for tropical regions that there is more temperature fluctuation during 24 hours than the average temperature fluctuation throughout the year.

Precipitation

Annual precipitation records in Colombia vary from about 150 mm in northern Guajira to more than 10,000 mm in Chocó (the most elevated annual precipitation known in the country and for the world is 11,770 mm at Tutunendó, Chocó, upper San Juan river). Unlike temperature, annual precipitations throughout the country vary in intensity throughout a typical year, causing rainy and dry seasons in many (but not all) regions of the country under the influence of the intertropical convergence zone.

Altitudinal thermic Levels	Isothermic Values	Surface (km²)	Synonymus
Hot	> 24°C	923,866.23, (80.92%)	"Tierra caliente". tropical.zone
Temperate	18-24°C	110,431.33, (9.67%)	"Tierra templada", subtropical zone premontane zone (HOLDRIDGE)
Cold	12-18°C	77,746.75, (6.81%)	"Tierra fría" temperate zone lower montane zone (HOLDRIDGE)
Subpáramo	6-12°C	27,559.15, (2.41%)	subalpine zone montane zone (HOLDRIDGE)
Páramo	1.5-6°C	2,027.15, (0.177%)	alpine zone subalpine & alpine (HOLDRIDGE)
Permanent Snows	<1.5°C	74.00, (0.006%)	

Table 1. Altitudinal thermic levels of Colombia (excluding the Archipiélago de San Andrés y Providencia) (fide HERNÁNDEZ-C. & DEFLER, 1989).

Gentry found a correlation of precipitation with plant diversity (GENTRY, 1988) and REED & FLEAGLE (1995) showed another correlation for primate species richness and precipitation up to about 2,500 mm, at which point it (according to their data) seems to stop increasing. From sites in eastern

Colombia with rainfall over 3,000 mm we can see that primate richness then tends to decrease (DEFLER, in press) or at least to level off at levels below that of the highest species richness. Because of the apparent dip in diversity above 2,500 mm a simple relationship of precipitation with primate diversity does not exist as long as adequate water sources are present at all times. The ecological consequences of drought exercise an effect on the vegetation cover, while the effect of vegetation differences on the primate community, whether via floral components, primary production, or both has been very poorly studied. An exception is STEVENSON´s (2001) analysis of fruit production with respect to primate richness. Nevertheless, it is clear that such relationships do exist in the tropics, although they need to be clearly measured (DEFLER, 1996a; DEFLER & DEFLER, 1996; FLEAGLE et al., 1999). Another poorly studied secondary effect of precipitation on the primate community is the variability along mountain slopes, where high precipitation and fog may affect the thermic tolerance of a particular primate species for an altitudinal zone. Species drop out along tropical slopes as altitude increases (DURHAM, 1971)

Edaphic Factors

Edaphic variations may determine whether some species will be present or absent at a particular site. Soil fertility and patterns of precipitation throughout the year may permit the development of high resource availability or the contrary, scarcity, this being reflected by phenological cycles and primary production and perhaps by key differences in some floral components (EISENBERG, 1979a; HUSTON, 1994). Floral characteristics may determine not only what primate species might be present in a region, but they probably determine some primate densities as well (DEFLER & DEFLER, 1996), although, again, the details of these ecological interactions need to be clearly demonstrated (MACARTHUR & CONNELL, 1966; HUSTON, 1994; FLEAGLE et al., 1999).

Competition

Confounding our understanding of the above factors, the poorly fathomed interactions of competition also play a role. Observations suggest that there are some primate species which make life difficult for other primates (EISENBERG, 1979) either directly by means of aggressive interactions or indirectly, harvesting resources more efficiently or before the other species gets to them. Some type of competition could probably exclude a species from a region (MACARTHUR & CONNELL, 1966; HUSTON, 1994) although proof is not easy to obtain. Various authors have inferred this type of competition affecting distributions of *Ateles* spp., *Lagothrix lagothricha* and *Alouatta seniculus* (IZAWA, 1976; DURHAM, 1975; FREESE *et al.*,1982; HERNÁNDEZ C. & COOPER, 1976), sympatric versus parapatric distributions of two species of *Cebus* (DEFLER, 1985) and perhaps *Cebus* versus *Saguinus*, among others.

Historic Factors and Pleistocene Refuges

Historic factors also have determined the distributions of primates, just as with other animals and plants, and this aspect should be considered. We know through the studies of HAFFER (1969, 1974, 1982, 1987a, 1987b), HERNÁNDEZ C. *et al.* (1992), VANZOLINI (1970, 1973), VANZOLINI & WILLIAMS (1970), BROWN (1976, 1977a, 1977b, 1979, 1982, 1987a, 1987b), AB'SÁBER (1977, 1982), BROWN & AB'SÁBER (1977), and VAN DER HAMMEN (1974, 1982) and VAN DER HAMMEN *et al.* (1982) among others that numerous climatic changes have occurred in South America, particularly during the Pleistocene, and these changes have very strongly affected the vegetation and the animals that existed in each region.

It is generally accepted that climatic changes were particularly strong during the last two million years of the Quaternary. Adverse changes in the form of marked diminution in precipitation during cold and dry glacial periods, repeatedly caused forests (and regions covered with other types of vegetation) to fragment and xerophytic vegetation to expand, leaving "refugia" of forest and other vegetation types during these critical phases of climatic fluctuations. Of course forest refuges were most important for forest organisms like primates, forest birds and other organisms adapted to forest, since they afforded habitat for which these organisms were adapted (HAFFER, 1974).

In these isolated forest fragments or forest refugia a differentiation of populations began to take place via normal evolutionary processes, so that a population of a species in one refuge differentiated distinctly from another population in another forest refuge. When the continental glaciers (which also were very developed in the Andean cordilleras) began to melt, and the climate began to improve, a subsequent increase in precipitation permitted an expansion of the forest vegetation that up to then had been confined to forest refugia. These refuges then became re-connected, and previously isolated populations of organisms came back into contact. By now many of these related populations may have been reproductively isolated and therefore different species.

This view of refuges as "species pumps" however has been more and more challenged due to molecular calculations now possible, which date many extant species to before the Pleistocene glaciations into the Pliocene. Also the location of some refuges has been questioned and ultimately direct fossil evidence from the upper Napo River indicates a late Pleistocene mammal fauna more typical of savannah/woodland, where the great Napo refuge has been hopothecized to have been located (WEBB & RANCY, 1996).

Figure 4 is an interpretation of the distribution of Colombian Pleistocene vegetation of about 20,000 years ago during the Wisconsin or Würm glaciation. It should be emphasized that sea level during this glacial period was about 185 m lower than it is today, and all of this water was locked up in the world's glaciers. This period resulted in the altitudinal levels of vegetation being much lower on the respective flanks of the cordilleras (PIELOU, 1979; SÁNCHEZ et al,1990). It is noteworthy that the Chocó forests may have been isolated from the forests of Paramillo-San Lucas, the Sierra Nevada de Santa Marta and the Catatumbo in the north as well as some areas in the south and in western Putumayo, areas in the trapezium and some other areas of less probability (according to BROWN, 1982, using lepidopterans as data) in the east. Many of these refugia may have been isolated from each other for long periods of time during glacial advances. Some very large regions that are covered in forest today were probably covered in savannas, Amazon caatingas (campinas or campinaranas) and other types of vegetation during this earlier time. Some of this vegetation was almost desert-like as is indicated by extensive dune systems that are found in the interior of what is now Amazonian forest (see map, Fig. 3).

HAFFER (1969, 1974) was the first to suggest Pleistocene forest refuges in South America. In some areas he discovered the concurrence of forest avian diversity centers, and he explained these centers using a theory of Pleistocene refugia. Since then a lot of evidence has accumulated demonstrating high diversity of other organisms (including primates) in many of the centers identified by Haffer or in new areas not mentioned by Haffer (Vanzolini, 1970, 1973; VANZOLINI & WILLIAMS, 1970; MÜLLER, 1973; PRANCE, 1973, 1982; BROWN, et al., 1974; BROWN, 1982; KINZEY, 1982; KINZEY & GENTRY, 1979).

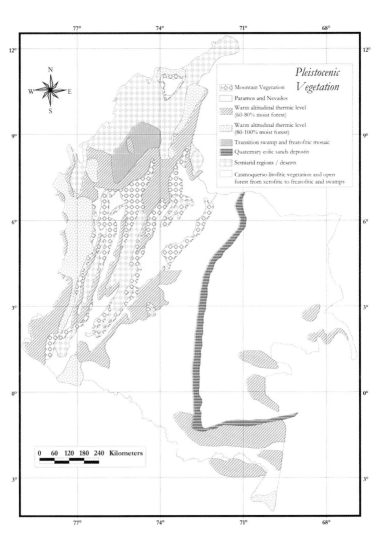

Figure 4. Interpretation of distribution of Colombian Pleistocene vegetation about 20,000 years ago (HERNÁNDEZ-C. & DEFLER, 1989)

Not only have Pleistocene refugia been one of the principal mechanisms evoked to explain the presence of specific regions with high diversity, but also refugia are suggested as one of the driving forces for biological speciation in the neotropics, since allopatric speciation generally is accepted as a principal mechanism for the evolution of species of vertebrates and of many plants. Allopatric speciation suggests that for the evolution of a new species, the populations of the originating species must be divided into two isolated populations for sufficient time for them to develop reproductive barriers. HAFFER (1967, 1974) suggested periods of 20,000 - 200,000 years for the evolution of a new species of forest bird. Pleistocene refuges then could play a role in isolating populations for sufficient time for such differentiation. VAN DER HAMMERN (1972, 1974, 1978, 1982), VAN GEEL & VAN DER HAMMEN (1973) and LIVINGSTONE & VAN DER HAMMEN (1978) supplied palynolgical data which supported the idea of vegetation changes that occured along with the glaciations.

Other authors, however, have not concurred with the above model, sustaining that there are other eco-geographic explanations for diversity centers and that other evidence has not been sufficiently examined (Endler, 1977, 1982; BENSON, 1982; COLLINVAUX, 1996). Also, new molecular evidence suggests ages for many birds and primates which exceed Pleistocene times, extending into the Pliocene (for example, *Ateles* species, COLLINS, 2001). This suggests that the refugia may not have been important as "species pumps", however, they surely protected taxa during adverse conditions.

Primate Diversity and Colombian Diversity

The Formation of Species

Along with a reproductive barrier, which is what defines a distinct species, changes in the manner in which the isolated populations exploit their habitat are what give origin to the exploitation of distinct niches so that two related species can exist sympatrically, not interfering excessively with each other. Ultimately a given species must be able to maintain itself as an interbreeding population and so the individuals of a species develop mechanisms to reduce competition with other species. The species may become a specialist in some aspect of resource exploitation, especially in the tropics where many species are packed together.

Allopatric speciation via refuges is not the only hypothesis for the origin of taxonomic diversity in South America. The origin of species via other vicariant effects producing isolated populations also seeks to explain allopatric speciation. Especially HERSHKOVITZ (1963, 1969, 1972b, 1976, 1997) has argued strongly for this viewpoint, evoking the changes in the courses of large rivers when necks of river meanders are cut and subsequent enclaves of populations are established on the opposite side of the river from the main population.

The concepts of parapatric and sympatric speciation may be valid, especially for organisms with little vagility (MAYR, 1971), Nevertheless, ENDLER (1977, 1982) suggests parapatric speciation for birds as well. The majority of biologists, however, see allopatric speciation as the major mechanism of speciation, although these questions are by no means settled.

The Diversity of Primates in Colombia

Colombia is probably sixth in the world in the number of species and subspecies of primates, and because of the various factors mentioned above, Colombian primates show various complex patterns of geographic distribution. None of the proximal causes of these factors are well understood since they probably involve many different interacting factors. Some tendencies have been recognized in distributions within the country, and these are discussed below.

Altitudinal Clines

Due to marked altitudinal differences found in Colombia, all of the forest slopes of mountains show clines in species richness diminishing with altitude. This is apparently due to the effects of temperature, since in Colombia the altitudinal thermogradient varies usually around 0.54-0.55°C for every 100 m (HERNÁNDEZ & DEFLER, 1989). This means, for example, that if one location at an altitude of 550 m close to the Cordillera Oriental in Putumayo registers a temperature of 24°C at 10 A.M., another location situated locally at an altitude of about 1,500 m would have a temperature that was 5.4-5.5°C (or 17.4-18.6°C) lower at the same time.

Such altitudinal variations excercise a marked influence on the presence or absence of a particular species in any particular altitudinal zone. This is explained by at least two reasons: (1) the primary altitudinal effects upon the forest vegetation (i.e. upon the vegetation community), which excercises affects on the resources available for the primates living there and (2) primary effects on the primates because of specific thermic tolerances.

Altitudinal Effects on Food Resources

It is generally accepted that the diversity of plants diminishes with altitude (PIELOU, 1979; MCARTHUR, 1965; GENTRY, 1990). This decrease in diversity has profound effects on the quantity of available biomass available for consumption, not only in terms of the diversity of the flora that is available for a specific diet but also in terms of primary production. DURHAM (1971) showed effects of this type in the populations of some primates in southern Perú, demonstrating a decrease in the size of the primate groups and in the densities of *Lagothrix lagothricha* and *Ateles b. chamek* at four different altitudes between 257-1,424 m.

HERNÁNDEZ C. & DEFLER (1985, 1989) discussed the direct or primary effects of altitude on primates, due to their thermic tolerances. For any altitudinal gradient, some species (usually the smallest and stenothermic species) disappear while the larger and eurythermic species persist. This accords with increased thermic tolerances of larger species, since the relation between the area of an animal and its mass increases in smaller animals so that smaller animals lose heat much more rapidly than do large animals at a rate described by the following equation $Log_{10}{}^t = -0.632685 + 1.35933/(log_{10}{}^p)$, where t = thermic tolerance and p = cubic root of the body weight, based on weights calculated by the authors for Colombian animals in HERNÁNDEZ-C. & DEFLER (1989). The fact that the small primate *Aotus* which are also active at night and persist to very high altitutdes of over 3200 m suggests important physiological adaptations in this genus.

In Colombia the small species of primates tend to have a very low thermic tolerance and they are considered to be stenothermic species. The eurothermic species usually are larger and are found in a greater altitudinal gradient. Table 2 shows estimates for thermic tolerance of Colombian species based on body weight and anual isotherms for the altitudinal limits know for the country.

Species	Altitudinal limits estimated	Thermic limits (Annual isothermas)	Thermic tolerance	Body weight males (g)	Body weight Females (g)	Mean in both sexes (g)
Cebuella pygmaea	ca. 80-500m	23.9-26.4 °C	2.5 °C	90 (N=1)	83.3-120 (N=2) (106.6)	101.1
Saguinus nigricollis	ca. 80-500m	23.9-26.4°C	2.5°C	425,500 (N=2) (462.5)	470 (N=1)	465.0
Saguinus fuscicollis	ca. 80-500m	23.9-26.4°C	2.5°C	385-562 (N=4) 461.7 ± 18.4	333-534 (N=4) (461.7 ± 22.3)	461.7
Saguinus inustus	ca. 100-300m	24.0-26.4°C 24.0-26.0°C	2.4°C	670 (N=1)	803 (N=1)	736.5
Saguinus leucopus	ca. 25-1500m	20.4-28.5 °C	8.1°C	456,532 (N=2) (494.0)	435-525 (N=2) (490.0)	487.0
Saguinus oedipus	ca. 0-1500m	21.6-28.0°C	6.4°C	458,490 (N=2) 474.0	466-579 (N=3) (505.6 ± 36.7)	489.8
Callimico goeldii	ca. 100-500m	23.8-26.2°C	2.4°C		630 (N=1)	630.0
Saimiri sciureus	ca. 80-1500m	21.0-26.4°C	5.4°C	835-1380 (N=9) (1082.0 ± 84.1)	595-1148 (N=5) (858.8 ± 106)	970.4
Aotus lemurinus (=A. zonalis, A. griseimembra)	ca. 0-3200m	9.5-28.5°C	19.0°C	608-1150 (N=7) (920.7 ± 79.9)	578-1050 (N=6) (859.0 ± 87.6)	889.8
Aotus brumbacki	ca. 400-1543m	19.4-26.0°C	6.6°C	(1 sin sexo anotado: 875)	455 (N=1)	665.0
Aotus vociferans	ca. 80-500m	23.9-26.4°C	2.5°C	568-800 (697.5 ± 24.0) (N=4)		697.5

Callicebus cupreus	280-500m	25.5-26.8°C	1.3°C			
Callicebus torquatus	80-500m	23.9-26.4°C	2.5°C	1232-1722 (N=5)	1030-1360 (N=9) (1265.1 ± 54.3)	1377.9
Pithecia monachus	80-500m	23.9-26.4°C	2.5°C	2106-3000 (N=4)	2101 (N=1)	2281.25
Cacajao melanocephalus	ca. 100-300m	ca. 24.0 -25-0°C	ca. 1.0°C			
Alouatta palliata	ca. 0-1400m	18.2-27.6°C	9.4°C	6000-8000 (N=5) (6600 ± 367.4)		6600.0
Alouatta seniculus	0-3200m	9.5-28.5°C	19.0°C	5000-12500 (N=8) (7540.3 ± 884.9)	4000-10000 (N=9) (6297.6 ± 697.3)	6918.95
Cebus albifrons	0-2000m	16.0-28.0°C	12.5°C	2650-5055 (N=8) 3415.3 ± 279.6)	1750-5008 (N=3) (2864.3 ± 1072.1)	3077.9
Cebus capucinus	0-2100m	15.7-28.0°C	12.3°C	3550-3708.9 (N=3) (3655.9 ± 52.9)	2500 (N=2) (2500.0)	3077.9
Cebus apella	80-2800m	12.5-27.9°C	15.4°C	2870-4215 (N=9) (3737.5 ± 197.2)	2040-3000 (N=5) (2327.0 ± 182.9)	3032.25
Lagothrix lagothricha	80-3000m	26.4-11.3°C	15.1°C	7868-11500 (N=2) (9684.0)		9684.0
Ateles ssp. (=*A. belzebuth*, *A. hybridus*, *A. geoffroyi*)	0-2500m	13.3-28.5°C	15.2°C	7875-8625 (N=2) (8250.0)	7500-10500 (N=7) (9150.7 ± 336)	8700.35

Table 2. Altitudinal thermic tolerances and weight in colombian primates (modified of HERNANDEZ-C. & DEFLER, 1989).

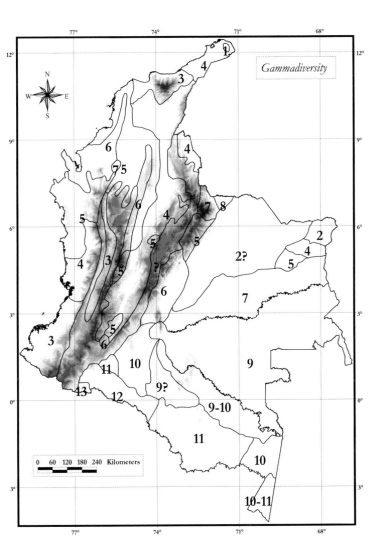

Figure 5. Distribution of primates in Colombia according to their gammadiversity. (modified from HERNÁNDEZ-C. & DEFLER, 1989).

Horizontal Clines

Horizontal clines are evident in various parts of the country. The map of Figure 5 shows some of the horizontal and vertical clines in Colombia. In the eastern lowlands there is a range of values from only two species in the north to around 12-13 sympatric species in the south. Additionally on the Pacific coast there is a less-exaggerated cline from the south, increasing to the north and being most elevated north of the equator.

It may be possible to correlate some of this increase in species numbers with an increase in the complexity of the vegetation or with a decrease in the duration of dry seasons. EISENBERG (1979) found a clear correlation between the diversity of primates and the total duration of the dry season during the annual cycle, which in parts of the Llanos Orientales can be as long as four months with little or no rain. GENTRY (1989) found a correlation between the increase in plant diversity and an increase in precipitation, although the correlation does not hold above about 2,500 mm. This correlation and that for precipitation above 2,500 mm and primate diversity shows the diversity leveling off and probably decreasing at higher precipitations, probably due to soil leaching and less biomass production (REED & FLEAGLE, 1995). This diversity tendency is what I find in eastern Colombia where precipitations vary from about 3,200 mm in Leticia to almost 4,000 mm north of the Caquetá River, apparently the result of soil leaching from excessive rains.

The lower Apaporis River in Vaupés have only eight sympatric primates with an annual precipitation of almost 4,000 mm. The forests north of Leticia and south of the Río Putumayo have only nine sympatric primates, coupled with an annual precipitation of about 3,400 mm. The most species rich site found by this author in eastern Colombia has been only 10 sympatric species on the Puré River, close to the Brazilian border, contrasting with sites with 12-14 species at 7°S in western Brazil by PERES (1997), in areas with less precipitation.

An additional factor of a reduction in soil fertility may be the main reason explaining these differences since the Estación Biológica Caparú, where the Vaupés data were collected, is found in a transitional zone where the poor soils of the biogeographic Province of Guyana interdigitate with the comparably more fertile soils of the biogeographic Province of Amazonas. The Apaporis River, close to the Caparú Biological Station is basically a mixture of blackwater and clearwater, which reflects the drainage of soils which are very infertile and include white sands (podsols), while the interfluvial zone between the Putumayo and Amazonas includes areas of soils that are more fertile than those from Vaupés. Of course the ultimate cause of the comparatively poor primate diversity is high rainfall.

The question arises whether edaphic reasons are enough to explain high diversity of primates in the Colombian Amazon, at least where there is sufficient precipitation during the entire year. Diversity in the piedmont forest of the Cordillera Oriental seems to be correlated with an increase in the development of such vegetation and of higher fertility closer to the mountains, perhaps allowing more niches. But this relationship needs to be measured quantitatively.

The zone identified as probably the most diverse in primate species in Colombia is also congruent with the hypothesized Napo refugium, where HAFFER (1967, 1974) found the highest avian diversity in Colombia and for the upper Amazon as well. However, the origin of these postulated refugia still is controversial, though there is no question that they are centers of high diversity.

Historic Origins for Geographic Distributions of Colombian Primates

HERSHKOVITZ (1963, 1977) suggested that the modern neotropical primate taxa in the Amazon had their origins in zones of peripherical uplands of the upper Amazon and that posteriorly these primates dispersed to the lowlands centripetally, strongly directed by the river courses which divided and cut off meanders in these rivers. We know that many modern taxa had ancestors living in what is now the upper Magdalena River basin in mid-Miocene times and that early on in this period there was no eastern Cordillera. The central Cordillera was the first to begin to form, permitting an ecological continuity throughout the upper northwestern proto-Amazonian area and probably blocked to the south by the proto-Amazon River which probably drained into the Pacific (KAY & MADDEN, 1997). Generally HERSHKOVITZ (1977) did not resort to Pleistocene refuges as a mechanism to explain differentiation, suggesting that river dynamics played a much more important part than did refuges. More recently AYRES & CLUTTON-BROCK (1992) discussed the effect of rivers on species range size in Amazonian primates.

Another older hypothesis for the differentiation of primates (and other flora and fauna) has been the argument that the majority of species and subspecies in the tropics had their origin in the Tertiary and are due principally to characteristics of continental zones and the sea, which repeatedly isolated and reconnected organisms that before had a wide geographic distribution. According to this view this allowed changes to develop in these populations of original species. The Hershkovitz (1977) view of the evolution of species is closely related to this vicariant model.

The third and currently most popular hypothesis is that of Pleistocene refuges, briefly described above. This model suggests that the Amazonian forest never has remained stable. It has changed considerably according to the reigning world climatic conditions. Since there have been strong climatic changes throughout the last 2,000,000 years of the Quaternary, there were alternatives in cold periods with low precipitation with warmer periods with high precipitation, which correspond to advances and retreats of glaciers.

Palynological research supports the idea of the occurrence of extensive changes in the vegetation. The maximum forest cover (which was more extensive than today's maximum) probably contracted many times, leaving dispersed enclaves or refugia where many forest organisms survived the drier times. In these refugia presumably such organisms differentiated until the next change in world weather patterns permitted the inevitable expansion and coalition of the forest and renewed contact between now related species. The conflicts in theories now have to do with the exact extent and locations of such refuges and vegetation changes were undoubtedly far more complex than we have imagined up to now as well as the putative ages of many species.

These competing hypotheses are often offered to explain speciation and even subspeciation, and each of the hypotheses uses allopatric speciation (well-described by Mayr, 1971), as the principal mechanism for the differentiation of the vertebrate species. It is to satisfy these conditions of isolation that hypotheses search for processes capable of isolating two or more populations of the same species from each other for enough time for a new species to arise. The requisites of the speciation process may have been filled via river barriers, refugia or other changes in the topography such as mountain ranges, etc.

Disagreements About the Allospeciation Process

Nevertheless, there is strong disagreement about the role these processes play in speciation, since because of the geological time involved such mechanisms are difficult to observe directly. Two categories of disagreement can be mentioned here. First, the possible existence of other processes of speciation besides allopatric speciation and second, the mechanisms in the development of subspecies.

Besides the suspicion of many that allopatric speciation is the principal mechanism for the evolution of two or more species from a parent species (MAYR, 1971; HAFFER, 1969, 1974; TEMPLETON, 1981), several other mechanisms of speciation have also been suggested which possibly could be important in the tropics: polyploidism, sympatric speciation and parapatric speciation. It is probable that these mechanisms of differentiation are more applicable to organisms which have reduced vagility, and because of this their importance to the majority of vertebrates would be reduced.

The Concept of the Species as Contrasted with the Subspecies Concept

Frequently refugia are evoked as the origin of many subspecies as well as the origin of species. Perhaps this is an error, since the two taxonomic levels (i.e. species and subspecies) represent processes which are quite distinct, and the differences are not sufficiently appreciated.

A species represents populations of an organism which are reproductively isolated from other populations. Woolly monkeys do not breed with spider monkeys when they are sympatric, thus each of these species exhibits reproductive isolation from the other. The development of reproductive isolation has been explained principally by the mechanism of geographic isolation, leading to evolutionary processes that result in any change which produces reproductive isolation. This is theoretically the role that the forest refugia fill for primates. A species is then a natural unit of evolution.

In contrast, a subspecies is an aggregate of populations that are phenotypically similar in a population, and such a subspecies is a geographic subdivision (subset) of the total distribution of the species. A subspecies is not necessarily an evolutionary unit as some imply. In many cases it is an artificial category which is always collective and possessing a taxonomic trinomial. The concept of subspecies makes life easier for the taxonomist, but in many cases it has no more importance, than to distinguish different populations. The tendency of some to equate subspecies to pronounced genetic differences actually subverts the original meaning of what a subspecies is or is not. Thus, the suggestion by COLLINS & DUBACH (2000a. 2001) and COLLINS (2001) that genetic evidence suggests the presence of only one southern subspecies of *A. geoffroyi* in Central America, due to the high overall similarity of haplotypes, seeks to redefine the meaning of subspecies, insisting that it be genetic. I am not certain that is a good thing, due to the many described subspecies based on pelage differences alone, which are certainly worth continuing to recognize and to conserve. A more profoundly natural classificatory system could easily group morphological subspecies according to their genetic differences, thus preserving both levels of information. The subspecies was never defined as a natural phylogenetic unit, but as a mnemonic taxonomy, recognizing phenotypic differences within a species.

There are differences in the evolutionary importance of subspecies. For example a subspecies that is geographically isolated from other populations of the species is evolutionarily more important since it is automatically an incipient species. This type of subspecies should be viewed as being more important than other subspecies, and should be accorded special conservation efforts, since this population is hypothetically on the road to the eventual development of reproductive barriers and will eventually become a separate species. Also subspecies which possess pronounced genetic differences may be of higher priority for preserving than subspecies which are geneticially less different from other subspecies.

HERSHKOVITZ (1988) criticizes the use of the concept of Pleistocene refugia especially as it is used to explain the origin of subspecies, principally, according to him, because the nature of the majority of mammalian subspecies is clinal, which is to say that the changes in the characteristics of the population occur in a series of contiguous populations in a continuous manner, a reflection of genetic flow in the population. Primatological examples of this form of subspecies are *Pithecia monachus monachus* and *Pithecia monachus milleri*; *Saimiri sciureus cassiquiariensis*, *Saimiri sciureus macrodon*, and *Saimiri sciureus albigena*; *Saguinus nigricollis nigricollis*, *S. n. graellsi* and *S. n. hernandezi*. There is no evidence in these subspecies of an interruption of gene flow which would be manifest in isolated populations.

FORD (1994) has recently found evidence of clinal variation in populations of *Aotus* made up of the species *A. lemurinus, A. brumbacki* and *A. vociferans*, while she found interrupted characters between this group and *A. trivirgatus*. Whether this proves the existence of a wide-spread *A. lemurinus* made up from the first three "species" listed above or these clinal characteristics have another interpretation remains to be seen. Certainly DEFLER *et al.* (2001) and DEFLER & BUENO (2003) emphasize the evidence of the karyotypes over that of the aparent clinal nature of some morphological characteristics, although this question of FORD´s (1994) needs much more analysis.

Nevertheless, there are examples of subspecies which exhibit abrupt character changes which might have had their origins in temporary isolation before reestablishing contact with other populations of the species. Perhaps various subspecies of *Saguinus fuscicollis* fill this description although not the subspecies *S. f. fuscicollis* and *S. f. fuscus*, which show clinal differences. Possibly *Callicebus cupreus* subspecies are examples of isolation since they exhibit some abrupt changes in coloration. However, it is worthwhile mentioning that the *Callicebus* have a social system of very short-distance dispersal, which probably greatly reduces gentic flow when compared to species which disperse great distances like *Alouatta seniculus* and *Ateles belzebuth*.

Lowering genetic flow probably permits phenotypes to develop and to stabilize, for example on the two sides of a river such as, for example, *Callicebus torquatus lugens* and *Callicebus torquatus lucifer* separated by the Caquetá/Japurá River. But *Callicebus toquatus lucifer* and *C. t. medemi* show an abrupt difference without being isolated from each other. Should a difference in hand color (white or black) be enough to distinguish different species, as argued by GROVES (2001)? Some abrupt color changes may actually be mediated by simple genetic differences and other color fields on *Callicebus torquatus lucifer* and *Callicebus torquatus medemi* are the same in the two taxa.

Since little is known about the genetics of many color changes, it is possible that many abrupt changes in phenotype are mediated by very simple allelic differences. In the case of the two subspecies of *Callicebus torquatus* this question could easily be studied, since populations of the two subspecies are probably in contact and may exhibit some simple Mendalian sorting.

Some Possibilities concerning the Development of Contemporaneous Geographic Distributions of Colombian Primates

It is not yet possible to identify the ancient origins of neotropical primate taxa due to the scarce fossil evidence, although we do know from the La Venta primate fauna that modern lines were probably well-established in the area of the upper Magdalena River valley (which probably did not exist as a river or a valley at the time) by the mid-Miocene, before the initiation of the Andean orogenesis, which began only about 5 million years ago in the Pliocene (KAY & MADDEN, 1996). More recent primate dispersal is very poorly understood, although it is possible to make some observations.

Cebuella pygmaea

This species probably originated in the upper Amazon, perhaps in a Napo refugium, then extending itself with very little differentiation to the Caquetá River, except for a population which managed to cross the upper Caquetá , presumably via the interruption of meanders in the river. The species then dispersed along the whitewater Caguán River, at least to the area of San Vicente de Caguán, perhaps managing to cross to the Guayabero River (also whitewater), although this needs to be confirmed with more observations.

Apparently the species could not establish itself along blackwater rivers for ecological reasons. Although this monkey is very small, its dispersal abilities must be very efficient since the homogeneity of its populations suggest a very strong genetic flow between populations.

Saguinus

In Colombia this genus, which principally is of the upper Amazon, has a very interesting geographical distribution, since there are trans-Andean taxa in northern Colombia which are widely separated from the principal distribution. Ecologically this genus can be called a pioneer genus, since it survives well in zones of secondary vegetation, permitting high population densities in disturbed forest.

One possible dispersal route for the Pleistocene ancestor of the three species of trans-Andean *Saguinus* (*S. geoffroyi*, *S. leucopus*, and *S. oedipus*) found in northern Colombia seems to be the piedmont of the Cordillera Oriental to the north, crossing the Cordillera to the west of Maracaibo Lake in Venezuela or via some low pass over the Cordillera Oriental, such as near Uribe, which is 500-600 m higher than contemporary members of these three transandeans are known to go today. There is no explication for the contemporary absence of *Saguinus* to the east of the Magdalena River where there is adequate habitat for this genus, although interestingly an excellent description of *S. leucopus* was published by Fray Diego for the right bank town of Timaná on the upper Magdalena River. Perhaps the specimens which Fray Diego saw were transported there or perhaps this was a remnant population that may no longer exist?

The differentiation of the three trans-Andean species from a southern ancestor probably occurred for *S. geoffroyi* in a Chocoan refugium, for *S. oedipus* in a refugium of Paramillo and for *S. leucopus* for a refugium in the Serranía de San Lucas (Nechí ?). These three species seem to be adapted to conditions that are a bit more xeric than those of the majority of *Saguinus*. It is interesting to note that the cis-Andean species *S. inustus* and *S. nigricollis* also have populations which survive in habitats which are drier than usual for the genus.

The origin of *S. inustus* could have taken place in any one of various small refuges that probably existed around serranías such as Chiribiquete (P.N.N. Chiribiquete), Naquén or Taraira, although the ecological requirements of this species need to be confirmed to know whether it tolerates such poor vegetation. The species is not present in the Amazon caatingal in the east of Guainía nor is it found in the right bank of the lower Guaviare (neither in secondary vegetation nor primary vegetation), according to inhabitants of the zone. This suggests some interesting limitations, since the species has been collected on the right bank of the Guaviare close to San José de Guaviare.

The origins of *S. nigricollis* and *S. fuscicollis* also are speculative and complex, because of several subspecies. Nevertheless, it is possible that the two species originated in a Napo refugium. The two species dispersed along the upper Caquetá River to the north, *S. nigricollis* reaching the Guayabero River as *S. n. hernandezi* while *S. fuscicollis* was unable to disperse to the east side of the Caguán River. The two are macrosympatric west of the lower Caguán River.

Aotus

This genus is in great taxonomic flux since HERSHKOVITZ' (1983) recent revision which underlined the need for karyological and morphological information necessary to have a clear understanding of the relations between populations. There are many well-founded speculations that can be mentioned. GALBRAITH (1983) proposed a hypothetical plan for the evolution of *Aotus* karyotypes which helps us to understand the possible relationships that exist. HERSHKOVITZ (1983) proposes an origin for *Aotus* from the north of the Amazon river, based on the idea that the grey-necked group (found north of the Amazon) is more primitive than the group to the south, which has orange-necks, according to the theory of the principle of metachronism in mammals (see, for example, HERSHKOVITZ, 1977 and below). The Napo refugium was probably important in speciation of *Aotus*, based on its apparent importance for so many other taxa and the fact that there were other now extinct species dating back at least to the Miocene to the west of the hypothecised refugium.

The species recognized as *A. lemurinus* (*sensu* HERSHKOVITZ, 1983) is made up of various distinguishable populations, including the Central American *A. l. lemurinus* which is distinguishable from the Colombian population *A. l. lemurinus* (which perhaps would best be called *A. l. zonalis* or *A. zonalis*) and whose origin is probably from the Chocoan refugium. The taxon *A. l. griseimembra*, which was important in malarial research, has perhaps two origins, one from the Caribbean coast (or from a refugium of Sierra Nevada de Santa Marta) and the other from higher altitude Andean habitats and perhaps requiring yet another taxonomic designation.

The karyologically distinct *A. brumbacki* originated perhaps in a Villavicencio refugium. This species is most closely related karyologically to *A. vociferans* of the Colombian Amazon, which probably had its origin in the Napo refugium. FORD (1994) compared cranial data for *A. lemurinus, A. l. griseimembra, A. brumbacki,* and *A. vociferans* and found that these northern (grey-necked) taxa presented a clinal gradient that would suggest that they are one species group. This species group contrasts with *A. trivirgatus*. It seems worthwhile here to closely examine the karyological behavior of these groups, in order to determine how reproductively isolated these populations are from each other, but it is our belief that the known karyomorphs of this species group would act as reproductive barriers (DEFLER & BUENO, 2003).

Without doubt there are still other species of *Aotus* that need to be recognized. We have synonimyzed the recently described taxon *Aotus*

hershkovitzi Ramírez-Cerquera *et al.* 2000 (2n=58) from the Cordillera Oriental with *Aotus lemurinus* (Defler *et al.*, 2001). A species from Maipures (in
Vichada) (2n=50) does not seem to be *A. brumbacki,* although the original
karyotypic preparatons are now lost. Torres *et al.* (1998) described yet
another karyomorph of *Aotus* with 2n=50, which also appears to be a
species other than *A. brumbackii.* There is still much research to be done to
be able to explain the phylogenetic relationships and systematics of this
interesting genus.

Of ecological interest is the wide range of thermic tolerance found in this
genus, since other Colombian primates in this size range do not reach such
varied altitudes. *Aotus* reaches altitudes of 3,200 m. Ecophysiological and
other adaptations have permitted dispersal of the genus throughout the
country to altitudes almost as high as *Alouatta, Cebus* and *Lagothrix* which
are animals considerably larger and diurnal, taking advantage of the warm
hours of the day. In contrast, *Aotus* is active at night, during the coldest
hours of the diurnal cycle. There are also areas in Colombia where the genus
is apparently absent, such as the northern parts of Vichada, north of the
Tomo river and perhaps in the eastern parts of Arauca and Casanare, reflecting
some other ecological limitations.

Callimico goeldii

This genus shows an interrupted distribution in the upper Amazon,
suggesting peculiar ecological requirements which are not easily filled. Rylands
(com. pers.) suggested that the geographic distribution of this species might
correspond to the western shores of an immense post-Pleistocene lake,
now extinct, recently suggested by Frailey *et al.* (1988) and Campbell and
Frailey (1984), based on studies of lacustran deposits. This species seems
to require a very particular, specialized habitat, since it is scarce where it is
found. To be able to say more about its geographic range the species'
ecological requirements will also have to be better understood.

Saimiri sciureus

The geographic distribution of this species is principally Amazonian,
which may by the origin of the species. Nevertheless, the species has
dispersed to the upper valley of the Magdalena, requiring an explication for

the dispersal over the Uribe pass of 1,600 m. The present distribution suggests the distribution occurred during a past epoch that was a bit more moist, when the limits of thermic tolerance were at a higher altitude (for *S. sciureus* only about 300 m higher).

Perhaps the dispersal of this species to the upper Magdalena river occurred during the last climatic optimum 5,000 years ago, although it is more probable that there had been sporadic connections during Pleistocene times, necessitating a Pleistocene refugium in the upper Magdalena valley, a possibility that has been suggested by other types of evidence. The fact that another species of *Saimiri* is clearly recognized from La Venta fauna suggests that *Saimiri* and perhaps *S. sciureus* have an ancient history in the region along the upper Magdalena River, perhaps this is the region from which the genus dispersed to the Amazon, instead of the other way around.

There are three recognized subspecies of *S. sciureus* in Colombia: *S. s. albigena* of the piedmont of the Cordillera Oriental and the upper valley of the Magdalena river, *S. s. cassiquiarensis*, probably east of the Apaporis river and extending to the Vichada river, and *S. s. macrodon*, south of the Apaporis River. These subspecies are strongly clinal and probably originated from a part of a continuous population, although others place their origens in refugia: *S. s. macrodon* from the Napo, *S. s. albigena*, perhaps from the upper Magdalena River and *S. s. cassiquiarensis* from somewhere in the eastern lowlands.

Callicebus

HERSHKOVITZ (1988) suggests an origin for *Callicebus* in the upper Amazonas, to the west of the contemporary species, from which the genus speciated as it dispersed down the sides of rivers. Possibly there was isolation via the occasional savannas which replaced forests, producing species and subspecies. HERSHKOVITZ (1988) suggests that *C. torquatus* (which he terms a «derived species», because of its size, color and karyotype) probably originated in the northwest Amazon.

Callicebus cupreus

The ancestors of *C. cupreus* presumibly originated in the area of the southeast Amazon, dispersing via gallery forests of the Amazonian tributaries (HERSHKOVITZ, 1988). A white frontal blaze on the forehead, marbled tail, rust-red feet and stomach and the phenolization of the limbs are shared between *C. c. discolor* and *C. c. ornatus*, which demonstrates an afinity and recent conection with a contiguous population that has been interrupted.

This species seems to be adapted to low forest. Often it is found in flood-plain forest in very thick vegetation, as well as in gallery forest in low areas. In Colombia there are apparently three isolated populations. The northernmost population *C. c. ornatus* is distributed principally north of the Guayabero River to the Upía River in Meta. Possibly this subspecies developed in the Villavicencio refugium. But it is also possible that this clinal subspecies was merely the end of a varied and larger population which became isolated, perhaps after increasingly moist conditions supported a rainforest that did not encourage this species, success, since the species niche suggests that it survives better in low forests such as might be found in floodplain or which is poorly stratified.

Callicebus c. ornatus seems to be isolated from another population of *C. cupreus* that was reported by MOYNIHAN (1976b) for southern Caquetá Department between the Ortequaza and Caquetá Rivers. This population is not yet described, but apparently it does not have the frontal blaze. Absence of the frontal blaze would imply that subspecific differences are discrete and not clinal and would also be a serious problem for HERSHKOVITZ' (1969, 1977) theory of pigmentation, since he states that a reversal in bleaching (the frontal blaze) is unlikely. Alternatively both *C. c. discolor* and *C. c. ornatus* may be derived from this southern Caquetá population.

The southern Caquetá population is isolated from another subspecies, *C. c. discolor* known south of the Guamués River in the department of Putumayo and extending throughout a great region at least to the Marañon river in Peru and possibly to the Colombian trapezium. Neither the population in the trapezium nor that between the Orteguaza and Caquetá Rivers have been confirmed by this author.

It would seem that an expanded and connected population of *C. cupreus* would have had to be united by appropriate habitat in the piedmont, i.e. a corridor of low, poorly stratified forest similar to that where the species is most commonly found today and distinct from the tall and exuberant forest which now fills in the gaps (or at least filled in the gaps until the contemporary destruction of that forest). Perhaps the extensive Pleistocene inundations permitted a successional gallery forest that was capable of supporting a continuous distribution of this species of *Callicebus* from the border with Ecuador up to the Upía River in Colombia. Or, contrary, the last dry phase, perhaps Wisconsin glacial, supported such a forest. Then, with the passage of time and changing conditions, ecological succession may have permitted a more developed forest (and more apt for *C. torquatus*) to replace the extensive *C. cupreus* habitat, leaving remnants of this species capable of sustaining itself. Nevertheless, some anecdotal evidence suggests that *C. cupreus* may be present in very low densities where the habitat permits in some other parts of the Colombian Amazon, but this needs observational confirmation.

Callicebus torquatus

This species seems to be adapted to high and well-developed rainforests (DEFLER, 1994a). It is found widely distributed throughout the Colombian Amazon up to the Vichada River. To the north of the Vichada River a population is found in gallery forests which reach the right bank of the Tomo River. Though this population is found also towards the source of both the Tuparro and Tomo Rivers, it is not found down river on the middle or lower parts of the Tomo, and this population extends on the Tuparro only to the middle Tuparro where it ends.

The gallery forest along Caño Hormiga and on the west side of Tuparro National Park, which contains the northernmost known Colombian population of *C. torquatus*, has a precipitation at Tapón of 2,363 mm, a bit more than that registered on the other side of the park at the mouth of the Tomo River (1,905 mm at Centro Administrativo) close to the Orinoco (where *C. torquatus* is not found). Tapón represents the eastern extent of the northernmost population of the species in Colombia (a bit further to the south the species extends more easterly along the Tuparro River). Perhaps during the time of the climatic optimal of about 5,000 years ago there was more forest between the middle Tomo and the middle Vichada, which would have been more apt for the species. Presently the land is 80-90% savanna, with great areas bereft of trees and few connections between gallery forests. The present population found near Tapón probably represent a remnant of what previously occupied a more forested zone.

KINZEY (1982) and KINZEY & GENTRY (1979) locate the origin of *C. torquatus* in an Imerí refugium, and they postulated subspecific origins in specific Pleistocene refugia. Nevertheless, their hypothesis does not agree with a more modern concept of *C. torquatus* subspecies, according to HERSHKOVITZ (1988, 1990). Also, it is probable that the Imerí refuge did not include land to the west of the Negro River, which represents the eastern edge of the geographic distribution for the species. Nevertheless, there is an island of elevated precipitation south of Mitú and north of the Caquetá River, which reaches about 4,000 mm. An annual decrease of precipitation in the Colombian Amazon of 1,500 mm would cause the savanization of great sectors of what is now Amazonian forest and would leave a forest island in this zone of high precipitation that could represent the area where *C. torquatus* developed, taking into account that, as part of its repertoire of habitats this species survives well in forests which grow over white sands. This region postulated as a possible refuge for *C. torquatus* contains a mosaic of soils of white sands and clays of various colors. As *C. t. lugens* exhibits a diploid chromosome number of 16 as opposed to a diploid number of 20 from more southerly populations of this complex. It may be that *lugens* was separated for some time from those with 2n=20. HERSHKOVITZ (1988, 1990), prefers to evoke fragments of gallery forests, detours of river beds, and interruptions of river meanders as mechanisms that are sufficiently capable of causing subspeciation and speciation of mammals, usually clinally and not discretely.

Pithecia monachus

The origin of *Pithecia* could have been from areas to the west of the upper Amazon, especially considering that *Cebupithecia* or *Mohanamico hershkovitzi* from the La Venta formations of the upper Magdalena River might have been ancestors of this genus. During the Miocene there was of course no barrier to what is the Amazon valley today. *Pithecia m. milleri* would have had to disperse to the north, crossing the upper Caquetá River above the mouth of the Caguán River . There seems to be an absence of pithecines between the Caguan and Apaporis Rivers, *Cacajao melanocephalus* being found east of the latter river, although more field work must be done to confirm this.

Cacajao melanocephalus

The geographic distribution of the scarce populations of *C. melanocephalus* seems to be connected ecologically to the use of igapó (inundatable blackwater forests). Recent censuses of primates in the department of Guainía on the banks of the Guaviare River, a whitewater river, confirm that the species is found only in the banks of blackwater affluent of the Guaviare and not in the actual Guaviare.

This species is closely related to *Cacajao calvus*, both species have the same chromosome numbers (2n=55, 56). *Cacajao melanocephalus* seem typically to exhibit long distance movements through upland forests when there is no food present there. The origin of this species and of *C. calvus* could have taken place in the banks of large rivers such as the Solimoês in the case of *C. calvus* and the río Negro in the case of *C. melanocephalus* where it became isolated on the right bank of the Río Negro. HERSHKOVITZ (1987a) mentions that *C. melanocephalus* is more primitive than *C. calvus* because of its pigmentation and hair present on the head. It is possible that eventually two subspeces might be found in Colombia (rather than only *C.m.ouakary*), since *C.m.melanocephalus* east of the Inírida river has no geographic barrier isolating Colombian and Venezuelan species and the subspecies west of the Casiquiare Canal in Venezuela is supposedly *C.m.melanocephalus*.

Cebus apella

This species has a geographic distribution that is the most extensive of any neotropical primate, although in Colombia the extent of its distribution is second to *Alouatta seniculus*, which is the most widely distributed Colombian species. Both *C. apella* and *A. seniculus* have been termed pioneer species, because of their considerable distribution which includes very diverse habitats. *Cebus apella* is, like *Saimiri sciureus*, established in the upper Magdalena River, but it is not found in the middle or lower parts of that river where it is replaced by its congener *C. albifrons*. The dispersal route between the upper Magdalena and the Amazon basin must be the 1,600 m Uribe pass between Meta and Huila.

In various isolated zones (Arauca, upper Cahuinarí River in the Amazon, lower Tomo in Vichada), *C. apella* has been completely replaced by *C. albifrons* producing an island population of *C. albifrons* surrounded by *C. apella*. The usual Amazonian situation is the presence of both sympatric species with *C. albifrons* at lower densities. However, in the upper Cahuinarí *C. apella* drops out, leaving comparatively high densities of *C. albifrons* in this white sand region (see DEFLER, 1985a); HERNÁNDEZ C. & COOPER, 1976). Since in the zones of sympatry of these two species known to the author, *C. albifrons* densities are much lower than *C. apella* densities, the suggestion is that *C.*

apella excercises an inhibitory effect upon *C. albifrons*, or simply that extensions of lowland rainforest are not the ideal habitat for *C. albifrons*.

Cebus albifrons

This is a species broadly and sympatrically distributed in the Colombian Amazon with *C. apella*. The species seems to replace *C. apella* totally in the upper Cahuinarí River. Since the upper Cahuinarí has extensive zones of white sands, apparently this species can exploit this habitat better than its widespread congener. Nevertheless, the two species are sympatric in some white sand areas east of the Inírida River, although the vegetation on that white sand is also very different from the upper Cahuinarí.

Cebus albifrons exhibits a stronger preference for xerophytic vegetation than does *C. apella*, which prefers more mesophytic vegetation (DEFLER, 1985a), suggesting that *C. albifrons* could have originated as a species in the rocky and sandy forests or the biogeographic Guyana Province. Nevertheless, *C. albifrons* has dispersed to the north of Colombia towards the Magdalena-Cauca Rivers and has reached the isolated Serranía de Macuira in the Guajira peninsula, perhaps following the same route as *C. apella* during a different epoch or perhaps following the piedmont of the eastern side of the Cordillera Oriental, diverging into several clinal subspecies along the way.

HAFFER (1974) compares the parapatric distribution of the related *C. albifrons, C. nigrivittatus,* and *C. capucinus* to superspecies complexes of birds, although no research has been done on the contact zone between these three species. As a superspecies complex these species ought to have some hybridization in narrow zones of contact (HAFFER, 1974; MAYR, 1971). HERNÁNDEZ-C. & COOPER (1976) mentioned a possible zone of hybridization between *C. capucinus* and *C. albifrons* which deserves to be studied in this respect.

Although *C. apella* seems to be more differentiated from the other three species of *Cebus*, a preliminary description of the zone of contact between populations of *C. albifrons* and *C. apella* in El Tuparro National Park suggests interesting interactions between these two species, although hybridization is not known. Similar diets seem to guarantee strong competative interactions between these two species. Future studies of the ecological relations between *C. apella* and *C. albifrons* should help clarify the interesting parapatric distribution between the two found in some places.

Cebus capucinus

This species probably originated in a Chocó refuge from where it spread northward in Meso-America and southward, along the coastal Chocoen plain. Populations of this species are poorly differentiated, and many do not fit into described subspecies because of this lack of differentiation (HERNÁNDEZ-C. & COOPER, 1976; COIMBRO-FILHO & MITTERMEIER, 1981).

HERNÁNDEZ-C. & COOPER (1976) describe a probable hybrid zone of *C. capucinus* X *C. albifrons* along the middle San Jorge River. This contact zone needs to be studied. *Cebus capucinus, Cebus albifrons,* and *Cebus nigrivitattus* (Venezuela and Guyana) form a superspecies that would benefit from comparisons and contrasts when finally enough observations of these species are available.

Alouatta seniculus

This is the primate with the most extensive distribution in Colombia. No clearly differentiated subspecies for *A. seniculus* have been described for the country, probably indicating very low variability and intense genetic flow, according to HAFFER (1974), who compares this primate with two species of cotingids (*Querula purpurata* and *Gymnodenus foetidus*) which are also widely distributed throughout the Amazon but with a very homogeneous population and no known subspecies. *Alouatta seniculus* has strong dispersal abilities. These monkeys are capable of crossing extensive treeless savannas, swimming wide rivers and surviving in zones of poor vegetation. These characteristics support the idea of strong genetic flow in the species, and such dispersal abilities easily explain the wide distribution, since broad rivers are no barrier for the species.

Alouatta seniculus is also a eurythermic species with an altitudinal range which varies between 0-3,200 m. This is the widest range of ecophysiological adaptations of all the Colombian species and is reflected in the species dispersal ability. Adaptations favoring wide dispersal ability are the ability to swim, lowest stable social unit being a pair of animals, a highly herbivorous diet, the habitual dispersal of both sexes, and perhaps other adaptations.

Alouatta palliata

This primate species has a limited geographic distribution in the western part of Colombia, although it is also widely found in Central America. This monkey is perhaps not as exceptionally adapted as *A. seniculus* to disturbed vegetation nor to long-distance dispersal, although frequently it is the last species found in patches of forest when the species is not being hunted out. It is possible that *A. palliata* is better-adapted to closed-canopy upland forest, when compared to *A. seniculus*, a species adapted to floodable forests, forest edge and successional forest. Sympatric populations of the two *Alouatta* species on the west bank of the Atrato River (Chocó) find *A. palliata* on hills and *A. seniculus* in the lowlands between the hills (HERNÁNDEZ-C. & COOPER, 1976; HERSHKOVITZ, 1949). It is possible that the species evolved in a Chocó refuge.

Lagothrix lagothricha

Lagothrix lagothricha is found throughout the great majority of Amazonian forest in Colombia, including the slopes of the Cordillera Oriental up to about 1,800 m altitude. This species was originally dispersed along the slopes of the Central and Eastern Cordillera above the Magdalena River, from Puracé in the south to the Serranía de San Lucas in the north (Bolívar). Nowadays the species is very scarce (although present locally) in the western slope of the Eastern Cordillera, which is an area of the country which has been radically altered by humans. Between the valley of the Páez River and the Serranía de San Lucas in the Cordillera Central no sign has been found of this primate, although at one time it would have had to be a continuous population in that region as well, since currently an isolated population is known for the San Lucas.

Lagothrix lagothricha probably originated in the western Amazon, presumibly in a Napo refuge. The two Colombian subspecies are strongly clinal with much color variation in both, making them at times rather difficult to distinguish using FOODEN'S (1963) revision.

Ateles

Ateles is widely distributed and a systematically difficult group, since there is much variation and hybridization. KELLOGG & GOLDMAN (1944) recognized four species. They illogically defined widely separated subspecies of one species intercalated with subspecies of another species. HERSHKOVITZ (1969, 1972a) and HERNÁNDEZ-C. & COOPER (1976) see *Ateles* as one (*A. paniscus*) polymorphic species.

Recently FROEHLICH *et al.* (1991) published a morphometric analysis based on 50 measurements of 284 specimens, from which they were able to produce a dendrogram of taxonomic relationships which supports a taxonomic revision that would recognize three (or perhaps four, according to FROEHLICH, per. com. 1994) species, located more logically than that of KELLOGG & GOLDMAN (1944). With this concept of *Ateles* all of the taxa of NW Colombia together with the Central American taxa form one species (*Ateles geoffroyi*), either including or not including *A. hybridus*. *Ateles belzebuth* (without *A. hybridus*) would be another distinct and clinal species which would include *A. b. belzebuth* (found in Colombia), *A. b. chamek* (found in Perú, Bolivia and Brazil) and Brazil's *A. b. marginatus* (*sensu* KELLOGG & GOLDMAN, 1944), while *A. paniscus* of the northern Amazon and Guyanas would be part of an independent species that is known to be the best-defined.

With the above hypothesized scheme Colombia has two (*A. geoffroyi*, including *A. hybridus*) or three (*A. geoffroyi*, *A. hybridus* and *A. belzebuth*) species. If we now include the comparative DNA work of COLLINS (1999) and COLLINS & DUBACH'S (2000a, 2000b, 2001) with their research on mitochondrial and nuclear DNA of *Ateles* we must conclude that the species status of *A. hybridus* is a monophyletic group "without clear ties to any other spider monkey clades". The species *Ateles hybridus* would include both *A. h. hybridus* (east of the Magdalena River into Venezuela) and *A. h. brunneus* in the northern parts of the Central Cordillera and endemic to Colombia. This is the best concept of *Ateles* systematics that has been suggested so far. It provides data which may indicate that *A. belzebuth* could have originated in the upper Amazon in some refugium (Napo ?) or via some vicariant event, giving rise to its dispersal into a sort of ring species showing clinal variation in skull morphology from one side to other of major rivers, even though these same rivers influence color patterns.

Using the above concept of *Ateles* it is possible to hypothesize that *A. geoffroyi* originated in some refugium in northern Colombia (Chocó ?) during a Pliocene glacial period, with later dispersal and differentiation via fluvial barriers, other vicariance events or refuges during the repeated wet-dry cycles of the Pleistocene.

Cladistically *A. hybridus* forms a well-defined group, but no data exists to examine populations on both sides of the Magdalena River, which should have been a rather effective barrier since the last glaciation. It is clear that the previous *"Ateles belzebuth hybridus"* does not belong within the *Ateles belzebuth* group, and has obviously been isolated from it for a long time. The nature of any contact zone between *Ateles belzebuth* and *A. hybridus* on the Cordillera Oriental of Colombia would be highly interesting to study.

These preliminary comments on the zoogeographic and systematic relationships of Colombian primates will hopefully provide a basis for future hypotheses and research that will allow us to understand more clearly how Colombian primate populations came to be distributed as they are. Most exciting may be the future possibility of understanding the basic ecological and historical factors which formed the present patterns of primate biodiversity in the country.

4. The conservation of primates in Colombia

Even though Colombia is one of the most primate-rich countries in the world (MITTERMEIER & OATES, 1985), this fact remains generally unappreciated within the country, and many primate populations, both of species and subspecies, continue to become increasingly threatened. Forest-living indigenous peoples and other inhabitants of the lowland forests, traditionally hunt primates for meat, and a considerable commerce arose in the 1960s and 1970s with their export for biomedical research. Primates are otherwise given little value in Colombia, and here I consider this unfortunate aspect of the modern world where our ignorance of such topics is not quite so pronounced as it was 500 years ago when our European forebears first arrived.

Because the original forests, once covering about 80% of Colombia's 1,197,000 km² of national territory, were so pervasive and overwhelming to the new European colonists, coming as they did from a more subdued European land, the great tropical *selva* and its mysterious inhabitants, including indigenous peoples and primates were seen only as a barrier to the subjugation of the land. This attitude, seen in all of the "discoveries" of the Americas, has contributed heavily to much unnecessary destruction of the natural environment, since the dominant land ethic has always been its conversion to crops and pasture for cattle and other livestock. This philosophy is a European legacy (see, for example, NASH, 1983), and although such a view has fed and enriched many people and permitted human dominion over vast expansions of land (for better or for worse), an often ignored consequence has been the precipitous loss of biodiversity that had otherwise sustained itself for millennia, much longer than the history of humankind. Biodiversity (including primates) has a value which is absolutely comparable in importance to crops and cattle, and this reality needs to be carefully considered by society.

Consequences of Ecosystem Degeneration to Humans

Usually the value of an organism is expressed in terms of its role in a functioning ecosystem. Since our world is made up of many, many ecosystems, we are beginning to realize that the health and well-being of one part (such as a city) depends on the well-being of another (the land, forest and the sea). This concept is intellectually accepted by most modern countries, and is endorsed in important international documents such as The World Charter for Nature of the United Nations General Assembly and the official Colombian report for the United Nations Conference on Development and the Environment of 1992 (REPÚBLICA DE COLOMBIA, 1992). Nevertheless, in practice there is a huge gulf between the official statements of our world's governments, and the realities of operating a government policy as if an ecologically balanced world did in fact matter.

There are many thousands of examples of ecosystemic disturbances in the world, and they are increasing. In Colombia, the destruction of the forests along the banks and the drainage slopes of the Río Magdalena, along with the river's contamination, has had enormous consequences for hundreds of thousands of people because of the loss of land for food production and the danger of flooding. The degeneration of watersheds in Colombia has greatly increased annual flooding and reduced the supplies of clean drinking water and potential electricity generation for millions of citizens; and this in a country with very high rainfall and immense hydroelectric potential.

The Colombian *páramos* are the key to fresh water, and the country is privileged with 57% of the entire extent of these peculiar Andean vegetation types. Yet *páramos* have been invaded and destroyed by cattle and colonists, ploughed under by potato farmers in every corner of the country, with no consideration for watershed preservation or for the survival of a unique ecosystem containing many endemic organisms. National parks in Colombia legally protect 10% of the nation, yet almost 30% of parks´ land is invaded by colonists or subject to other claims. Personnel of the national park system have never received the national support they need to accomplish their duties, subject as they are to excessive budgetary constraints and lack of political will.

In many parts of Colombia despite legal prohibitions, there are still no actual consequences for trapping or shooting primates, birds, or any other animal, including those of species which are endangered. These products of millions of years of natural selection and evolution are devalued and deprecated on all sides, with the attitude that somehow the products of the natural world have little value and are, at any rate, there to be consumed at the pleasure of any human being. Somehow the destruction of nature has become a biblical right.

Let us then ask ourselves the question, what actual value does an organism such as a monkey have? Why worry about conserving it?

Assessing the Value of an Organism

One of the traditional ways of assigning a value to an organism is to consider it within the framework of its natural ecosystem. Although some primates such as the Anubis baboon (*Papio anubis*), the gelada baboon (*Theropithecus gelada*), and the vervet (*Cercopithecus aethiops*) in Africa live in open, terrestrial habitats, and others such as the Japanese macaque (*Macaca fuscata*), the Barbary ape (*Macaca sylvanus*), the Hanuman langur (*Semnopithecus entellus*), and the golden monkey (*Rhinopithecus roxellanae*) in Asia can be largely terrestrial, occupying temperate forests, most primates are dependant on tropical forests. All Neotropical primates are totally dependent on tropical or (some few) subtropical forests. Although the function of primates within a natural ecosystem is far from being well understood, many ecologists have stressed their importance as seed-dispersers (VAN ROOSMALEN, 1985; TERBORGH, 1983; ESTRADA & FLEMING, 1986; DEFLER & DEFLER, 1996; LAMBERT & GARBER, 1998; STEVENSON, 2000). Within the species-rich tropical forests where primates are most common, they are for this reason major contributors to the maintenance of plant diversity.

As an example, a troop of woolly monkeys (*Lagothrix lagothricha*) annually disperses the seeds of 200-300 species of plants throughout their home range (DEFLER, 1989c; DEFLER & DEFLER, 1996; PERES, 1993a, 1994a; STEVENSON, 2000), while other sympatric primates greatly increase this list. Another example of important seed dispersers are the highly frugivorous spider monkeys (*Ateles*) (VAN ROOSMALEN, 1985; DEW, 2002a; STEVENSON *et al.*, 2002). This vital role is not sufficiently appreciated, since without them and other seed dispersers, the extraordinary diversity of these forests is diminished (ESTRADA & FLEMING, 1986; RAEZ-LUNA, 1994).

Besides contributing to the well-being of their tropical forests, and thus to our own by helping maintain these diverse ecosystems, there are other values to be considered. Renewable resources can be classified as having direct and indirect values (MᶜNᴇᴇʟʏ *et al.*, 1990): as commodities (direct) or as amenities, their existence improves our lives (indirect); and moral/ethical terms allow for the rights other life forms to exist (Nᴏʀᴛᴏɴ, 1988). As commodities primates provide meat and are used as pets and in biomedical research. However much these uses may be deplored by some, the use of the larger primates species as a source of protein by indigenous hunters and some colonists is of no small economic importance; these are human populations which may measure its cash income at the bottom of the national economic ladder. Because of low levels of protein biomass and the difficulty of its detection, an Amazonian hunter often kills any game animal that crosses his path, even though some are preferred over others (Rᴇᴅꜰᴏʀᴅ & Rᴏʙɪɴsᴏɴ, 1991), unfortunately including the larger primates (Rᴏʙɪɴsᴏɴ and Rᴀᴍíʀᴇᴢ, 1982; Rᴏʙɪɴsᴏɴ & Rᴇᴅꜰᴏʀᴅ, 1991, 1994; Mɪᴛᴛᴇʀᴍᴇɪᴇʀ, 1987b). Indirectly, primates play a fundamental ecological role as part of the tropical forest, and their value can be considerable for wildlife observation and ecotourism.

Direct values can often be assigned using market place pricing, but do not take into consideration those values which are not as yet widely recognized, and monetary values change: a resource's worth today is not a true reflection of its value in the future. Indirect values are even more difficult to calculate, although many believe that they are ultimately the most important for humanity. We may eventually fully understand the ecological role of organisms such as primates in an intact ecosystem, but equally our society must work to develop a full appreciation of any organism as a unique representative of a process of evolution every bit as ancient and unique as our own. It seems probable that future generations on this planet will be both dismayed at the number of organisms that we let slip into the oblivion of extinction and will be deeply thankful for what we of this generation were able to save.

Why are Colombian Primates Threatened?

Habitat loss and environmental degradation are by far the most important reasons that primates are today so threatened in Colombia. The impact of hunting and capture for pets has increased enormously where forests are destroyed and fragmented, diminishing the populations, restricting their capacity to disperse, and facilitating access to hunters (now, sadly, the situation over enormous areas of the Colombian Amazon; DEFLER, 2003). But because of their very low population numbers, the capture of one of the more endangered Colombian primate, the cotton-top tamarin (*Saguinus oedipus*), has a disproportionate effect on the species when compared to such as the woolly monkey (*Lagothrix lagothricha*) which, although hunted for meat and as pets, still maintains much higher populations over a wider range than do *S. oedipus*..

Habitat Loss

When the first European colonists arrived in the New World and began their exploration of what is now Colombia, about 80% or 800,000 km² of land was under forest, most of this lowland tropical moist forest. At that point in the history of Colombia the human population probably did not exceed one million Indigenes, presumably living in a more balanced and sustainable state with their natural resources than their European and African successors.

A sustainable co-existence of people and wildlife in tropical lowland forest demands a low density of human beings, such as that existing 400-500 years ago in Colombia. Being human, even Indigenes overstep the carrying capacity of the land when they became concentrated as they were in certain areas of Meso-America or along the banks of the Amazon. Primates were extirpated on the Caribbean islands of Jamaica and Hispaniola even before European colonization, most likely due to the exploitation of indigenous peoples and the higher susceptibility to extinction of island populations (NILSSON, 1983; DIAMOND, 1984).

European colonization in Colombia began in earnest along the Caribbean lowlands. This region was covered by dry deciduous forest along with some very large savannas, and it was there that the agricultural population was concentrated, subsequent to the eradication of the tropical forests. The second great focus of Colombian colonization was in the Bogotá savanna where human population densities are the highest in Colombia. Later, human settlements were established at Medellin and Cali. The highlands were particularly affected due to a climate that was more amenable to human activity, including western agricultural practices. Today the greatest concentrations of humans and of environmental degradation are in the Caribbean lowlands and in the Andean highlands. Only in the last 50 years or so have major inroads been made in the lowland tropical forests, the ecosystems with the highest primate diversity (HERNÁNDEZ CAMACHO & DEFLER, 1985, 1989).

Because of the human demographic explosion and social carelessness, most natural ecosystems in Colombia are becoming drastically degraded and reduced. This occurs because of outright destruction, for example by cutting down a forest, by resource exploitation that results in the destruction of key species or in perturbations of ecosystems such as through pollution and exotic and invasive species, or simply through development projects that disregard the environmental damage they cause.

About one-third of Colombia's forests have now been destroyed, leaving perhaps 494,000 km² (43% of the country) reasonably intact (REPÚBLICA DE COLOMBIA, 1992), although the state of some of this forest is debatable. Most of this destruction has been for growing crops and for creating pastureland for colonists rather than through lumbering, but this may soon change. About 95% of the Caribbean lowland forest has been cut in the geographic range of the endangered cotton-top tamarin (*Saguinus oedipus*), which until recently has been one of the highest conservation priorities amongst Colombia's primates, partly because this small primate is found nowhere else outside of Colombia.

Many of the remaining forests in Colombia may be largely devoid of the larger primates and other mammals, as noted for the Peruvian forests (FREESE *et al.*, 1982), for Brazilian forest (PERES, 1990; 1991c) and at least for certain parts of Colombia, like parts of the Chocó (HERNÁNDEZ-CAMACHO & COOPER, 1976; VARGAS, 1994a; H. RUBIO, pers. comm.). Many forests have apparently been hunted out or disturbed in some manner so that, although the trees are left standing, large vertebrates are largely absent (DEFLER, in press).

Habitat loss in the north and north-central parts of the country due to colonization has also affected the populations of another endemic primate, the white footed or silvery tamarin (*Saguinus leucopus*), which are currently in serious decline, since original habitat for this monkey is the least extensive of any *Saguinus* species. The lemurine night monkey (*Aotus griseimembra*) and the spider monkeys, *Ateles hybridus brunneus* and *Ateles hybridus hybridus* are also restricted to northern Colombia, and are likewise critically endangered due to the widespread loss of their forests, as well as hunting. *Ateles* species are a target of meat hunting in Colombia, wherever they occur.

Hunting Pressure

Where the geographical range of a species or subspecies is small, as it is in the threatened primates mentioned above, habitat loss has a relatively greater impact, since the original population numbers and habitat are already limited. Widespread and large primates, such as the spider monkeys and woolly monkeys, become vulnerable because their very size makes them a target for hunters, and their population recovery time is very long due to a low birth rate. Thus, Colombian populations of black spider monkey (*Ateles geoffroyi rufiventris*) and variegated spider monkey (*Ateles hybridus*) are under pressure from hunting in almost all parts of their ranges, and it is becoming exceedingly difficult to observe these primates outside of large national parks. *Ateles* is the genus most at risk in Colombia due to the fact that human settlement has been particularly intense within their ranges, even in the Amazon.

National laws (Decree 1608, of 1978) permit only subsistence hunting of primates in Colombia, but these laws are generally not enforced, because of a lack of the appropriate authorities in most parts of the country. Unfortunately many people in Colombia, as in other South American countries, enjoy shooting animals, sometimes only for recreation! Most human indigenous populations hunt any animal large enough for the pot even in protected areas, since present laws permit "traditional" indigenous hunting most anywhere, in and out of national parks and reserves. Hunting pressure increases when primate populations are already severely reduced because of extensive habitat destruction. Earlier, population densities of indigenous peoples were always low and animal populations had few other human pressures. But as primate habitat decreases and hunting technologies change from blow-guns to shotguns, risk to primate populations increases dramatically (RAEZ-LUNA, 1993). For example, extensive vertebrate censuses in Utria National Park indicated an absence of traditional game species, except in some very few nuclei (H. RUBIO, pers. comm.), and populations of *Ateles* had been perceived by several park chiefs to have decreased due to hunting in both the Katios and La Macarena National Parks (D. PINTOR & C. MURILLO, pers. comm). Otherwise intact forests west of and inside the Orquideas National Park were said by colonists to be devoid of large primates due to indigenous hunting pressures of nearby communities

(VARGAS, 1994). Such anecdotal evidence needs to be investigated with systematic censuses.

The Impact of the Jaguar and Ocelot Skin Trade on Primates

Previously jaguar and ocelot were commercially hunted in the Colombian Amazon, and in order to attract the cats many animals were shot, including thousands of primates that were used as bait. It was common for a group of men to pursue a group of monkeys until they were all shot for bait, and many areas alongside rivers became devoid of monkeys, even in quite isolated parts of the Amazon (DEFLER, 1983c). Fortunately both skin hunting and primate hunting are now illegal, the key to enforcement being the international CITES agreement which discourages skin hunting by prohibiting commerce in spotted cat skins between countries.

Primates as Pets

Although prohibited by law in Colombia, the use of primate pets is widespread, but it is biased towards certain species. There is an illegal but active trade in both Geoffroy's tamarin (*Saguinus geoffroyi*) and the silvery-brown bare-face tamarin (*Saguinus leucopus*), some of which appear for sale on the streets of Bogotá and Medellin or find themselves in the illegal export market. VARGAS (1994) reported commercial trapping of *Saguinus geoffroyi* near Murrí (Antioquia) in a newly discovered extension of its geographic range east of the Río Atrato. A recent animal price list from Miami listed supposedly captive bred *S. oedipus* at US $1000 each and *S. geoffroyi* at US $1500. These prices suggest that it is commercially viable to risk trading in illegally imported individuals via Panama or northern Colombia. *Saguinus leucopus* on the streets of Bogotá are often drugged to make the animals appear tame. Some have dyed manes and other portions of the body in an apparent effort to make the individuals appear more exotic. Entire groups of *Saguinus* are often captured using a group of several traps with a caged animal of the same species in the center of the traps to attract them. When the group comes to the vocalizations of the captive animal they find food in the surrounding traps, enter them and are caught. However, this traffic may be diminishing in Colombia due to more active enforcement, although there is little data to back this up (DEFLER & RODRÍGUEZ-M, unpublished).

Other larger primates are usually acquired as young animals when still being carried by their mother, when she is shot for food. Hunters search out the mothers carrying infants. In the Colombian Amazon *Lagothrix* babies are obtained in this way and are probably the most coveted as pets, but spider monkeys (*Ateles*), capuchin monkeys (*Cebus*), sakis (*Pithecia*) and titi monkeys (*Callicebus*) are also popular. Presently the pet trade has the greatest impact on the three northern species of tamarin (*S. geoffroyi, S. leucopus, S. oedipus*), as well as on *Lagothrix* and *Ateles*. Anytime a mother primate is hunted to obtain her infant, severe damage is done to that social grouping, since the future breeding potential of the mother is also lost.

Commercial Pressures

Prior to 1974 the exportation of Colombian primates constituted a significant pressure on primate populations. In 1970, Colombia reported the export of 15,000-20,000 primates (COOPER & HERNÁNDEZ-CAMACHO, 1975; MACK & MITTERMEIER, 1984), probably only a portion of what was actually exported, and only a small part of the actual numbers captured. Many hundreds die before being shipped. The trade was banned through the Natural Resources Code of 1974 and INDERENA Resolution 0392.

Although this commerce in primates is now illegal, there are still demands within and outside of the Colombian government for a resumption of this pernicious trade. Recent arguments are even cloaked in conservation rhetoric, saying that putting a commercial value on wildlife is the only solution to its eventual conservation. Thus, market forces are evoked as the ever-popular solution to another difficult problem, while failing to recognize that the uncritical acceptance of market forces as the prime mover in the world today is a simplistic reaction to complex issues. Market forces have always resulted in immense environmental damage to world and to Colombian ecosystems and resources, since cost calculations do not include all the elements, and adequate control to direct and organize such commerce is an impossibility.

As a counterargument to such pressures for commercialization, Colom-bians must recognize natural resources as being held in common for all citizens – present and future; and that using market forces on natural products which are not produced by human beings ought to be very carefully controlled, since commercializing any natural product will result in its being heavily exploited by all, with little regard for the many environmental costs which may never be recognized by the consumer.

Societal Pressures which Threaten Primates

Views towards nature are too often fearful and uninformed, and many people perceive the forest as a barrier to "progress", defined only as "development" and ignoring the possibility of cultural attitudes towards natural resources as sources of pride and as purveyors of many indirect values (NASH, 1982). Our legacy of the deprecation of nature belongs in the past. Education must include an orientation towards the beauty and health of a society that preserves and appreciates its natural assets alongside appropriate development. National pride in the possession of tracts of natural ecosystems with all of their organisms intact can be taught and nurtured, and healthy national parks should be a source of national pride and of regional financial profit due to the outsiders attracted to the region. Ultimately the values of natural resources such as plants and animals and their conservation must be based on a value system which does not depend so much on money as it does on other human values concerning respect for life, for evolution, and for our place on this planet, which ultimately must be less dominant and pervasive. The present world emphasis on market forces has no regard for the resouces to be consumed for the market place.

Poverty does not allow many people to look at a tree without thinking of firewood or to see a large mammal without feeling hunger. If the people of a nation do not have a way to feed, cloth, and educate their children, it is difficult to expect them to appreciate a national park or a tract of forest. Appropriate development must proceed hand-in-hand with a national and international conservation movement. The establishment of a local park requires local education and an improvement in local living standards, especially if the park land removes previous economic possibilities. Present efforts to incorporate human needs with resource conservation will continue to have many conflicts, since there is no way that a burgeoning population cannot leave a huge impact. We can only seek to reduce the impact that is inevitable.

Services such as basic education, an adequate diet, and affordable health care (including family planning) are the keys to a society that is able to bank some of its natural ecosystems in national parks, reserves, and forests for the future. National survival depends on all of these elements, and none of them can be neglected without further impoverishment.

Conservation Priorities for Colombian Primates

Our Colombian national committee for the IUCN classification of mammals considers the two subspecies of *Ateles hybridus* as the most endangered taxa in Colombia with a CR classification (RODRÍGUEZ-M. *et al.*, unpublished). Following *Ateles hybridus*, we rate *Ateles geoffroyi rufiventris* as EN and 12 other taxa including 11 species as VU. We believe that *S. oedipus* should be lowered from EN to VU after having reassessed this endemic species. We also redefine *Aotus griseimembra* from EN to VU. *Ateles geoffroyi griscesens* is probably EN or CR, but it needs to be confirmed for its presence in Colombia and assessed.

Taxons	IUCN INTERNATIONAL (Hilton-Taylor, 2000 (Hilton-Taylor and Rylands, 2002 www.redist.org)	COLOMBIAN IUCN CATEGORY (Rodriguez *et al.*, 2004)	% of total distribution represented by the Colombian population	Notes
Data Defficient				
Ateles geoffroyi grisescens	EN/DD	DD	9	
Critically Endangered				
Ateles hybridus hybridus	CR	CR A3 a,c,d	40	**(1)**
Ateles hybridus brunneus	CR	CR A3 a,c,d	100	**(1)**
Endangered				
Ateles geoffroyi rufiventris	CR	EN A2 a,d	90	**(1)**
Vulnerable				
Callimico goeldii	NT	VU D2	20?	
Saguinus oedipus	EN	VU B1a,c	100	**(2)**
Aotus griseimembra	EN	VU C1	70	**(3)**
Aotus brumbacki	VU	VU A2c	100?	**(4)**
Aotus zonalis	VU	VU C1	80	**(5)**
Aotus lemurinus (=*A. hershkovitzi*)	VU	VU C1	70?	**(6)**
Saguinus leucopus	VU	VU A2c	100	

Callicebus cupreus ornatus	VU	VU B1b(iii)	100	
Callicebus cupreus discolor	LC	VU B1b(iii)	5	
Pithecia monachus milleri	VU	VU A2c	100	
Ateles belzebuth belzebuth	VU	VU A2a,c,d	10-15	
Alouatta palliata aequatorialis	LC	VU A2a,c,d	50?	
Lagothrix lagothricha lugens	VU	VU A2a,c,d	95-100	**(7)**
Near Threathened				
Lagothrix lagothricha lagothricha	LC	NT	50	
Cacajao melanocephalus ouakary	LC	NT	35	
Cebus albifrons cesarae	NT	NT	100	
Cebus albifrons yuracus	DD	NT	10?	
Cebus albifrons malitiosus	NT	NT	100	
Cebus albifrons versicolor	DD	NT	80 ?	
Least Concern				
Callicebus torquatus medemi	LC	LC	100	**(8)**
Alouatta seniculus seniculus	LC	LC	20	
Aotus vociferans	LC	LC	50?	
Callicebus torquatus lucifer	LC	LC	40	
Callicebus torquatus lugens	LC	LC	40	
Cebuella pygmaea	LC	LC	20	
Cebus albifrons albifrons	LC	LC	30	
Cebus apella apella	LC	LC	?	
Cebus capucinus	LC	LC	45	**(9)**
Pithecia monachus monachus	LC	LC	15	
Saguinus nigricollis nigricollis	LC	LC	20	
Saguinus nigricollis graellsi	LC	LC	20	**(10)**
Saguinus nigricollis hernandezi	LC	LC	100	
Saguinus fuscicollis fuscus	LC	LC	90?	
Saguinus geoffroyi	LC	LC	40	
Saguinus inustus	LC	LC	50	
Saimiri sciureus albigena	LC	LC	100?	
Saimiri sciureus cassiquiarensis	LC	LC	40?	
Saimiri sciureus macrodon	LC	LC	60?	

Table 3. Conservation priorities of Colombian primates, based on new IUCN Red List conservation categories (IUCN, 2000; Hilton-Taylor and Rylands, 2002; Rodríguez-M. *et al.*, in prep.)..

Notes:

(1). In agreement with the cladogram drawn by FROEHLICH *et al.* (1993) using morphometrics, his personal comments, and COLLINS & DUBACH´S (2000a) phylogenetic analysis of mitochondrial and nuclear DNA and COLLINS' (1999) comments, this paper recognizes *Ateles hybridus* as a separate species containing the subspecies *A.h.hybridus* and *A.h.brunneus* and we consider all *Ateles* west of the Cauca River to belong to *Ateles geoffroyi*. Actually there is no confirmation of the presence of *A. g. grisecens* (=*A. f. grisecens*) in Colombia. Its presence in the country was suggested by KELLOGG & GOLDMAN (1944) and by HERNÁNDEZ-C. & COOPER (1976), but there are no specimens or observations to prove it. If they are there, the population would be small and extremely endangered. Comments made by colonists near the northern parts of the Serranía de Baudó region talk of two "types" of *Ateles,* one in the lowlands (which is definitely *A. g. rufiventris*) and another different one above 500 m – 600 m altitude (RODRÍQUEZ-M., *verbatim*). This is the only real suggestion that this taxon might actually be present in Colombia.

(2). In 2000 a Colombian national mammal committee was formed in order to classify all Colombian mammals using the latest IUCN (2000) criteria. This was particularly appropriate, since the authors of the IUCN criteria suggested that classifications using the same international criteria should be made on national and regional levels. The new, upgraded IUCN (2000) classification uses additional species characteristics such as the capacity to adapt to new habitats or to degraded or intervened habitats, among others. At the last meeting of our national mammal committee, we discussed the necessity of re-classifying 15 taxa of primates, elevating the category of 10 of these and lowering the category of 5, among which we lowered *S. oedipus* downward to VU instead of EN, based on the observations of RODRÍGUEZ-M. (*verbatim*) and Hernandez Camacho, as well as considerations and evidence in northern Colombia that the populations of wild *S. oedipus* do not meet the population criteria for EN and that it is more appropriately a «VU» species. Here we reclassify *S. oedipus* VU, due to its improved population numbers over the past twenty years, and the species failure to qualify for the newest EN classification criteria. This change is not yet an official classification as mandated by IUCN (2000), although documentation in the form of a reassessment has been submitted (DEFLER & RODRIGUEZ-M, unpublished).

(3) Because of chromosome differences, this taxon may best be considered a separate species from *A. lemurinus* (DEFLER *et al.*,., 2001; DEFLER & BUENO, 2003).

(4) *Aotus brumbacki,* because of wide chromosome differences from *Aotus lemurinus*, cannot be a subspecies of *Aotus lemurinus* as suggested by GROVES (2001). This taxon is undoubtedly reproductively isolated prezygotically.

(5) Because of chromosome differences, this taxon is a separate species from *A. lemurinus* (DEFLER *et al.*, 2001; DEFLER & BUENO, 2003).

(6). We believe *Aotus hershkovitzi* to be a synonym for *Aotus lemurinus*, which we restrict to the Andean highlands (DEFLER, BUENO & HERNÁNDEZ-C., 2001). This taxon is undoubtedly a separate species from *A. griseimembra, A. zonalis,* and *A. brumbacki* because of pronounced chromosome differences (DEFLER & BUENO, 2003).

(7). DEFLER (1996b) analyzes the basis of why this taxon should be classified as VU instead of CR, *contra* RYLANDS *et al.,* (1995), who in turn followed a previous classification of DEFLER (1994) before the latest reassessment.

(8). GROVES (2001) classifies this taxon as a full species, but we feel that this is doubtful. Actually GROVES (2001) elevates the following in Colombia to full species (*Callicebus ornatus, Callicebus discolor, Lagothrix lugens, Saguinus graellsi,* and *Callicebus medemi*) and he eliminates *Cebus albifrons cesarae, Cebus albifrons malitiosus,* retaining only *C. albifrons versicolor* for central and northern Colombia. He considered *Cebus capucinus* as monotypic in agreement with HERSHKOVITZ (1949) and HERNÁNDEZ-CAMACHO & COOPER (1976) and he considered *Aotus brumbacki* as a subspecies of *Aotus lemurinus* and accepted *Aotus hershkovitzi.* This document does not evaluate his classification, since it is out of the scope of this book. However, in many instances we take exception as described for each taxon. VAN ROOSMALLEN *et al.* (2002) also consider all *Callicebus* variants in Colombia as species, a view that DEFLER *et al.* (2003) cannot accept. Our preference, for want of any new convincing evidence, is to follow HERSHKOVITZ (1990) for all Colombian *Callicebus,* although new karyological information for *C. torquatus lugens* shows that the species is probably a complex of species.

(9) HERSHKOVITZ (1949), HERNÁNDEZ-CAMACHO & COOPER (1976) and GROVES (2001) all agreed that there is not enough variation in this species to merit the distinction of subspecies, so it appears monotypic here.

(10) HERNÁNDEZ-C. & COOPER (1976) and RYLANDS *et al.* (2000) classify this taxon as a separate species, but such a designation requires proof that it is sympatric with some other population of *Saguinus nigricollis* and we leave it here as a subspecies of *S. nigricollis.* A collection of *S. nigricollis nigricollis* and deposited in the Instituto de Ciencias Naturales de la Universidad Nacional de Colombia, without specific collection data but supposedly from the left bank of the Río Putumayo may be in error (HERSHKOVITZ, 1977).

(11). *Aotus nancymai* and *Aotus nigriceps* observed in captivity by P. HERSHKOVITZ & J. HERNÁNDEZ-C. in Leticia and reported in DEFLER (1994b) have never been confirmed in the wild for Colombia and a traffic in *Aotus* does exist in Leticia from the other side of the Río Amazonas to Colombia for sale to the local primate lab from where they are erroneously treated as being from Colombia (since it is illegal to purchase fauna from either Brazil or from Peru). It is most probable that the specimens examined by HERSHKOVITZ & HERNÁNDEZ-C. were purchased animals from the other side of the river. *Aotus trivirgatus* listed in DEFLER (1994b) is probable but not confirmed for Colombia; three *Aotus* from widely separate sites in the eastern Colombian Amazon, examined by DEFLER and confirmed by HERSHKOVITZ (*verbatim*) have phenotypes agreeing with the description for *A. trivirgatus* (HERSHKOVITZ, 1983; HERSHKOVITZ, pers. com.). Karyotypes still urgently need to be determined for this phenotype and for *Aotus trivirgatus. Saguinus labiatus* and *Saguinus tripartitus* reported in RODRÍGUEZ *et al.* (1995) are possible, judging from their known areals but must be confirmed via field studies, which have never been done. HERNÁNDEZ-C. & COOPER (1976) accept *Saguinus graellsi* as a valid species, but its validity must be confirmed via field studies. *Cacajao calvus* is known in Brazil, not far from the border with Colombia and may have historically been present in Colombia in Caño Uacarí upriver from Leticia, but there are no recent observations. If all the above were to be confirmed as Colombian species the total number would go up to 33 species of primates in Colombia. Should some of these species be confirmed in the near future, it is likely that they would be classified as endangered within the country, due to very small populations (esp. *A. nancymai, A. nigriceps, S. tripartitus, S. labiatus* and *Cacajo calvus*.)

Colombian Conservation Priorities of Primate Taxa

Hooded spider monkey, *Ateles geoffroyi grisescens* (=*A. fusciceps grisescens*) – DD

Considered endangered globally, there is no recent information regarding the presence and extent of occurrence of *A. geoffroyi grisescens* (=*A. fusciceps grisescens*) in Colombia. Observations in the middle Baudó have indicated that this subspecies is present only in the immediate portion of the Sierra de Baudó abutting Panamá and not the entire range indicated by HERNÁNDEZ-CAMACHO & COOPER (1976). Colombian populations of this taxon need confirmation and management, since it would be certainly the most endangered spider monkey in the country. As soon as its national presence can be confirmed a management plan would need to be formulated. The forests where it is believed to occur are relatively intact and there is little settlement in the area. Unfortunately violence in the zone make field work impossible at this time.

Variegated spider monkey, *Ateles hybridus* – CR

Both subspecies of the variegated spider monkey, *A. h. hybridus* and *A. h. brunneus* are considered to be at extreme risk in Colombia. *A. hybridus* has been considered subspecies of both the white-bellied spider monkey, *A. belzebuth* and Geoffroy's spider monkey, *A. geoffroyi* (KELLOGG & GOLDMAN, 1944; FROEHLICH *et al.*, 1991). However, FROEHLICH (pers. comm., 1995) commented that it is very probably a "good" species, a view strongly supported by his cladogram of *Ateles* species (FROEHLICH *et al.*, 1993). Further work by COLLINS (1999) and COLLINS & DUBACH (2000a) confirmed that it is a separate species. *A. h. brunneus* has not been widely recognized as a valid taxon. It is a Colombian endemic found in a recognized dispersal center (the Nechí). Both these spider monkeys are severely threatened with fragmented habitat, although *A. h. brunneus* has a large possible habitat in the Serranía de San Lucas. Census work is urgently needed, and local populations and their status to be clearly identified and actively managed. *Ateles h. brunneus* is not known to occur in any protected area, and its total population size is probably much smaller than that of *A. h. hybridus*, which is more widely distributed in small remnant population fragments in both Colombia and Venezuela. The only stronghold of *A. h. brunneus* is in the Serranía de San Lucas in southern Bolívar, as such it is an important site for the establishment of a national park, which would also protect a number of other endemics to this dispersal center or refugium. The area has, however, suffered civil unrest for years and census work there would be hazardous. Guerilla groups have placed anti-personnel mines in some parts of the mountain range.

Colombian black spider monkey, *Ateles geoffroyi rufiventris* – EN

Apart from *A. hybridus* all spider monkeys are considered to be at risk in Colombia, either endangered or vulnerable, due to both habitat destruction and hunting. Populations in the National Natural Parks of Los Katios and Orquideas are reportedly extremely low now because of indigenous hunting pressure (N. VARGAS, pers. comm.; H. RUBIO, pers. comm.). At Los Katios National Park the population density of *Ateles* may be decreasing (D. PINTOR, pers. comm.). It is not easy to see this species, due to a continuing decline in numbers. Censuses and monitoring of all the Colombian spider monkeys throughout their ranges are vitally needed, especially in national parks. Problems of hunting need to be identified and countered in the parks, and the indigenous rights to continue hunting endangered taxa needs to be reconsidered. *A. g. rufiventris* is heavily hunted, its populations are fragmented, and its situation seems precarious at best. Censuses attempted to date have rarely located any groups, even in national parks. *Ateles geoffroyi* may be one of the first taxa to go extinct in Colombia (along with *A. hybridus*) because of hunting now even more than habitat loss. Neither the prohibition of hunting nor educational efforts are easily implemented in the isolated areas where they occur and because many areas are off limits to officials and conservationists. The continued violence in rural areas is probably directly affecting all primate species, and *Ateles* is by far the most threatened because it is favored game in areas which have suffered and are suffering, heavy deforestation.

Cotton-top tamarin, *Saguinus oedipus* – VU

Saguinus oedipus is one of the best examples in Colombia of a partial success story for an endangered species. Many cotton-top tamarins were exported during the 1970's, and a number of colonies have been successfully established in other countries. A significant portion of the species' habitat has been modified by agriculture and for cattle ranching. However, secondary forests are still widespread in the Magdalena valley, the Montes de María and areas extending to the eastern Andean Cordillera, which maintain healthy populations of the species.

Today we belive that most of the traffic in *Saguinus oedipus* has stopped, not only because of its prohibition but also because of the various awareness campaigns that have been carried out by the government agency INDERENA (now replaced by the Ministry of the Environment) which initiated the first conservation efforts on its behalf. The species is able to survive in isolated fragments of forest and shows a remarkable capacity to tolerate great weight loss during food shortage in the dry season (J. V. RODRÍGUEZ-M., pers. obs).

Cotton-top tamarins are able to colonize new areas when given the chance, as demonstrated by introduced populations in the Yotoco forest (Chocó), around Cali (a population that the Colombian primatologist Carolina Gomez has been unable to confirm to be still extant), and in the Tayrona National Natural Park. These populations were established through the haphazard liberation of captive specimens by their owners or confiscated animals by government officials. Ironically the present activities of the guerilla and the paramilitary have left extensive areas of small holdings in Sucre, Córdoba and Bolívar without their owners, and this national disaster has had positive effects on Colombian fauna, including *Saguinus oedipus* (J. V. RODRÍGUEZ-M., pers. obs.; RODRÍGUEZ-M. *et al.* in prep.). A fair captive population is maintained and registered by studbook in North America, Great Britain and Europe (MAST & FAJARDO, 1988; MAST *et al.*, 1993).

Although some surveys for this species were undertaken a few years back (BARBOSA *et al.*, 1988) and more recent field work has been undertaken at the Colosó (Sucre) primate facility and in Sucre (SAVAGE, 1988a, 1989a; SAVAGE & GIRALDO, 1990; SAVAGE *et.*, 1990a) with an effort to develop a local conservation program, much work remains to be done. The Colosó primate research facility in Bolívar department, where the current major effort to study the species is located, was closed to researchers because of civil unrest caused by guerillas and paramilitary groups several years ago, but a research group from the area is currently planning more primate studies. Anne Savage and collaborators have continued their immense long-term effort of developing an environmental education plan via work in zoos, and an alternative site in northern Sucre with the idea of making the community aware of the necessity of not using this and other endangered species as pets.

The species continues to be captured in small numbers locally as pets, and enforcement of Colombian laws to protect the species is complicated by many social factors. Nevertheless, both police and guerillas often enforce the laws prohibiting the species´ capture, a measure of the success of the conservation program. A reassessment of the species conservation category has recently been written and is being considered by the IUCN committee (DEFLER & RODRIGUEZ-M., unpublished).

Necessary Conservation Actions

A recent review article by MAST *et al.* (1993) suggest several action categories needing to be addressed for a more aggressive conservation effort for *S. oedipus*; these are listed below with commentary and could apply equally well to the very endangered *Ateles* taxa.

Conservation in situ: identification of key sites, creation of public and private reserves. Although some censuses have been carried out, broad areas of its putative geographic distribution have not been investigated for the presence of this and other primates. Fieldwork and primate inventories are needed in the upper Río San Jorge and around the Paramillo National Park where the most extensive forests remain. The southwestern limits to the species' range are still poorly known and it would be of interest to locate a possible zone of contact with a recently observed population of *Saguinus geoffroyi* to the east of the Río Atrato (VARGAS, 1994a). Other isolated populations throughout the species' range also need to be mapped, to provide for a gazetteer of the extant populations. J. V. RODRÍGUEZ-M. (pers. comm.) found an apparently healthy population along the left bank of the Río Magdalena between Carmen de Bolívar and the Zambrano region (Bolívar). Populations such as these should be given special protection. Because of his finding in the Zambrano area, Rodríguez-M. feels that the species is best classified as VU, while also emphasizing the excellent results of the conservation efforts initiated through the government and the work of Proyecto Tití. Populations located in municipal watersheds or other forests (including private forests) could be touted as special resources to be proudly protected, since this is a unique Colombian species. The Ministry of the Environment and local semi-autonomous corporations need to become more involved in the preservation of this endemic species via the establishment of appropriate reserves and conservation programs.

Environmental education, law enforcement, and ecotourism. Educational programs at the national and local level are vital for the conservation of this species. Local people should be convinced that it is not acceptable to trap or buy these primates as pets. Conservation education needs to be included in school curricula, and communities close to wild populations of *S. oedipus* must be made aware and proud of the importance of their local population. Local populations could be made into tourist attractions in this way, including the participation of conservation and wildlife clubs. Publications, videos and workshops should be made available to the local people. Environmental education programs should be extended to the local authorities to encourage a stricter application of the wildlife laws protecting the species, and should run parallel with the development of ecotourism programs for the forests and these highly attractive primates.

Forest restoration. Forest regeneration programs are few and far between in Colombia. Because natural vegetation is so important for water conservation and because a group of this primate species does not require excessive space, there are probably many possibilities for local communities and even private individuals to support a forest regeneration programs that would insure the survival of a small local population of *S. oedipus.*

Lemurine night monkey, *Aotus lemurinus* – VU

Much of the habitat of this night monkey is congruent with widespread human disturbance. It is probably not as threatened as *A. griseimembra* (see below), but censuses of their populations and habitat are needed to provide a more accurate picture of its status.

Night monkey, *Aotus griseimembra* (=*A. l. griseimembra*) – VU

A. lemurinus appears to be highly vulnerable in Colombia, in part due to habitat loss and in part due to its use as a model for malaria research, which results in enormous numbers being taken from the wild. STRUHSAKER *et al.* (1975) found it to be extremely rare in a past census. GREEN (1976), however, was able to locate populations in the Sierra San Lucas. This area is not protected, however, and could easily be destroyed should the local problems of civil unrest be solved, stimulating renewed colonization of the area. Populations recognized as *A. griseimembra* are genetically heterogeneous and may actually require more than one taxonomic designation. The lowland populations may be distinct from *A. griseimembra* known from more montane areas. Likewise, it is possible that genetically distinct populations of *Aotus* in the north of Colombia may actually be an unrecognized sibling species. These night monkeys are in danger of becoming extinct even before we understand their systematic relationships. Further research is necessary to provide a full understanding of the geographical distributions of a number of the Colombian night monkeys, including *Aotus lemurinus, A. brumbacki,* and *A. vociferans*. The protection of the forests in northern Colombia harboring these night monkeys, and a national ban on any collections of *A. griseimembra* for research in malaria (other less threatened Amazonian populations have been identified which are also susceptible to malaria) are vital steps for its effective conservation.

Brumback's night monkey, *Aotus brumbacki* – VU

Except for the region around Villavicencio, Meta, the geographic range of this species is unknown. Although classified as vulnerable, it is possible that a Data Deficient classification is more appropriate. The known part of this taxon's geographic range is congruent with much human activity, but if it reaches south to the Río Guayabero (which has not been confirmed) it would have a healthy population protected by one or even two large national parks.

The taxonomy of the night monkeys, a difficult and controversial genus, is still confused, despite Hershkovitz' synthesis published in 1983. FORD'S (1994) recent morphometric analysis points to an alternative view of *Aotus* in this part of Colombia as one of clinal species. DEFLER *et al.* (2001) believe, however, that the three subspecies of *A. lemurinus*, including *A. l. zonalis* of the Andean highlands and Chocó [not recognized as a valid taxon by HERSHKOVITZ (1983)], should be elevated to species status. DEFLER *et al.*, (2001) argue *Aotus hershkovitzi* described by CERQUERA in 1983 is, on the other hand, no more than a junior synonym of *Aotus lemurinus*.

Goeldi's monkey, *Callimico goeldii* – VU

This species is naturally sparse throughout its geographic range and has been registered from only six sites in the Colombian Amazon. Two are in national parks, although one locale in the Cahuinarí National Park [reported to HERSHKOVITZ (1977) by the Colombian herpetologist Federico Medem], has never been precisely located. Being rare, its conservation and management requires a detailed knowledge of the precise location of all known populations, so that each can be protected. Since four of the six populations are found in generally unperturbed primary forest, the pressures upon them are generally not considered to be severe. Nevertheless, the apparently special habitat requirements needed by this primate may mean that any site is fundamentally at risk for one reason or another. This question must remain open until further field studies are accomplished focusing on this monkey.

Silvery-brown bare-face tamarin, *Saguinus leucopus* – VU

This small endemic primate does not occur in any park or reserve, while its entire range is located in a region of intensive colonization and habitat destruction. A basic conservation program for this species must include detailed ecological field studies, censuses, and the establishment of adequate reserves (VARGAS & SOLANO, 1996b). Municipal watersheds would make for ideal local reserves, as long as the tamarins are protected from trapping and sale as pets. Local educational campaigns should convince that this beautiful small Colombian mammal should be strictly protected wherever it occurs. The creation of the proposed Serranía de San Lucas National Park needs to be supported so that populations of this, as well as of *Ateles hybridus brunneus* and other endangered species of the Nechí refugium may be conserved for the future.

Ornate titi monkey, *Callicebus cupreus ornatus* – VU

This small endemic titi monkey is known to occur only in a small area of intense colonization and widespread forest destruction. Perhaps the best populations of this primate are located in and adjacent to La Macarena National Park, so there is some hope that populations there and in Tinigua National Park can constitute a safe nucleus. However, habitat destruction is widespread in this part of La Macarena and enforcement of national laws is very difficult, especially since this area is subject to civil unrest and heavy insurgent activity. Censuses and local environmental education campaigns are probably the best tools at the moment to insure its survival. Convincing local farm and ranch owners of the importance of protecting a portion of their forests and woodlots would go a long way towards protecting this primate.

Titi monkey, *Callicebus cupreus discolor* – VU

Although this titi monkey is not considered threatened throughout most of its range in Ecuador and Peru, the Colombian portion is very small and at risk due to the destruction of its forests through agriculture, colonization, oil exploration, and drug activities. For this reason its national status is considered to be vulnerable. The Colombian Amazon between the Ecuadorian border and the Río Guamués is part of the Napo refugium, and the most biodiverse region within the country, giving fieldwork, made difficult by the presence of guerillas, an extremely high priority. There is an urgent need for a national park in the south-west of the Department of Putumayo. This is recognized by the Colombian government, but any action by the appropriate authorities would be opposed by insurgents and coca growers. There is no defined colonization front in the region, only widespread, isolated, and illegal coca farms which leave no part of the forest intact. Recent attempts by the government to eradicate these farms has led to confrontation, and ensures that any conservation attempts on a national scale will be even more difficult. Pressure from the United States government is resulting in widespread spraying to eradicate the illegal coca, and the likely result will be more environmental degradation and damage to the natural ecosystems surrounding these fields. There is tentative evidence that other patchy populations of this or another form of *Callicebus* may be found in other parts of Putumayo or in the southern portion of the Department of

Caquetá, and censuses are needed. MOYNIHAN (1976a) mentioned one of these populations for southern Caquetá, at Valparaiso, between the Ríos Orteguaza and Caquetá. Titi monkeys he observed there were not attributable to either *C. c. ornatus* or *C. c. discolor*, but since then no qualified observations have confirmed his report.

Miller's monk saki, *Pithecia monachus milleri* – VU

Much of the known distribution of this taxon is congruent with colonization. Fortunately La Paya National Park should help to protect a healthy population and it may also eventually be found between the Ríos Caguan and Yarí. Its classification as vulnerable may be due only to the little information we have about its range and status, and it is possible it may be more correctly assigned to the category of "near threatened" in the future.

White-bellied spider monkey, *Ateles belzebuth belzebuth* – VU

This is probably the most endangered taxon in the Colombian Amazon (DEFLER, 1989, 1994; DEFLER *et al.*, 2003). Geographic distributions published by HILL (1962), HERNÁNDEZ-CAMACHO & COOPER (1976) and KONSTANT *et al.* (1985) greatly overestimated the presence of this taxon in Colombia, and its range is mainly congruent with heavy colonization in the Colombian Amazon and piedmont forest to the north. Although it is known to occur in at least three national parks (La Macarena, Tinigua and Picachos), pressures from colonization around these parks make it necessary for a stricter control of hunting. Unfortunately management in this region is greatly complicated by the presence of insurgents. Another large block of forest between the heavily colonized Ríos Caguán and Yarí to the east evidently has a good population of this spider monkey. Only the comparative isolation of this forest will keep this population intact for the near future, but any development which makes it easier for colonists to move into this region will quickly endanger them.

Colombian woolly monkey, *Lagothrix lagothricha lugens* – VU

As a species *Lagothrix lagothricha* (=*L. l. lagothricha*) is classified as LC (Least Concern) (HILTON-TAYLOR, 2000). Three of the four subspecies (which are being called species by some taxonomists) are considered LC by the latest IUCN (2000) classification. *L. l. lagothricha* seems to have widespread and healthy populations. Although RYLANDS *et al.* (1995) classified *L. l. lugens* as Endangered, it is best characterized as Vulnerable since it is still found in a number of large protected area and relatively widespread piedmont forest in Colombia (DEFLER, 1996b). All populations of *L. l. lugens* are threatened by nearby human populations. Although it is known to be present in six or seven national parks, the sizes of the populations are not known. Censuses are urgently needed and local threats need to be evaluated. Because it may be a Colombian endemic (possibly there may be populations in Venezuela), it is important to emphasize the responsibility of Colombian conservationists in carrying out an effective conservation program. The establishment of a proposed San Lucas National Park would hopefully protect the northernmost, apparently isolated population, but preliminary work on this project has not yet begun, partly because of problems explained above for *Ateles hybridus brunneus*.

Equatorial mantled howling monkey, *Alouatta palliata aequatorialis* – VU

Although the mantled howling monkeys, which occur as far north as southern Mexico, are considered LC on a global level, *A. palliata aequatorialis* in Colombia is threatened because of widespread hunting by Afro-Colombian and indigenous people. Population estimates are not available, and censuses are an urgent priority. Recent faunal evaluations in the Utría National Park found that they were extremely rare there, and that they had been more common in years past (H. RUBIO, pers. comm.). Widespread hunting in the Chocó is probably the greatest pressure, aside from habitat loss. Historical populations as far east as the area around Cartagena have now been totally replaced by *Alouatta seniculus*.

Primate Conservation in Colombia

In general terms, the conservation of primate species requires several steps: any one may comprise the main thrust of a conservation research project based around one or several species and subspecies.

Censusing. Knowledge of the location and size of viable populations of the threatened species is fundamental, as is the assessment and monitoring of populations in protected areas. The protected areas themselves should be evaluated for threats to their integrity and management. Parks with no personnel offer no or little protection only. Census techniques are described and discussed in NRC (1981) and elsewhere, but even intensive observations (searches) in different parts of a protected area will usually generate important data.

Basic natural history, ecological and taxonomic studies. Extinction risks vary greatly with the biology of each species. Informed management requires a detailed knowledge of the species' natural history, behavior and ecology. The information needs to be pertinent to the population in question, studies carried out elsewhere may not reflect the same ecological conditions. It is also important to obtain a good understanding of the local human population's relation to the species in question, such as whether it is hunted or trapped and to what extent. New techniques in karyology and molecular genetics are necessary to give us a better understanding of systematic relationships and differences, and consequently the diversity we need to conserve. It is likely that this type of research will uncover differences between populations which today are considered to be of the same taxa. This type of research with *Aotus* is especially urgent and is expected to point our more variation than has so far been recognized.

Protected areas. Although Colombia is fortunate in having a fine park system, not all ecosystems and species are protected and there are many problems, such as inholdings and land claims as well as budgetary constraints and too few personnel. More protected areas are required, while improving the protection available to those already established. I encourage all Colombian citizens and others interested in the conservation of Colombian fauna and flora to support the Colombian park system fully and to insist that the park system be given a budget adequate to its needs. Also, it is important to promulgate research within the protected areas to provide a better understanding of the biodiversity they contain. National parks and reserves are like a national land bank where many, many organisms are saved for the future. As the future comes upon us these organisms and the land itself will be increasingly valuable to the nation and to the world.

At the present time, the central control of national parks is under question by the Colombian government, which in its effort to cut spending, sees the transfer of parks to regional entities as a solution. Most environmentalists in Colombia are vehemently opposed to a de-centralization of national parks, since it is well-understood that such a step will lead to local abusive control by political authorities who have no idea about national resources and interests. The development of national parks in Colombia now has involved 30 years of difficult and underfunded growth, which has led to the development of competent authorities and a pool of knowledge wielded for the development and protection of parks in the country. A transfer of parks to regional control would be a national mistake with grave future consequences for the integrity of the park system and would lead to continued degeneration of the park units which all have fought to preserve.

Environmental education. It is vital that all Colombians understand the importance of the conservation of the rich biodiversity found in this country. Educational programs, books, films and other devices can be used to tell Colombians and foreign friends about the flora and fauna, and specifically about the primates in the country. Every time a reserve or a national park is established an educational campaign should be organized to explain to the local people about its value and meaning. Any conservation program ought to have an educational element, and local support should be actively sought.

Users of this guide are urged to contact the author about sightings of Colombian primates, especially when they are of threatened taxa. Information about primate populations observed within parks and reserves would be much appreciated. In many cases the local inhabitants and even park officials fail to understand the importance of particular primate populations which, after being identified, may form the basis of a local education program.

General considerations and specific topics of neotropical primate conservation can be found in KONSTANT *et al.* (1985), MITTERMEIER (1986a, 1986b, 1986c, 1986d, 1987a, 1987b, 1988, 1991), MITTERMEIER & CHENEY (1987), MITTERMEIER *et al.* (1976, 1978 1993, 1994, 2000), MITTERMEIER & COIMBRA-FILHO (1983), MITTERMEIER & OATES (1985), RAEZ-LUNA (1993, 1994), ROBINSON & RAMIREZ (1982), RYLANDS *et al.* (1994, 1996/1997, 1997), TERBORGH (1983, 1986a), THORINGTON (1978a, 1978b). For discussions on aspects of primate conservation in Colombia, I refer the reader to Alderman (1989), BARBOSA *et al.* (1988), DEFLER (1989c, 1994b), FAJARDO (1988), FAJARDO & CLAVIJO (1988), HERNÁNDEZ & DEFLER (1985, 1989), MAST & FAJARDO P. (1988), MAST *et al.* (1993), SAVAGE (1988), SAVAGE & GIRALDO (1990), SAVAGE *et al.* (1990b), AYRES & MARTINS (1990), CEBALLOS & NAVA-

RRO (1991); HELTNE & THORINGTON (1976) among others. GROVES (1993, 2001), RYLANDS *et al.* (1997, 2000), SILVA JR. (2001) and VAN ROOSMALEN *et al.* (2002) provide the most recent taxonomic listings of New World primates, although their views on what is a species may not be adopted in this guide.

Cebuella pygmaea

5. The phylogeny of the platyrrhines and their classification

Modern analyses of molecular data have created a revolution in the way that we understand the phylogenetic relationships of the Colombian platyrhines. The first information on aligned sequences of nuclear DNA of the epsilon-globulin for 16 genera of platyrrhines was published by Schneider et al. (1993). Following this several papers appeared, some of which involved neotropical genera (SCHNEIDER et al., 1995, 1996, 1997; CANAVEZ et al., 1999a; VON DORMAN et al., 1999; GOODMAN et al., 1998). Other researchers have concentrated on particular clades (CANAVEZ et al., 1996b; CHAVES et al., 1999; MEIRELES et al., 1999 ; PORTER et al., 1997a, 1997b ; PORTER et al., 1999 ; PASTORINI et al., 1998 ; TAGLIARO et al., 1999) and a synthesis of work accomplished (RYLANDS et al., 2000).

Classical taxonomic schemes of the last century based principally on comparative morphology, organized the neotropical primates in 2-3 families which in the case of Colombian primates included the Cebidae (*Saimiri, Callicebus, Pithecia, Cacajao, Aotus, Cebus, Ateles, Alouatta* and *Lagothrix*) and the Callitrichidae (*Cebuella, Saguinus* and sometimes *Callimico*), although HERSHKOVITZ (1977) preferred to classify *Callimico* apart in its own family, Callimiconidae.

The importance of recent molecular analyses lies in the clear separation of three evolutionary branches of platyrrhines, the atelines, the cebids and pithecids, and these are classified as families (Atelidae, Cebidae and Pitheciidae) in agreement with morphological and molecular analysis (SCHNEIDER & ROSENBERGER (1996). Nevertheless, there are still difficult questions to be resolved with reference to the relationships of *Aotus, Cebus* and *Saimiri* which RYLANDS et al. (2000) partially resolve.

The Cebidae are the most complex group, though the relation of *Cebuella, Saguinus* (and the other callitrichids) to the *Saimiri* and the *Cebus* has been accepted for various years (ROSENBERGER, 1981; RYLANDS et al., 1995; GROVES, 2001). In the new nomenclatural arrangement the callitrichids become a subfamily (Callitrichinae) in the Cebidae, together with the subfamies Cebinae (including the *Cebus*) and the Saimirinae (including all of the *Saimiri*).

The phylogeny of the platyrrhines and their classification

According to molecular criteria the *Aotus* have been considered a subfamily of the Cebidae by SCHNEIDER *et al.* (1993, 1995, 1996) and PORTER *et al.* (1997), but using morphological criteria the *Aotus* have been grouped with the *Callicebus* in the family Pithecidae (ROSENBERGER, 1981; SCHNEIDER & ROSENBERGER, 1987) making it difficult to place them in an appropriate phylogeny. The difficulty evidently is that the genus is ancient, found in the La Venta fauna of the Upper Miocene of Colombia and retains evidence of its relationship to all of these groups, which ultimately goes back to a small founder population from Africa. Nevertheless, it is important to remember that *Saimiri* is also an equally ancient genus compared to other platyrhine genera. In this book I follow RYLANDS *et al.* (2001) and place the *Aotus* in their own family Aotidae.

The Atelidae have been accepted for some time as a clade which is divided into two subfamilies (Atelinae: *Ateles, Brachyteles, Lagothrix, Oreonax* and the extinct taxa *Caipora*† and *Protopithecus*† and the Alouatinae: *Alouatta* and the extinct genus *Stirtonia*†) (FLEAGLE, 1988; FORD, 1986; ROSENBERGER, 1981; RYLANDS *et al.*, 2000; GROVES, 2001).

In the past decade the relationship of the *Callicebus* to the pithecids has become well-established via the work of SCHNEIDER *et al.* (1993, 1996, 1997), PORTER *et al.* (1997, 1998), VON DORNUM & RUVOLO (1999), HOROVITZ *et al.* (1998) and CANAVEZ *et al.* (1999) and this has resulted in the tendency to divide the Pitheciidae into two subfamilies, the Pitheciinae and the Callicebinae (GROVES, 2001) or sometimes into three with the Aotinae, but here we consider the *Aotus* as belonging to their own natural group (Aotidae).

Using the above information, based on the various studies using molecular and morphological comparisons I use the classification seen below. Though GROVES (2001) presented various arguments modifying the traditional nomenclature presented below, I continue with traditional terms which were later ratified by BRANDON-JONES & GROVES (2001) after the publication of GROVES (2001) book.

106 • Primates of Colombia

Order. **Primates**

Infraorder. **Platyrrhini**

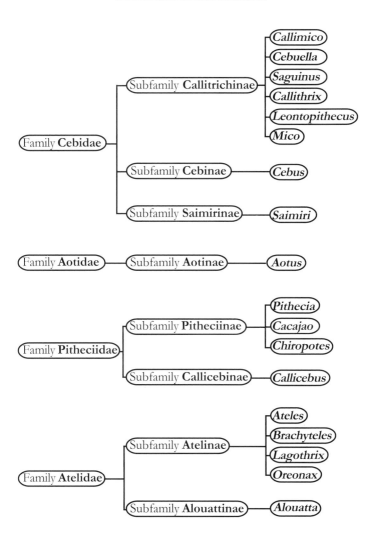

6. How to use this field guide

This book is dedicated to one of the most important groups of mammals in Colombia, the primates. The country is one of the richest in the world for diversity of primates with 43 taxa (species and subspecies), including 27 species. The goal of this guide is to provide all interested readers with up-to-date information on the state of these our nearest relatives in the Animal Kingdom, providing data on their taxonomy, distribution, behavior, ecology and conservation status.

The first chapters offer a general introduction to the history, zoogeography and conservation of Colombian primates with the aim to guide the reader through an appropriate context and provide it with an improved capacity for appreciating the importance of these animals, as well as understanding their fossil history and providing some aspects of their distribution. I also include, questions yet to be answered about their distribution and other diverse aspects. I believe that readers could contribute important information to these themes and hope that readers will contact me with new information.

Fundamentally, the conservation of primates has been taken into account in this book. One of the great challenges of the past century and the present one is beginning to actively and effectively safeguard our planet´s biodiversity, including all the plants, animals, microorganisms and the ecological processes of which they are part. Originally the book was to be published in Spanish because of the ample need which exists in Colombia and the rest of Latin America for Spanish language information, especially since most of the species found in Colombia are also found in many other neotropical countries. Indeed, an earlier version of this book was published in Spanish in 2003, but the decision was taken by Conservation International to also publish this information in English and thus cover a much wider audience, adjusting both editions to the Tropical Field Guide Series of Conservation International.

This chapter contains a practical and simple graphic key to each species, based on morphological and color aspects and silhouette, which should help the reader to identify Colombian primates quickly and easily. Pagination provides the option of going directly to the text, distribution map and color plate of the species which a reader might wish to consult. The phylogenetic order adopted in the present volume follows a nomenclatuaral order very recently adopted by some (particularly RYLANDS *et al.*, 2001) based on new comparative molecular data, which is permitting a revolution in how we see they phylogeny of the primates. The concepts and considerations made by this author with respect to these changes are listed in various sections called **Taxonomic Comments,** which accompany each species and genus description and which try to present a "state of the art" view, in some cases pointing out the need for more research.

Certainly the heart of the book is the section which presents the detailed information on each species, arranged as species grouped around each genus, with some short comments (sometimes detailed comments in the case of *Aotus*, *Ateles* and *Callicebus*) on each genus (and family). If there are subspecies recognized then I included a key for the identification of the subspecies. A very large effort has been made to provide a bibliography of field work that is as inclusive as possible up to the end of 2003, to which I continued to add references in 2004 up to the point where the text had to be given to the editor. Selected publications on some captive and laboratory research are also included, if in my opinion it would clarify details not studied in the field. For example, much of the research of Epple particularly on *Saguinus fuscicollis* needed to be done in a controlled setting and it explains much about hormone mediated behaviors of this and other species. These references especially relating field work to each species, but also partially covering fossil history have been the result of an exhaustive search up to the most recent dates possible. Although I have not examined every citation, I have read many, and mean for this exhaustive list to be a tool for those interested in these primates. This is the place to begin for any student new to the field.

I have made an effort to provide as many Colombian names as possible for each species, including both "Spanish" and indigenous names and of course I have included the most common English, German, Portuguese and French names as they were available to me. The Colombian names are usually the greater part of the list, since these animals are referred to and discussed by many country people in their respective idioms. This section bears analysis as being more valuable than the European language names, since the local names have been invented by people who know the animals intimately, not by a white European scientist looking at a skin in a European or American museum. I continue collecting indigenous names, since they are a reflection of those people´s cultural attitudes towards each primate and seem to be particularly valuable for an understanding of the animals´ natural history.

The description of each genus and species is accompanied by illustrative material in the form of color plates and pen and ink drawings of silhouettes and positions that complement the text in order to acquaint the reader with the animal. The description first tries to give a general idea of the size and weight and lists any sexual differences. It then lists the colors of pelage and skin in various parts of the body, especially those characteristics which are evident to an observer in the field, as well as individual variations. A detailed map of the distribution of the species in Colombia is also included, based on geo-referenced collection data from the four main collections in the United States and Great Britain, as well as the two main collections in Colombia, supplemented by field observations by myself and a few other data given to me by Colombian field biologists. The text is accompanied by a glossary which includes many of the terms in the text which may be only familiar to the specialist.

The different sections in the text for each species are standarized and include: **Taxonomic Comments**, including often a short history of different concepts and ending with the taxonomy accepted in this book, usually (but not always) based on the work of Anthony Rylands and other members of the Primate Specialist Group of the Survival Service Commission of the International Union for the Conservation of Nature; **Common Names**, including Colombian "Spanish" (which in many cases are adapted from an indigenous name), English, other European languages, Brazilian Portuguese and; **Indian Languages**, including all names that I could either personally collect or find in the literature from Colombian indigenous groups; **Identification**, as described above; **Geographic Range**, with details of the Colombian distribution and a general statement of the extra-Colombian distribution and including the known habitat preferred by the species; **Natural History,** including an explicit overview of all field work published on the species, with a description which includes average group size, home range or territory size, calculated densities, activity budget, diet, reproductive characteristics, infant care, vocalizations, inter- and intraspecies interactions and some displays; **Conservation State**, a description of the national and international classification according to the IUCN categories; and finally

Where To See It, including some suggestions as to where one might be able to observe the species.

The last block of text in the book includes the color plates which illustrate all of the Colombian species and subspecies known for the country. Also included here are the distribution maps, showing the historic distributions for all species (though some parts are probably now unoccupied by the species), including subspecific distribution and some question marks indicating a lack of knowledge for their presence or not. Then comes a rather extensive glossary, as described above and finally the extensive bibliograpy ends the book.

Dorsum

Mantle

Frontal stripes

Saddle

Temporal stripes

Malar stripe

Ventrum

Key for rapid identification

Large body, more than 50 cm head-body length and a weight 6-12 kg or more. Prehensile tail naked callosity on ventrum. Tail longer than body. Diurnal.

Slender appearance, weight 6-12 kg or more; with arms, legs and tail elongated. Hands without a thumb.

Body pricipally black. *Ateles geoffroyi*.

Plate 8 • Page 339 • Map 16

Body black with yellow or whitish stomach, at times the interior part of the legs are also yellow. *Ateles belzebuth*.

Plate 8 • Page 331 • Map 15

Dorsum and flanks of the body dark brown, contrasting with the ventrum which is light brown. *Ateles hybridus*.

Plate 8 • Page 347 • Map 14

Robust appearance, 6-12 kg or more with the limbs moderately elongated. Tail long, prehensile with hairless callosity . Pelage moderately long or short with variable texture. Hands with thumb.

Woolly Monkeys, Genus **Lagothrix**

Color usually olivaceous brown to dark brown, almost black or greyish with a silvery aspect; sometimes with abundant and very long dark chest hair that contrasts with the rest of the body. *Lagothrix lagothricha*.

Plate 9 • Page 357 • Map 17

Robust appearance, 4-12 kg with moderately elongated limbs. Tail is long and prehensile, pellage is long and rough. Throat inflated, especially in males. Females notable smaller than males.

Howler Monkeys, Genus **Alouatta**

Generally black but often with a lighter brown saddle. *Alouatta palliata*.

Plate 10 • Page 370 • Map 18

Generally chestnut red with patches of golden yellow on back and on tail. *Alouatta seniculus*

Plate 10 • Page 384 • Map 19

Robust appearance 2.5 – 4.5 kg with limbs moderately long. Tail prehensile, without callosities and totally covered in fur. Pellage rough and moderately long.

Dark brown limbs and tail, light brown body, sometimes with a head tuft which may be bifurcate in females and young males. Old males may be almost without hair on head. *Cebus apella*.

Plate 5 • Page 216 • Map 12

General coloration blackish with whitish shoulders and chest and black head pelage giving appearance of a cap. *Cebus capucinus*.

Plate 5 • Page 227 • Map 11

General coloration light brown or dark brown overall, sometimes with forearms and legs colored reddish or yellowish and a cap usually a darker brown. *Cebus albifrons*.

Plates 5, 6, 15 • Page 207 • Map 10

Body of moderate size, 30-50 cm head-body length and 1-3 kg. Tail is not prehensile and is long or short. Diurnal, crepuscular or nocturnal.

Body covered with abundant, long and thick hair, sometimes with a disorded aspect; diurnal; tail long or short and more than 2 kg in weight.

Body sometimes has a chubby appearance due to its habitat of erecting its long body hair when it is nervous. Otherwise, calm, it appears very slim. The long hair is thick and lightly wavy. Head hair directed forward up to forehead where it becomes very short over the entire face, giving an odd appearance of the animal with a wig. The underparts have much less hair than the dorsum. *Pithecia monachus.*

Plate 13 • Page 278 • Map 22

Body long and slim, tail very short and not prehensile with long, smooth body hair, 2-3 kg, males heavier than females; head hair is characteristically directed foreward, falling over the forehead. *Cacajao melanocephalus.*

Plate 14 • Page 287 • Map 23

Body covered with moderately abundant, fine and moderately short hair giving appearance of slender elegance. Diurnal. Tail long and not prehensile. Weights 1-2 Kg.

The pelage color is black or reddish black with white collar, with or without white hands or grey agouti with reddish limbs and circling the face. *Callicebus torquatus.*

Plate 11 • Page 314 • Map 20

General color greyish agouti with whitish tail, whitish band across the forehead, reddish tones on limbs and on ruff surrounding the face. *Callicebus cupreus.*

Plate 12 • Page 306 • Map 21

Body covered with abundant short hair, moderately fine giving a slender appearance. Face with attractive, complex design; diurnal or nocturnal; tail long and not prehensile; weight around 1 kg.

Diurnal; face with white mask, nasal area black, forehead and crown a contrasting grayish brown. *Saimiri sciureus.*

Plate 4 • Page 237 • Map 9

Nocturnal; tail long and non-prehensile; eyes very large; head with black stripes on the face and crown and usually with white areas over the eyes. Usually weighs around 1 kg. Genus *Aotus*.

The interscapular region with no crests or whirls.

Ventrum of arms and legs completely brownish like elk or moose like the thorax or with orange or creme extending to the stomach and the mid part of the arms and legs (rarely to the ankles). Dorsal pelage short and thick or long and loose; coloration of the dorsum and body variable from light orange tones to dark grey; sometimes a wide dark dorsal stripe, coffee or orange colored; the head stripes may be united or un united behind.

Dorsal hair long, more or less 2.6 – 3.6 mm; limbs may be blackish due to interspersed hairs (especially those from the Cordillera Occidental) or not (especially in the north of the Cordillera Oriental). Two basic color phases (sometimes in the same group); one corresponds to a dorsum of greyish light brown and the other to a dorsum of reddish brown; ventrum almost always light yellow. (karyotype 2n=58; Defler *et al.*, 2001). *Aotus lemurinus*.

Plates 7, 15   •   Page 262   •   Map 13

Dorsal hair short and plastered against the body ; dorsal color brownish or brownish yellow; ventrum light yellow; dorsal part of hands and feet usually light brown. (karyotype 2n=52, 53, 54). *Aotus griseimembra.*

Plates 7, 15   •   Page 262   •   Map 13

Appearance similar to *Aotus griseimembra* but with the hands and feet having blackish appearance due to interspersed hairs. (karyotype 2n=55,56) The interscapular region has hair formed into a crest. *Aotus zonalis*.

Plates 7, 15 • Page 262 • Map 13

The interscapular region with fur forming a crest.

Interscapular crest with hair directed away from the center and to the sides; gular gland large (5 cm) and with circling hair extending away from the center; head design with black temporal stripes which extend from the exterior angle of the eye caudaly, converging but not uniting except for disperse black hairs. The crown between the temporal lines is beige; black, well-defined malar stripes and an extesion of the temporal stripes; supraorbital patches of cream, sides of head, neck below the ears colored greyish to cream, finely speckled with brown (karyotype 2n=50). *A. brumbacki;*

Plate 7 • Page 267 • Map 13

The interscapular region forms a whirl (vortex).

Temporal stripes close to each other, almost united behind the head; malar stripes well-defined or absent; slender tufts in the feet which do not extend to the nails; animals of small size, tail around 340 mm (300-363 mm). (karyotype 2n=46, 47, 48). The interior parts of the extremeties are orange or beige and the same as the chest and stomach, extending to the wrists and ankles; the dorsal pelage is short, thick and dominantly grey in the upper parts, sometimes brownish beige with dark narrow band mid- dorsally which contrast strongly with the orange-brown of the neck and stomach. *A. vociferans.*

Plate 7 • Page 269 • Map 13

Interior part of the extremeties with orange or beige as the chest and stomach, extending to the wrists and ankles; pelage of dorsum short, thick, dominantly grey in the upper parts.

Temporal stripes do not unite; darker central dorsum area as undefined stripe; no interscapular whirl (unknown karyotype). *A. trivirgatus*

Plate 7 • Page 273 • Map 13

Small body around 300 – 300 mm head-body and between 400 – 700 g.
Attractive colors; long tail, non-prehensile, diurnal.

Body principally lustrous black; head with a large black mane; third molar present. *Callimico goeldii.*

Plate 1 • Page 126 • Map 1

Color generally blackish with brown undertone; cheeks and genitals are white in high contrast to rest of body.

Without fur and contrasting white. *Saguinus inustus.*

Plate 2 • Page 169 • Map 4

Generally colored black, chestnut or olive or a combination. *Saguinus nigricollis.*

Plate 3 • Page 179 • Map 8

Body usually dark brown with a saddle of clearer color beige and brown. Naked, black face, highly contrasts with rest of body which has white or reddish ventrum and legs, striped brown-beige dorsum and dark tail. *Saguinus fuscicollis.*

Plate 3 • Page 150 • Map 7

Face naked and black in high contrast to the rest of the body which is white in varius parts of the pelage.

Dorsal pelage principally with intermixed dark chestnut and white lines, white stomach, dark brown tail; head with a modest crest of hairs in the form of a crest in the center of the crown. *Saguinus geoffroyi.*

Plate 2 • Page 163 • Map 3

Dorsal pelage principally café of various tonalities, white belley, reddish tail proximal half and blackish the distal half. Head with a full white mane. Body grey-brown with ends of hair white; hands and feet white, tail dark brown at times with a white tail tip. *Saguinus oedipus*.

Plate 2 • Page 188 • Map 6

Body brown with points of hairs white-tipped, hands and feet white, tail brown sometimes with a white tip. *Saguinus leucopus*.

Plate 2 • Page 173 • Map 5

Body very small, 130-160 mm head-body with a tail of about 200 mm and weight 85 – 140 g. Diurnal. Long, non-prehensile tail with bands of black and cinnamon. *Cebuella pygmaea*.

Plate 1 • Page 135 • Map 2

Family **Cebidae**

   The classical meaning of Cebidae has undergone a revolutionary change in the last few years as comparative morphology and molecular research have more and more clearly demonstrated the phylogenetic connections of *Callimico, Saguinus, Cebuella* (as well as *Mico, Callithrix* and *Leontopithecus*) to *Cebus* and to *Saimiri*. Because *Saimiri* is apparently an ancient taxon, recognizeable even in the mid-Miocene, this genus shows other relationships as well, but its relationship particularly to *Cebus* has become clear.

Early on it was clear that *Saimiri* and *Cebus* were closely related (POCOCK, 1925; ROSENBERGER, 1977) and ROSENBERG (1979) pointed out that relatives of these two groups were probably relatives of the marmosets, since the entire complex shares absent third molars, enlarged canines, somewhat enlarged anterior premolars, foreshortened faces, gracile zygomatic arches and shallow, open glenoid fossae for articulation with the mandibular condyle (ROSENBERG, 1981:22). These characters contrast sharply with the atelines and other living primates, suggesting a "natural group", which because of priority can be called the Cebidae. This view then sees the callitrichines as a highly derived group of cebids.

It has been clear for a long time that the callitrichines were monophyletic and could be diagnosed by "clawed digits, reduced hallux and pollex and a modified form of incisal occlusion" (ROSENBERG, 1981:23). There was great resistance against the idea that these were derived characteristics, perhaps because of the weighty authority of people like HERSHKOVITZ (1977) and others, who considered them primitive and so looked for an origin other than the cebids.

The pioneering work of SCHNEIDER *et al.* (1993, 1995, 1996) first clearly supported ROSENBERG'S (1981) morphometric interpretation of the relationship of the callitrichines (including *Callimcio*) to *Cebus* and *Saimiri* via closely related DNA sequences, much of which suggested that *Aotus* was also related to this clade. Other morphological evidence (ROSENBERG, 1981; SCHNEIDER & ROSENBERGER, 1996) and molecular sequencing evidence (GOODMAN *et al.*, 1998) saw *Aotus* as being more related to the atelids or to the pitheciines. Finally, mt DNA RFLP variation found a close relation of *Aotus* to *Cebus* and neither of these to *Saimiri* (RUÍZ-GARCIA & ÁLVAREZ, 2003). The lack of agreement for the phylogenetic position of *Aotus* and its ancient lineage (at least to the mid-Miocene; chapter 2) has caused me to place it into its own family while its phylogentic position is further analyzed. This view is supported by GROVES (2001).

Therefore, a preponderance of morphological and molecular evidence seems to group the callitrichines with *Cebus* and *Saimiri* in their own family. This family can be divided into subfamilies as follow: Callitrichinae (*Callimico, Cebuella, Saguinus* in Colombia), Cebinae and Saimirinae.

Goeldi´s Marmoset

FAMILY **CEBIDAE**
Genus *Callimico* MIRANDA-RIBEIRO, 1912

This is a monotypic genus and until recently it was considered to belong to the monotypic family Callimiconidae because of its unique dental formula and other characteristics However, molecular analyses done by SCHNEIDER *et al.* (1996), SCHNEIDER & ROSENBERGER (1996), GOODMAN *et al.* (1998) and morphological research by ROSENBERGER (1981) have demonstrated that this genus is closely related to other callitrichids such as *Saguinus, Callithrix, Mico, Cebuella* and *Leontopithecus* and that this group is closely related to *Cebus* and *Saimiri* with a common origin. Therefore, this genus is considered to be part of the Callitrichinae in the family Cebidae, along with Cebinae and Saimirinae (ROSENBERGER, 1981; ROSENBERGER *et al.*, 1990) and they are classified in this manner here. Nevertheless, there still are other taxonomies used.

Goeldi's Marmoset

Callimico goeldii (THOMAS, 1904)

Taxonomic comments

Plate 1 • Map

This species had been problematic with respect to its phylogenetic position. HERSHKOVITZ (1977) had placed it in a separate family on the basis of its dental formula (36), including the third molar that is lost in the callitrichines and the birth of one offspring instead of two. It has also been grouped with *Callithrix, Saguinus* and *Cebuella* and even placed in the Cebidae (in the classic sense). However, the bulk of evidence now places it with the callitrichine primates in the cebid family along with *Cebus* and *Saimiri*. Chromosome comparisons with *Saguinus* and *Callithrix* (DUTRILLAUX *et al.* 2, 1988; MARGALIS *et al.*, 1995) suggest that *Callimico* probably had a common origin with the other two genera, but that now *Callitrhrix* and *Saguinus* are much more related to each other. No subspecies are known (ROSENBERGER, 1981; RYLANDS *et al.*, 2000; GROVES, 2001).

Common names

Colombia: *chichico negro* around Puerto Umbria in Putumayo; German: *Springtamarin*.

Indian languages

Dohogo (Huitoto); *júhusaryje* or *cibominajuubari* (also used for *Saguinus* sp.) (Muinane); *baisisi* (Siona).

Identification

Callimico goeldii adults weigh around 393-670 g and measure 210-310 mm head-body and 255-324 mm tail length. The hands and feet have claws except for the hallux. These animals have thick, glossy black hair of about 1-

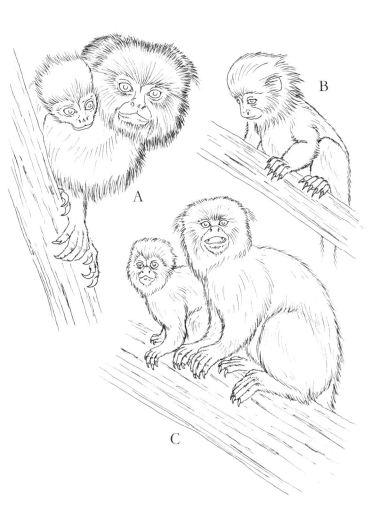

Figure 6. Various positions of *Callimico*. a) Carrying the infant; b) Lone infant waiting; c) Male in aggressive attitude, protecting the infant.

Figure 7. Various views of *Callimico*. a) Face. b) Clinging. c) Vigilant. d) Jumping.

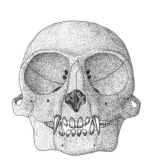

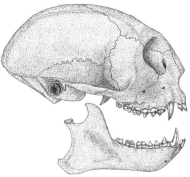

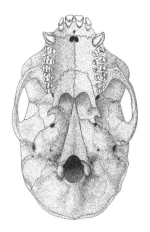

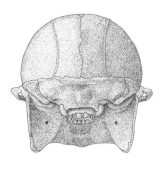

Figure 8. Skull of *Callimico*.

2 cm covering the head, dorsum and lateral parts of the body. There is a ruff of longer hair around the neck, giving a lion-like appearance. The skin of the face is also black, although somewhat lighter around the eyes, cheeks and external ear margin, while the rest of the skin of the body is pale white. The species has a dental formula including three molars instead of the usual two for the callitrichines, for a total of 36 teeth.

Geographic range

In Colombia *Callimico goeldii* has been collected from various localities from the base of the Cordillera Oriental of the Andes in Putumayo department to the mouth of the Cahuinarí river, a major right bank affluent of the Caquetá. In no place is the species common, and it is possible that eventually it will be detected further east between the Caquetá and Putumayo Rivers as well as in the Colombian trapezium. Outside of Colombia the species has been recorded in Amazonian Bolivia and Perú (but not Ecuador although probably it is found in Ecuador as well at upper Amazonian sites.

All known specimens were collected in upland forests that were flat or had low hills. It has also been mentioned as being found in "degraded" forest, secondary growth, sometimes with bamboo growing on very damp ground. Nevertheless, in upland rainforest the animals move at low heights of around 3 m, only climbing into the canopy to forage (MOYNIHAN, 1976a; POOK & POOK, 1979b; IZAWA, 1979b; HELTNE *et al.*, 1981; HERSHKOVITZ, 1977; HERNÁNDEZ-C. & COOPER, 1976).

Natural history

This species has been studied in the field by POOK & POOK (1979a, 1979b, 1981, 1982), IZAWA (1979b) & MASATAKA (1981, 1983a & b), WHITTEMAZE (1970), CHRISTEN (1994, 1998, 1999, 2000) and CHRISTEN & GAISSMANN (1994) in Bolivia. It has also been studied by PORTER (2000a, 2000b, 2001a, 2001b, 2001c, 2001d, 2001e, 2002), PORTER *et al.* (2001), PORTER & CHRISTEN (2002), BUCHANAN-SMITH *et al.* (2000), HANSON & PORTER (2000), CHRISTEN & PORTER (1999), FERRARI *et al.* (1999) and HARDIE (1998) in Bolivia and by GARBER & LEIGH (2001), CALOURO *et al.* (2000), GARBER (2000), GARBER & REHG (1998) and REHG (2003) in Brazil. It was observed more casually by MOYNIHAN (1976a) and HERNÁNDEZ-C. & COOPER (1976) in Colombia. Posture and locomotion was studied by DAVIS (1994). MASATAKA (1983) studied vocalizations in the wild.

Group size averages 5-6 individuals: one group contained 8 (IZAWA, 1979b, HELTNE *et al.*, 1981). Captive studies suggest that within the group one male and female are bonded, so that a natural group is probably a family group made up of the adult pair and its young. One study showed a home range of 33-60 ha (POOK & POOK, 1981). One group traveled an average of 2 km per day. Densities seem generally to be extremely low and groups well-separated from each other at about 1 group/km^2 (POOK & POOK, 1981).

Callimico goeldii uses much clinging and leaping to progress on small-stemmed and low vegetation. Favored height from ground is 2-3 m. When traveling higher they use more conventional quadrupedal locomotion, but to move from one tree to another they often descend to do a horizontal leap. Groups sleep in tangled vines at around 10-15 m in height (IZAWA, 1979b) and dense thickets of low vegetation (POOK & POOK, 1981).

Individuals consume a variety of soft fruits and insects (especially grasshoppers), which at times are plucked from the forest floor (HELTNE *et al.*, 1981). LORENZ (1971) reported that captive but wild-caught *Callimico* killed and ate small snakes, lizards and frogs as well as young mice and chicks. Some fruits eaten were *Cecropia morassi, Cecropia* sp., *Pourouma* sp., *Piptadenia* sp. *Clarisia racemosa* and *Parkia* sp. *Callimico* drink open-standing water in captivity but whether on the ground or in the trees has not been observed.

The highest ranking female initiates sexual behavior. Gestation is about 151-159 days, based on seven pregnancies in captivity (LORENZ, 1972), but the range is probably 150-165 according to HELTNE *et al.*, 1981; HEINEMANN, 1970). Sexual cycles were reported with a range of 21-24 days (n=13)(LORENZ, 1972). Birth season in Bolivia is September-November, during the early part of the wet season (POOK & POOK, 1981). Behavioral and hormonal aspects of reproduction are discussed by JURKE (1996) and JURKE *et al.* (1995).

Callimico goeldii gives birth to one infant which is carried by the mother ventrally at first, gradually shifting the infant to the back during two weeks. The infant is then gradually carried more and more by the father, who finally only gives it over for nursing (HELTNE *et al.*, 1973, 1981; LORENZ & HEINEMANN, 1967).

The 1st day infants have open, cloudy blue eyes with a naked ventrum and short black dorsal hair. At 2-3 days the infant is curious about surroundings and during the first week the mother carries the infant exclusively about 40% of the time on the ventrum. During the second week the mother also is the exclusive carrier with the infant on ventrum about 20% of the time. Third week, the mother begins attempting to remove the infant, which wails in distress, causing the father to carry it. This week the father carries 87% of the time, the mother 13% of the time. Fourth week the same division of labor exists between the father and the mother. Fifth week, father carries infant about 96% of the time. Sixth week, infant makes short hops, begins to walk between parents and show independence. Seventh week, increasing independence. Eighth week, infant independent of parents and exploring. Tenth week, only mounts father when frightened; otherwise he causes infant to dismount using sharp bites to the ankle.

Adult young probably must separate from the groups as there is evidence of aggression between same-sex animals. Group members groom each other, perhaps the females more than the males (LORENZ & HEINEMANN, 1976). The species exhibits group cohesion, often traveling in a close bunch and usually within 5 m of each other.

Some calls mentioned by HERSHKOVITZ (1977) and taken from POOK & POOK (1981) and WHITTEMAZE (1970) are listed as follow: (1) *bark, tchuk, chok* - low-pitched and throaty response to unusual noise; (2) *long-call, tsee-call, contact-call, shrilling* – series of piercing calls given with open mouth after hearing recordings of other *Callimico*; (3) *twitter, feeding call* – bird-like twittering when feeding or satisfied; (4) *scream* – distress call; (5) *lost call* – progression towards longer notes (perhaps the same as contact call); (6) *trill* – extreme excitement; (7) *whistle, shriek* – long, monosyllabic note as alarm; (8) *eek-eek* – succession high-pitched, voiced coughs with aggressive context.

A list of some displays described by POOK & POOK (1981) and WHITTEMAZE (1970) follows: (1) *rapid tongue flicking* – aggressive context; (2) *marking* – both with sternal glands and urine; (3) *tail ruffling, swaying* – while emitting *tschuk* danger call; (4) *tail curling* – under the body when relaxed; (5) *arched back display* – non-aggressive interaction of adults while nuzzling and walking repeatedly past each other and jumping over each other (not clear if in a sexual context); (6) *urine tail marking* – while resting over the tail; raises up brushed colied tail back and forth over ano-genital region, moistening tail.

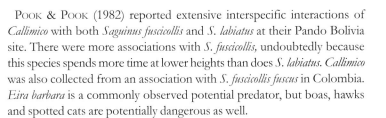
Pook & Pook (1982) reported extensive interspecific interactions of *Callimico* with both *Saguinus fuscicollis* and *S. labiatus* at their Pando Bolivia site. There were more associations with *S. fuscicollis*, undoubtedly because this species spends more time at lower heights than does *S. labiatus*. *Callimico* was also collected from an association with *S. fuscicollis fuscus* in Colombia. *Eira barbara* is a commonly observed potential predator, but boas, hawks and spotted cats are potentially dangerous as well.

Conservation state

This species is listed in Appendix I of CITES and VU in both the international and national IUCN classification, mostly based on its scarcity in its natural habitat, including scarcity of information in Colombia. Surveys of this species are needed in Colombia in order to establish the location of populations and to confirm the extent of the presence of this species in La Paya and Cahuinarí Natinal Parks where the species has been observed before. If good populations of this species can be identified then special efforts could be made for their protection. Especially if a healthy population is detected within a national park species protection should be easier. If populations are found outside the parks it may be necessary to seek to establish a reserve especially to protect *Callimico*. Any incipient commerce in this species should be immediately controlled by the authorities.

Where to see it

This primate is not easy to view in Colombia. The animal is not very well-known even in regions where some have been sighted by biologists over time. Although it has been years since the sighting by Moynihan (1976a), it seems that searching for this species near Puerto Umbría on the Mocoa-Puerto Asís road (Putumayo department) would be the best possibility, since these animals are the most accessible at that place. They were seen in second growth in abandoned plantations mixed with Asiatic bamboos. Other possibilities would be to go by boat downriver from Puerto Asís on the Putumayo River to the Quebrada (creek) La Hacha, one of the western limits of La Paya National Park and travel up this creek as far as possible, looking for *Callimico*. The author would be grateful for reports of any Colombian sightings of this rare primate species.

Pygmy Marmoset

FAMILIY **CEBIDAE**
Genus ***Cebuella*** GRAY, 1866

Cebuella is considered monospecific with two subspecies recognized (*C. p. pygmaea* and *C. p. niveiventris*) (VAN ROOSMALEN & VAN ROOSMALEN, 1997). This very small primate has long been regarded by many as at most a subgenus of *Callithrix*, based on both morphological and molecular grounds (BARROSO, 1995; BARROSO *et al.*, 1997; MOREIRA, 1996; SCHNEIDER *et al.*, 1996; SCHNEIDER & ROSENBERGER, 1996; TAGLIARO *et al.*, 1997; PORTER *et al.*, 1997a &

b; CANAVEZ *et al.*, 1999), but the division of *Callithrix* into the *Callithrix* of the Atlantic coast and the Amazonian *Mico* argues for maintaining *Cebuella*, because of the small size and tree-gauging specialization. The animals are slightly larger than *Microcebus* or mouse lemurs, among the smallest of primates and *Cebuella* is certainly the smallest anthropoid primate in existence.

Pygmy Marmoset

Cebuella pygmaea (SPIX, 1823)

Plate 1 • Map 2

Taxonomic comments

Neither HERSHKOVITZ (1977) nor HERNÁNDEZ-CAMACHO & COOPER (1976) registered subspecies although recent work recognizes two, *Cebuella pygmaea pygmaea* being the only subspecies found in Colombia (VAN ROOSMALEN & VAN ROOSMALEN, 1997).

Common names

Colombia: *chichico* in the río Putumayo; *leoncito* in the departments of Putumayo and Amazonas; *piel roja* and *mico de bolsillo* popularized by animal traffickers.

Indian languages

Túmi (Huitoto); *caspicara chichico* (Ingano); *upicha* (pl. *jupíchiu*) (Miraña); *jibidlaje* (Muinane); *tsünt-sificú* (Okaima); *punk a sisi* (Siona); *ípoa* (Yucuna).

Identification

This is the smallest neotropical primate. SOINI (1988a) weighed 63 adults that weren't pregnant and obtained weights varying between 85-140 g, with an average weight of 119 g. The females were 12% heavier than the males. The head-body measurement of these animals varied between 331-362 mm with an average of 339 mm. The tail measured around 200 mm. These monkeys have an agouti dorsum with an ocher-white tail. The tail may be faintly to conspicuously ringed and the urogenital area in both sexes is covered in black hair.

Geographic distribution

Cebuella is well-known in the Colombian Amazonian region south of the Río Caquetá. There are records of the species' presence on the upper Río

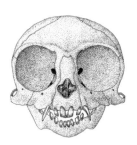
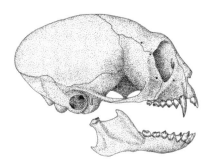
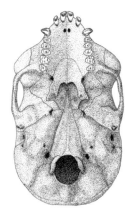
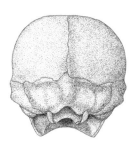

Figure 9. Skull of *Cebuella*.

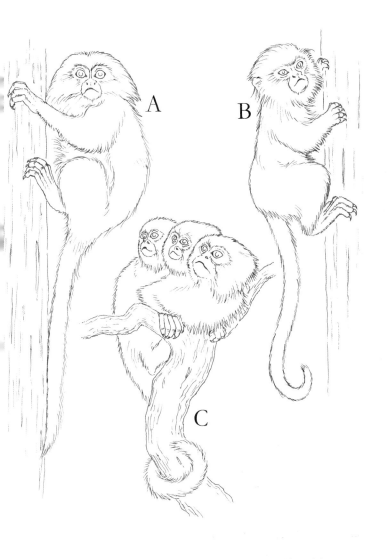

Figure 10. Various positions of *Cebuella*. a,b) Climbing positions. c) Mother with two infants.

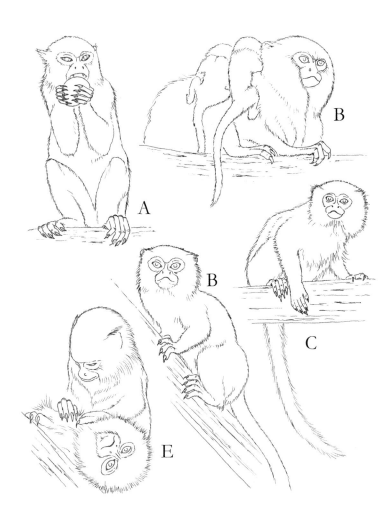

Figure 11. Various positions of *Cebuella*. a) Holding fruit. b) Infants holding onto mother. c-d) Resting adults. e) Social grooming.

Guaviare and Izawa (1975) observed the species on the Río Caguán a little to the south of San Vicente de Caguán. A captive specimen supposedly from the Caño Morrocoy on the right bank of the Guayabero about 2 km from the village of Refugio (also known as La Macarena) just south of the Sierra La Macarena is the best evidence for this region. But also the species was listed in research on the eating habits of the Nukak Macu Indians from the right bank of the Río Guaviare, downriver from the mouth of the Río Ariari (Politis & Rodríguez, 1994), perhaps the only report in the literature of this primate as human food. This locality and the others need to be confirmed.

Outside of Colombia *Cebuella* is found to the south of the Amazon river in hygrophytic forests of low lying areas including eastern Ecuador and Peru and western Brazil to the western bank of the Río Madeira (van Roosmalen & van Roosmalen, 1997) and the banks of the Río Orthon-Mamupiri or Río Madre de Dios in northern Bolivia. In Colombia *Cebuella* inhabits mature inundatable and non-inundatable forest. Near La Pedrera, Amazonas in Colombia the species can be seen in low trees at the very edge of the Río Caquetá. Also, it is common on the edges of the Río Puré (south of La Pedrera), Colombia. Close to Puerto Asís in the forest remnants a group was observed in a large *Parkia* sp. (guarango) tree. Soini (1989) found the species in floodable forest (during three months a year). He maintained that the forests on the edges of rivers had the highest populations. These primates seem to need border habitats and for this reason they easily colonize areas influenced by humans like forest edges around pasture-land. They are often observed close to human settlements.

Natural history

The majority of the information in this species account comes from the following studies concluded in Peru by Soini (1982, 1986b, 1987a, 1987b, 1988, 1993, 1995c), Bennett *et al.* (2001) and Heymann & Soini (1999). Other Peruvian studies were accomplished by Snowdon, (1991a, 1993b), Snowdon & Cleveland, (1984); Snowdon & Hodun, (1981); Snowdon & Pola, (1978) Ramírez *et al.* (1977), Castro & Soini (1977), Fess (1975a), Freese *et al.* (1977). Terborgh (1983) published short notes for Peru. For Ecuador De La Torre & Snowdon (2002), Snowdon & De La Torre (2002), De La Torre (2000), De La Torre *et al.* (2000), Zingg & Martin (2001), Youlatos (1999a) have been published on the species. For Colombia MOYNIHAN (1976a, 1976b), HERNÁNDEZ-CAMACHO & COOPER (1976), HERSHKOVITZ (1977) IZAWA (1975, 1976) published some notes for the species. Good reviews of the species were published by RYLANDS (1993,1997, 2001) and by TOWNSEND (2001).

The group size varies between 2-9 independently locomoting animals, carrying one or two infants. These groups are usually composed of a reproductive female, her mate and their young for usually the last four births. Sometimes a third associate adult that is not one of the pair's young is included in the group. Rarely two groups associate for a few hours forming a group of 10-15 individuals.

These groups occupy home ranges which vary from around 0.1 - 0.5 ha with an average of around 0.3 ha (n=7). These small home ranges are either exclusive or overlapping and have a core area of around 20-40% the size of the total home range. *Cebuella* have a day range of 200-300 m. Two measurements of day range were of 280 m and 300 m. A solitary individual, however, had a trajectory of about 850 m for an entire day.

In SOINI's study from the Río Saimiri the density for this species was 51-59 independently locomoting individuals/km^2, although strictly speaking, the densities at the river-edge were 210-233 independently locomoting individuals/km^2. In Colombia on the Puré and Purité rivers the densities are much lower and are correspondingly difficult to measure.

An activity budget calculated for a group of this species was about 41% resting, 11% moving, 32% exudate feeding and 16% insect-eating (total for eating 48%). A sexually active pair invested 68% of their time resting and grooming each other and only 30% of their time eating. Locomotion of this primate is actually very similar to a squirrel, including numerous stops and starts. This freezing behavior probably is an anti-predator strategy, since these small animals are difficult to see when they aren't moving. The movement consists of frequent vertical hanging and jumping as well as quadrupedal running and jumping (Moynihan, 1976b). Habitually and perhaps in order to hide from observers (predators?) they often move to the opposite side of the trunk or branch from where they are first observed. The majority of the groups sleep literally piled on top of each other at about 7-10 m above the forest floor on the proximal end of branches near the trunk or in epiphytes or a terminal bunch of vegetation.

Cebuella pygmaea consumes arthropods and exudates of some tree species as well as a few fruits, buds, flowers and some nectar and occasionally some vertebrates. The exudates are the resource most frequently consumed by these animals. The arthropods in the diet consist of insects and spiders; although orthopterans are the most-preferred. Among the insects, butterflies are also included, lured by the exudates and which the *Cebuella* catch and consume. The exudates are obtained via specialized feeding behavior, which is made possible thanks to especial adaptations of the incisors, utilized to rasp and dig holes in the tree-bark of the chosen trees. The holes cause sap to run within the holes, making it available as food. A group concentrates this activity usually in the principal feeding tree and, thanks to detection of the perforations in the bark, they can be detected by human observers. The temporal nature of feeding on exudates was discussed by ZINGG & MARTIN (2001).

A list of preferred exudate feeding trees (*fide* SOINI, 1982, 1987a) follows here: 87 *Vochysia lomotaphylla* (Vochyciaceae); 10 *Spondias mombin* (Anacardiaceae); 8 *Inga* spp. (Leguminosae); 6 *Campsiandra laurifolia* (Leguminosae); 7 *Swartzia* sp. (Leguminosae); 4 *Parkia oppositifolia* (Leguminosae); 2 *Croton cuneatus* (Polygonaceae); 2 unidentified sp. (Polygonaceae); 1 *Qualea amoena* (Vochyciaceae); 1 *Cassia* sp. (Leguminosae); 1 *Terminalia* sp. (Combretaceae), 1 unidentified sp. (Sapotaceae). VAN ROOSMALEN & VAN ROOSMALEN (1997) observed the use of *Enterolobium schomburgkii* (Mimosaceae), *Inga edulis* (Mimosaceae), and *Ficus guianensis* (Moraceae) as exudate sources on the banks of the Río Madeira.

The following lists lianas used as exudate sources (SOINI, 1982, 1987a): 27 *Gnetum* sp. (Gnetaceae); 9 *Entada polystachys* (Leguminosae); 6 *Maripa* sp. (Convolvulaceae); 5 *Dioclea* sp. (Leguminosae); 3 *Cheiloclinium* sp. (Hippocrateaceae); 2 *Acacia riparia* (Leguminosae); 1 *Bauhinia* sp. (Leguminosae); 1 unidentified (Compositae); 1 *Doliocarpus* sp. (Dilleniaceae);1 *Coussapoa* sp. (Moraceae).

The rest of the sources were of lesser importance for the *Cebuella* diet. HERNÁNDEZ-CAMACHO & COOPER (1976) and MOYNIHAN (1976a, 1976b) reported that in the department of Putumayo the *guarango* (*Parkia*) were extremely important for this primate as an exudate source and that the two species formed an ecological association. In the upper Río Purité of the Colombian trapezium *Cebuella* seems to be often associated with a species of caracoli; *Anacardium* sp. (Anacardiaceae).

MOYNIHAN (1976a) observed *Cebuella* using the available resources of *Parkia* sp. (Leguminosae), *Matisia cordata, Cedrela odorata* (Meliaceae) and *Inga* sp. (Leguminosae). During his studies SOINI (1982, 1987a) observed *Cebuella* eating fruits on only five occasions. These belonged to the genera *Ficus* (Moraceae), *Pourouma* (Moraceae), *Cecropia* (Moraceae) and an unidentified red berry. Whenever they were available the fruits of *Ficus* and *Pourouma* were used and usually during various days. *Cebuella* licks the raindrops from the surface of the vegetation and obtains the humidity of the various exudates, which it eats.

Cebuella reproduces very early in life from about 18 months of age, although usually reproduction is not until they reach 24-42 months. Estrous appears to continue despite the birth of the young, prolonging for about nine days afterwards, but estrous behavior extends for another six days of estrus, from approximately day 13 to about 18 days after parturition. Only one female of the group becomes pregnant at a time and one or two young are born to her.

The male consort shows much interest in the female when she is in estrous; he follows her constantly, smelling and licking her genitals. Body contact increments and tongue displays are very evident. CHRISTEN (1974) and CONVERSE *et al.* (1995) observed females of a captive group raising their tails and exposing their genitals to the male. In order to copulate the males grab the female's waist and both remain standing on the substrate or he grabs the legs of the female with his feet. Copulation is very short and when completed the two may remain very close, grooming each other and continuing with the activities they had been engaged in before (LARSON *et al.*, 1982).

There seem to be two annual birth peaks. In almost two-thirds of the births the females give birth to two young (more frequently at night) and only one young the rest of the time, after a gestation of 137-138 days. Repeated births after six months occur in about 85% of the individuals and there are rarely triplets. Neonates weigh only 13-15 g and are born with a fuzz of hair colored agouti lemon yellow with a gray head. At the end of the first month these colors have changed and the animals have the same color as the adults.

The majority of the members of a group participate in the care and transport of the infant, which is carried constantly for the first 1-2 weeks. Afterwards, the infant is placed in a protected place, usually in the foraging tree, while the adults feed. Soini (1982, 1987a) noticed that after the twentieth day the associated adult male was the one which most often carried the infant, while the presumptive father continues completely occupied with the estrous mother. Before 20 days the young brother or sister carries the infant up to about 61% of the time. Soini's (1982, 1987a) observations suggest that males other than the father could be the principal carriers of the infants. Even the brother of six months carries the infant until he no longer can lift it.

Towards the end of the third month the infant may cease nursing completely and begin consuming insects before reaching four weeks of age. When they beg for insects they are frequently given the insects by other members up to about 5-6 months. Around 6 weeks of age the infants begin to feed themselves with exudates which they lick from the holes used by the adults; it is not until they have completed 5 months that they are capable of excavating their own holes. Soini (1987b) studied dentary development of this species as a method to estimate their age.

The reproductive female seems to dominate the other members of the group, while the reproductive male is dominant over all of the males of the group. Grooming is a very important social activity which consumes 9-20% of the time budget of daily activities (Soini, 1988a; Ramírez et al., 1977).

Play behavior is important between infants, juveniles and subadults and consists of rough and tumble play between 2-4 animals, which occurs during the principal periods of rest. Grooming is important for all members of the group, although adults groom more than the other members. Christen (1974) showed that dominant males and females groom other adults equally while subordinate females groom dominant females more than dominant females groom subordinate females.

Soini (1988a) lists 28 visual and olfactory patterns of expression in *Cebuella pygmaea*. Some examples are as follow: genital exposure is common among the adults and subadults of both sexes; this is observed when the animals are confronting each other from a distance, raising their tail, fluff their hair up (piloerection) and arch their back and expose their genitals, at times dribbling urine from their genitals. This type of genital display seems different from the genital display of females directed towards males during the estrous cycle, since these females do not arch their back nor do they fluff up their tail hair. Tongue thrusts may be associated with sexual context between males and females or less frequently in situations which involve gestures of sympathy and in light conflicts which happen between individuals. Body rocking and balancing of the body is a display observed in situations of extreme excitation during mobbing o during play sessions. Spasmodic tail wagging is observed when the animals are frustrated or anxious.

CHRISTEN (1974) and POLA & SNOWDON (1975) studied the vocalizations of this species in captivity. Some of the vocalizations have extremely high frequencies, being high enough to be heard by avian predators according to SNOWDON & HODUN (1981). SOINI (1988a) lists 29 distinct vocalizations, although many are intergradations which are dependent on the level of excitation. Predators of this small primate are as follow: *Cebus apella,* the ulamá or *Eira barbara* (GALEF *et al.,* 1976), cats, raptors and snakes. Domestic cats also kill and eat these small monkeys when the two species live close to each other (DEFLER, pers. obs.). This species has been observed close to *Callicebus cupreus, Saimiri sciureus, Saguinus fuscicollis* and *Saguinus mystax* with little interaction.

Conservation state

Cebuella pygmaea is not considered in danger of extinction, although there is a low level of illegal trade as pets in Colombia. Although commercialization and primates as pets is prohibited in the country, it is easy for a traveler in an airport to hide these animals on their body in order to transport the animals to Bogotá where they can sometimes be found for sale, even though the majority die very quickly of the cold. The species is classified in Colombia as LR in the IUCN system.

Where to observe it

This species probably is most easily observed close to the visitors' center in the Amacayacu National Natural Park, upriver from Leticia, where the park employees can indicate to the visitor where the trees most used by the monkeys are located. In the villages alongside the Río Caquetá and in other small towns on the banks of the Putumayo the local people very often know the location of a local group of this primate and can indicate its location to the visitor.

Tamarins

Taxonomic comments

This polytypic genus contains 13 species, according to HERSHKOVITZ (1977), although Rylands *et al.* (2000) prefers to recognize 14-15 species, including the separation of *Saguinus geoffroyi* and *S. oedipus* and the separation of *S. graellsi* and *S. nigricollis*. Using molecular analysis, the strong monophyletic origin of *Saguinus, Cebus, Saimiri, Mico,*

Callimico, Callithrix and *Leontopithecus* has been recognized and this group is now placed into the family Cebidae with subfamilies Cebinae, Callitrichinae, and Saimiriinae (GOODMAN *et al.*, 1998; ROSENBERGER, 1981; SCHNEIDER *et al.*, 1996, SCHNEIDER & ROSENBERGER, 1996). Nowadays it is universal to recognize *S. geoffroyi* and *S. oedipus* as separate species and it is anticipated that more subspecies of *Saguinus fuscicollis* may eventually be raised to species status.

The recognition of *S. graellsi* is more controversial and depends in part on whether *Saguinus nigricollis nigricollis* actually does exist in Putumayo, sympatric with *S. graellsi*. The presence of *S. n. graellsi* (*S. graellsi*) in Putumayo is supported by specimens in Colombian collections and observations of MOYNIHAN (1976). Specimens of *S. nigricollis nigricollis* marked "Putumayo-Caquetá" or just "Putumayo" are also known and one specimen is marked specifically "Quebrada La Hacha", a known site in Putumayo province. In this book, the author recognized the existence of six Colombian species, although a future resolution of the taxonomic position of *graellsi* could change this.

Common names

Common names in Colombia are *chichico* in the Colombian Amazon; *titi* in the rest of Colombia; German: *Perückäffchen*. Etymology: derived from a word in Tupi-Guaraní meaning "small monkey".

Tamarins (*Saguinus, Leontopithecus*) are distinguished in part from marmosets (*Callithrix, Mico, Cebuella*), by their more developed lower canines, which are much larger than their incisors and which possess a thicker layer of enamel on the lingual side of the incisor, compared to marmosets.

Identification

These small primates with striking pelage design usually weigh around 500 g and have a total head-body length of around 180-300 mm. The tail is around 250-450 mm, completely furred and non-prehensile. The hands and feet have claws instead of nails and their dental formula is I2/2, C1/1, P 3/3, M 2/2. It is likely that future revisions of this genus will recognize more species, especially within the *S. fuscicollis* complex.

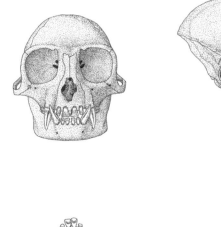

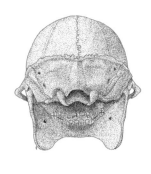

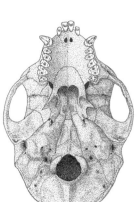

Figure 12. Skull of *Saguinus oedipus*.

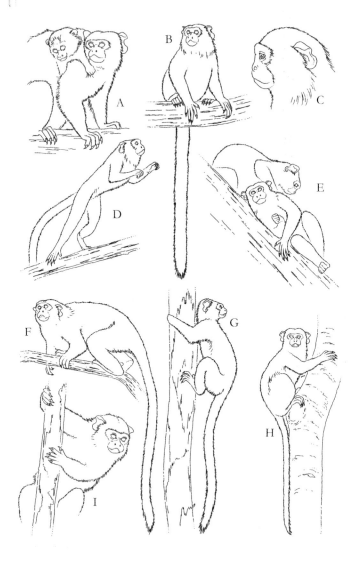

Figure 13. Various positions of *Saguinus*. a) Mother with very young infant. b-c) Relaxed observation. d) Jumping e) Grooming. f) Standing still. g) Clinging while watching. h-i) Watching.

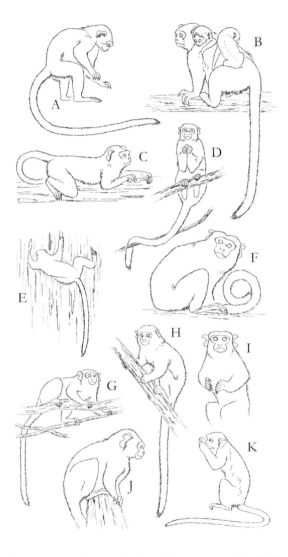

Figure 14. Various positions of *Saguinus*. a) Searching. b) Mother with two infants. c) Marking with pubic gland. d) Manipulating object. e) Hunting arthropods. f-j) Resting. k) Object manipulation.

Geographic Distribution

Saguinus monkeys are basically Amazonian, although there are three trans-Amazonian species in northern Colombia and Panama: *S. oedipus, S. geoffroyi*, and *S. leucopus*. All other Colombian *Saguinus* are found in the Colombian Amazon.

Saddle-back Tamarin

Saguinus fuscicollis (SPIX, 1823)

Plate 3 • Map 7

Taxonomic comments

Only one of the 12 currently recognized subspecies of the saddleback tamarin is known for Colombia, Lesson's saddleback tamarin, *Saguinus fuscicollis fuscus* (LESSON, 1840). *Saguinus* populations near San José del Guaviare (Río Guaviare), east of the Cordillera Macarena may, however, belong to a distinct form which has yet to be described (HERNÁNDEZ-CAMACHO & COOPER, 1976), although there is no confirmation via a specimen of their presence and *S. inustus* (not known to tolerate other species of *Saguinus*) has been collected immediately north of San José on the south bank of the Guaviare River, opposite the mouth of the Ariari River). *S. f. fuscus* is the most northern of the saddleback tamarins, and HERSHKOVITZ (1977), who considered coat color and MOORE & CHEVERUD (1992) who studied the facial morphology concluded that it is one of the most primitive of the tamarins, nearest to the hypothetical ancestor. CHEVERUD & MOORE (1990) concluded that differences between the recognized forms of saddleback tamarins are consistent with their ranking as subspecies; small founding populations which have budded off a parent population, each invading a new interfluvial basin and then remaining isolated. However, the same authors (MOORE & CHEVERUD, 1992) later indicated that the primitive *S. f. fuscus* may be better considered a separate species due to the similarities in facial morphology with *S. leucopus*. The morphological difference between *S. f. fuscus* and other *S. fuscicollis* subspecies is of the same order as that which separates the cotton-top tamarin (*S. oedipus*), GEOFFROY's tamarin (*S. geoffroyi*), and the silvery-brown bare-face tamarin (*S. leucopus*) (MOORE & CHEVERUD, 1992). Interestingly, a molecular genetic study by CROPP *et al.* (1999) found that *S. f. fuscus* grouped (strongly) with *S. nigricollis* rather than with other *S. fuscicollis* subspecies, and they also suggested that it be considered a distinct species.

Common names

Colombia: *bebeleche* (many parts of its geographic range) *chichico* or *chichico bociblanco* (along the lower Río Caquetá), *micoblanco* (piedmont of Amazonian Cauca); Brazil: *Sagüi, sauim, sagüi-de-boca branca;* French: *Tamarin à tête brune.*

Indian languages

Abujijiyo (Cubeo), *yana chichico*, (Ingano, referring to the dark *Saguinus, fide* J. V. RODRÍGUEZ-M., 1993), *juusari* (Muinane), *ya/rih*, (Tikuna), *jipoa* (Yucuna; refers to the subspecies *S. f. fuscus* on the south bank of the Río Caquetá).

Identification

Adult saddleback tamarins range from 342-411 g in weight depending on the subspecies (HEYMANN, 1997). Their head and body length is 175-270mm (average 222 mm), while their tail measures 250-383 mm (average 322mm) (HERSHKOVITZ, 1977). Thirteen subspecies were recognized by HERSHKOVITZ (1977), each being distinguishable by their coat colors. For *S. f. fuscus* the key features are the black crown, the mantle on the upper third of the body which is ochraceous orange ticked with black, and the orange arms, rump and thighs. The saddle across the middle of the back is a marbled, dark agouti. The face and crown are black with gray hairs around the mouth (thus, the local Spanish name *bebeleche*). The underparts and the undersurface of the tail are reddish brown, and the dorsal surface of the tail is black. The black-mantle tamarin, *S. nigricollis nigricollis*, has a uniformly black mantle, and Graell's black mantle tamarin, *S. n. graellsi*, has a reddish brown agouti or blackish agouti color, with the lower rump, thighs and underparts dominantly buffy red-brown. The mottled-face tamarin, *Saguinus inustus* is largely black, with a black face variously mottled with white or pinkish blotches.

The markings of the confirmed subspecies (*S. f. fuscus*) and one which might eventually be found in Colombia (*S. f. fuscicollis* and *S. f. tripartitus*) are distinguished as follows:

• The mantle is chestnut red and extends along the upper third of the body and the arms; the barred designs below the shoulders is contrasting and obvious; the buttocks and thighs are lighter and reddish; the face and crown are black..*S. f. fuscicollis.*

• Golden-yellow mantle in contrast with a black head; mid- to lower back (saddle) is barred with chestnut red and whitish. Also a clearly visible grayish band between the eyes...*S. f. tripartitus.*

Geographic range

The distribution of *S. fuscicollis* in Colombia is not very well understood. The range of *S. f. fuscus* is defined by HERNÁNDEZ-CAMACHO & COOPER (1976) as the area south of the Río Caquetá, and north of the Río Putumayo. Additionally these authors write that it extends north of the upper Caquetá along the right (west) bank of its tributary, the Río Yarí, and in the north and east of the upper Yarí, eastward to the area around San José del Guaviare on the south bank of the Río Guaviare, about 1 hour by boat downstream from the confluence of the Ríos Guayabero and Ariari, to the headwaters of the Ríos Vaupés and Apaporis in the southern department of Meta. To the east from there it is replaced by *Saguinus inustus* (see DEFLER, 1994b). However, there are no supporting data for large sections of its proposed range. HERSHKOVITZ (1979) recorded the presence of *S. f. fuscus* west of the Río Caguán, north of the Río Caquetá, and between the Ríos Caquetá and Putumayo, extending someway into Brazil, between the Rios Solimões and Japurá and around La Pedrera, south bank of the Río Caquetá in Colombia. The easternmost locality known is the Rio Tocantins, a north bank tributary of the Rio Solimões (HERSHKOVITZ, 1977), and I have seen them on the lower Rio Puré (Purué), Brazil, a right bank tributary of the Rio Japurá. A specimen has also been collected (in the ICN collection) from the south bank of the Guayabero River, west of the village of La Macarena, but east of that point there are no collections to support the extension to the vicinity of San José de Guaviare shown in HERNÁNDEZ-C. & COOPER (1976).

MOYNIHAN (1976a) observed *S. f. fuscus* on an island in the Río Guineo, a tributary of the upper Caquetá, and around the localities of El Pepino and Rumiyaco, near Mocoa, forming as such the westernmost known localities of its range. He also observed it around Valparaiso, Caquetá between the Orteguaza and Caquetá Rivers, where there are voucher specimens collected.

Nowhere is *S. fuscicollis* known to be sympatric with *S. inustus*. *S. f. fuscus* is, however, sympatric with *S. nigricollis graellsi*, and its range overlaps partially with *S. n. hernandezi* east of the upper Río Orteguaza (HERSHKOVITZ, 1982b) and on the upper Río Guayabero, where the Colombian biologist Nancy Vargas studied (and collected a voucher specimen) *S. n. hernandezi*. South of the Río Caquetá, *S. f. fuscus* is sympatric with the pygmy marmoset, *Cebuella pygmaea* and possible with *S. nigricollis nigricollis* and *S. n. graellsi*. In Brazil its range is poorly known, but its geographic range superimposes that of the red-bellied tamarin, *Saguinus labiatus thomasi*, at least in the region of the Rios Tocantins and probably the Purué (Puré).

Outside of Colombia, *S. fuscicollis* extends to the south in the upper Amazon in Ecuador and Peru to Bolivia (Río Mamoré), and east as far as the west bank of the Rio Madeira. From there, it occurs south of the Rios Solimões and Rio Japurá (HERSHKOVITZ, 1977; RYLANDS, 1993a). However, within this vast region, there are large areas for which information is completely lacking regarding the presence or otherwise of the species. Aditionally it is probable that some of these subspecies will eventually be raised to species status.

Wherever saddleback tamarins are sympatric with one of the moustached tamarins (*S. mystax*, *S. labiatus* and *S. imperator*), groups of each species tend to travel and forage together for much or even all of the day. They form so-called "mixed-species groups". This phenomenon is well-reviewed by BUCHANAN-SMITH (1999) and HEYMANN & BUCHANAN-SMITH (2000), who discussed the ecological and behavioral differences between the saddlebacks on the one hand and the moustached tamarins on the other, which allow for this close association. The authors also described the evidence regarding the benefits that individuals of each species accrue in terms of safety from predators, foraging efficiency and resource defense (from neighbouring groups). Aspects of these mixed-species associations are also discussed in the section Natural History. *S. f. fuscus* is not sympatric with any of the moustached tamarins, but its geographic distribution overlaps with *S. nigricollis graellsi* between the Ríos Putumayo and Caquetá, west from Puerto Leguizamo, and with *S. nigricollis hernandezi* between the Ríos Orteguaza and Caguán (HERNÁNDEZ-CAMACHO & COOPER, 1976; HERSHKOVITZ, 1982b) as well as on the right bank of the upper Río Guayabero. Associations of the sort described for other saddlebacks and the moustached tamarins have not been observed, but although Moynihan (1976) indicated that *graellsi* and *fuscicollis* were in fact mutually exclusive (where one was, the other was absent), HERNÁNDEZ-CAMACHO & COOPER (1976) reported mixed groups of *graellsi* and *fuscicollis* around Puerto Asís on the Rio Putumayo (at the mouth of the Río Guamués), as well as along the Río Guamués. No evidence is available regarding mixed groups of *S. f. fuscus* and *S. n. hernandezi*.

HEYMANN & BUCHANAN-SMITH (2000) cover the key aspects of ecological, behavioral and morphological dissimilarities between the saddleback and moustached tamarins which allow for their sympatry. One of these is body-size - the moustached tamarins are larger (GARBER & TEAFORD, 1986; HEYMANN, 1997). *S. n. graellsi* and *S. f. fuscus* show no significant difference in size, and no evidence has come forth that they form mixed-species groups in the same way the saddlebacks and moustached tamarins do (HEYMANN & BUCHANAN-SMITH, 2000). The species which associate also differ in foraging substrata, the type of foraging manoeuvres used for locating and capturing prey, and prey size and colour (HEYMANN & BUCHANAN-SMITH 2000:175). HEYMANN (1997) suggested that *S. nigricollis* and *S. fuscicollis* are too similar, particularly with regard to animal prey foraging, for mixed species groups to form and predicted that they use different habitats, and would not be found to associate, nor even occur in the same forests. HEYMANN'S (1997) arguments, based on the similarity of *S. f. fuscus* and *S. nigricollis* which would deny them being syntopic, are further reinforced by CROPP *et al.* (1999) who reported that they are phylogenetically more closely aligned than *S. f. fuscus* is to other *S. fuscicollis* subspecies. Field research in the interfluvium of the upper Ríos Putumayo and Caquetá, and surveys along the Rios Orteguaza and Caguán are needed to understand the circumstances of this enigmatic co-occurrence of *S. nigricollis* and *S. f. fuscus*.

Natural history

There has been a considerable amount of field research on wild populations of saddle-back tamarins, but mainly in Perú, Bolivia and Brazil. In Perú CASTRO R. (1990), CASTRO & SOINI (1977), SOINI (1981, 1986b, 1987c, 1990b, 1995c), TERBORGH ((1983), TERBORGH & STERN (1987) studied *S. f. weddelli* in Manu National Park. This study was amplified by TERBORGH & GOLDIZEN (1985) and GOLDIZEN & TERBORGH (1986, 1989), concentrating on the pairing process. GOLDIZEN (1986, 1987a, 1989, 1990) and GOLDIZEN *et al.* (1996) studied social relations and the transport of infants within a polyandrous group. HARDIE (1998, 1996), HARDIE & BUCHANAN-SMITH (1997, 1998), POOK (1977), POOK & POOK (1982a), YONEDA (1981, 1984a, 1984b) and BUCHANAN-SMITH *et al.* (2000) examined the adaptive basis of reproductive strategies, taking as base information collected during 13 years in Manu National Park. Other recent studies on diet have been accomplished by BICCA-MARQUES (2000), BICCA-MARQUES & GARBER (1999, 2000, 2002a,b, 2003), GARBER & BICCA-MARQUES (2002), CALEGARO-MARQUES *et al.* (1995), GARBER (2000), GARBER & LEIGH (2001), PORTER (2001a, 2001b). GARBER & LEIGH (2001) studied postural differences in Brazil. BEJARANO (1981) studied the species´ ecology in Bolivia.

CRANDLEMIRE-SACCO (1988) studied sympatry of this species with *Callicebus moloch*; GARBER (1988a, 1988b, 1990, 1991a, 1991b, 1992, 1993a, 1993b, 1994, 2000) studied spatial memory, patterns of foraging and positional behavior. HEYMAN (1990, 1992, 1993, 1996a, 1996b) worked on associations of this species with *Saguinus mystax* and with the hawk *Harpagus bidentus*. OVERSLUIJS-VASQUEZ & HEYMANN (2001) analyzed predation by the eagle *Morphnus guianensis* on two *Saguinus* including *S. fuscicollis;* HEYMANN & SMITH (1999), HEYMANN *et al.* (2000a, 2000b) and SMITH (1999, 2000a, 2000b) studied predation and infanticide. PERES (1991a, 1992a, 1992b, 1993b, 1993c, 1996) in Brazil studied the consequencies of joint territoriality with *S. mystax*. TERBORGH (1990) commented on the mixture of this species with various species of birds.

S. f. fuscus remains largely unstudied, but it is reasonable to suppose that its ecology and behavior will be very similar to other subspecies unless this taxon proves to be a separate species. They are found in a wide variety of habitats, in secondary vegetation and isolated forest patches and tall *terra firme* forests, up to about 700 m above sea level along the Andean foothills (HERNÁNDEZ & COOPER, 1976; MOYNIHAN, 1976), but are generally not found in flooded forests (PERES, 1997). They tend to travel and forage well below the forest canopy. YONEDA (1984b) recorded that *S. f. weddelli* in Bolivia spends most of its time moving and foraging less than 10 m above the ground, and this has been found to be true of other subspecies studied. They sometimes go to the ground to search for prey in the leaf litter. CASTRO & SOINI (1977) describe locomotion in the species as consisting of quadrupedal walking and running, bounding or galloping but also leaping between terminal branches. Vertical clinging and leaping is common, especially when foraging for animal prey (Castro & Soini, 1977; Garber, 1992).

Saddle-back tamarins travel in groups of around 5-10 individuals (Freese 1975; Janson & Terborgh, 1985a,b; Peres, 1991a). Of 62 different groups counted by Soini (1993), the average group size was 5.8, with a mean of 3.2 adults per group. They are comprised of a breeding pair and their offspring; the larger groups sometimes having one or more extra adult males or females. Population densities recorded for the species vary widely. Freese *et al.* (1982) estimated densities ranging from a low of 2.4/km² (*S. f. leucogenys* at Panguana, Río Yupichis, Peru) to 29/km² (*S. f. lagonotus*, Río Itaya, Peru) over 11 sites in Peru and Bolivia. Pook & Pook (1981) provided an estimate of 33/km² at Triunfo, south-west of Cobija, Bolivia (*S. f. weddelli*), and Peres (1997) recorded densities ranging from 20 to 68/km² at 11 different sites in Brazil (including three different subspecies; *fuscicollis*, *avilapiresi* and *melanoleucus*) and densities of *S. f. fuscus* of about 8 individuals/km² on the upper Purité (Purueté) in Colombia (Defler, in prog.).

Reported home ranges vary from about 15 to 40 ha (Pook & Pook, 1981; Soini, 1987c; Terborgh, 1983), but Peres (1992b) reported that a mixed-species group of *S. f. avilapiresi* and the red-capped moustached tamarin, *S. mystax pileatus*, occupied an area of 141 ha. Day ranges are usually more than 1 km, and they have been recorded traveling as far as 2100 m (Garber, 1993a; Peres, 1992b; 1987c; Terborgh, 1983; Yoneda, 1981).

The monkeys are active for rather less time per day than most primates. Groups leave their sleeping sites quite late in the morning, shortly after sunrise (later still when there is rain or cloudy conditions), and retire well before sunset. Two studies have made time budget estimates for the species, with some divergent results as follows: Soini (1987c) estimated that individuals of a group he studied spent on average 45% of their day foraging for small animal prey, 6% traveling, 14% feeding on fruits and other plant foods, and 32% resting and grooming. These figures of course give only a rough idea of the way they apportion their day, which of course varies in different sites with the weather in different seasons and whether they are traveling with one of the moustached tamarins in a mixed-species group (Soini's group were not). Peres (1991a), watching *S. f. avilapiresi* in a mixed-species group with the red-capped moustached tamarin, *S. mystax pileatus*, recorded overall time budgets of 33% moving, 23% resting, 25% feeding, 14% foraging and about 6% in social behaviors, including interactions with neighboring groups.

When sleeping at night callitrichids become extremely torpid. While sometimes they enter holes in trees, in the open they sleep in tight huddles which look like a termite nest rather than small animals— an excellent predator defense strategy (SOINI, 1983; TERBORGH, 1983). HEYMANN (1995) made a particular study over 35 days of the sleeping sites used by *S. f. nigrifrons* at the Quebrada Blanco in Perú. He identified five types: crowns of *Jessenia bataua* palms (also recorded by PERES, 1991), tree hollows, dense tangles of vegetation, the crotches of tree branches, and open horizontal branches. Sleeping sites were largely concentrated in the central parts of the home range which do not overlap with those of other groups. In his study, PERES (1991a) found tree hollows to be important, but SOINI (1987c) never saw his study group using them during five years (other groups in his area were seen to use them). HEYMANN (1995) found the animals would sometimes use the same site for up to six nights in a row.

Saddleback tamarins eat a wide range of small ripe fruits from trees, lianas and epiphytes, as well as nectar, and gums and exudates from tree-trunks, lianas and leguminous seedpods. Rarely they eat seeds from unripe fruits and fungi. They drink water from tree holes as well as from the surface of vegetation (SOINI, 1983). Small ripe fruits are predominant in the plant part of the diet. SOINI (1987b) recorded a saddleback tamarin group eating fruits of nearly 60 plant species. PERES (1993b) recorded 141 species over 13 months, and in any one month they eat fruits of as many as 46, which are generally indehiscent and often yellow or red. They eat the fleshy mesocarps of large-seeded fruits, small, many-seeded fruits, such as berries, and the arils of some dehiscent fruits, such as *Inga* (Mimosaceae) and *Souroubea guianensis* (Marcgraviaceae). In Peres' study, key families included Moraceae, Leguminosae, Guttiferae, Linaceae and Sapotaceae.

Exudates and nectar can also be important items in the diet, especially when fruits are scarce. PERES (1993b) observed them eating gums from a bromeliad, and from the bark of 11 tree species. Notable is their consumption of the gum produced from the hanging seedpods of the giant legume trees, *Parkia pendula*, *P. nitida* and *P. oppositifolia* (Mimosaceae) (GARBER,

1993a,b, SOINI, 1987b; PERES, 2000). Gums are also exuded from trunks and branches due to abrasion and windstorm injury, insect perforations and holes gouged by squirrels and especially pygmy marmosets, *Cebuella* (SOINI, 1987b, 1993). Saddle-back tamarins will sometimes bite at the bark to get at the gum, but do not gouge for it systematically as do the pygmy marmosets. HEYMANN & SMITH (1999) examined the daily pattern of gum feeding in *S. f. nigrifons* and the moustached tamarin, *S. mystax*, and found that they visited gum sources more frequently and spent more time feeding on them in the afternoon. Gum is difficult to digest and requires fermentation, and they suggested that this is a means by which the tamarins can maintain it for longer in their intestinal tract to allow for fermentation. Nectar can be an especially important food at times of fruit scarcity. PERES (1993b) noted their avid consumption of the nectar of *Symphonia globulifera* (Guttiferae), and in Perú, JANSON *et al.* (1981) found that they are probably important pollinators when licking nectar from the flowers of *Combretum fruticosum* (Combretaceae).

The other major component of the saddleback tamarin's diet is small animal prey. The monkeys spend a considerable part of the day foraging in holes in tree trunks, turning over leaves, examining curled-up leaves, exploring crevices, stalking, pouncing, and rummaging through vine tangles and leaf litter. Their principle prey are insects, and especially orthopterans, mainly katydids (Tettigonidae, especially Pseudophyllinae, but also Phaneropterinae, Letroscolidinae, Copiphorinae, and Agraeciinae), are of particular importance (GARBER, 1993; NICKLE & HEYMANN, 1996; Peres, 1993b; SOINI, 1987b). CRANDLEMIRE-SACCO (1988) studying *S. f. weddelli* in Tambopata, Perú, estimated that 73-78% of the animal prey they eat were orthopterans. Estimates for the contribution of katydids to the animal prey part of the diet range from 61% (*S. f. weddelli* in Manu, Perú [TERBORGH, 1983]) to 67% (CRANDLEMIRE-SACCO, 1988), to as much as 82% (*S. f. avilapiresi*, Rio Urucú, Brazil [PERES, 1993a]). Short-horned grasshoppers (Acrididae), stick insects (Phasmidae), crickets (Gryllacrididae), romaleid grasshoppers, and wingless locusts which mimic stick insects (Proscopidae) are also eaten, along with cockroaches (Blattodea), preying mantis (Mantidae), army ant (*Eciton burchelli*) larvae, cicadas, moths, caterpillars and lepidopteran egg cases, and spiders. Detailed lists of invertebrate prey items can be found in NICKLE & HEYMANN (1996) and SMITH (2000b). These monkeys also eat small lizards, frogs and, rarely, nestlings. SOINI (1987b) recorded *S. f. illigeri* eating geckos, *Gonatodes humeralis*, and the anole lizard, *Anolis [Norops] fuscoauratus*. HEYMANN *et al.* (2000), comparing vertebrate predation by *S. mystax* and *S. fuscicollis*, found that the saddleback tamarins tended to eat more lizards from the understorey and forest floor, while the moustached tamarins caught more frogs in the foliage of the upper strata. This difference is related not only to the heights in the forest in which the two tamarins tend to forage, but also their distinct foraging methods. Lizards recorded as prey items included members of the Iguanidae (*Anolis*), Teiidae (*Kentropyx pelviceps*), Scincidae (*Mabuya nigropunctata*), and Polychrotidae (*Anolis [Norops] fuscoauratus* and *Anolis [Norops] nitens*). The monkeys also ate hylid frogs (*Osteocephalus* sp. and *Phyllomedusa* sp.).

Saddle-back tamarins usually forage for these prey items low down, in the forest understory, mostly below 10 m above the forest floor (BUCHANAN-SMITH 1990, 1999; PERES, 1993c; YONEDA, 1984b), and their foraging techniques contrast with all other members of the genus (GARBER, 1993a,b). They are very manipulative, searching and probing in closed microhabitats where the prey is concealed, evidently in preference to the more open substrates such as leaf surfaces. This contrasts with the moustached tamarins with which they are sympatric over much of their geographic distribution, which forage at higher levels (above 10 m) and search more for cryptic but exposed insects in daytime resting places, on or under leaves (BUCHANAN-SMITH, 1990; GARBER, 1993a,b; NICKLE & HEYMANN, 1996; PERES, 1993b; YONEDA, 1984a, 1984b). Travel at these low levels in the forest involves much more vertical clinging and leaping between relatively large supports, and the moustached tamarins which tend to use higher levels of the forest use measurably more quadrupedal walking and running along smaller, oblique and horizontal branches and vines (BUCHANAN-SMITH, 1999; YONEDA, 1984b). Considering both their techniques and the levels of the forest in which they forage, it is evident that the moustached and saddle-back tamarins exploit very different animal prey (BUCHANAN-SMITH, 1999; HEYMANN & BUCHANAN-SMITH, 2000; NICKLE & HEYMANN, 1999; PERES, 1993b; YONEDA 1984b).

HEYMANN (1990a), working at the Quebrada Blanco Biological Station, studied the reactions of mixed-species groups of *S. f. nigrifrons* and *S. m.*

mystax to avian predators and other birds which elicited alarm reactions. The saddle-back and moustached tamarins responded to each others alarm calls, which were given at a rate of about one every two hours. Birds which caused alarm included vultures, forest falcons (*Micrastur*), black-collared hawks (*Busarellus nigricollis*), double-toothed kites (*Harpagus bidentatus*), anis (*Crotophaga ani*), herons, caciques (*Cacicus cela*), toucans (*Rhampastos*) and parrots. Double-toothed kites are insectivorous, capturing their prey on the wing. They tend to follow monkey groups, including capuchin monkeys and squirrel monkeys as well as tamarins (EGLER, 1991). They evidently do so to take advantage of the insects flushed by the foraging monkeys. This association has been reported for mixed-species groups of *S. f. nigrifrons* and *S. m. mystax* by HEYMANN (1992). Although the presence of the double-toothed kites, which follow groups for more than two hours at a time, is generally ignored by the tamarins, they sometimes attempt to drive the kite off; but when the kites fly up they cause the tamarin to alarm call. The tamarins showed an overall increase in their rate of alarm calling when the kites were following them. Similar but briefer associations have also been observed for nunbirds (*Monasa morphoeus*) and the great jacamar (*Jacamerops aurea*) (HEYMANN, 1992). HEYMANN (1990a) observed an attack by an ornate-hawk-eagle (*Spizaetus ornatus*). While in most cases alarms involved moving down in the forest or moving fast from the periphery of a tree to its trunk, the (unsuccessful) attack by the hawk-eagle caused the animals which were exposed feeding in the forest canopy, to literally fall out of the tree. For some days afterwards the tamarins of both species were still nervous, evidenced by a higher than usual alarm call rate, and they also spent more time than was typical in the lower levels of the forest. OVERSLUIJS-VASQUEZ & HEYMANN (2001) recorded the predation of an infant saddleback tamarin by the Guianan crested eagle (*Morphnus guianensis*). Snake-mobbing was recorded by BARTECKI & HEYMANN (1987). The saddle-back group were seen to approach and mob (giving a specific trill call) two mating boas (*Corallus enydris*, Boidae) hanging from a liana, some tamarins going as close as 1.5 to 2 m.

Marmosets and tamarins are cooperative breeders – the male(s) and other group members help carry the (usually) twin infants, and may also provision them with food following weaning (FEISTNER & MCGREW, 1989). Saddleback tamarin groups are generally made up of a breeding pair and their offspring, but often with extra adults who may not be related. As in all tamarins and marmosets, the dominance of the breeding female is maintained through an extraordinary mechanism of physiological (suppression of ovulation) and behavioral reproductive inhibition of other subordinate females of breeding age (ABBOTT *et al.*, 1993). Although long believed to be an effectively monogamous mating system, recent evidence has shown that, at least in some cases, the breeding group is cooperatively polyandrous (two adult males breeding with the same female) or even may be made up of two reproductively active females with one or more males (TERBORGH & GOLDIZEN, 1985).

The sexes are similar in size and pelage color. The females have an estrus cycle of 15.5 days (PRESLOCK *et al.*, 1973), and gestation is about 140 days with the female giving birth (in about 80% of instances in one study) to twins (SOINI, 1981), although in many cases one of them may succumb. Most births occur during a birth season which is a 2-3 month period early in the rainy season, when fruit is most abundant, but SOINI (1987b,c) found that they could occur at any other month of the year as well. When food is abundant, either seasonally or in particularly productive forests, they may produce offspring twice a year, but usually only once. The infants are carried, and largely dependent till at least two months old. They are born relatively large, and for the mother the energetic demands of foraging, traveling, carrying and suckling the young, are believed to be the reason she needs help (TARDIF *et al.*, 1993). It is probable that a critical number of helpers are necessary for the survival of the young (GOLDIZEN & TERBORGH, 1986), and the occurrence of polyandry may result from the need for helpers in the absence of older offspring (SOINI, 1987).

Coat color in the neonate is different from the adults, the saddle being absent and replaced with black, which changes to a more adult coloration within a month. The adult face pattern does not appear until the 7-9th week. Deciduous teeth are present by the 2nd week and the first permanent teeth by the 4th week and is completed by the 10th week. Canine growth requires five more weeks. The vulva becomes adult-like by the 13-15th week and testes reach adult size by the 15th week, marking the transition to adulthood (SOINI, 1983).

Saddleback tamarins are very social, and can often be seen sitting together on a sunny branch and grooming each other. They communicate to some extent through facial expressions and posture, but mainly through their calls and by scent marking. Facial signals include grimacing and baring the teeth in threat, and tongue-flicking, a rapid rhythmic protrusion of the tongue, typical of all tamarins and seen in a variety of situations, when grooming, threatening or in sexual solicitation. The most distinct vocalizations, often the first clue before seeing them in the forest, are the long calls they use to maintain contact. They are especially frequent in the early morning, but can be heard throughout the day, especially when the group is on the move, and when they are in a state of excitement, for example during an encounter between neighboring groups. Long calls are quite high in frequency and use short syllables strung together (SNOWDON and SOINI, 1988). They are divided into two types, *loud long calls* used for territory defense and long distance contact, and *soft long calls* for cohesion in the group. HODUN *et al.* (1981; see also SNOWDON, 1993a) demonstrated subtle but distinct differences between the saddleback tamarin subspecies in their typical long calls. MOODY & MENZEL (1976) list and describe 16 different vocalizations for *S. fuscicollis illigeri*.

All the marmosets and tamarins have cutaneous glandular fields in the anogenital, suprapubic and some also in the sternal areas. They use these glands to mark objects with oily scent, using quite distinct postures and movements (see LINE DRAWING). Male and female scents can be discriminated, as well as dominance and submissiveness. The sexual state of the female is assessed by chemical scents and, at least in the case of *S. fuscicollis*, it has been shown that suprapubic marking has a *territorial function* (EPPLE *et al.,* 1993; BARTECKI & HEYMANN, 1990; SNOWDON & SOINI, 1988). *S. fuscicollis* have glands in all three regions, but HEYMANN (2001) found anogenital marking, or *sit-rubbing* to be the most frequently performed. Suprapubic marking, or *drag-marking* (the individual sprawls along a branch and drags its pubic region forward), is also quite common, and sometimes up to five individuals mark at the same time, or sequentially, at the same site (HEYMANN, 2001).

Conservation status

S. fuscicollis fuscus is not threatened in Colombia, especially since it is able to live in degraded and secondary vegetation close to human beings and is native to the Colombian Amazon, which has not been widely deforested. It is easy to observe close to human settlements where secondary vegetation is growing, and as long as such areas are not converted into fields and pasture the species will probably persist within the country. It occurs in La Paya National Natural Park (422,000 ha) and in the Cahuinarí National Natural Park (575,000 ha) (DEFLER, 1994; POLANCO-OCHOA *et al.*, 1999). It is listed on Appendix II of the CITES, along with all other low-risk category primates. Its commerce is forbidden within Colombia and is classified as of LC in the *2003 IUCN Red List of Threatened Species* (HILTON-TAYLOR, 2000).

Where to see it

S. fuscicollis is very easy to observe in the secondary vegetation around garden plots near any small Amazonian village between the Ríos Caquetá and Putumayo. Often a local person will be glad to point a group out. The species is also very common on the upper Río Cahuinarí river, but this is a little remote for most people to travel. It is probably best to hire a local person to guide potential observers to some nearby gardens. It is easily observed along the air strip and in local gardens in La Pedrera, Amazonas.

Geoffroy´s Tamarin

Saguinus geoffroyi (PUCHERAN, 1845)

Plate 2 • Map 3

Taxonomic comments

HERSHKOVITZ (1977) did not accept this species as valid. Instead he assigned it as a subspecies of *S. oedipus (geoffroyi)*. HERNÁNDEZ C. & COOPER (1976), MITTERMEIER & COIMBRA-FILHO (1981) and HONACKI et al. (1982) and GROVES (1993) distinguished two separate species *S. geoffroyi* and *S. oedipus*. Hybrids of the two species have been produced (HERSHKOVITZ, 1977), although it is not known whether backcrosses were fertile. SKINNER (1986a, 1986b), using measurements found in HERSHKOVITZ (1977), did a morphometric comparison of bare-faced, trans-Andean *Saguinus*, *S. geoffroyi*, *S. leucopus*, and *S. oedipus* and found *S. oedipus* was closer to *S. leucopus* than it was to *S. geoffroyi*. NATYORI & HANIHARA (1988, 1992) found that an analysis of multivariance of dental morphology showed, despite similarities between *S. oedipus* and *S. geoffroyi*, the relation was less than that between *S. fuscicollis* and *S. nigricollis*. MOORE & CHEVERUD (1992) applied a multivariate analysis and found that *S. oedipus* and *S. geoffroyi* differed morphologically on the species level and found that these differences were not clinal but rather discontinuous. RYLANDS (1993) discussed all of the above evidence, supporting the two species concept rather than one. Comparisons of vocalizations and displays should be examined with rigor. Possible zones of contact or overlap ought to be studied for details of the relationship of these two taxa, so that more corraborating evidence for the two species concept is made available, but this view has been widely accepted.

Common names

Other English: *Panamanian tamarin; Red-crested bare-faced tamarin.* Colombia: *tití* and *bichichi* in the Department of Chocó; German: *Geoffroyi-Peruckäffchen.*

Indian languages

Tit (pronounced as in Spanish):(Cuna); *bichichi* (name in Emberá, Indian reservation in the upper Baudó, Coroboro Valley, upper Bojayá, upper Río Buey), Chocó.

Identification

Individuals of this species measure 225 -240 mm body length and 314 – 386 mm tail length and weigh an average of 486 g in the males (n=486) and 507 g in the females (n=41). Individuals possess a wedge of white over the top of their head, contrasting with dark on either side. The dorsum is a streaked agouti and the ventrum, arms and legs are white or (quite often in Colombia) sulfur-yellow. The rather hairless face is dark-pigmented, while the tail is dark brown.

Geographic range

In Colombia *Saguinus geoffroyi* is found from the Panamanian frontier south, probably to the Río San Juan on the Pacific Coast. Previously the eastern limit was thought to be the entire Río Atrato, but a recent primate census conducted by VARGAS (1994a) has located good populations of the species near Orquideas National Park, near the village of Mandí (Antioquia) at 1000 m altitude. The eastern and southern limits of this species need to be established via census work. Besides Colombia the species is found throughout Panama and reaches the southern tip of Costa Rica.

This species seems more abundant in secondary forests in new secondary vegetation mixed with cut forest, and it seems less abundant in primary forest. The species habitually forages on the forest floor and seems to prefer edge habitat (DAWSON, 1979).

Natural history

The most extensive field study was accomplished by DAWSON (1976, 1977, 1979; DAWSON & DUKELOW, 1976). HLADIK & HLADIK (1969) studied its feeding ecology on Barro Colorado. Later, GARBER (1980a, 1980b, 1984, 1991a, 1991b, 1994, 1995) studied the species' locomotor behavior and feeding ecology. GARBER & BICCA-MARQUES (2002) studied foraging and locomotion and RASMUSSEN (1998) studied the changes in spatial use of animals accustomed to people. LINDSEY (1980) and SKINNER (1985) both did field studies of the species as well. The species has unfortunately not been studied in Colombia. An evaluation of the state of the Colombian populations is vitally needed as a first effort towards its conservation.

The average group size is usually 5-7 with a range of 3-9 (DAWSON, 1976; LINDSAY, 1980). These groups maintain and defend a territory by vocal and physical means which may vary from around 9.4 ha (GARBER, 1980a,b) to 26 ha and 43 ha (DAWSON, 1979). Groups exist in densities that vary from 3.6 individuals/km^2, 4.7 individuals/km^2 to 5.6 individuals/km^2 on Barro Colorado (EISENBERG, 1979) to at least 20-30 individuals/km^2 where habitat is more apt (DAWSON, 1976). Day range is an average of 2061 m (DAWSON, 1976, 1979).

No activity budget has been calculated for this species as yet. Locomotor behavior was studied by GARBER (1980a, 1980b, 1991a). He found that, although the species was able to cling to the bark of large tree trunks, they preferentially ascend rather than descend, and they prefer to leap from tree to tree from thin terminal branches. In general, GARBER (1980a, 1980b) found that the use of the tégula for clinging was quite different from the squirrel *Sciurus*, with *S. geoffroyi* avoiding vertical supports during travel.

GARBER (1980a) calculated by time sampling a diet for *S. geoffroyi* made up of of 38% fruits, 14% exudate and 40% insects and 8% other. GARBER (1980a) also estimated that 98% of exudate feeding was on *Anacardium excelsum* (Anacardiaceae). While most feeding of *S. geoffroyi* occurs on thin supports less than 10 cm circumference, an exception is the exudate feeding which occurs while hanging from trunks, which are often greater than 100cm in circumference. HLADIK & HLADIK (1969) recorded 59 species of plants in the diet, and they estimated the diet at 30% insects and other animals, 60% fruits and 10% green plant parts. This diet contained an estimated 16% animal protein (mostly orthoptera and coleoptera), 5% vegetable protein, 9% lipids, 29% reducing sugars, 7% cellulose and 34% of mainly indigestible components. The sap of the tree *Enterolobium cyclocarpum* was consumed a great deal in this study, which took place on Barro Colorado Island. DAWSON'S (1976) study estimated fully 49.5% insects (70% of which were grasshoppers) and 50.5% fruits. These proportions changed throughout the year, according to annual fluctuations of availability. *Saguinus geoffroyi* drank water on Barro Colorado from the corollas of the flowers of *Ochroma limonesis* (Bombacaceae) (HLADIK & HLADIK, 1969), but based on observations of other tamarins, they probably occasionally use tree holes as well.

Saguinus geoffroyi exhibits a birth peak between April and June, although a few births may occur any month of the year (DAWSON & DUKELOW, 1976). Though some *Saguinus* (*S. mystax*, SOINI & SOINI, 1982) exhibit seasonal changes in testes size, this species apparently does not (DAWSON & DUKELOW, 1976). Estrus cycles are an average of 15.5 days. Gestation period in this species is unknown but probably similar to *S. oedipus*, whose gestation is about 145 days (HARVEY *et al.*, 1987). Parents seem to share equally in infant care in this species (MOYNIHAN, 1970). SKINNER (1986a) reported nine interbirth intervals varying from 154 days to 540 days (average=311 days). Females which gave birth to twins had the longer interbirth intervals.

Births occur mostly in April-May in northern Colombia whereas infants were born on Barro Colorado in captivity in May-June (HERSHKOVITZ, 1977). Infants are born fully furred and weigh around 40-50 g. and represent about 10% of the average adult female's body weight. Neonates exhibit different pelage characteristics from the parents, having black bodies with a beige blaze on the black fur and a face with sparse white fur that highlights the cheeks. The tail is solid black, becomming proximally rufous with maturity. The dorsal agouti pattern begins to develop in the third week in the shoulder region, then slowly spreading caudally.

Both sexes carry and groom infants, although males perform this and other parental care behavior (except for cleaning the infant) significantly more than females. These parental behaviors include allogrooming, protection and social feeding. Sometimes the carrying parent bites their infant, especially if it is active. This causes the infant to scream, resulting in the other parent approaching and taking the infant. Nips and tugs at infant appendages cause it to transfer to the other parent. Little carrying is observed by older siblings, and infants prefer carrying by adults rather than by their siblings. Siblings often socially feed the infants.

The infant first begins locomotion at 2-5 weeks and reaches 50% at about 5-23 weeks but usually at about 5-10 weeks. Infants begin eating solid foods from 4-7 weeks. First allogrooming is usually at about 4-5 weeks and solitary play at 6-10 weeks. They engage in social play at about 7-21 weeks. At 10-18 weeks young are totally independent and they are completely weaned at about 15-25 weeks.

In her captive study SKINNER (1986a) found that males more often approached females and groomed them than the converse. Grooming scores were significantly higher when females were not pregnant. Additionally, males begged more to females than females to males, although females had significantly more success begging. Pregnancy increased success and frequency of begging in females.

Some behavior patterns noted by SKINNER (1986a) are listed as follow: (1).*Ruff Flash* (Rufous Ruffle, MOYNIHAN, 1970) - After approaching, either sex turns face away simultaneously piloerecting and displaying the rufous-colored neck ruff. (2). *Foot Display*.- Males sitting on a branch stretched one or both legs forward, displaying large feet for several seconds before grooming his thigh. (3). *Vertical Stretch* - Both sexes; stand bipedally on toes and stretch with arms over head, arching back and pressing chest against vertical surface. Individual then autogroomed while sitting and partner often approached and *Groom Begged*. (4). *Groom Beg* - Both sexes; approach anterior end of mate, crouch low, often directly beneath mate's head displaying nape. If object turns away partner persists. Male usually responds to female's *Groom Beg*, but female shows unpredictable response. (5). *Tail Coil* (*Upward Tail Coiling*, Moynihan, 1970) - Both sexes; regular coiling up of tail into coil between legs during rest, grooming, or huddling. (6). *Food Beg* - Both sexes; partner anteriorly approaches mate with food item; repeated squeaking by both; if feeding animal doesn't flee, begger sits beside it vocalizing and tries to take food which may or may not be given up. (7). *Tongue Display* - Both sexes; extend tongue about 2 cm, curled up at end and vibrated toward conspecific during courtship as solicitation; used also in play-chase and agonistic interaction. (8). *Marking* - Both anal and supra-pubic gland fields deposited over substrates and investigated by other member of pair. (9). *Cuddling* - One partner touches the other for periods of 30 seconds or more, using a significant portion of the body, usually the trunk.

ANDREW (1963), EPPLE (1968) and MOYNIHAN (1970) studied vocalizations in this species. MOYNIHAN'S (1970) scheme includes the following vocalizations: (1). *Plantive Whistles* (a). *Long Whistles* - Loud, plaintive, prolonged in notes of 2-4; long-distance signal when separated from group, in response to separated animals and when defending territory. (b). *Twitters* - Moderately soft, high-pitched notes in series of 4-10 notes; sing notes uttered in short whine; low intensity directed towards potential predators, intraspecies disputes as alarms, greeting among captive animals. (2). *Sharp Notes* (a). *Trills* - Moderately loud series of 4-12 rapidly sounded short notes; includes loud ultrasonic components to 24 kc; used reacting to potential predators and hostile situations. (b). *Loud Sharp Notes* - Variable series of sharp tsit sound or single loud, sharp notes loud and higher ultrasonic components; socially contagious and used in reaction to potential predators, intraspecific disputes, escape movements. (c). *Soft Sharp Notes* - Single or short variable series of *tsit* notes identical with a & b with ultrasonics up to 44 kc; use not known since it is only heard in captivity. (d). *Sneezing Sharp Notes* - Sneezelike single note or «tschuck» which may be the same function. (3). *Rasping* (a). *Long Rasps* - Loud, harsh, prolonged screech.

Conservation state

We have very little knowledge of the conservation state of this species in Colombia. Nevertheless, internationally and nationally *S. geoffroyi* is classified LC, though in Coombia there is actually little information available to support this classification. The species has a very restricted geographic range and there is active forest reduction in almost all parts of their geographic range. It is fortunate that this small primate seems to prefer disturbed secondary vegetation. Colombia's Katíos National Park protects a population of this species, but secondary vegetation, municipal watersheds and other managed zones should also provide adequate habitat for this primate if it is not hunted or trapped out. Besides habitat destruction, live trapping and the pet trade are probably one of the greatest pressures on *Saguinus geoffroyi*, and these activities should be controlled.

VARGAS (1994a) recently reported active trapping and selling of this species west of the Atrato River, where she discovered a new extension in the population of this primate. The extent of such commercialization is completely unknown and ought to be elucidated. The details of the area containing *S. geoffroyi* east of the Río Atrato also needs to be defined and studied, since in this region or to the north there may be contact with *S. oedipus*. Censuses and studies of this primate in Colombia are urgently needed.

Where to see it

This species is common in and around Katíos National Park. It is also easy to observe along the Atrato river, although not in flooded forest.

Mottled-faced Tamarin

Saguinus inustus (SCHWARTZ, 1951)

Plate 2 • Map 4

Taxonomic comments

This species has no described subspecies although another subspecies may exist between the northern and southern parts of its range (HERNÁNDEZ-C. & COOPER, 1976). There is very little material in collections to ascertain subspecies and the species is very poorly known.

Common names

Colombia: *mico diablo*, *diablito* and *tití diablito* in the department of Vaupés; *chichico negro*, *hueviblanco* in the Caquetá River; Brazil: *sagui* .

Indian languages

Macuchi (Carijona); *abujíjiyo* (Cubeo); *usica* (Macuna); *nosí* (Macú, *fide* Capitán Francisco Yujup-Macú, 1992); *otsari* (Miraña, *fide* Napoleón Miraña, 1985); *dyafíixu* (Okaima); *teu teu* (Puinave); *pisácaca* (Tanimuca); *usica* (Tucano); *pijerú* (Yucuna).

Identification

Saguinus inustus is basically a black or very dark-brown animal with white hairs covering the genitalia (thus, the local common name *hueviblanco*). The animals have a mantle of lighter-brown hair covering the dorsum, which in appropriate light contrasts with the darker pigmentation of the limbs and tail, which appear black. The muzzle has a white patch between the upper lip and the nostrils. The face is unpigmented except for some blotches (thus, the name mottled-face tamarin). This is the darkest of all the *Saguinus*. Head/body measurements are about 233 mm (208-259; n=11) and tail length averages 366 mm (330-410; n=11) (HERSHKOVITZ, 1977). Females have large and conspicuous pinkish colored genitals, the most developed of the species in this genus making it easy to distinguish the females from the males in the field.

Geographic range

In Colombia *Saguinus inustus* is found north of the Caquetá River and east of the Yarí River. The Savannas of Yarí probably separate *Saguinus inustus* from *Saguinus nigricollis* and *Saguinus fuscicollis*, which are known from above Angostura I on the Guyabero River. The northern limit seems to be the Guyabero/Guaviare Rivers, although recent field work suggests that the species is unkown on the lower Guaviare and Inírida Rivers and is absent east of the lower Inírida River (DEFLER, unpublished data). Outside of Colombia the species is known in Brazil between the Río Negro westward to the frontier with Colombia.

This small primate has been observed in closed-canopy rainforest, but it is much more common around the secondary vegetation of Indian gardens, where it enters the gardens to take advantage of certain human crops as well as plant species which grow well in such disturbed vegetation (DEFLER, 1983b; 2003).

Natural history

The only studies of this species have been remedial efforts to study a population of *S. inustus* around a local Indian community on the lower Caquetá River, near La Pedrera in Amazonas department (PALACIOS *et al.*, 2003). A few observations were made previously by DEFLER (1983b, 2003) and other brief observations by PALACIOS *et al.* (in press).

Observed groups have been made up of about 7 animals, although larger groups are said to be seen from time to time. Three group counts were 3, 7 and 8 and a home range was calculated at 35 ha (PALACIOS *et al.*, 2003). Densities have not been calculated, although anecdotally it seems to this author that densities are higher around the secondary vegetation of Indian settlements than in primary forest. After eleven years working at a site of about 10 km² in Vaupés no groups have ever been observed in this primary forest vegetation.

A day range of one group was an average of 961 m (750-1100 m; n=5) (PALACIOS *et al.*, 2003), but no time budget has been calculated. This small monkey uses quadrupedal locomotion frequently as well as leaps and clinging to the sides of the trees. These animals descend to the ground to reach isolated blocks of forest. This species seems to have a similar diet to many other species of *Saguinus*. Two garden species which greatly attract them are *Pourouma cecropiifolia* (Cecropiaceae), which is very common in Indian gardens and highly appreciated by these monkeys which carry the seeds in their mouths, dispersing them from the mother-tree. Another species which is eaten about six months of the year except July-August-September and

January-March (when there are none) is *Inga edulis* (Leguminosae), while other *Inga* and other fruits are sought in the nearby forest. Developing small fruits begun to be eaten in October and April (there are two crops every year), and the animals continue using them until they ripen in June and the crop is finally used up. This species licks vegetation and drinks from tree holes, although when they are eating *Pourouma* they probably do not drink because of the juice of the fruit. The short study by PALACIOS *et al.* (2003) identified as part of the diet *Inga edulis, I. leptocarpa, I. pilosula, I. thibaudiana, I. yasuniaria* (Mimosaceae) and three other species of *Inga*; *Pouroma cecropiifolia* and *P. tomentosa* (Cecropiaceae); *Pouteria guianensis* (Sapotaceae) and two additional species from this genus; *Couma macrocarpa* and *Lacmellea* cf. *arborescens* (Apocynaceae); *Tapirira guianensis* (Anacardiaceae); *Rollinia mucosa* (Annonaceae); *Mendoncia ovata* (Acanthaceae); *Buchenavia* sp. (Combretaceae); *Abuta grandifolia* (Menispermaceae) and a species belonging probably to the Quiinaceae. They have also been observed eating small spiders, orthopterans and ant larva while foraging in secondary vegetation (rastrojo) of Indian gardens (PALACIOS *et al.*, 2003).

There seem to be two birth seasons when infants are seen, a main season in March-April and another minor season in September. Little is known about infant development, although within six weeks the infants seem to be locomoting independently. Since some groups of 10-20 are seen, groups may actually forage side-by-side for awhile before separating again. The vocalizations are similar to many other *Saguinus* with many bird-like twitters. They include a long call, alarm call, contact calls and others. This small species is apt to be preyed upon by many hawks, small mammals, and snakes, especially since its routes to reach food trees in the Indian gardens leave the groups very vulnerable because of the short, bushy and sparse vegetation.

Conservation state

This primate is not considered to be endangered and it should do well as long as the indigenous people tend gardens in the way that they do, allowing cut-over areas to go grow back to forest vegetation. Indians usually ignore this small primate. There seems to be little pressure on this monkey on Indian lands (which make up 60-70% or this primate's Colombian range). The species seems to be almost a commensal of Indians. It is classified LC by the IUCN and by Colombia.

Where to see it

Saguinus inustus can be seen on the north bank of the Caquetá river around the gardens of any Indian community. Potential observers should contact the local chief (capitán), asking permission and paying a guide to help look for the animals. It would be very inexpensive to pay several youngsters to search for this small monkey.

White-footed Tamarin

Saguinus leucopus (GÜNTHER, 1876)

Plate 3 • Map 8

Taxonomic comments

The species has been described as monotypic (HERSHKOVITZ, 1977). It is possible that two subspecies exist, since two specimens collected around Mariquita at the southern extreme of the known range differ in the tone of their pelage from other known examples (HERNÁNDEZ C. & COOPER, 1976). This species is most closely related to *S. oedipus* and *S. geoffroyi*, probably sharing trans-Andean ancestral stock. A morphometric comparison of the three species using data from HERSHKOVITZ (1977) suggested that *S. leucopus* and *S. geoffroyi* were more closely related than either species was to *S. oedipus* (SNOW, 1986).

Common names

Other English: Silvery brown bare-face tamarin. Colombia: *tití* and *tití gris* throughout their geographic range in north-central Colombia; *mico tistís* in Antioquia.

Identification

These *Saguinus* weigh about 462 g (n=8; HERNÁNDEZ & DEFLER, 1989). Body length is around 230-250 mm with an average tail length of about 380 mm. The dorsum of this monkey is a pale silvery brown with lighter streaks throughout. The ventrum is ferruginous while the tail is brown with a white tip, although in some individuals a white blotch may be also seen further up on the tail, and occasionally there is no white tip at all. The face is almost without hair, being thinly furred with fine white hairs. Between the ears and along the neck is a thick ruff of brown hair, while the hands and the feet are white.

Geographic range

Saguinus leucopus is an endemic species in Colombia. It is found in northeast Antioquia (Cáceres, Valdivia and the valley of the middle Magdalena River) in the departments of Antioquia, Caldas and the north of Tolima (at least to the neighborhood of Mariquita). The general limits of this species are the eastern banks of the lower Cauca river, the west bank of the middle Magdalena river (including all of the largest islands in the river) and the foot of the Cordillera Central up to about 1500 m altitude. The range is the smallest of all the *Saguinus* species.

GREEN (1978) believed that this species most often used the low and mid-canopy. The species is found in tropical dry forest, tropical humid forest, very humid tropical forest and very humid premontane *(sensu* HOLDRIDGE), and the habitat includes primary and secondary forests, including isolated forest relicts (which probably become overpopulated from the refugees fleeing from the heavy colonization activity in the areal of distribution of this species. CALLE (pers. com.) reports that these animals are common in forest of very steep slopes, which have been unable to be utilized by colonists, and this may be one conservation tactic for the future, if such forests can be preserved.

Natural history

The field data in this report derive from the work of CALLE (1992c) and of VARGAS & SOLANO 1996a, 1996b) the last two of whom recently did some censuses around the Miel river. Recently POVEDA (2000) completed six months of work with an urban and forest group of the species near Mariquita. BLUMER & EPPLE (unpublished) did a short captive study with a group of three male and one female individuals and POVEDA *et al.* (2001) reported some of her data. CUARTAS-CALLE (2001) provided new distribution data of observed groups.

The groups were usually made up of 3-9 individuals (average=4.6; n=42 groups), although occasionally solitary animals or temporary associations of 14 or more individuals have been observed. CALLE found the larger groups had a predictable territory, while smaller groups were highly unpredictable. Perhaps this was due to the probable packing of many animals together from the surrounding destroyed forests, leaving smaller groups and individuals without territories. POVEDA (2000) observed seven groups varying in size from 2-12 with an average group size of 6.6.

POVEDA (2000) calculated a home range of 17.7 ha during six months of observation near Mariquita. She also observed an urban group which used only 0.73 ha. CALLE (1992c) found very high densities of *S. leucopus* in a small forest of her study area (82 animals/km²) along the Miel river in Antioquia, most probably due to the active destruction of the many colonists in the region. BERNSTEIN *et al.* (1976a) reported 1-4 individuals/km² for northern Bolívar.

POVEDA (2000) calculated day ranges for two days of 1848 m and 1851 m and she calculated an activity budget of 38% foraging and feeding, 9% resting, 34% travel, 8% observation, 10% social activities, and 1% non-social activities. This species, like most of the Callithrichinae, are quadrupedal, using many leaps. The monkeys also often proceed in a quadrupedal-suspensory mode, clinging to an above branch with all four limbs, while they proceed hanging below the branch. Much genital marking is accomplished while hanging thus upside down.

S. leucopus like other *Saguinus*, eat mainly soft fruits and insects. POVEDA (2000) calculated a diet consisting of more than 70% consumption of ripe fruit of which only 12 species of plants were utilized as listed below. In an urban group observed by POVEDA (2000) the animals subsisted on fruits that had been planted by residents of the neighborhood in their backyards and they consisted of 48.47% mango (*Mangifera indica*, Anacardiaceae), 15.29% zapotes (*Matisia cordata,* Bombacaceae), 11.47% fruits provided by residents, 7.34% papaya (*Carica papaya*, Caricaceae), 4.13% guava (*Psidium guajaba,* Mirtaceae), 3.67% rubber, 1.53% avocado (*Persea gratissima,* Lauraceae), 1.38% plantain (*Musa sapientum,* Musaceae), 1.22 % pomarosa (*Eugenia jambos,* Myrtaceae), 0.61% oranges (*Citrus aurantium,* Rutaceae), 0.31% carambolo (*Averrhoa carambola,* Oxalidaceae), 0.31% coconut (*Cocos nucifera,* Palmae), and 0.31% *Hibiscus sp.*, Malvaceae).

The species eaten by the *Saguinus leucopus* in the woods adjacent to Mariquita during the six months observations of POVEDA (2000) and percent of feeding time spent eating them are as follow: *Cecropia peltata* (Cecropiaceae) 35% fruit; *Talisia* sp. (Sapindaceae) 24.2% fruit; *Protium* sp. (Burseraceae) 15.8% fruit; *Sorocea sprucei* (Moraceae) 12.4% fruit; *Rollinia edulis* (Annonaceae)

5% fruit; *Trichospermum mexicanum* (Tiliaceae) 1.7% flowers; *Tetrochidium aff. echeverianum* (Euphorbiaceae) 1.7% fruit; *Pera arborea* (Euphorbiaceae) 0.8% fruit; *Didymopanax morototoni* (Araliaceaea) 0.8% bark; *Byrsonima spicata* (Malphigiaceae) 0.8% fruit; *Tococa sp.* (Melastamataceae) 0.8% unidentified; *Zanthoxylum sp.* (Rutaceae) 0.8% bark.

Young animals were reported by HERSHKOVITZ (1977) for May-June, and again (second season) in Oct.-Nov. reported by VARGAS & SOLANO (1994, 1996a, 1996b). BLUMER & EPPLE, (unpublished) and EPPLE (1968) divided vocalizations of this species into 13 call types, based on physical characteristics as follow: (1). *Contact Calls* - used to maintain or monitor contact. (a). *Monosyllabic contact calls* given in close visual contact - *low amplitude peeps and twitters* with perhaps some sexual dimorphism in frequency, which may be a duet for cementing the pair bond. (b). Monosyllabic contact calls given when out of visual contact - *cheeeh* given as long-distance contact call and to maintain intergroup spacing (EPPLE, 1968), usually given in series of 2-4 calls. At low intensity *chee* may separate from *eeeh*, the *chee* being used during agonistic interactions. (2). *Intergroup Agonistic Calls* (a). *Chee* a component of the long-distance call, but used in a series of 1-3 in agonistic encounters. (b). *Rapid Trills* are more variable than most calls and may reflect various emotional states in intergroup interactions. (3). *General Arousal Calls* - situations of high excitement (a). *Chirps* attract attention of group members to cause of arousal. (b). *Chucks* directed at cause of arousal or given in long series and undirected at disturbance. (4). *Investigatory/Inquisitory Calls* - Tonal whistles directed at unfamiliar stiumuli. (a). *Whistles-* almost always preceded by a short monosyllabic call such as a *peep* or a *chirp*. (b). *Sliding Whistle (Type A)* - greatest degree of downward sloping at beginning of call. (c). *Sliding Whistle (Type B)* - responding to predators or unfamiliar animal. (5). *Threat/Mobbing Calls* - given when angry by being threatened or startled or defending food source. (a). *Chat-* most common of this class and perhaps the lowest excitement level. (b). *Chatter-* extremely high threat motivation. (c). *Tsik* - highest excitement levels in presence of predator; part of mobbing response.

BLUMER & EPPLE (unpublished) suggest that the vocalizations of this species show that *S. leucopus* is not as related to *S. geoffroyi* and *S. oedipus* as those two species are related to each other. Further, the vocalizations support HERSHKOVITZ ' (1977) suggestion that *S. leucopus* may be closer to the basic stock which gave rise to the three trans-Andean species.

One frequent display seen in captive animals is the *tongue flick*, where the tongue is extended exaggeratedly forward out and back with rapid movements with the tip curled up at the end, when an animal wishes to solicit play (DEFLER, pers. obs.) or in an agonistic context (BLUMER & EPPLE, unpublished). HERSHKOVITZ (1977) reported the same display when he surprised a wild, adult male, so it would seem that the display occurs during different states of excitation. *Urine washing* is performed, somewhat like *Saimiri*, seemingly in an agonistic situation. Piloerection is similar in context to other *Saguinus*. *Agonistic lateral head shaking* was rarely seen. Arch posture is common in an agonistic context accompanied by piloerection and splaying of the shoulders to make the body appear larger. Genital display is performed by the dominant animals towards subordinants.

Saguinus leucopus does both ano-genital and genital marking similar to *S. geoffroyi* and *S. oedipus*. Also, sternal marking is performed, rubbing the sternal region over a substrate. EPPLE & LORENZ (1967) suggest that this may function as socio-sexual cues. A *urine kick wash* is used during agonistic displays, involving urinating on the homolateral hand and foot while they are lightly rubbed together, then thrusting the foot up toward the armpit and out 7-8 times (BLUMER & EPPLE, unpublished). This may be repeated on alternate sides of the body. The behavior has been observed in *Saimiri, Cebus capucinus* and other cebids and has been discussed by CANDLAND *et al.* (1980) as to its possible functions.

Conservation state

Saguinus leucopus has been placed in Appendix I of the CITES agreement, and the IUCN classifies it as VU as does the Colombian IUCN classification of endangered primates. During the X Congress of the International Primatological Society in Japan in 1990, naturalistic studies and conservation of this species were declared to be an international priority, along with a few other species.

The species is highly vulnerable, since its limited distribution (the most limited of all *Saguinus* species) is found in a zone of very active colonization and currently the animals are often sold as pets on the streets of Bogotá and Medellín. One marketing trick is to bleach out areas of these monkeys' fur (such as their top-knotch) so as to claim that they are very rare and from some other region of the country than their actual provenance. One vender claimed to the author that his *Saguinus leucopus* was a rare monkey from near Leticia, thinking perhaps that Leticia, being the furthest point he could think of, would be unknown to others. Another strategy is to drug the wild-caught animals, so as to claim that the individuals for sale are tame. The species is not protected in any reserve in Colombia.

The species still has numerous populations at several sites such as La Miel River, although many of the areas are being rapidly deforested and a large hydroelectric dam is being constructed there (CALLE, 1992). Reports of high densities in one census are actually results of a crashing population with many individuals seeking refuge in the censused forest.

Live trapping and commerce is extremely destructive to this and any other *Saguinus*, since the multiple-trap technique of trapping entire groups is known to the animal traffickers. With this technique, a captive animal is surrounded by other baited traps into which the individuals of the attracted group enter (see, for example, NEYMAN, 1977 for a description of the technique).

Captive breeding and captive studies ought to be undertaken while there are still plenty of animals in commerce. These individuals could be acquired legally through actions of The Ministry of the Environment. A program to establish reserves, to educate the populace and to reintroduce is vitally needed, although reintroduction may only supply more commerce until such commerce is severely punished.

Where to see it

Saguinus leucopus is easily viewed in the countryside surrounding the Miel River, near the right bank of the Magdalena River. Several pairs are observable in the trees surrounding the Medellín Zoological Park; this being the easiest way to see the species. This monkey is still fairly common in many areas, but tree removal is quickly lowering population densities.

Black-mantle tamarin

Saguinus nigricollis (SPIX, 1823)

Plate 2 • Map 5

Taxonomic comments

HERSHKOVITZ (1982b) divided this species into three subspecies: *S. nigricollis nigricollis*, *S. n. graellsi* (Graell's black-mantled tamarin) and *S. n. hernandezi* (HERNÁNDEZ-CAMACHO's black mantled tamarin). All of them occur in Colombia. HERNÁNDEZ-CAMACHO & COOPER (1976) and DEFLER (1994b) considered *S. n. graellsi* to be a separate species (= *S. graellsi*) because they believed it to be sympatric with *S. nigricollis* around Puerto Leguízamo, on the north back of the Río Putumayo. This was based on a specimen of *Saguinus nigricollis nigricollis* in the collection of the Instituto de Ciencias Naturales (ICN) which is registered as having been collected in the "Quebrada [creek] El Hacha", a left bank affluent of the Río Putumayo by H. GRANADOS & H. ARÉVALO in 1960 and on the presence within this collection of various specimens of *S. nigricollis nigricollis* marked as a collection locality only "entre el Caquetá y Putumayo" (between the Caquetá and Putumayo Rivers). These specimens would indicate that the subspecies occurs to the north of the Río Putumayo. HERNÁNDEZ-CAMACHO & COOPER (1976) also wrote, however, that the *S. n. nigricollis* of the region was actually an animal approaching the appearance of a dull-colored *S. f. fuscus*. PHILIP HERSHKOVITZ (pers. comm. to the author) believed that the H. GRANADOS/H. ARÉVALO specimen was actually a mistake; and later classifications accepted HERNÁNDEZ-CAMACHO & COOPER's (1976) dull-colored *S. f. fuscus* as just that, *Saguinus fuscicollis fuscus*, but a bit duller than animals further east of the piedmont. *S. f. fuscus* is well-collected between the Caquetá and Putumayo from the piedmont to the Brazilian frontier, and this author is familiar with it along the lower parts of the Rio Puré (Purué) in Brazil, as well.

It is clear from four Colombian specimens that *graellsi* is present between the Caquetá and Putumayo; three of them were collected at 900 m from the upper Río Guamués ICN3554, ICN3562, ICN3563), south of where it was commonly observed by MOYNIHAN (1976a) close to Puerto Umbría, north of the upper Río Putumayo. Geographically *S. n. graellsi* overlaps with the saddle-back tamarin, *S. fuscicollis fuscus*. MOYNIHAN (p.35), however, did not observe them together in the same locality. At Santa Rosa on the Rio Guamués he found *graellsi*, but not *S. fuscicollis*, and the local Indians were not familiar with it. However, HERNÁNDEZ-CAMACHO & COOPER (1976) reported that mixed troops of *graellsi* and *fuscicollis* could be commonly be observed around Puerto Asís on the Putumayo (at the mouth of the Río Guamués), as well as along the Río Guamués.

The fourth specimen of *graellsi* collected in Colombia (ICN3556) was collected by GRANADOS & ARÉVALO in 1960 and is registered in the ICN as coming from "El Encanto, Caquetá". However, no "El Encanto, Caquetá" exists, although certainly "El Encanto, Amazonas" does exist, downriver from where the two collectors had been working. It is probable that this specimen was in fact collected around El Encanto, Amazonas at the mouth of the Cará–Paraná, and that the collectors´ notes were mistaken in registering the Department of Caquetá instead of Amazonas. Additionally, because of HERSHKOVITZ' belief that a mistake had been made with the *S. n. nigricollis* specimen, I speculate further that in fact the El Encanto *graellsi* may in fact have been collected in the Quebrada El Hacha, upriver from Puerto Leguízamo and below Puerto Asís, and that it was likely that the *S. n. nigricollis* specimen was purchased around El Encanto, or indeed anywhere along the river. It occurs downriver on the right bank of the Río Putumayo (the westward extent is unknown in Perú) at least to the Río Yuvineto (Perú), not far (downstream) from Puerto Leguizamo (AQUINO & ENCARNACIÓN, 1994). *S. n. graellsi* can be found opposite El Hacha (upriver) and around Puerto Asís (HERSHKOVITZ, 1977; DEFLER, pers. obs.). It would not be the first time that specimens and data were mixed up and inadequate notes padded out from memory, but these details are going to have to be determined in the field, difficult at the present because of insecurity and violence in the entire region.

Surveys between the middle reaches of the Ríos Putumayo (both banks) and Caquetá are needed to determine better the distributions and any sympatry, since the observation of *nigricollis* and *graellsi* in the same forests would be needed to confirm *graellsi* as a good species. Here I continue to recognize *graellsi* as a subspecies of *S. nigricollis* in agreement with HERSHKOVITZ (1977) until such time as field work can clarify the situation.

Common names

Other English: Red and black tamarin. Colombia: *Leoncito* (upper Río Putumayo), *bebeleche*, *bociblanca* (Colombian Amazon). Ecuador: *Chichico*. Brazil: *Sagüi-de-manto-preto*.French: *Tamarin rouge et noir*.

Indian languages

Pronounced as in Spanish: *jití* (Huitoto), *tiiru* (Ocaima), *nea sisi* (Siona) (*fide* RODRÍGUEZ-M. *et al.*, 1995)

Identification

In general, this small monkey (both sexes weigh around 500 g) is lighter in color (reddish, red-black or buff) on the rear half of the body, with a blackish nape and mantle on the upper back, at least when considering *S. n. nigricollis* and *S. n. hernandezi*. Their ears are black and naked, and they have a black face, but with white hairs broadly outlining the nostrils and mouth. Infants do not have the typical white moustache, but small patches of white hairs on each cheek. Juveniles gradually lose them, and beyond 14 months old, the subadults are grayish around the muzzle, gradually becoming whiter as they grow older. *S. n. graellzi* of course are fundamentally differently marked, as they are basically agouti-brown (or reddish-brown) animals, darker along the nape and shoulders than on the lower parts of the back. HERSHKOVITZ (1982b, p. 647) distinguishes the three subspecies as follows:

• *S. n. nigricollis*with nape and mantle nearly or entirely uniformly blackish (eumelanin), the mantle usually tapered behind and extending to lower back; sacral region and dorsal surface of tail with proximal 3-5 cm marbled or striated reddish (pheomelanin) and blackish; ventral surface of tail base, rump, and legs dominantly or nearly entirely reddish or mahogany, remainder of tail   blackish; neck, chest, and belly blackish, often mixed with reddish or mahogany.

• *S. n. graellsi*with nape and mantle blackish or dark brown (eumelanin) agouti, the mantle more or less square behind and not extending beyond midback; remainder of back, sides of body, thighs, legs, and proximal 5-16 cm of dorsal and ventral surface of tail buff (pheomelanin) agouti; remainder of tail blackish; neck, chest, and belly varying from dominantly buffy agouti to dominantly blackish.

• *S. n. hernandezi* with nape and mantle nearly uniformly blackish, the mantle tapered behind to midback with the blackish continued, usually as a mid-dorsal band or stripe, across lower back and proximal portion of tail; sides of lower back mixed blackish and orange (pheomelanin), sides and ventral surface of proximal 5-10 cm of tail orange agouti, remainder of tail blackish; neck and chest dominantly orange agouti, belly mixed orange and blackish.

Some have argued that *graellsi* is too dissimilar to *S. n. nigricollis* and *S. n. hernandezi* to belong to that group, and that, being geographically located between them, it is difficult to see *graellsi* as being related. However, I have no problem understanding the pelage changes as being the result of metachromic evolution as described by HERSHKOVITZ (1969, 1977). *S. n. graellsi* has by far the most primitive color with its predominant agouti hair pattern. HERSHKOVITZ (1968, 1977) has clearly shown that agouti is a precursor to melanistic (black) and pheomelanistic (reddish) color patterns found in other forms, including the other two subspecies, due to increased eumelanin in the growing hair to achieve black tones or to the deposition of pheomelanin in the growing hair to achieve reddish tones, thus achieving striking patterns. According to the principle of metachromism, these changes follow orderly pathways, which could derive the *nigricollis* and *hernandezi* color patterns from an agouti ancestor.

Geographic range

In Colombia, *S. nigricollis nigricollis* is common in the Colombian trapezium, between the Ríos Amazonas and Putumayo. *S. n. graellsi* is known from Puerto Leguizamo or perhaps from El Encanto (between the Caquetá and Putumayo Rivers) westward, probably, to the base of the Cordillera Oriental (MOYNIHAN, 1976a; HERNÁNDEZ-CAMACHO & COOPER, 1976). It is possibly found eastward between the Ríos Caquetá and Putumayo from Puerto Leguízamo as well, because of the El Encanto specimen discussed above. *S. n. hernandezi* occurs on the Río Peneya, an affluent of the northern bank of the Río Caquetá, between the Ríos Orteguaza and Caguán, and from the south bank of the Río Guayabero, near La Macarena. The entire range of this subspecies may include the area west of the Río Yarí (and the Llanos de Yarí) and north of the Caquetá, perhaps to the eastern Cordillera of the Andes, although further studies are needed.

Outside of Colombia, *S. n. nigricollis* extends into Brazil and Perú, between the Ríos Putumayo-Içá and Solimões (=Amazonas), and north of the Río Napo, west as far as the Ríos Lagartococha and Güepi, which form the frontier with Ecuador (Aquino & Encarnación, 1994). *S. n. graellsi*, ranges south of the upper Caquetá, across the upper Río Putumayo, into Ecuador where it occurs along the upper reaches of the Ríos Aguarico, Coca, Napo, and Pastaza (but not between the Napo and Curaray in the Yasuní National Park, domain of the golden-mantle tamarin, *S. tripartitus*). In Perú, Aquino & Encarnación (1994) placed *S. n. graellsi* between the Ríos Napo and Nanay. Hershkovitz (1982b) recorded two localities for *S. n. graellsi* at the mouth of the Río Curaray (a south bank affluent of the Napo), but a recent survey by Heymann *et al.* (2002) found no evidence for it along either bank of the river. There is no indication that *S. n. graellsi* occurs along the middle and lower reaches of the Ríos Tigre and Pastaza in Perú, although Hershkovitz (1982b) suggested it might extend as far south as the north bank of the Río Marañón. Further surveys are required to establish its true range in Perú.

Natural history

Typical of all the tamarins, *S. nigricollis* occupies both primary humid forests and secondary forests. All species show a propensity for dense, secondary vegetation, possibly due to increased protection from predators, but also the abundance of animal prey and preferred fruits. In the Amacayacu National Park, *S. n. nigricollis* is evidently more common in advanced secondary vegetation. Along the Río Purité, in the colombian trapezium, they were frequently seen foraging for insects in the riverside vegetation, just as the sun had begun heating the low bushes and trees along the watercourse (Defler, 1994d). *S. n. nigricollis* has never been studied in the wild, but De la Torre (1994, 1996) and De la Torre *et al.* (1992, 1995) carried out a field study of *S. n. graellsi* in the Cuyabeno Faunal Production Reserve in Ecuador, and they found it to prefer terra firme forest over the various types of flooded forests found there. *S. n. hernandezi* has been studied in humid lowland forests on the Río Peneya by Izawa (1978a), and later by Vargas (1992, 1994b) in the northernmost part of the species' range on the south bank of the Río Guayabero, in highly seasonal dry forest and spiny-leaved scleromorphic scrub, known as *arrabal*. Tamarins on the Río Guayabero spent much of their time either in the *arrabal* or in edge habitat between forest and pasture, rivers or lakes. They especially preferred the river edge in the dry season due to the more humid conditions there, allowing for a greater abundance of insect prey.

Saguinus nigricollis groups are small; from 4 to 11 individuals. Ten groups counted by IZAWA (1978a) on the Río Peneya averaged 6.3 individuals. To the north on the Río Guayabero, six groups averaged 7.5 individuals, ranging from 4 to 21 (VARGAS 1994b). TOKUDA (1968) reported a rather high group average of 13. They are generally made up of a paired male and female and their offspring, and while larger groups may include unrelated males and females, usually only one female breeds, as has been found in *S. fuscicollis* by GOLDIZEN (1987b).

These small groups live in partially overlapping territories, which in the Peneya area varied from about 30 to 50 ha; Izawa's particular study group ranged over 41 ha. In similar forest in the Cuyabeno reserve, the home range of the *S. n. graellsi* group studied by DE LA TORRE ranged from 42 ha in the wet season to 55 ha in the dry season. Within these ranges, groups may travel as much a 1 km during the day. Interestingly, IZAWA (1978a) occasionally observed 3-5 neighboring groups forming aggregations of up to 40 (usually 10-20) tamarins, which stay together for 1-2 days, ranging over a larger area of 100 to 150 ha. DE LA TORRE *et al.* (1995) also observed *S. n. graellsi* forming these aggregations, which were peaceful and sometimes involved up to five neighboring groups foraging together. Aggregations such as these have also been recorded for woolly monkeys, *Lagothrix lagothricha*. Two or three subgroups from two or three different troops join up and travel together, and it is evident that many of the individuals are quite familiar with each other. These aggregations are undoubtedly important for females, and possibly also males, to migrate to other groups. Female woollies generally do not breed in the groups they were born in.

IZAWA (1978a) calculated an ecological density for primary forest of 19-24 individuals/km² of *S. n. hernandezi* and an overall crude density throughout his study area of about 10-13/km². DE LA TORRE *et al.* (1995) found a similar density for *S. n. graellsi* in Cuyabeno, at 22-33 individuals/km². DEFLER (in prep.) calculated crude densities for *S. n. nigricollis* at four sites along the upper Río Purité (Pureté) (Departamento de Amazonas) ranging from 4 - 15 individuals/km², mostly in upland primary forest. It is possible that tamarins reach higher densities in secondary forest than in primary forest (SNOWDON and SOINI, 1988), but this has yet to be demonstrated with hard data. In the Amacayacu National Park near Leticia there is a good population of *S. n. nigricollis* in both primary and secondary forests, and a comparison of densities in both types of habitat would be of considerable interest.

S. n. hernandezi studied by VARGAS (1994b) in the northern dry forests, fed heavily on fruit in the mornings, from about 06:00 to 08:00 hr, and again in the afternoons from about 14:00 to 16:00. They also foraged for small animal prey in the mornings, and commonly rested at mid-day from 12:00 to 14:00. She found, however, that there were distinct seasonal differences in the time the tamarins dedicated to the different activities. Comparing the wet season with the dry season: In the dry season, the percentage time spent feeding dropped from 24% to 15%; they spent more time foraging (increased from 18% to 25% in the dry season); rather less time moving about (12% of their time in the dry season compared to 15% in the wet season); more time resting (30% of their time compared to 17% in the wet season); and less time in social interactions (only 14.5% of their time in the dry season compared to 26.5% in the wet). Averaged over the year they fed on fruits for 17.5% of the day, foraged for small animal prey for 18%, moved about during 15%, rested for 9%, and were occupied in social and sexual behavior 21% of the day (VARGAS, 1992). In the richer, more humid forests in Cuyabeno, DE LA TORRE *et al.* (1995) found *S. n. graellsi* to have similar time budgets, but also with some seasonal changes. In the dry season, they were feeding for 27% of the day, foraging for 35%, resting for 21% and traveling during 21% of the day. In the wet season they foraged more (38%), rested more (30%) and traveled more (26%), but fed for considerably less time (6%). Contrasting with many other forests, in Cuyabeno, it seems that food abundance for the tamarins was higher in the dry season. Both *S. n. hernandezi* and *S. n. graellsi* were found to spend much of their time low in the dense understoreys, from the ground to 10 m, especially when foraging or resting. They do, however, use all levels of the forest, and at times go right up into the canopy at 20 more meters or to feed on fruits.

Their most common form of locomotion involves walking and, when faster, bounding, classified as arboreal quadrupedalism, but they also frequently jump from tree to tree, often between vertical supports, especially when foraging, when they otherwise move around more slowly and stealthily (IZAWA, 1978a). They usually sleep along branches in dense vegetation and vine tangles. VARGAS (1992, 1994b) observed *S. n. hernandezi* sleeping on the spathe bracts of tall palms and on large diameter trees with tangled vines.

The diet of *S. nigricollis* is made up of small, soft fruits, such as those of *Pourouma* (Cecropiaceae), *Pouteria* (Sapotaceae), and *Protium* (Burseraceae), flowers and nectar, and invertebrates, and small animal prey, including especially arthropods such as caterpillars, stick insects, ants, termites, beetles and spiders, as well as small lizards and frogs. Most of their time they hunt large grasshoppers of the families Tettigoniidae and Acrididae, catching about 5-6 a day. Food is shared between mothers and their infants, but, although sometimes individuals beg and try to steal them, grasshoppers are not shared between adults. Although tamarins are not specialized in gouging with their teeth as the marmosets are, the four lower incisors fit tightly next to each other and resemble an adze. At Caparú Biological Station I have seen *S. n. nigricollis* scraping the thin-barked branches of a lemon tree. SOINI (1993) has reported saddleback tamarins, *S. fuscicollis*, scraping with their lower jaws at gums exuded from the holes gouged by pygmy marmosets in the same fashion. They have not been seen to drink, but they probably make use of any arboreal water sources they find (tree holes, bromeliads), as do other species of *Saguinus*. A free-ranging black-mantled tamarin at Caparú Biological Station often lapped rainwater from leaves, branches and twigs. The soft fruits they eat generally provide sufficient water.

Tamarins have a gestation period of about 5 months, and in captivity they are able to give birth at intervals which are not much longer than this, due to them having post-partum estrus. However in the wild, their capacity to obtain enough food to sustain pregnancy and lactation evidently constrains the number of births each year. *S. nigricollis*, at both the Peneya and Cuyabeno study sites, would seem to be in privileged forests in terms of year round food availability – in both sites females produce twins twice a year. On the Río Peneya, birth seasons are spaced six months apart (IZAWA, 1978a). Neonates appeared in December and, by indirect evidence, around June. Of eleven December births observed by IZAWA (1978a), four were twins and another seven were singletons. Mating behavior was observed in January. In Cuyabeno, DE LA TORRE *et al.* (1995) recorded births in January towards the end of the dry season and, for some other groups at least, again in the middle of the wet season, in June.

During the first two weeks, the mothers are very shy and the infants are completely dependant, clinging to mother's fur, and usually on her back even from early on. After 2-4 weeks the infants take their first steps away from mother while she is resting. Play begins at this time, with jumping, leaf plucking, and twig shaking. During the 4th-6th weeks the infants begin foraging alongside the mother and beg food from other individuals, and move about exploring leaf surfaces and tree holes. At 5-8 weeks the infants begin moving about on their own, only mounting the mother when a swift group movement is necessary. They also indulge a lot more in solitary play and rough-and-tumble. After about 8 weeks the infants can be considered juveniles and independent.

A tongue display is often used in play and appeasement behaviors: The tongue is fully extended straight out from the mouth and slightly curved up at the tip. Another tongue display is used in sexual behavior, where the tongue is extended up over the nose and a slight chopping motion is made with the mouth, while the corners of the mouth are stretched in a grimace. Both of these displays were commonly used by a lone free-ranging female *S. nigricollis* towards other primate species at Caparú Biological Station (pers. obs.). Tail curling and pubic marking is common in the species, their social and communicative functions being evident, but poorly understood. They also mark branches with drops of urine. Vocalizations are similar to those described for *S. fuscicollis*.

Predators on this small species are many. IZAWA (1978a) observed the barred forest falcon (*Micrastur ruficollis*) with a dead tamarin. De la Torre *et al.* (1995) noted that the entire group would descend when individuals gave alarm calls on seeing such species as the white hawk, *Leucopternis albicollis*, the red-throated cara-cara, *Daptrius americanus* and the ornate hawk eagle, *Spizaetus ornatus*. The tayra, *Eira barbara*, has been observed to hunt these monkeys (GALEF *et al.*, 1976). Snakes and small cats are also predators.

VARGAS (1992) observed *S. n. hernandezi* traveling behind and below a mixed group of capuchin monkeys, *Cebus apella*, and squirrel monkeys, *Saimiri sciureus*, although they showed caution and avoidance when the capuchins approached them.

Tamarins

Conservation status

Saguinus nigricollis is not threatened in Colombia nor in other parts of its range in Perú, Ecuador and Brasil. It is classified as of LC in both the national and global IUCN classifications (HILTON-TAYLOR, 2000).

Where to see it

Saguinus nigricollis nigricollis is very easy to observe around the visitors center of the Amacayacu National Park, upriver from Leticia, and it can be located in wooded areas on the edge of the town. It is a very common primate in other parts of the Colombian trapezium as well. *Saguinus n. hernandezi* can be observed up above Angostura 1 (a rapid) of the Guayabero River, but it is apparently not as easy to see as *S. n. nigricollis*. Finally, *S. n. graellsi* is common along the road in secondary vegetation between Mocoa and Puerto Asís in Putumayo and along the Guamués River, west of Puerto Asís.

Cotton-top Tamarin

Saguinus oedipus (LINNAEUS, 1758)

Plate 2 • Map

Taxonomic comments

The major revision of the genus was done by HERSHKOVITZ (1977); he recognizes two subspecies: *S. o. geoffroyi (=S. geoffroyi)* and *S. o. oedipus,* endemic to Colombia. HERNÁNDEZ C. & COOPER (1976), GROVES (1993) and MITTERMEIER & COIMBRA-FILHO (1981) recognize these two populations as diferent species and *S. oedipus* as an endemic Colombian species distributed between the Magdalena and Atrato Rivers. Research on this question is needed, even though most distinguish *oedipus* as being a separate species from *geoffroyi*. Reproductive research and comparisons of vocalizations as well as comparisons of other behavioral traits could be a guide for the determination of the relation between these two taxa. Hybrids are known, nevertheless, the fertility of such hybrids has not been documented.

SKINNER (1986a, 1986b) did a morphometric analysis of the two species and *S. leucopus*, using data found in HERSHKOVITZ (1977), and she found that there was a greater morphometric relation between *S. leucopus* and *S. geoffroyi* than with *S. oedipus*. HANIHARA & NATYORI (1987) found that an analysis of multivariance of dental morphology showed that, despite similarities between *S. oedipus* and *S. geoffroyi*, the relation was less than that between *S. fuscicollis* and *S. nigricollis*. MOORE & CHEVERUD (1992) applied a multivariate analysis including discriminate function and cluster analysis and found that *S. oedipus* and *S. geoffroyi* differed morphologically on the species level and that these differences were not clinal but rather discontinuous. RYLANDS (1993) recently discussed all of the above evidence, supporting the concept of two species rather than one. All of the above *Saguinus* (except for *S. fuscicollis* and *S. nigricollis*) as well as *S. bicolor* are called "bare-faced tamarins" (HERSHKOVITZ, 1977).

Common names

Colombia: *titi cabeza blanca*, *titi*, *titi blanco*, *titi leoncito* and *titi pielroja* in all of its geographic distribution in northeast Colombia.

Identification

Saguinus oedipus is a small tamarin that weighs 416.5 g (n=10) in the wild (SAVAGE, 1990) while in captivity they weigh an average of about 566 g. The length of the head and body is around 225-240 mm and both sexes are rather similar, except for the tail, which seems to be longer in the female (HERSHKOVITZ, 1977). The species' most distinguishing characteristic is the topknotch and mane of the head, both white. The majority of *S. oedipus* seem to have totally white bellies, forearms, hands and posterior limbs; one museum specimen and various captive specimens from the San Jorge River are sulfurous yellow rather than white in the same body parts mentioned above. Also, various specimens from Zambrano, Bolívar have also been observed with a yellowish coloration (pers. com., J. V. RODRÍGUEZ). Another characteristic that varies from individual to individual is the tone and the extent of the rufous color on the thighs and the proximal region of the tail.

Geographic range

Saguinus oedipus is endemic to Colombia. It is found from the region of Urabá in the northwest in Antioquia to the south, at least to the León

River, as well as in the Departments of Córdoba, Sucre, and northern parts of Bolívar and Atlántico. The eastern limit is the western banks of the lower parts of the Magdalena and Cauca Rivers and these limits extend to north-central Antioquia. The species is not found on Mompós Island, where it is replaced by *S. leucopus*.

Saguinus oedipus is found in rainforest, deciduous forest and secondary growth, but it is not found in xerophytic forests. Neverthless, some of the deciduous forests where it is found lose almost all of their leaves at certain times of the year (pers. com., J. V. RODRÍGUEZ). The highest altitudes known for this species are about 400 m, but it is possible that the limits are superior to this in the upper valley of the Sinú river.

Natural history

NEYMAN (1977, 1978) studied social behavior, diet and home range of *S. oedipus* for two years in the north of the Department of Sucre. During the course of one year she identified, marked and followed six different groups. Another research project examined the conservation status of the species and evaluated possible reserves as well as extending our knowledge of the geographic distribution and the reproductive biology of the species (SAVAGE, 1988, 1989a, 1989b, 1990, 1995a, 1995b, 1995c, 1996a, 1996b, 1997, 2002; SAVAGE & BAKER, 1996; SAVAGE & GIRALDO, 1990; SAVAGE *et al.,* 1986, 1987, 1988, 1990a, 1990b, 1990c, 1993, 1996a etc.). Another field study was reported by LINDSAY (1980). A review article by SNOWDON & SOINI (1988) about the *Saguinus* also includes much information about *S. oedipus*.

This species has been extensively studied in captivity, in part because it has been used as an animal model for various biomedical research projects (as, for example, research on colitis and on colon cancer) and because of being maintained in various countries such as the United States, Great Britain and Germany. Ample laboratory research has helped to define the characteristics of this species with respect to vocal and chemical communication, reproduction, social development and social interactions, and these are important contributions to understanding natural populations of the species: **Social Behavior**, WOLTERS (1980); CLEVELAND & SNOWDON (1982, 1984); DE LA OSSA *et al.* (1988); FRENCH *et al.* (1983, 1984); FRENCH & SNOWDON (1981); FRENCH & CLEVELAND (1984); McGREW & McLUCKIE (1986); MOORE *et al.* (1991); MOYNIHAN (1970); MUCKENHIRN (1967); OMEDES & CARROLL (1980); PRICE (1992a); TARDIF & RICHTER (1981); TARDIF (1984); WELKER & LEHRMAN (1978); TARDIF *et al.* (1990) ; **Vocalization** (ANDREW (1963) ; EPPLE (1968) ; McCONNEL & SNOWDON (1986); MUCKENHIRN (1976); SNOWDON *et al.* (1983); **Reproduction** (BRAND (1981a, 1981b); FESS (1975b); FRENCH (1982); FRENCH *et al.* (1984b); FRENCH & CLEVELAND (1984); HAMPTON & HAMPTON (1965, 1977); JOHNSTON *et al.* (1991); KIRKWOOD (1983); KIRKWOOD *et al.* (1983) ; SNOWDEN *et al.* (1985) ; TARDIF *et al.* (1984, 1990, in press) ; TARDIF & COLLEY (1988); TARDIF (1984); **Locomotion**, WELKER *et al.* (1980); **Activity Cycles**, PRICE (1992a); **Reintroduction**: PRICE (1992b); PRICE *et al.* (1991); Scent Marking, FRENCH (1982), FRENCH & CLEVELAND (1984), FRENCH *et al.* (1984).

In natural habitat this primate is observed in groups of 2-9 animals. There can be more males than females; six groups were composed of an average of 2.7 males and 1.7 females (n=77). SAVAGE (1989) observed two groups which increased to 9 members during the two years. One group consisted of three adult males and three adult females and two unsexed young. Apparent polygamy was detected in one group observed with two pregnant females (SAVAGE, 1989a), but so far, no androgyny has been confirmed, though some groups do have more than one adult male.

The groups maintain fixed territories, which they defend physically and with vocal displays. NEYMAN (1977) calculated three territories measuring 7.8 ha, 7.8 ha, and 10 ha., which overlapped with neighboring territories approximately 20-27%. SAVAGE (1990, in press) reported home ranges of 12.4 ha and 10.5 ha. In one field study the population showed a density around 30-120 individuals/km^2 after mapping various home ranges (NEYMAN, 1977) and the day range of one group was estimated at 1500-1900 m.

Time budgets calculated by SAVAGE (1990) for two groups were quite different from each other as follow: 44%/31% forage, 37%/29% rest, and 19%/40% moving. Locomotion and posture have not been specifically studied, except for the captive study of WELKER *et al.*(1980). Like other *Saguinus*, the animals are quadupedal walkers, runners and gallopers. They jump from one terminal branch to another and often cling from a vertical position.

Sleeping sites are in the upper canopy (13.5 - 20 m in Neyman's study), where they sleep on a wide branch or forking branches, often with little vegetation cover above them. Sometimes they use a cover of lianas and branches and occasionally they use a dense mass of vines (NEYMAN, 1978). NEYMAN's group often entered the tree chosen for sleeping at 1630 and the animals were sleeping by 1830.

This species is similar to other *Saguinus* and consumes small fruits and arthropods, flowers, leaves, nectar and gums of trees (especially *Enterolobium cyclocarpum*) and occasionally small vertebrates such as frogs and lizards. NEYMAN (1977, 1979) reported 48 species of fruit belonging to 28 families, listed as follows by the order of their importance value, according to the number of species chosen per family: Moraceae (7); Flacourtiaceae (5); Sapindaceae (4); Sterculiaceae (3); Anacardiaceae (2); Bombacaceae (2); Leguminaceae (2); Myrsinaceae (2); Rubiaceae (2); Annonaceae (1); Araceae (1), Apocynaceae (1); Bignonaceae (1), Boraginaceae (1); Capparidaceae (1); Elaeocarpaceae (1); Guttiferae (1); Malpighiaceae (1); Marantaceae (1); Meliaceae (1); Myrtaceae (1); Phytolaccaceae (1); Piperaceae (1); Rosaceae (1); Sapotaceae (1); Ulmaceae (1); Urticaceae (1); Verbenaceae (1). In the same study the most popular fruits were *Inga punctata* (Leguminosae), *Ficus* sp., *F. palmicida* (Moraceae), *Anacardium excelsum* and *Spondias mombin* (Anacardiaceae) and *Quararibea* sp. (Bombacaceae). SAVAGE (1990) reported exudate feeding from the tree *Anacardium excelsium* and she listed 28 species of plant foods from 19 families.

Foraging takes place in mid-canopy (4.5-13.5 m in Nehman's study) in northern Sucre. The sharing of foods seems common, especially fathers sharing with their young ones and older siblings with younger siblings (FEISTNER & PRICE, 1990). Because of the seasonal scarcity of food, RODRÍGUEZ M. (pers. com.) found that the body weights varied as much as 30% between rainy and dry seasons.

There is evidence that *S. oedipus* practices facultative polyandry as does *S. fuscicollis* and some other members of the genus, especially since more than one male is often found in the small family groups (SAVAGE *et al.*, 1996). However, SAVAGE (1990) found that most groups were made up of monogamous pairs. She found two pregnant females in one group. SAVAGE (1990) found that unpaired females did not cycle, and FRENCH *et al.* (1984) found that females living in their family group did not cycle for up to 40 months. The estrus cycle of a paired female is around 15.5 days, according to PRESLOCK *et al.* (1973), and a birth season has been reported in March-May at the beginning of the rainy season, with another birth season six months later in September-November, although SAVAGE (1990) was unable to confirm a second birth season in northern Sucre and also found births during other times of the year. The gestation period is 125-140 days. PRICE (1992a) showed that the females slow down during the latter part of their pregnancy while the nursing mother eats twice as much food and forages twice as much as before. This indicates that there is strong selection pressure for a communal system of rearing infants.

SAVAGE (1990) observed 4 sets of twins born in their natural habitat. During the first month the newborns are marked differently from the adults and their ventrum is naked. CLEVELAND & SNOWDON (1984) found that from the first week of life the adult male and the infant's brothers carry the newborn more than the mother. SAVAGE (1989a, 1990) observed that initially after birth both parents, extra females and the eldest animals took care of the babies equally with the juveniles and young taking some part later after the first three weeks as the new infants matured. In one group the extra female carried the infant for the first four weeks as much as the mother of the infant carried it, but all other group members were also involved. By the 4-5 week the infant was locomoting beside the care-giver, and by week 9 the infant locomoted 50% of the time independently. At 14 weeks the infants were independent and did not need to be carried. There were no sex differences in carrying the infant in this study. During the first month the infants were carried at all times and little by little the babies began to locomot independently until by 14 weeks they were totally independent. At week 7 the first social play was observed while at week 8 the infants began begging for food from other group members (SAVAGE, 1990). In captive groups the male carries the infant more, and juvenile females begin carrying infants during the second week. The infants are independent from week 9-11.

FRENCH & SNOWDON (1981) reported that intrusions of unknown animals of either sex caused the females of the paired animals to mark substrates with their suprapubic glands and to threaten from a distance. The males were aggressive towards the intruding males and showed little reaction to intruding females. Males did not mark.

HAMPTON *et al.* (1966) describes grooming in a group; this is similar in both sexes and is carried out principally towards the posteriors and tails of the receivers. OMEDES & CARROLL (1980) found that grooming lasted 5-65 seconds. SAVAGE (1990) observed that males were more likely to initiate contacts, huddling and social sniffing. She also found that the level of social interaction was much less in the wild than with captive animals and that females rarely scent-mark in free-ranging groups, even though in captivity they mark 20-30 times per 30 minute period. Groups in the wild have one adult member as a sentry while other group members are occupied. Group members defend their territory using adult and subadult bluff charges at territory boundaries. Subadult males were observed to emigrate between groups.

CLEVELAND & SNOWDON (1982) did an extensive study of the vocalizations of this species, some of which are listed below and ammended from another study by SNOWDON & SOINI (1988): (1). *Type A Chirp* - During mobbing or to preferred food source; wide open mouth, rapid erratic movement toward and away from focal object with piloerection. (2). *Type B Chirp* - To humans or familiar objects with fixed stare and frown. (3). *Type C Chirp* - During approach to food or object to be explored with mouth slightly open and cautious movement. (4). *Type D Chirp* - When possessing food or object in (*Hooked Chatter*) response to approaching begging infants with stationary posture. (5). *Type E Chirp* (*Chirp Chatter*) - As startle response to sudden stimulus with scanning or rapid flight with some piloerection. (6). *Type F Chirp* (*Chirp Trill*) - During intergroup antiphonal calling of Normal Long Calls to audible nongroup vocalizations with slow upward scan in stationary position and minimal piloerection. (7). *Type C Chirps* - In relaxed environment with head tilted at object. (8). *Type H Chirp* (*Chirp Trill*) - To novel stimuli at close proximity while staring with mild piloerection and mild approach. (9). *Squeal, Twitter* - During actual physical contact or wrestle play by passive participant crouched with crown smoothing. (10). *Slicing Scream* - During mobbing to sudden stimuli and to some preferred food sources with wide open mouth, staring, rapid erratic movement toward and away from focal object with piloerection. (11). *Chevron Chatter* - To potential threats of humans at close proximity (capturing) with head flicking, tonguing, piloerection and some evasive locomotion. (12). *Type A Trill* - During antiphonal long call interactions prior to atempted play mounts with tonguing at partner; with slow approach and moderate piloerection. (13). *Type B Trill* - While approaching and retrieving infants and while carrying them when the infants vocalize or try to get off, accompanied by rapid head-tilting and tongue flicking at infant with some piloerection. (14). *Type C Trill* - To signal the end of a nursing bout by the female while pushing at infants head or nipping it. (15). *Squeak* - During vigilance, foraging and investigation with increased locomotion, piloerection and arousal. (16). *Initially Modulated Whistle* - During low arousal with members with semi-open mouth, stationary. (17). *Terminally Modulated Whistle* - During resting contact, grooming, nursing, retrieving and carrying infants and showing semi-open mouth, stationary and relaxed. (18). *Flat Whistle* - While scanning and following type F chirps or while staring following type H chirp trills during moderate to high arousal with piloerection, scanning or staring. (19). *Ascending Descending Multi-Whistles* - In huddles, nursing, retrieving or carrying infants while resting, tail looping, semi-open mouth with some tongue and head flicking if directed at infants. (20). *Quiet Long Call (Partial Quiet Long Call)* - In huddles, nursing to audible calls from unfamiliar animals and with semi-open mouth, slight piloerection and and stationary or slow movement with scanning. (21). *Rapid Whistle Before Antiphonal Call Play* - Approaching to mount in with intense arousal; stationary, staring , pilo-erection, frowning and some tonguing and head flicks. (22). *Normal Long Call* - During isolation to distant non-group; *Combination Long Calls, Normal \ Long Calls, Type F Chirps* with wide open mouth, moderate piloerection, continuous scan and stationary position. (23). *Multi-level Large or Small Modulation* - During low arousal with group members but not when resting, grooming or nursing and with semi-open mouth, stationary, relaxed. (24). *Inverted U+ Whistle* - As alarm to faint acoustical stimuli after an alert has been signaled. (25). *Combination Long Call* - During play, isolation, or when socially disturbed with continuous vocalization, rapid erratic movement, variable piloerection often scanning. (26). *Type F Chirp+ Whistle* - By individuals less confident than when giving *Normal Long Calls*; response to *Combination* or *Normal Long Calls* and in isolation and the same as *Normal Long Calls* with increased movement. (27). *Squawk* - By recipient of aggression as appeasement gesture and given by some as grooming invitation while crouching or with evasive locomotion away from aggressor. (28). *Sneeze* - After eating, drinking, sniffing or after rubbing nose on substrate or with hand. (29). *Scream* - When other attempts to steal food and with rapid head flicks, salivation, rapid evasive action from pursuer while holding food.

Callitrichins possess specialized glands for olfactory communication. These are found on the sternum, the suprapubic and anal region (HERSHKOVITZ,

1977). FRENCH (1982) described some marking episodes of *S. opedipus*. Anogenital marking was more frequent than suparpubic marking. Sternal marking was not observed. Anal marking is observed before and after copulation, and suprapubic marking is observed when unknown animals appear. EPPLE (1978a, 1978b) and EPPLE & CERNY (1979) showed that this marking behavior is under gonadal control in *S. fuscicollis*, and there is no reason to think that the situation would be different for other *Saguinus*, including for *S. oedipus*.

Visual communication doesn't seem to be as important as chemical and vocal communication, principally due to the dense nature of the vegetation in the habitat. *Saguinus oedipus* is known for using various displays which are socially important. *Tongue flicking* is common during grooming and reproduction. *Head down* is used to appease others. *Crouch with crown smoothing* is a reaction of alarm. *Tail looping* is used to maintain physical contact with other members of the group. *Mobbing* is an aggressive display of all members of the group together.

Predation of this species probably occurs via avian predators and perhaps some snakes. MOYNIHAN (1970) reported the predation of an individual by a tayra (*Eira barbara*). NEYMAN (1977) observed agressive behavior of *Cebus capucinus* towards *S. oedipus* which makes this *Cebus* a suspected predator of *S. oedipus*.

Conservation status

Saguinus oedipus is considered by the Colombian committee to classify conservation status of the country´s mammals to be VU, due to a recent reanalysis of its status (DEFLER & RODRÍGUEZ-M., unpublished manuscript), although this downgrade has not been as yet officially accepted by the international chair. The CITES convention places this specie in Appendix I. The reassessment is as follows:

The conservation status of *Saguinus oedipus* and its analysis according to IUCN criteria has been changing since the time of uncontrolled harvesting of the species and its first natural history studies by NEYMAN (1977, 1978). HERNÁNDEZ-CAMACHO & COOPER (1976) and COOPER & HERNÁNDEZ-CAMACHO (1975) have described how during the 1960's and early 1970's the species was the object of exaggerated exploitation, estimating that the unregulated numbers easily reached 20,000 – 30,000 individuals. In 1970 over 2,000 animals were exported and it is thought that at least that number was probably lost due to deaths and lack of data (COOPER & HERNÁNDEZ-COOPER, 1975).

The establishment of Inderena (Instituto Nacional de los Recursos Naturales Renovables and the Environment) in 1969, marks the beginning of some control over this illegal business, due to the establishment of appropriate legal controls which sought to regulate exportation via new regulatory laws such as *Resolución 574* (July 24, 1969) permanently forbidding the hunting of *Saguinus oedipus* due to its endangered status and *Resolución 392* (April 18, 1973) prohibiting the capture and commercialization of all nonhuman primates in Colombia. These laws were sustained and amplified by the establishment of the *Código Nacional de los Recursos Naturales Renovables y del Ambiente* (*decreto Ley 2811* of 1974) and the established regulatory legislation for colombian fauna (*decreto Ley 1608* of 1978)

There is no doubt that the impact of this illegal and uncontrolled traffic, driven by high international prices, was a very significant pressure on natural populations, since the number of captured animals was much greater if one takes into account that usually the methods of capture and transportation generated many dead animals, which added to the ongoing habitat deterioration that was resulting in severe fragmentation of habitat, resulted in the first Colombian species to be considered threatened by extinction by the International Union for the Conservation of Nature (IUCN, 1977) and by Inderena. It is important to recognize that even though the illegal commerce in this primate species in Colombia has never completely disappeared, the birth of an environmental institution and the formulation of legal regulations permitted the beginnings of control and the implementation of processes of conservation of threatened species, which on the international scene was successfull due to Colombia's signing of the CITES treaty, ratified as law via the *Ley 17* of 1981, resulting in the almost complete cessation of illegal exportation of this primate to other countries.

At various levels these activities have generated a national consciousness about the importance of this primate species as a national emblem, given that it is a Colombian endemic from north-western Colombia and it is threatened. The entire conscious awakening led to the establishment of a small 1000 ha reserve emblematic of the species, El Santuario de Fauna y

Flora de los Colorados, close to the small city of San Juan Nepomuceno in the department of Bolívar. Later the Centro de Investigaciones Primatológicas de Colosó, close to the slopes of the Serranía de Montes de María in the department of Sucre was developed, in recognition of the species' presence in the area and the first studies that had been accomplished nearby (NEYMAN, 1977, 1978). Further studies of the species were begun parallel to this, education campaigns were initiated nationally by Inderena, some of which have continued as private initiatives. The best example of this is *Proyecto Tití* initiated by A. Savage and Colombian collaborators such as H. GIRALDO (SAVAGE, 1989b, 1995a,b, 1996; SAVAGE & GIRALDO, 1990). This *Proyecto Tití* program continues today with great success from its present headquarters at El Ceibal, Municipio de Santa Catalina, departament of Bolívar, close to Cartagena (SAVAGE *et al.* 1986, 1987, 1988, 1990c, 1996a, *etc.*).

We believe that *Saguinus oedipus* is an example of a successful conservation effort. In this context CI-Colombia in agreement with the Ministerio del Medio Ambiente, under the leadership of the Institute von Humboldt, a semi-autonomous corporation and another five institutions, with the support of the scientific community, organized the study and analysis the Colombian biota in order to make recommendations for its management and conservation. In the case of mammals, the national committee for the categorization of Colombian mammalian species, was organized and directed by J. V. RODRÍGUEZ-M, and this particular primate species´ case was widely and extensively debated; this group of specialists had wanted to objectively and justly focus the few existing resources on the conservation of species and groups of species that require immediate attention, considering the growing number of threatened species in the country. Therefore, we analyzed the national conservation status of all Colombian species, especially in light of new IUCN criteria, the object being to identify taxa which might more correctly belong to a different higher or lower category (IUCN, 2000). In the case of *S. oedipus*, the conservation status no longer seems much different from that of *S. leucopus,* another Colombian endemic that has been considered VU since its inclusion in the IUCN list. It seems to us that the efforts of past years, including all actions taken, have had a positive impact on the future survival of *S. oedipus* and, analyzing the evidence and instrinsic characteristics of the species convinces us that *S. oedipus* should be relegated to the category VU.

Considering the past 40 years of decade after decade of changing conservation status for *S. oedipus,* we see that during the first two decades the population decline was very dramatic and population levels became quite low, parallel with extensive habitat fragmentation, particularly due to cattle ranching, the preponderant economic activity of the region. Nevertheless, it is important to realize that in the last two decades the situation began changing with a diminishment of animal traffic, even though ironically the few confiscated individuals began arriving at zoos, especially in Cali and Baranquilla (14 animals in 1999, SAVAGE com. pers.). These zoo arrivals indicated that there was still a moderate illegal traffic in the species, and also it suggested that such traffic would need to depend on some important natural populations, which permit such traffic. Also, we can conclude that the greater part of the specimens arriving at these zoos were taken there by citizens who are increasingly rejecting such illegal traffic in fauna.

We would like to underline that many of the statements made in the analysis are derived from unpublished observations of the second author of this position paper since 1974, when the biology of this species was first becoming known by JVRM. First, via the collection of a specimen of *S. oedipus* from the left bank of the Río Atrato within the Los Katios National Park in the area of the Tumaradó swamps, José Vicente Rodríguez Mahecha made an addition to the known geographic distribution for the species. Other observations were made by him from the south or right bank of the Rio Atrato, as well as observations welling from a project organized for an evaluation of the state of *Saguinus oedipus* in 1987, directed by José Vicente Rodríguez Mahecha while he worked as Head of Fauna at Inderena during more than ten years, plus other field experience with the species and observations while working in northern Colombia in the field with the Colombian national parks section of Inderena. JVR's experience with the species was considerably expanded as a result of his participation in a project for the evaluation of colon cancer in this primate species, which necessitated visits to the field in Bolívar department during about 10 additional years. This and other elements motivated the preliminary discussions that have permitted us to conclude that the species should be categorized as VU at this time, according to the criteria of the IUCN (2000). But first, we would like to list several specific points which have bearing on the conservation status of *Saguinus oedipus.*

The species has been legally protected in Colombia since 1969, but in practice few large-scale efforts have been taken. Population numbers are unkown, but the majority of forests where it was formerly distributed have been cut, and remnant populations are small. Between 1960-1975 probably about 30,000-40,000 animals were exported to the United States, not to mention the numbers to other countries. All exportation from Colombia has been illegal since 1974, but some illegal exportation continues. One technique for exporting the animals has been to send the animals to Panama in order to acquire exportation permits there, a process that was particularly easy when General Omar Torrijos was in power. But traffic continues at a lower level. Some Colombian cargo airlines still export fauna illegally to Panama, (and perhaps to Miami) since there is often little control when air cargo flights leave the country.

Other threats are the capture of this primate for the Colombian pet trade, as well as continuing habitat loss and hunting (RAMÍREZ, 1985). *Saguinus oedipus* is still often found maintained in cages in households in northern Colombia with the concommitant loss to the small reproductive population. Since there is so little habitat for the species', future losses to the breeding population may be critical. Conservation units have been established in the area of the geographic distribution of this species, such as Paramillo National Park, the Sanctuary of Los Colorados, and the Reserva Forestal Cerro de Coraza-Monte de Marja, but more are necessary and an increase in physical protection in these units is absolutely necessary (see, for example, FAJARDO P. & DE LA OSSA, 1994).

Ann Savage and her collaborators developed class-room lectures on the forest and wildlife of the region of Colosó with the aim of teaching conservation of these resources, especially of *S. oedipus*. This group developed special teaching techniques which seemed to be rather successful for the region. Also, they produced videotapes of their project, using video footage of wild animals.

MAST *et al.* (1993) reviewed all aspects of the conservation of *S. oedipus*, including suggestions for amplifying such efforts. Among the suggestions made were the following: (1) a list of key sites which must be preserved; (2) an increment of educational efforts in the conservation of this species; (3) an increase in public awareness; (4) increase efforts for legal regulation; (5) increase efforts for allowing regeneration of forests; (6) make more efforts for transplanting the species into other appropriate habitats; (7) make efforts for developing a type of ecotourism which generate earnings from the viewing of wildlife.

The range of efforts listed above for conserving this valuable part of Colombian fauna could provide graduate thesis projects for many Colombian or non-Colombian students who would like to help in this important effort.

Where to see it

One of the best places to observe this endangered species is around the primate research center near Colosó where Ann Savage did her observations. These animals are accessible and somewhat protected. If small populations of this monkey are found at other sites by the reader, the author would like to know the full details of the location, size of the population and other particulars. It is urgent that all extant populations be protected for the future.

Capuchins

FAMILY **CEBIDAE**
Genus **Cebus** ERXLEBEN, 1977

Taxonomic comments

Of the seven species currently recognized by many primatologists and based mostly on the HERSHKOVITZ (1949) review, three are found in Colombia: *Cebus apella, Cebus albifrons* and *Cebus capucinus.* Comparative molecular analyses provide strong evidence for a monophyletic origin of this genus with *Saguinus, Callimico, Cebuella, Mico* and *Saimiri* so that all of these may be placed in the family Cebidae with subfamilies Cebinae, Callitrichinae and Saimirinae. SILVA (2001, 2002) recently studied the taxonomy of *Cebus* and divided them into the untufted *Cebus* with four species and subgenus *Cebus* (*Cebus capucinus, Cebus albifrons, Cebus olivaceus* and *Cebus kaapori*) and the *Sapajous* subgenus including seven species *Cebus (Sapajou) apella, Cebus (Sapajou) macrocephalus, Cebus (Sapajou) libidinosus, Cebus (Sapajou) cay, Cebus (Sapajou) xanthosternos, Cebus (Sapajou) robustus* and *Cebus (Sapajou) nigrittus*. On the level of Colombia the species identities remain the same irrespective of what taxonomic scheme is adopted for the entire genus. Nevertheless, there is good evidence for distinguishing *Cebus apella* from the "untufted" *Cebus* in Colombia.

Common names

Colombia: *maicero*; (Spanish generally) *capuchino*; Quechua *machín*; Brazil *macaco*; German *Rollaffe, Rossschwanzaffe, Wickelaffe*; French *sajou, sapajou*; Italian *scimmie cappucine*; Swedish *schmauz, capucin-apar*.

Indian languages

Sai, cai, caitaia, caiarara (Tupi-Guaraní) were apparently general names in the Amazon region.

Etymology

According to HILL (1960:322), *Cebus* comes from a transliteration of the Greek word which Aristotle used to refer to certain primates with long tails, but particularly *Cercopithecus*. ERXLEBEN (1777) first used the word κήβοι, to refer to neotropical primates, later written κήπος, transliterated *cephus* and applied to *Cercopithecus*.

The species of this genus have homogeneous morphological characteristics and behavior. They usually weight about 2-4 kg, they have prehensile tails with lengths in relation to their body length somewhat less than that of the atelids (*Ateles, Lagothrix, Brachyteles* and *Alouatta*), although their tails are furred ventrally rather than having a naked callosity like the atelids. Generally their body morphology is similar to that of Old World primates. The texture and color of fur varies widely, according to the species. The *Cebus* are omnivorous and are very intelligent and very inventive and adaptive while

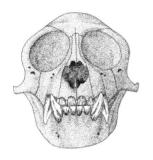

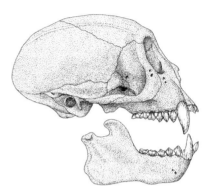

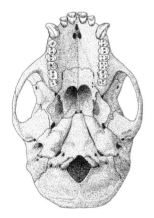

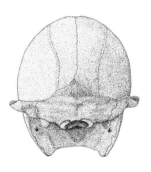

Figure 15. View of skull of *Cebus albifrons*.

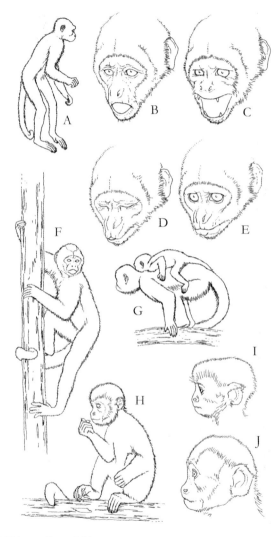

Figure 16. Various positions of *Cebus*. a) Bipedal position. b-e) Various facial gestures. f) Climbing. g) Carrying infant. h) Feeding. i) Infant´s face. j) Adult´s face.

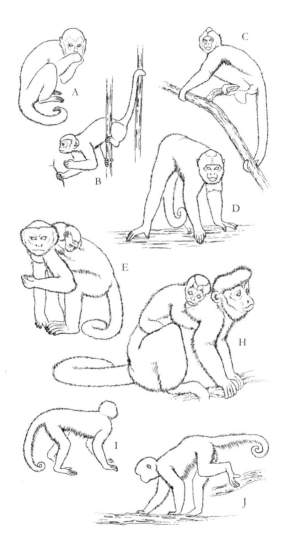

Figure 17. Various positions of *Cebus*. a) Submission. b-c) Uses of the tail. d) Threat. e) Carrying newborn infant. f) Mother carrying older infant. g) Stand. h) Quadrupedal locomotion.

foraging on different substrates and manipulating different food resources. They move about walking and running quadrupedally and frequently execute jumps between trees, although they commonly forage and move about on the ground as well. They live in social groups made up of various adult males and females with their young.

This genus has the widest geographic distribution of any neotropical primate with the exception of *Alouatta*. To the north they first appear in Honduras, extending throughout Central America and most of South America to northern Argentina, including the Pacific and Caribbean coasts and inter-Andean valleys up to 2500 m altitude as well as some islands, including Gorgona (Colombia), Margarita (Venezuela) and Trinidad.

White-fronted Capuchin

Cebus albifrons (HUMBOLDT, 1812)

Plates 5, 6, 15 • Map 10

Taxonomic comments

This species has had problems with its name, description and type locality. The holotype does not exist and the original description of V. HUMBOLDT (1812) describes an animal that is much darker (grayish) than those that exist close to the type locality, and the description includes a dark tail tip, a character that is completely unknown in any population of the species. Additionally, the animal which v. Humboldt examined was a tame animal in Maipures, where the species is not found. The closest population is about three kilometers to the north, on the other side of the Tuparro river (DEFLER & HERNÁNDEZ-C., 2002).

DEFLER & HERNÁNDEZ-C. (2002) established a neotype from the population that was called *Cebus albifrons albifrons* by HERNÁNDEZ C. & COOPER (1976) and the type locality was fixed by these authors as Maipures. Another problem has been that the taxon *C. a. unicolor* described by SPIX (1823) and further defined by HERSHKOVITZ (1949) was indistinguishable from *C. a. albifrons*; the two are synonymous (HERNÁNDEZ-C. & COOPER, 1976; DEFLER & HERNÁNDEZ-C., 2002). HERNÁNDEZ C. & COOPER (1976) describe eight subspecies for Colombia (including *C. a. albifrons* and *C. a. unicolor*), but it seems likely there are five, since *C.a.versicolor* includes *C. a. pleei* and *C. a. leococephalus* and *C. a. unicolor* is synonymous with *C. a. albifrons*. *C. a. yuracus* has been synonymized with *C. a. cuscinus* by GROVES (2001).

Capuchins

Common names

Colombia: *Mico, macaco, mono blanco* and *caraira* in the vecinity of Leticia; *mico tanque* or *tangue* in the watershed of the Caquetá River; *maicero, maicero cariblanco* and *mico cariblanco* in regions outside of the Amazon; *mono blanco* along the Orinoco river (origin Venezuela); *carita blanca* and *mico bayo* in the Departments of Magdalena, César and the southeast of Bolívar; *machín* in the Valle of the middle Magdalena. French: *sajou pieds dorés, sajou front blanc*. German: *Schrabrackenfaun* (VON PUSCH), *Weisstirn-kapuziner*. Brazilian portuguese: *caiarara branco*.

Indian languages

(Pronounced as in Spanish): *Takke* (Andoke); *tawachí* (Carijona); *jalio, jario* (Curripaco, *fide* Amazon 1992); *huaja* (Cubeo); *vánali* o *papábë* (Guahibo); *tiyo* (Huitoto); *tanke* (Ingano); *moi* (Macú, *fide* Capitán Fransico Macú of the lower Apaporis River); *gaqué-josé* (Macuna); *ouvapavi* (Maipures, *fide* v. Humboldt); *jumopa* (Miraña, *fide* Napoleón Miraña, 1985); *jímuhai* (Muinane); *uuri* (Okaima); *jichú* o *ji'ichu* (Piaroa); *yurac-machín, yana-machín* (Quechua); *bo také* (Siona); *tanqueboiya* (Tanimuca); *tou* (Ticuna); *seupeupunimi* (Tucano); *sowarama?* (Tunebo); *poiñ, poi* (Yucuna); *shirdë* (Yukpa).

Identification

The males of this species usually weigh an average of 3415 kg (n=8) and the females an average of 2864 kg (n=3), although a male on the Mirití-Paraná weighed 5.5 kg (DEFLER, 1983b). This primate is usually the color of maroon-white or palomino and creamy white. Below are descriptions of the known subspecies for Colombia.

(1) *Cebus albifrons malitiosus* (ELLIOT, 1909) is characterized by a color that is rather dark brown over almost the entire body with yellowish shoulders. "Pale area of front less extensive, upperparts and limbs paler than in *hypoleucus*. Cap Prout's Brown, median dorsal region Cinnamon Brown, forearm and foreleg not markedly contrasting in color with back and sides of body; hairs of belly and chest Ochraceous-Tawny to Cinnamon-Brown and silvery; contrasting pale area of front extending well over upper surface of shoulder and inner side of upper arm" (HERSHKOVITZ, 1949).

(2) *Cebus a. cesarae* (HERSHKOVITZ, 1949) is very light in color and quite well-defined as a subspecies. "The cap is Cinnamon or Snuff-Brown; median

dorsal region, forearm and foreleg with orange tones and contrasted with sides of back and trunk; hairs of belly and chest ochraceous-orange to pale ochraceous-buff and silvery; contrasting pale area of front extending over variable amounts of upper surface of shoulder and inner side of upper arm" (HERSHKOVITZ, 1949).

(3) *Cebus a. versicolor* (PUCHERAN, 1845) is a complex which includes dark populations and lighter populations and probably includes the subspecies *C. a. leucocephalus* (GRAY, 1865) and *C. a. pleei* (HERSHKOVITZ, 1949). HERSKOVITZ' (1949) description of *C. a. pleei* is of a very reddish animal, particularly in its limbs and *C. a. versicolor* is an animal similar to *C. a. pleei* with a lighter red. *C. a. leucocephalus* was described as a dark brown animal with reddish tonalities in the hind legs. Nevertheless, HERNÁNDEZ C. & COOPER (1976) discuss evidence that the three subspecies: *C. a. leucocephalus, C. a. pleei* and *C. a. versicolor* could be subsumed into one subspecies *C. a. versicolor*, since the variations (*C. a. pleei* and *C. a. leucocephalus* along with *C. a. versicolor*) seem to be found in a very well-defined zone and even in the same groups, close to Barrancabermeja on the eastern bank of the middle Magdalena river in the Department of Santander. This strongly suggests that the dark phase (*C. a. leucocephalus*) and the light phase (*C. a. pleei*) are extremes of an intermediate (*C. a. versicolor*). This possibility needs to be evaluated, comparing individuals of various critical areas, particularly the western bank of the middle Magdalena and the Colombian watershed of Lake Maracaibo.

One population of very pallid coloration is found in Arauca, the northern part of Boyacá and the eastern part of Norte de Santander and probably represents *C. a. albifrons*.

(4) *Cebus a. albifrons* is found in eastern Vichada, close to the type locality and was defined by v. HUMBOLDT (1812) using an animal maintained by humans (and a pig) in the village of Maipures. The original description of v. HUMBOLDT (1812) described an ashy gray animal with a black tail tip, characteristics that are not typical of any known population of *C. albifrons*. The *C. albifrons* located three kilometers to the north of Maipures are very light colored animals with yellowish or reddish tones, very similar to the population of Arauca. A description of the recently established neotype is

practically the same as *Cebus a. unicolor* (SPIX, 1823) white-fronted capuchin which is also very light colored with yellowish tones and it seems clear that it is a synonym of *C. a. albifrons* (DEFLER & HERNÁNDEZ-C., 2002).

(5) *Cebus a. yuracus* (HERSHKOVITZ, 1949) (GROVES, 2001, has synonymized this taxon with *cuscinus*) is another subspecies found south of the Guamués river and colored a light brown.

Several years ago some specimens of *Cebus* were preserved from the Baranquilla market, which supposedly had come from the middle valley of the San Jorge River. It is difficult to determine whether these are *C. capucinus* or *C. albifrons*. Intermediate characteristics include a dark crown that is high and removed from the forehead. Also, the white parts in the face are more distinctively bald and the outside parts of the arms and legs are lighter; all of these characters suggest *C. capucinus*. The body is darker and more uniform than in individuals of *C. a. malitosus*. Also, some *C. a. versicolor* (*pleei* type) seen in the market at Magangu, and probably captured in the lower Cauca river, also show similar tendencies to the above, except that there are is no increase in the dark pigmentation. Based on these observations and on various intermediate specimens colored dark brown of *C. capucinus* from northern Colombia, it seems possible that an investigation of the contact zone between *C. capucinus* and *C. albifrons* ultimately "could show that these forms are conspecific" (HERNÁNDEZ-C. & COOPER, 1976). Another critical zone for this analysis is an area in northeast Ecuador where *C. a. ecuatoriales* (ALLEN, 1914) and *C. capucinus* are found, although neither sympatric distributions or intergradation have as yet been determined.

Geographic range

In Colombia *Cebus albifrons* is found from the northern slopes of the Sierra de Santa Marta to the south, in the valley of the Magdalena River to an as yet undefined point in the Department of Tolima and in the valley of the lower Cauca River, to the eastern parts of central Antioquia and the southern parts of Sucre to the west. In Guajira the species is found to Riohacha and an isolated population is apparently found in the Serranía de Macuira, though this needs confirmation. The species is also found along the slopes of the Serranía de Perijá and the Cordillera Oriental. To the east of the Cordillera the species is found in Norte de Santander, western Arauca, in eastern

Vichada between the Meta and Tuparro Rivers, and then south of the Vichada River, although east of the Ariari River, not including the Ariari itself. It is not known whether the species is found in the rather extensive forests of the upper Manacacías River in Meta. South of the Guayabero and Guaviare River *C. albifrons* is found throughout the Amazon. The species reaches an altitude of 1500-2000 m in the Department of Tolima.

Outside of Colombia *C. albifrons* is found from the Cordillera de los Andes throughout eastern Ecuador, Perú and northern Bolivia to the Tapajós River in Brazil, south of the Amazon River. North of the Amazon River the species is found in the southern parts of the Venezuelan Federal State of Amazonas and in northern Brazil between Colombia and the Branco River.

Natural history

Cebus albifrons is found in a variety of forest types. DEFLER (1985a) showed that the species in Vichada exploits a more xeric habitat in terms of drainage, in comparison with *C. apella*, which tends to be found in forests that are more mesophytic. *Cebus albifrons* often is found in flooded forests in contrast to *Cebus apella*. *Cebus albifrons* survives well in forests growing over white sand and in forests of *high caatinga* growing in the rocks and gravel at the foot of mesas.

This species has been studied in Colombia by DEFLER (1979a, 1979b, 1980, 1982, 1985a) and in two different sites in Perú by SOINI (1986a), and TERBORGH (1983) and in Trinidad by PHILLIPS (1998a, 1998b, 1999; PHILLIPS & NEWLON (2000, 2001); PHILLIPS & SHAUVER (2001) and PHILLIPS & ABERCROMBIE (2003).

In eastern Vichada these monkeys are found in large groups of around 35 individuals, while to the south in closed forest (perhaps as a result of competition with *C. apella*) they have average group sizes of around 8-15 individuals. A group in Vichada used a home range of about 120 hectares (DEFLER, 1979a), while TERBORGH (1983) found a home range of more than 150 ha. Near the type locality in gallery forest and islands of forest in Vichada, the species has an ecological density of around 30 individuals/ km^2 (DEFLER, 1979a). In forests with closed-canopy in Colombia and in southern Vichada, many areas have very low densities. Around the lower

Apaporis river (Vaupés), for example, densities are less than one individual/km² and the size of the groups is around 15 individuals. Calculated densities at four sites on the Purité River in the Colombian trapezium, where *Cebus apella* is not found, varied from 4.4 – 15.8 individuals/km², whereas on the Puré River where *Cebus apella* is sympatric the calculated densities at one site were about 1 individual/km² (DEFLER, in prep.). Low densities found in many parts of the Colombian Amazon make it difficult to detect the presence of the species.

TERBORGH (1983) calculated an average of 1800 m for the day range of a group and he calculated the following time budget of the study group in Manu National Park, Perú (18% rest, 21% travel, 22% feeding on plant material and 39% feeding on insects; total feeding 61%). These primates are primarily quadrupeds, although they utilize a great variety of gallops, jumps, falls and climbing. During certain times of the year *C. albifrons* is extremely terrestrial, especially when there is a scarcity of available fruits and the troop must search for arthropods in the dry leaves of the forest floor. In some parts of the Llanos Orientales where this species is found, these primates often walk over the grassy savanna between forests, leaving well-beaten trails. *Cebus albifrons* in Vichada use preferential trees for sleeping at heights of 25-30 m. The palm *Attalea regia* is often used for sleeping in this zone.

All species of *Cebus* tend to have a rather similar diet in broad terms; they are omnivores, eating fruits and small invertebrates, small vertebrates and birds´ eggs, which they forage at all levels of the forest, frequently descending to the forest floor. In northern Colombia during the dry season when there are few fruits to be found, *C. albifrons* spends more than half their time on the ground, searching for and capturing small prey. These animals are extremely good at manipulating objects and spend a great deal of time examining dry leaves from which they collect invertebrates (for example small beetles and ants eggs) from rolled up leaves. These primates hunt frogs and drink the water which accumulates in the spaces between the bracteoles of the common plant *Phenakospermum guianense*, where frogs hide. Hunting amphibians seems to be a cultural phenomenon which the members of each group learn. *P. guianense* is commonly present in large, dense stands in some types of forest.

In Manu National Park the animal material in the diet includes frogs, lizards, small mammals and birds eggs as well as many invertebrates, including orthopterans, lepidopterans and hymenopterans (especially ants and wasp larvae). In the Pacaya-Samiri National Park SOINI (1983) saw these monkeys eating tent caterpillars. TERBORGH (1983) identified 73 species of plants from 33 families consumed by this primate. The Moraceae was the most important family by a wide margin, counting the number of species (17) eaten, equivalent to 23.3% of all plant species consumed. Importance values for plant families consumed by *Cebus albifrons* in one study are as follow (TERBORGH, 1983): Moraceae (17. 23.3%); Leguminosae (5. 6.8%); Araceae (4. 5.5%); Bombaceae (4. 5.5%); Palmae (4. 5.5%).

DEFLER (1979a) collected 40 species of plants from 23 families eaten by this primate in Vichada with the following importance values, according to species consumed per family: Arecaceae (7); Moraceae (6); Chrysobalanaceae (3); Leguminosae (3); Passifloraceae (2); Bromeliaceae (2); Burseraceae (2); Bombaceae (1); Celastraceae (1); Connaraceae (1); Euphorbiaceae (1); Lecythidaceae (1); Maranthaceae (1); Melastomataceae (1); Anacardiaceae (1); Myrtaceae (1); Annonaceae (1); Musaceae (1); Apocynaceae (1); Orchidaceae (1); Araceae (1); Rubiaceae (1); Bignoniaceae (1).

In terms of importance value, palms are highly valued by all species of *Cebus*. In El Tuparro National Park the lovely and rather common palm *Maximiliana regiae* was a key species for this primate, the nuts being a principle food (DEFLER, 1979a). In Manu National Park in Peru the palms *Astrocaryum* and *Scheelea* were the most important palm genera, but perhaps not at the same level as *Maximiliana* in El Tuparro. Also, at Manu various species of *Ficus* were very important to *C. albifrons*; this emphasis on *Ficus* was not observed in the El Tuparro study, although this study did not include an entire year. Nevertheless, research on other species suggests the importance of palms as key species and the lack of importance of *Ficus* in such habitats as gallery forests in the llanos of Colombia and Venezuela, contrasting with their high importance in more fertile habitats like Manu.

Cebus albifrons takes advantage of almost any water source, drinking water from tree holes when available, but also drinking from brooks and springs when necessary. During the driest season in Vichada the group studied by DEFLER (1979a, 1979b) went to the ground every day to a water seep from under a huge boulder, which was the only water source available in their home range.

This species of *Cebus* is polygamous. The male mounts the female, holding her legs with his hind feet and copulates with her for a few minutes. Although the time of gestation is unknown, it is probably around 160 days like *C. apella*. Usually one infant is born. Observations of a newborn in Tuparro National Park showed the process by which the newborn discovered the appropriate position for riding on the mother. Newborns ride oriented sideways over the mothers' shoulders, but during the first days the baby holds on to any part of the mother such as the base of the tail, the tail, the legs, and the arms before discovering and learning that the position over the shoulders is best and most secure. After several weeks the baby makes the change from the sideways position over the shoulders to riding on her back.

All the members of the troop are interested in the newborn, and they take advantage of any opportunity to examine and look at its genitals if the mother permits it. With time the baby begins to climb up on other members of the troop, including the adult males who are interested in protecting the little one. Playing behavior is principally with one companion, and all members of the troop from the alpha male, the mother and all young members of the group solicit play with the young one.

Adult males are notably tolerant of each other in the group, but they are very aggressive towards males of other groups. DEFLER (1979b) described some intergroup aggressive behaviors in El Tuparro, which resulted in one group fleeing towards the central parts of their territory. Alpha males seem to exercise a control position at the center of the group, since all members are extremely conscious and alert to his location, and they all observe his reactions. If the alpha reacts with intense fear or panic or if he pays close attention to something, all members of the troop react similarly. The presence of adult males seems to lend psychological support to the smaller adult females. DEFLER (1979b) noticed that more timid females often became quite aggressive towards him when a male appeared on the scene, although the females often needed to press up against the flank of the male for reassurance.

Vocalizations are variable and some are listed as follow (DEFLER, 1979b): (1) *ua* - a soft bark given repeatedly and used by all members of the group when danger is perceived; (2) *ya* - excited animals around the alpha, towards alpha and towards perceived danger; (3) *eh-eh* - threat towards potential danger, but especially adult females; accompanied with open mouth showing teeth (OMT); (4) *sqeaky hinge* - threat given especially by young animals; (5)

squeal - conflict within the group during a fight; (6) *whistle* - conflict in the group of a young animal; (7) *ahr* - a lost animal; others answer this call, apparently to direct it back to the group; (8) *uh!uh!uh!* - a common vocalization during feeding which may allow contact to be maintained and show general contment; (9) *uch!uch!* - an animal trying to keep up with the group; (10) *warble* - young animals establishing contact or coming close to an adult; (11) *purr* - close and pacific contact; (12) *chirriar* -pacific interaction of young ones during play.

Perhaps the most important display is the behavior of breaking branches, which all members of the group effect. Even infants break small branches (or twigs), letting them fall to the soil, but the most spectacular is the alpha male who chooses large, dry branches which he hits with his hands and feet in spectacular jumps, so that they fall. Usually such branches make a tremendous sound as they fall through the other branches, and the members of the group become very excited and chatter loudly. This behavior is quite commonly discharged towards an observer when the animals have lost their fear. One surmises that this behavior might drive a ground predator away from the immediate area.

These primates frequently travel with *Saimiri sciureus* and at other times with *Cebus apella* and *Alouatta seniculus*. The small hawk *Harpagus bidentatus* often accompanies these monkeys, exactly as it does other species of primates. *Cebus albifrons* feel threatened by avian predators, and they are very vigilant around any large bird of prey. In Vichada, DEFLER (1979, 1980) observed *Eira barbara*, *Boa constrictor* and the raptor *Spizaetus ornatus* trying to capture the monkeys, and these primates always were cautious around large birds. After detecting the *Eira barbara* and *Boa constrictor* the members of the troop showed little fear or caution, even though these animals had been trying to capture the monkeys. The most common behavior after detecting a potential ground predator is *ya-ya* vocalization and branch breaking over the head of the potential predator, similar to the display of the primate *Lagothrix lagothricha*. In contrast, after being frightened by a male *Spizaetus ornatus* the monkeys screamed intensely only once, then hid quietly, some descending to the ground to sneak away.

Conservation status

Cebus albifrons is adaptable, has a wide distribution and probably is not yet endangered in Colombia on the species level, which is LC. Nevertheless, it

Capuchins

is probable that some subspecies are under considerable pressure. For example, *C. a. cuscinus* is found only is a small part of the SW Colombian Amazon and is classified NT for the country (DEFLER, 1996a). We need to census the various subspecies and to clarify the taxonomy in order to evaluate properly the situation within the country. This species is found within 10-15 National Parks and it is probably not excessively hunted (Defler, 1994b). Also, it survives well in secondary vegetation close to human beings. *C. a. cesarae, malitiosus* and *versicolor* are classified NT in Colombia.

Where to see it

Cebus albifrons albifrons is very common in the eastern half of El Tuparro National Park. It is less common but still not too hard to find in Amacayacu National Park. *Cebus a. yuracus* or *cucinus* is known south of the Putumayo River, but this author does not know a particularly good place to see it. *Cebus a. versicolor* is widespread on the middle-Magdalena river and is observable in preserved woodlots of protected fincas. *Cebus a. malitiosus* is easy to observe in Tayrona National Park, east of Santa Marta. *Cebus a. cesarae* can be located in the Sierra de Perijá, east of Valledupar, César, although the author has not observed this subspecies in its natural habitat.

Tufted Capuchin

Cebus apella (LINNAEUS, 1758)

Plate 5 • Map 1

Taxonomic comments

If the HILL (1960) concept of this species were followed, it would be necessary to accept a number of sympatric subspecies of *Cebus apella* whose populations have not even been identified as being in Colombia. Phenotypical variations related to sex and age contribute to the problem of identifying subspecies. It seems parsimonious to recognize only one subspecies, *C. a. apella*, for Colombia and for northern South America, until someone completes a modern revision (HERNÁNDEZ-C. & COOPER, 1976).

Common names

Other English: Brown Capuchin. Colombia: *maicero* or *mico maicero; mico cornudo* or *maicero cornudo, cachón, cachudo* or *maicero cachón* (because the adults and especially females and young males have a characteristic tuft on the head); *macaco* or *macaco prego* (Leticia but of Brazilian origin); *mico* or *mono negro* (near the Orinoco River in the Department of Vichada); *maicero negro* in PutumayoBrazil: *macaco prego;* German: *Faun, Faunaffe, brauner Rollaffe;* French: *sajou brun;* Dutch: *capucijnaap;* Swedish: *schmaus;* Suriname: *kesi-kesi.*

Indian languages

Meeku (Carijona); *pove* (Curripaco, *fide* AMAZON 1992); *tangue* or *taque* (Cubeo); *pababë* (Guahibo); *jóma* (Huitoto); *comendero* (Ingano from the Aguas Claras-Cusumbo community s. Caquetá, *fide* J. V. RODRÍGUEZ-M., 1993); *to-otsípa* (Miraña, *fide* NAPOLEÓN MIRAÑA at Mariamanteca); *wai* (Macu-Yujup of the lower Río Apaporis, *fide* CAPITÁN FRANCISCO MACÚ, 1992); *gak, gaqu* (Macuna); *chullili* (Muinane); *wua p* (Nukak Macú, *fide* POLITIS & RODRÍGUEZ, 1994); *sooriiyo* (Ocaima); *tsic* (Piaroa); *bu* or *bup* (Puinave); *tankiya* (Tanimuca); *taicure* (Ticuna); *akenisami* (Tucano), although *aggu* (*fide* Olalla for the Tucano Indians from specimen tickets at UNIFEM; *sibroa* (Tunebo); *calapichi* (Yukuna); *sulihry* (Yuri) (pronounce as in Spanish).

Identification

This medium-sized primate with prehensile tail weighs around 2.5 – 3.5 kg, the males (3.7 kg, n=9) being larger than the females (2.3 kg, n=5) (HERNÁNDEZ-C. & DEFLER, 1989). The head-body length is 350-490 mm and the tail is about 380-490 mm. The generally light yellowish-brown, light brown to reddish-brown and sometime duller brown body has dark-brown to blackish legs, arms and tail. Usually the animals have a dark cap composed of erect hairs, which sometimes form ridges on each side of the crown. Some adult males neither have an erect cap, nor do they have much hair on their black-skinned face and head. Such hairless-faced males, throw their massive masseter muscles into such relief that their aspect is actually fierce and even diabolic. The males of this species develop a sagittal crest in contrast to the other species of *Cebus*. Color variants and subspecies ought to be revised.

Geographic range

In Colombia *C. apella* is found in the entire Colombian Amazon and in all of the eastern lowlands from the piedmont forest (at least to 1300 m above sea level) east of the cordilleras, with the exception of some eastern areas such as eastern Vichada (DEFLER, 1985a) and the upper Cahuinarí River where the species is known to be absent and replaced by *C. albifrons* and much of the Colombian trapezium where the species also seems to be mostly absent. The species is also found in the upper valley of the Magdalena River in Huila up to 2700 m (in the region of San Agustín) and in the region of Tierradentro in Cauca up to altitudes of about 2500 m (close to Inzá). A specimen marked "Tolima Mountains", collected in 1900 by White, was probably collected in Huila (e.g. specimens of *Lagothrix lagothricha lugens* collected by White and marked "Tolima" included the designation 2°20´N, which would place them in the Department of Huila of today. It is possible that *C. apella* and *C. albifrons* could have marginal sympatric contact or parapatric contact in the Department of Tolima. In Vichada *C. apella* shows parapatric distribution with *C. albifrons albifrons*.

Cebus apella has a larger geographic range than any other neotropical species of primate. Besides Colombia, it is found in the Venezuelan Amazon and Guyanas to the Amazon River. Also, it is found south of the Amazon River in eastern Ecuador, eastern Peru and Bolivia, northern Paraguay and Argentina (GIL & HEINONEN, 1933) and almost all the rest of Brazil.

The species is found throughout a very broad range of habitats from deciduous gallery forest of the Llanos Orientales to humid non-deciduous forest as well as secondary growth. This species can be termed a habitat generalist. In some parts of Colombia they become serious agricultural pests, feeding on corn (thus *maicero*), but also sugar cane, cacao and other fruit trees. Unlike *C. albifrons, C. apella* groups do not like to move and forage over flooded forest, and DEFLER (1985a) demonstrated a further habitat difference between these two species of *Cebus*: *Cebus apella* choses more mesic environments in semi-deciduous forests of the Llanos Orientales, while *C. albifrons* utilizes more of the thorny *Bactris* palm forest which grow on dry, sandy stream beds. Furthermore, in the Colombian Amazon *C. albifrons* is more apt to be found in forests surrounding rocky hills and in forest growing on white sand, while *C. apella* choses the more diverse and fertile forest. In Colombia *C. apella* have been found as high as 2700 m in the Department of Huila near San Agustín and up to 2500 m in Cauca Department near Inzá (HERNÁNDEZ-C. & COOPER, 1976). Also *C. apella* is found in interrupted or isolated forest in secondary growth, in mountainous areas (cloud forest) and in mangroves (FERNANDES, 1993a). They also cross open land, passing from one segment of forest to another. *Cebus apella* frequents the middle and lower strata of the forest, using medium-sized arboreal supports (FLEAGLE *et al.*, 1981).

Natural history

This species has been studied at far more sites than the other three species of *Cebus*, although it was only recently that it has been studied intensively. The following is a list of most of the field studies to 2003: **ecology** (BROWN & COLILLAS, 1984; CAUSEY *et al.*, 1948; DEFLER, 1985a; DI BITETTI, 2001; FERNANDES, 1991, 1993a; V. KÜHLHORN, 1943; FLEAGLE *et al.*, 1981; FRAGASZY *et al.* 1990; HIRSCH 2000; JANSON, 1990b, JANSON & V. SCHAIK, 1993; KLEIN & KLEIN, 1973a, 1973b; LYNCH & RIMOLI, 2000; PRANCE, 1980; STEVENSON *et al.*, 2002; TERBORGH, 1983; TORRES DE ASSUMPCAO, 1981; MANTILLA & BARRIOS, 1999); **diet and feeding** (BOLEN, 1999; BROWN *et al.*, 1986; BROWN & ZUNINO, 1990; CHALUKINA, 1985; DIBITETTI, 2002, 2003; DI BITETTI & JANSON, 2001; DRAPIER *et al.*, 2003; FERREIRA *et al.*, 2002; GALETTI, 1990; GALETTI *et al.*, 1994; GOMEZ, 2003; IZAWA, 1978b, 1979a, 1990b; IZAWA AND MIZUNO 1977; JANSON (1985a, 1985b, 1986a, 1988, 1990a, 1990b, 1998; JANSON & BOINSKI, 1992; JANSON & TERBORGH, 1979; JULLIOT & SIMMEN, 1998; MENDES *et al.*, 2000; MITTERMEIER & V. ROOSMALEN, 1981; PERES & V. ROOSMALEN, 2002; PHILLIPS *et al.*, 2003; PODOLSKY, 1990; SIEMERS, 2000; STEVENSON *et al.*, 2000; STRUHSAKER & LELAND, 1977; THORINGTON, 1967; VISALBERGHI, 1990, 1993; VISALBERGHI & ANTINUCCI, 1986; DE WAAL *et al.*, 1993; WESTERGAARD *et al.*, 1998; ZHANG, 1995; ZHANG & WANG, 1995; **social behavior** (BOINSKI *et al.*, 2003; DEFLER, 1982, 1985b; CASTILLO, 2001; HARE *et al.*, 2003; HIRSCH, 2002; IZAWA, 1980, 1988a, 1988a, 1988b, 1990a, 1992, 1994a, 1994b, 1997; JANSON, 1982, 1983, 1984, 1985a, 1985b, 1986a, 1986b, 1988, 1994, 1999, 1990a, 1990b; JANSON & VAN SCHAIK, 2000; JANSON & WRIGHT, 1980; NISHIMURA *et al.*, 1995; PHILLIPS & SHAUVER, 2001; PHILLIPS *et al.*, 2003; RIMOLI & FERRARI, 1997; V. SCHAIK & V. HOOF, 1983; V. SCHAIK & NOORDWIJK, 1989; V. SCHAIK & HORSTERMANN, 1994; STEVENSON *et al.*, 1994; TERBORGH, 1986c; THIERRY *et al.*, 1989; VISALBERGHI & ADDESSI, 2003; DE WAAL & DAVIS, 2003; WEIGEL, 1974; WELKER *et al.*, 1987, 1990, 1981; WELKER & LÜHRMANN, 1978; WELKER, 1979a, 1979b); **sexual behavior** (CAROSI & VISALBERGHI, 2002; DI BITETTI & JANSON, 2001; LYNCH, 1998); **conservation**

Capuchins

(BROWN, 1989; CHIARELLO, 2003; HERNÁNDEZ-C. & DEFLER, 1989; MARTÍNEZ *et al.*, 2000; SANZ & MARQUEZ, 1994); **infant development** (CALLE, 1992a, 1992b; ESCOBAR P., 1989a, 1989b, 1990; FRAGASZY, 1990; HARE *et al.* 2003; IZAWA, 1989b, 1990a; NOLTE & DÜCKER, 1959; VALENZUELA,1992, 1993, 1994); **seed dispersal** (STEVENSON *et al.*, 2002; IZAR & SATO, 1997; **distribution** (BOHER-BENTTI & CORDERO-RODRÍGUEZ, 2000). general (DI BITETTI *et al.*, 2000; FREESE & OPPENHEIMER, 1981; HERNÁNDEZ-C. & COOPER, 1976; JANSON, 1986a; ROBINSON & JANSON, 1986; SOINI, 1986a, THORINGTON, 1968; VISALBERGHI *et al.*, 2003 ; WESTERGAARD, 1994; WESTERGAARD & SUOMI, 1994a, 1994b. ZHANG, 1995); tool use (MOURA, 2002, 2003; BOINSKI, 2000; BOINSKI *et al.*, 2000 ; CLEVELAND *et al.*, 2003; CUMMINS-SEBREE & FRAGASZY, 2001; RESENDE & OTTONI, 2002 ; ROCHA *et al.*, 1998 ; WESTERGAARD *et al.*, 1997, 2003); **locomotion** (WRIGHT, 2001, 2003).

Some pertinent and selected research in captivity is listed as follows: **feeding** (ADDESSI & VISALBERGHI, 2002a, 2002b); **early infant development** (FRAGASZY, 1990; NOLTE & DÜCKER, 1959); **social aspects** (CAROSI *et al.*, 2001; DE WALL *et al.*, 1993; DOBRORUKA, 1972); WESTERGAARD *et al.,* 1999; **sexual aspects** (CAROSI & VISALBERGHI, 2002); **infant interaction with mother** (WELKER *et al.*, 1993; NOLTE & DÜCKER, 1959); **social structure of group** (LYNCH, 2002; WELKER *et al.*, 1990); **facial expressions** (WEIGEL, 1974); **tool use and comparison with other species** (CHEVALIER-SKOLNIKOFF, 1989; FERNANDES, 1991; OTTONI & MANNU, 2001; VISALBERGHI, 1990, 1993; VISALBERGHI & ANTINUCCI, 1986; WESTERGAARD & FRAGASZY, 1987).

Group sizes in the Llanos Orientales are 8-9, including one adult male, several females and young (DEFLER, 1982). In Tinigua National Park near the Duda River groups have been as large as 23 with an average number in a group of 16 and exhibiting an age-graded male group with various females and young (IZAWA, 1980, 1988b; STEVENSON *et al.*, 1992). JANSON (1985a) found groups of around 10 at Manu National Park. At the Klein study site on the north bank of the Guayabero River group sizes ranged from 6-12 (KLEIN & KLEIN, 1973a). At SOINI´s (1986a) study site in the Pacaya-Samiria National Reserve in northern Perú, ten groups ranged from 6-11 individuals with a mean of 8.7 animals per group. The range of these various observations is 6-23 individuals per group.

Home ranges vary according to the quality of the habitat. In El Tuparro National Park one home range was 90 ha (DEFLER, 1982). At Tinigua National

Park another home range was about 158 ha (STEVENSON *et al*, 1992) while the home range of a third group at Manu National Park was 125 ha (JANSON, 1985a) and another home range was 80 ha (TERBORGH, 1983) in the same area, while an average of seven home ranges in northern Argentina yielded 161 ha (BITETTI, 2001). One group north of Manaus in Brazil showed a home range of about 900 ha (SPIRONELO, 1987, 1991). Some of these same data give a population density of about 16 individuals/km^2 at Tinigua National Park (STEVENSON *et al.*, 1992), while the Manu National Park site had about 40 individuals/km^2 (JANSON, 1985a, 1985b). The shorter study in El Tuparro National Park in gallery forest yielded about 15-17 individuals/km^2 (DEFLER & PINTOR, 1985), while SOINI (1986) calculated an ecological density of 8-10 animals/km^2 for Pacaya-Samiria (Perú). Other densities are 8 individuals/km^2 at the Estación Biológica Caparú and 5.8 individuals/km^2 at the Río Puré (DEFLER, in prog.). Density data here range from 5.8-40 animals/km^2.

BITETTI (2001) observed agresive inter-group relationships with only partial overlap of home ranges, while mostly peaceful interactions were observed by DEFLER (1982), TERBORGH (1983) and JANSON (1986a,b).

Day ranges of this species were calculated at 2,070 m by TERBORGH (1983), at Tinigua such ranges varied between 370-2300 m, with an average of 1667 m (STEVENSON *et al.*, 1992). A time budget of *C. apella* in Manu in Perú was as follows: 12% rest, 21% travel, 16% feeding on plant material, 50% feeding on insects (66% total feeding), At Tinigua National Park in Colombia another time budget was 62% feeding, 27% moving, 7% resting and 4% other activities (STEVENSON *et al.*, 1992).

Cebus apella utilize a wide variety of locomotive behavior, although they are primarily quadrupedal locomotors. They also employ much galloping, running, jumping and climbing to get about and commonly they descend to the forest floor to search for invertebrates and lizards and to drink from streams and springs when there is no water available in the trees. Their prehensile tails allow secure feeding positions similar to that of the atelines and the other *Cebus,* although they used their tail much less than atelines and the tail is used much more frequently in feeding postures than in locomotion (YOULATOS, 1999). ZHANG (1995) found the majority of sleeping sites in French Guyana to be in the tops of the palm *Jessenia bataua*, which were concentrated in a particular area of the home range. The species uses this frequently as a sleeping site in eastern Colombia as well (DEFLER, per. obs.).

Cebus apella can be considered omnivorous. Because these animals spend fully 50% of their time feeding on insect prey, this is obviously an extremely important part of the diet, and such invertebrates include many lepidopterans, hymenopterans (especially ants), and orthopterans (TERBORGH, 1983). Vertebrate prey is also hunted and consists of frogs, lizards, birds, small mammals and eggs (GALETTI, 1990; TERBORGH, 1983; IZAWA, 1979a, 1990b). IZAWA (1978b) describes the goal-directed behavior of *Cebus* hunting frogs hidden in the interior of bamboo segments. The species has also been observed exploiting oysters from mangrove swamps (FERNANDES, 1991), just as *C. capucinus* (HERNÁNDEZ-C. & COOPER, 1976).

During two months of observations IZAWA (1979a) observed the use of 22 species of fruits from 11 families as listed here: Moraceae (9 species); Palmae (6); Mimosaceae (3); Lecythidaceae (2). This group ate 1 species from each of the following families: Sterculiaceae, Cyclanthaceae, Annonaceae, Clusiaceae, Simaroubaceae, Euphorbiaceae, Musaceae and Bignoniaceae. The species most intensively used in the IZAWA (1979a) study are as follow: *Pourouma lawrencei* (Moraceae); *Pseudolmedia laevis* (Moraceae); *Grias haughtii* (Lecythidaceae); *Enterolobium schomburgkii* (Mimosaceae); *Inga* sp. (Mimosaceae); *Amphilophium pannosum* (Bignoniaceae); *Astrocaryum chambira* (Palmae); *Jessenia bataua* (Palmae).

In the TERBORGH (1983) study 16% of time spent feeding on plant foods inlcuded 100 species of plants from 35 different families. The following lists the five most important families in terms of number of species chosen and these represent 51% of the total diet in terms of species chosen: Moraceae (21 species); Palmae (10); Leguminosae (9); Anonnaceae (6); Sapotaceae (5).

TERBORGH (1983) underlined the importance of palms and of *Ficus* in the diet of this species in Manu National Park, and he identified the palm *Astrocaryum* as a critical resource during the dry season, when there is little else to eat. The technique of processing the hard nuts of these palms was well-described by STRUHSAKER & LELAND (1977) and by IZAWA & MIZUNO (1977) and consists of hitting the seeds repeatedly against branches and biting them until opened. *Ficus* is also of great importance to this primate at Manu during the wet season.

The importance of *Astrocaryum* is the same on the lower Apaporis of Colombia (DEFLER, 2003) as is *Astrocaryum* at Manu, and one wonders what *Cebus* would do without that poor season food. Yet in the Llanos Orientales *Astrocaryum* is largely absent north of the Vichada River. The *C. apella* in that region seem to rely heavily upon pith eating of fronds from *Scheelea* palms and meristeme-eating of fronds pulled from small *Bactris* palms, although the palm *Jessenia bataua* is also important to fill the gap left by the absence of *Astrocaryum*.

SOINI (1986a) analyzed 96 feeding observations as follow: 71% fruit feeding; 16% leaf, shoot, meristem and petiole feeding; 7% seed predation and 3% pith feeding. In the same study the animals also ate larvae and the chrysalis of wasps, ants and ant eggs as well as large land snails.

STEVENSON *et al.* (1998, 2000) lists 126 species of fruits consumed by *C. apella*. In the same study the *Cebus* spent 14.2% of their time feeding upon *Astrocaryum chambira* and 12% of their time feeding on *Jessenia bataua*. The Palmae were the most important in terms of species consumed equaling 29% of the total, followed by the Moraceae (20%) and then the Burseracaea, Mimosaceae and Tiliaceae. PERES (1991b) describe seed predation on *Cariniana micrantha* (Lecythidaceae) by this primate. This species drinks water from a wide variety of sources, taking water from tree holes, bromeliads, rain drops on vegetation and descending to the forest floor to drink from brooks and springs.

It seems that in many groups there is only one male breeder, although all adult females breed. In the largest groups the reproductive male may change from time to time. PHILLIPS *et al.* (1994) as well as CAROSI & VISALBERGHI (2002) describe sexual behavior in this species. Other males present in the group do not contest the right of the dominant male (JANSON, 1984). Females show an estrus period of 18 days (WRIGHT & BUSH, 1977). Females solicit sex and at least 20 behavioral patterns are associated with sexual behavior (CAROSI & VISALBERGHI, *op cit.*). The mounting male either stands directly on the substrate or grasps the hind legs of the female in his feet. Gestation is 160 days (HARVEY *et al.*, 1987).

SOINI (1986a) observed births during all months except the dry season (October-June at Pacaya-Samiria Biological Reserve), while IZAWA (1990a,f) reported more births during the dry season, which at Izawa´s study site on the Duda River is much more pronounced than at SOINI´s study site in Perú.

Newborns are behaviorally very altricial compared to most other primates and they are unable to do much but cling to their mothers, while development of prehensility and postural control proceeds during the first two months (FRAGASZY, 1990). The development of postural control and locomotion is longer in this species than in *Saimiri*, *Papio* or *Macaca* (FRAGASZY, 1990a). Infants ride on the mother, first over her shoulders then, later, on her back, but they begin to become independent at 6-7 weeks, completing the process during the 5th or 6th month (FREESE & OPPENHEIMER, 1981; ESCOBAR, 1989a). VALENZUELA (1992) found that an infant began its independent locomotion in the wild during the first 3-4 weeks of life and first independence was initiated by the mother, who left the baby in tangled vegetation for short periods of time. During this first period of independence, other members of the group, especially older juveniles, also began carrying the infant. The mother carries the infant transversally, across the shoulders for the first two weeks, at which time the baby begins to orient itself more and more longitudinally in a mounted position, as well as further back towards the middle of the back. Infants nurse until about 6 months to 1 year. Infants and young play constantly either alone or with their friends, and play involves much chasing, biting and tail pulling. Sexual play also involves mounting and thrusting upon the mounted of either sex (ESCOBAR, 1989a).

ESCOBAR (1989a, 1989b) studied four infants between 3-9 months. At three months the four monkeys moved independently much of the time (varying from about 30% - 100% of locomotion), although they continued to nurse from the mother and did not become totally independent from being carried until 6 months of age. During feeding the alpha male often protected these young. The young begin early to investigate solid foods and at six months are able to feed on most solid foods eaten by the troop. Mother-infant contact is not as strong as in many other species and a dyad-specific call is used by both the infant and the mother in order to promote nursing. During resting, contact with the mother increases and the mother constantly grooms her infant.

IZAWA (1980) and JANSON (1985a, 1985b) studied social interactions in the species. JANSON (*op cit.*) concentrated on agonistic interactions and he found differences in the rate of feeding, according to social status. The dominant animals enjoyed a diet with 20% more total energy in comparison with the subordinates, but also the animals which received little aggression had a diet high in energy. A male which emigrated from the group had the lowest energy consumption in the group; levels of intragroup competition were ten times higher than levels of intergroup competition.

Adult males spend a lot of time on the periphery of the social group; they eat less and are more vigilant than the females. The males detect predators more frequently than the females. The high risk of predation may be an important selective pressure on the development of a dominant male with a long life (VAN SCHAIK & VAN NOORDWIJK, 1989).

JANSON (1984, 1986b) hypothesized that the sexual system of *C. apella* and of *C. albifrons* could have determined the social system shown by each species. He found differences in the size of the food patch exploited by the two species: *C. apella* feeds from food patches much smaller than those utilized by *C. albifrons*, assuring that *C. apella* group members are on the average much closer to one another than members of a group of *C. albifrons*. Because of this, the alpha male *C. apella* is able to offer benefits to the females and young of this group and this makes it likely that the alpha male will be the father of the females´ infants. This also assures much more tolerance of the dominant male towards the females (and his young) in the small tree being utilized (JANSON *op. cit.*).

FREESE & OPPENHEIMER (1981) describe vocalizations as follow: (1) *fuh* – loud and long, used by all when trying to localize the group (1-3 repitions per call); (2) *mik* – infants and juveniles when afraid, in order to attract mother (2-3 repitions per call); (3) *whine* – all members use for maintaining contact in group; (4) *peep* – short repeated call commonly used by infants to maintain contact with mother; (5) *grunt* – low-pitched pulsed call usually with headshake given by female with infant as friendly appeasment; (6) *flutelike* – rare, high-pitched call given by adult male when approaching estrous female; (7) *igk* – common, high-pitched trill given by younger or subordinate animals showing mild fear; (8) *kecker* – common broad-frequency pulsed loudcall given by all showing intense fear or when chasing or threatening; (9) *scream* – common broad-frequency loud call given by all to show extreme fear; (10) *tooth grind* – less common grinding of molars by adult male as threat; (11) *iku* – less common low frequency call given by adult to give alarm; (12) *ika* – common long series of staccato calls (bark) given by all as alarm. DiBitetti (2003) shows that these capuchins have food associated calls which all members of the troop understand.

Cebus apella frequently moves and forages in company with *S. saimiri sciureus* for periods of up to several days (Thorington, 1967, 1968; Tokuda, 1968; Klein & Klein, 1973a; Hernández-Camacho & Cooper, 1976; Mittermeier & Coimbra-Filho, 1977). During 243 observations of *C. apella*, the species was seen with *S. sciureus* 74% of the time in the Parque Nacional Pacaya-Samiria (Soini, 1986). In Surinam, during 50 observations of *C. apella* and 30 observations of *S. sciureus*, 29 observations were of the species in association (36% of total observations)(Fleagle & Mittermeier, 1981). Also, *C. apella* commonly forages and travels with *Lagothrix lagothricha* and occasionally they are seen with *C. albifrons* and with *Cacajao melanocephalus*. The two hawks *Harpagus bidentatus* and *Leucopternis albicollis* are often observed in association with *C. apella*, alghough the first hawk is be far the most commonly seen.

Conservation state

With its enormous geographic range this species is one of the least endangered of neotropical primates. Nevertheless, it is important to investigate and identify differences in populations at a subspecific level, and especially to determine karyological and phenological differences, since some of these subsets of the species are probably endangered and may prove to be species in their own right rather than populations of *C. apella*.

This species survives well in proximity to human beings, as long as it is not hunted. If they are persecuted, however, they are very quiet and wary, capable of hiding in the tops of trees (such as palms) for long periods of time. In Colombia this species is protected in at least 11-14 national parks. Its wide distribution means that it is one of the least endangered of the Colombian primates (Defler, 1994b, 2003; Defler *et al.*, 2003). It is currently classified LC in Colombia (Rodríguez *et al.*, in press).

Where to find it

Cebus apella is one of the most common species of primate in the Colombian Amazon in most regions (not, however, in the upper Cahuinarí River or in the Colombian trapezium, where it seems to be almost entirely absent) and in the Llanos Orientales. Local people can often be counted on to point out a group. They are very common in the forests near the cabaña Tapón in western El Tuparro National Park, although that is a very remote site. Often they are to be found around small towns in the western Llanos, but near people they learn to be very shy and to hide very effectively, due to human persecution.

At the Estación Ecológica in Tinigua Nacional Park there are several completely tame groups which come for hand-outs and are so accustomed to being observed that they are probably the easiest of all the sites to see the species. However, I strongly recommend coordinating any trip to this site with professor Carlos Mejía or with professor Pablo Stevenson both of Los Andes University (Bogotá), since there are some special problems involved in traveling to this site. This is not a difficult species to locate in most of its home range, either with a guide or without.

White-faced Capuchin

Cebus capucinus (LINNAEUS, 1758)

Plate 5 • Map 11

Taxonomic comments

Three subspecies have been defined for Colombia: *C. c. capucinus* (LINNAEUS, 1758); *C. c. curtus* (BANGS, 1905); and *C. c. nigripectus* (ELLIOT, 1909). However, these are neither recognized by HERSHKOVITZ (1949), HERNÁNDEZ-CAMACHO & COOPER (1976) nor by MITTERMEIER & COIMBRA-FILHO (1981), due to the great variability found even in the same group. In this book I also shall not consider subspecies.

Common names

Other English: capuchin monkey; Colombia: *mico negro* in the departments of Córdoba, Sucre and Norte de Bolívar; *maicero cariblanco*, *carita blanca* or *mico* in the departments of Antioquia (the Urabá region), Córdoba, Sucre, Norte de Bolívar and Atlántico; *machín* in the departments of Sucre and northern Bolívar; *mico maicero*, Pacific coast region; German: *Kapuziner*, *Weischulteraffe*, *Weiskehlkapuziner;* French: *sai gorge blanche, sajou capuchin.*

Indian languages

Missura (Embera, Indian reserve in upper Baudó, Córdoba valley, upper Bojaya, upper Río Buey; *aisur* (Emberá Chamí).

Taxonomic comments

Although HERSHKOVITZ (1949) first considered *Simia hypoleuca* as a subspecies of *Cebus albifrons*, later he emmended this interpretation, proposing "that the name be disposed of as an unavailable synonynm of *C. albifrons*" (HERSHKOVITZ, 1955). Nevertheless, CABRERA (1957) recognized *C. albifrons hypoleuca* as a subspecies with synonmy, including *C. malitiosus* (*C. a. malitiosus*). A careful examination of V. HUMBOLDT´S description of *Simia hypoleuca* demonstrates conclusively that it is conspecific with populations of *C. capucinus* with a brownish tone which covers much of the body, similar to the specimens from the valley of the upper Río San Jorge. Also, *C. capucinus* is known as much in the valley of the Río Sinú (in the region of Montería), as in the Tolú region not far from Puerto del Zapote in the lowlands of the coast of the Gulf of Morrosquillo. These populations can be confused with particularly dark populations of *C. albifrons* from the valley of the mid-Magdalena.

MITTERMEIER & COIMBRA-FILHO (1981) erroneously reported that HERNÁNDEZ-C. & COOPER (1976) had chosen to synonimize *C. capuchins* and *C. albifrons*, because of the presence of hybrids between populations of these two species in northern Colombia, when actually HERNÁNDEZ-C. & COOPER (1976) only pointed out the possibility of their being conspecific but treated the two species separately in their article. It should be pointed out that hybridazation, if it is present, does not necessarily argue for one species rather than two since even a hybrid zone can restrict gene flow between two populations such that they maintain their integrity (MAYR, 1971).

Identification

These animals are of medium size with a head-body length of 330-450 mm and a tail length of 350-550 mm. They weigh between 1.5 – 4 kg, the males being larger than the females. The tail is strongly prehensile and black, although its ventrum is sometimes brown. Hands and feet are blackish while the head is yellowish-white and black with a dark, black cap on the crown that gives rise to a comparison to a monk´s tonsure. The face is pinkish with widely scattered white hairs. Chest and throat are pale yellowish and the blackish belly is thinly-haired. Subspecific differences have been described on the basis of color of the light parts (from nearly pure white to white, strongly diffused with tawny ochraceous), the color of the chest and belly (black to black or dark drab strongly grizzled with light-yellowish white), the extension of a yellowish latero-ventral stripe to the genital region, sometimes including the medial thighs (as sparse yellowish hairs), the degree of development of yellow in the basal portion of the ventral tail hairs, and the development and coloration of the frontal tuft in females, although these characteristics are so variable as to make subspecific designation difficult or impossible.

Geographic distribution

In Colombia *Cebus capucinus* is found from the Panama frontier southward along the Pacific coast and the western slope of the Cordillera Occidental de los Andes (up to about 1800-2000 m altitude), the Island of Gorgona offshore from southern Cauca department, in the valley on the west bank of the upper Cauca River (between the cordillera Occidental and Central), and in the department of Antioquia (especially in the Urabá region). East of the frontier with Panama the species is found in the lowlands and the slopes of the Cordillera Occidental up to the Río Magdalena and to the central and lower parts of the Río San Jorge. *Cebus capucinus* is not found in the aluvial lowlands of the Pacific coast of the department of Valle (*i.e.* mangroves, forest of *nato* (a fresh-water swamp dominated by the *nato Dimorphandra oleifera* - Caesalpiniaceae and the lowlands of mixed forest) but the species appears in the piedmont (HERNÁNDEZ-CAMACHO & COOPER, 1976). Outside of Colombia, *C. capucinus* is found from Panamá to Honduras and to north-western Ecuador.

This monkey seems to prefer primary forest or advanced secondary forest, but it is also found in remnant degraded forests and in forests with large concentrations of palms, especially *Scheelea magdalenica*. The species is also at home in forests flooded with fresh-water.

Natural history

Several field studies have been published, although the species has never been studied in Colombia. FREESE (1976a, 1976b, 1977a, 1977b, 1978, 1983) studied *C. capucinus* in Costa Rica. OPPENHEIMER (1967a, 1967b, 1968, 1969a, 1969b, 1973, 1977a, 1977b, 1982, 1996) and OPPENHEIMER & LANG (1969) studied the species on Barro Colorado. HLADIK & HLADIK (1969) and HLADIK *et al.* (1971) included this species in their analysis of food habits and nutrition. More recently CHAPMAN (1986a,b, 1987, 1988b, 1989b, 1990a), CHAPMAN & CHAPMAN (1990b), CHAPMAN *et al.* (1989a), CHAPMAN & FEDIGAN (1990) did comparative studies of this species with *Ateles geoffroyi* and *Alouatta palliata* in Costa Rica. Other observations were published by Boinski (1988b, 1993), CHEVALIER-SKOLNIKOFF (1989, 1990), FEDIGAN (1983, 1990, 1993), FEDIGAN *et al.* (1985), FONTAINE (1980), GEBO (1992) on locomotor and postural behavior, MASSEY (1987) surveyed this and two other Central American species, MOSCOW (1987) studied movements and diet, ROSE (1992, 1994a, 1994b, 1999) studied sex differences in the diet, TOMBLIN & CRANFORD (1994) studied niche differences between *Alouatta palliata* and *Cebus capucinus*, WARKENTIN (1993) describes an association of this species with the raptor *Accipiter striatus* while BAKER (1996, 1999) discussed the use of medicinal plants by these animals. A census of this and the other sympatric primates was published by LIPALD (1988, 1989).

Lately PERRY & ROSE (1994) studied meat eating in the species and PERRY (1996) described intergroup encounters, while LONGINO (1984) analyzed anting behavior and PHILLIPS (1994, 1995) studied resource use. BUCKLY (1983) did his Ph.D. research on this species in northern Honduras, but this work has apparently not been published. A review article written about all *Cebus* species (recognized at the time) by FREESE & OPPENHEIMER (1981) included information up to about 1978.

Other research has been concerned with the following: **vigilance behavior** (ROSE & FEDIGAN, 1995; FEDIGAN *et al.*, 1996, JACK, 2001); **births** (FEDIGAN & ROSE, 1995); **vocalizations** (BOINSKI, 1996; BOINSKI & CAMPELL, 1995, 1996; GROS-LOUIS, 2001, 2002); **population dynamics** (OPPENHEIMER, 1996; FEDIGAN & JACK, 2001); **fur rubbing** (BAKER, 1996, 1999); **predation** (ROSE, 1997); **social interactions** (PERRY, 1997, 1998a, 1998b, 1999; MANSON *et al.* 1999; BOINSKI, 2000); **use of space** (HALL & FEDIGAN, 1997); **spatial memory** (GARBER & PACIULLI, 1997); **posture** (BERGESON, 1997a, 1997b; BERGESON *et al.*, 1976; PANGER, 1999a,b); **tool use** (PANGER, 1998); **hand preference** (PANGER, 1998); **feeding** (BERGESON, 1998); **behavioral ecology** (ROSE, 1999); **object manipulation** (PANGER, 1999); **tail use** (GARBER &

Rehg, 1999); **population density** (Pruetz & Leasor, 2000); **group size** (Chapman & Chapman, 2000; Burger, 2001); **foraging** (Panger *et al.*, 2002); **aggression** (Leca *et al.*, 2002).

Group size seems to vary from a few animals to around 20, depending on the quality of the habitat. The average size on Barro Colorado was 10-15 (Oppenheimer, 1969a) but the average size in drier forest of Santa Rosa National Park in Costa Rica was 15-20 (Freese, 1976a). Groups are made up of a few males and at least twice as many females, plus young.

Home ranges average around 80 ha on Barro Colorado (Oppenheimer, 1967a) and about 50 ha in Santa Rosa National Park, Costa Rica (Robinson & Janson, 1987). Crude densities of about 5-7 individuals/km² or ecological densities of 18-24 individuals/km² were measured for Barro Colorado (Robinson & Janson, 1987) and in Santa Rosa National Park ecological densities reach about 30 animals/km² (Freese, 1976a). Day ranges usually vary from 1-3 km per day and usually around 2 km (Freese, 1976a).

Oppenheimer (1968) calculated a time budget for the species as follows: 28% foraging, 47% moving, 14% resting, 8% allo-grooming and 3% play. These animals spend a great deal of their time moving about and foraging on the ground, as well as in trees up to the emergent crowns. Possessing a prehensile tail is very important to them in foraging, since prehensile tails seem to be primarily a feeding adaptation, used particularly in below-branch feeding and foraging and as leverage in above-branch prey extraction (Garber & Rehg, 1999). Youlatos (1999b) analyzed three other *Cebus* monkes and their use of the tail and found similar results. These animals utilize a great variety of locomotive behaviors such as walking and running (they are primarily quadrupedal), galloping, climbing, springing, jumping and bipedal walking for a few steps (on the ground) (Oppenheimer, 1968). Panger *et al* (2002) were able to demonstrate the presence of different foraging techniques in different groups of capuchins, making this species the only non-ape primate which exhibits different traditions at different sites.

Although the species is well-known as a tool-user (especially in captivity), a specific study focused on object-use in three troops failed to find the range of tool use observed in captivity (Panger, 1998). These primates sleep rather high in trees in densely-leaved parts of the tree (Oppenheimer, 1968). Manson & Perry (2000) found elevated rates of self-directed behaviors such as self-scratching and autogrooming when in proximity to conspecifics, irrespective of rank, suggesting that self-directed behavior may be indicative of anxiety in this species as it is in catarrhine primates.

Cebus capucinus are considered omnivores. Their diet was analyzed by HLADIK & HLADIK (1969) by weight into 20% animal prey, 65% fruit and 15% green material. On Barro Colorado the diet included 95 fruit species and the branch parts, buds and flowers of 24 species which are broken down according to number of species chosen per family (OPPENHEIMER, 1968): Leguminosae (10); Rubiaceae (7); Palmae (5); Sapindaceae (5); Flacourtiaceae, Moraceae, Euphorbiaceae (4); Musaceae, Anacardiaceae, Convolvulaceae (3); Tiliaceae, Guttiferae, Violaceae, Lecythidaceae, Myrtaceae, Sapotaceae, Apocynaceae (2); Araceae, Solanaceae, Gnetaceae, Bromeliaceae, Lacistemataceae, Polygonaceae, Nyctaginaceae, Menispermaceae, Rosaceae, Myristicaceae, Connaraceae, Rutaceae, Simaroubaceae, Burseraceae, Bombacaceae, Sterculiaceae, Dilleniaceae, Passifloraceae, Melastomaceae, Bignoniaceae (1).

Invertebrate prey taken by this primate includes ticks, spiders, grasshoppers and katydids, walking sticks, termites, spittlebugs (Cercopidae), Fulgoridae, beetles, wasp larvae, larvae and pupae of lepidopterans. Vertebrates such as birds eggs, birds, small mammals (for example coati babies, *Nasua nasua*) and small lizards are also opportunistically consumed (OPPENHEIMER, 1968).

FREESE (1977a) observed *C. capucinus* eating insects 30% of the time during the dry season and up to 51% of the time during the wet season. Another unpublished work analyzed feeding behavior of this primate (BUCKLEY, 1983) and PANGER *et al.* (2002) have demonstrated some foraging traditions comparing different study sites. Likewise, some other social conventions seem to be traditional as well (PERRY *et al.* 2003). ROWELL & MITCHELL (1991) compare seed dispersal of this primate to guenons. WEHNCKE *et al.* (2003) studied seed dispersal patterns of this monkey on Barro Colorado Island, showing substantial movement of seeds up to 844 m from the fruit tree. Over four months this included at least 95 species of fruits of which 67 species passed intact through the gut. This species of primate drinks from tree holes and also descends to drink from small creeks, springs and waterholes.-

In Central America this species seems to be a seasonal breeder with the births occurring during the dry season of December-April (FREESE &

OPPENHEIMER, 1981). Polygamous mating takes place after a series of chases by both male and female through the trees, with a curious dancing about the female by the male, including much chirping vocalization preceeding the mount. Copulation lasts around 2 minutes and during the mount the male may either stand upon the substrate or grasp the female's rear legs with his feet. Gestation apparently lasts about 5-6 months; HARVEY *et al.* (1987) reports 160 days for *C. apella*.

One infant is born and carried almost constantly by the mother, oriented across her shoulders for around 6 weeks. After being carried across the shoulders the infant orients itself lengthwise along the mother´s back. The infant may begin to leave the mother briefly for gradually lengthening periods of time after about 7-8 weeks. By the fifth or sixth month the infant usually locomotes independently. Nursing may last for 6-12 months (FREESE & OPPENHEIMER, 1981). Babies and young spend a great deal of time playing, while the mother forages or rests. The play is either solitary or with others. Young may either chase, do mock battle, tail pulling or they may imitate social and sexual behaviors which become important later (FREESE & OPPENHEIMER, 1981).

Other members of the group are always interested in the new infant, although the mother controls access to it, gradually loosening control over time. Many members of the troop may carry the infant, including adult males. Other members show heightened vigilance (ROSE & FEDIGAN, 1995). Infant handling by females other than the mother was studied by MANSON (1999), who wished to test the various hypotheses used to explain the behavior and this researcher concluded that infant handling my be used to test bonds between females, who require coalitions, since females tended to handle the infants of females which they groomed and formed coalitions.

GROS-LOUIS *et al.*(2003) reported violent group attacks on single males similar to chimpanzee attacks, which resulted in deaths and injuries. PERRY *et al.* (2003) studied the origin and use of social conventions in groups of capuchins in Costa Rica, concluding that such behavior such as handsniffing, sucking of body parts and three types of games were used to test the quality of the social relationships in the particular group and that a particular social convention rarely lasted over ten years.

Some vocalizations of *Cebus capucinus* are listed by OPPENHEIMER (1968) as follow: (1) *chirp* – a low intensity sound when in physical contact, by young animals when hugging; (2) *yip* – near or in physical contact and of medium intensity by all ages, but especially of youngsters of a few months of age to solicit maternal grooming and nursing; (3) *huh* – at medium intensity in visual contact by infants and juveniles to express hesitancy to move; (4) *arrawh* – out of visual contact and sometimes initially out of auditory contact at high intensity by individual or group of individuals lost from contact with main body or group; (5) *purr* – in physical contact at low intensity by individuals in hugging contact which may indicate friendly intentions (or contentment); (6) *trill* – near or in physical contact at medium intensity given by individual which follows, approaches or remains with an older and more dominant animal; (7) *whistle* – at medium intensity in situations which also invoke trills and which may indicate fear on the part of the vocalizer; (8) *gutteral chatter* – near or in physical contact and at medium intensity by adult females with recently born infants on their backs; (9) *scream* – at high intensity by an animal being attacked by a conspecific or other animal; (10) *gyrrah* – at high intensity by presence of another species on the ground (e.g. human, boa, tapir); (11) *bird gyrrah* – same call as above but given only once rather than several times when a fast-moving bird goes by; (12) *chortle* – at high intensity by old males and adult females who were threatened by holding their ground; (13) *group cohesion call* – at medium intensity via a series of *yips* and *huhs* while individuals are eating and which enables maintainance of contact with other members and may advise of food.

On Barro Colorado an adult male *Ateles geoffroyi* commonly traveled with the local *Cebus capucinus* group, an interaction that is reportedly common (BERNSTEIN, 1964; OPPENHEIMER, 1968). Each of the two species show social interest in the other, even resulting in grooming. Hardly any interations have been recorded between *Cebus capucinus* and *Alouatta palliata* except in food trees, where some physical threats of the *Cebus* take place towards the howlers (OPPENHEIMER, 1968).

Cebus capucinus have been observed preying upon coatimundi (*Nasua nasua*) babies and other small mammals. They react to the danger cries of *Tayassu tajacu* by climbing higher into the trees and both *Tayassu tajacu* and *Dasyprocta punctata* are attracted by the *Cebus* feeding, since they know that there will be fallen fruit (OPPENHEIMER, 1968). The small double-toothed kite (*Harpagus bidentatus*) commonly travels with *Cebus capucinus*, just as it does with many other actively foraging primate species (GREENLAW, 1967; FONTAINE, 1980) and apparently the white hawk *Leucoptenis albicollis* also uses the foraging monkeys to its advangtage, to hawk prey. WARKENTIN (1993) describes a foraging association of this monkeys with sharp-shinned hawks (*Accipiter striatus*).

Conservation state

This primate species is not considered to be endangered by the Ministerio de Medio Ambente in Colombia and is classified LC nationally and internationally and it is included in Appendix II by the international CITES agreement. Nevertheless, there are no recent data about the state of the populations in Colombia, and a consideration of the reality of the subspecific taxa should be undertaken in order to evaluate the conservation state of the species on that level. Contact zones between *C. capucinus* and *C. albifrons* ought to be studied on a genetic level to clarify the nature of the reported hybridization between the two. Additionally, censuses should be undertaken to evaluate the conservation health of this species in Colombia.

Where to see it

Cebus capucinus can be observed in Los Katios National Park in the Chocó. It is also found in nearby Colosó (Sucre) in the Primate Center. Probably the easiest population to observe is found on Gorgona Island in Gorgona National Park, since many of these animals are accustomed to human beings, but the species is widespread in the Pacific lowlands.

Squirrel Monkey

FAMILY **CEBIDAE**
Genus *Saimiri* VOIGT, 1831

Both HERSHKOVTIZ (1984) and THORINGTON (1985) have revised this genus as well as COSTELLO *et al.* (1993) and GROVES (2001). HERSHKOVTIZ ´(1984) revision has perhaps been more widely accepted, and it describes four species (*S. boliviensis, S. sciureus, S. oerstedi* and *S. ustus*). AYRES (1985) described one more species, *S. vanzolini* for Brazil. For Colombia the Hershkovtiz revision includes one species (*S. sciureus*) and three subspecies (*S. s. albigena, S. s. cassiquiarensis* and *S. s. microdon*).

HERSHKOVITZ´new revision means that many populations previously called *Saimiri sciureus* are different species. For example TERBORGH´S (1983) often-quoted work on the ecology of five New World primates presents data on *S. boliviensis,* although in the book the species is called "*S. sciureus*".

THORINGTON´S (1985) revision of *Saimiri* recognizes only two species: *S. sciureus* and *S. madeirae.* In this revision *S. sciureus* is divided into four subspecies: *S. s. sciureus, S. s. cassiquiarensis, S. s. boliviensis* and *S. s. oerstedii.* The HERSHKOVITZ (1984) revision seems to have taken priority over THORINGTON´S, but the case cannot be considered closed, and more karyotypic and other information is required.

COSTELLO *et al.* (1993) used a combination of genetic and behavior to propose uniting all of the taxa of the continent into one species, *Saimiri sciureus.* Nevertheless, BOINSKI & CROPP (1999) working with mtDNA and CROPP & BOINSKI (1999) with two nuclear genes and one mitochondrial gene concluded that the taxonomy of HERSHKOVITZ (1984) reflected the most probable scheme, and suggestions by RYLANDS *et al.* (2000), considering all of the above information, supported by GROVES (2001) agreed that recognizing the five species mentioned from the HERSHKOVITZ (1984) was the most acceptable. RUÍZ-GARCÍA & ÁLVAREZ (2003) found distinct haplotypes of mtDNA in each of the three subspecies.

The genus epithet *Saimiri* is derived from the Tupi *sai-mirim* or *çai-mirín,* where *sai* is a qualifier given to many species of monkeys and *mirim* signifies small (HILL, 1960).

Squirrel Monkey

Saimiri sciureus (LINNEUS, 1758)

Plate 4 • Map 9

Taxonomic comments

See under *Saimiri.* If popular revisions continue in the vein of insisting that all subspecies should be elevated to species status, then Colombian species may total three, instead of one in the future. Actually, RUÍZ-GARCÍA & ÁLVAREZ (2003) have found different mtDNA haplotypes for each of the Colombian subspecies, demonstrating some genetic differences, although this is congruent with the typical polymorphic species, which is the way this author prefers to view *Saimiri* in Colombia.

Common names

Colombia: *tití* in the Llanos Orientales, the Caquetá region and the upper Magdalena River valley; *vizcaino* en the department of Caquetá and the upper Caquetá and Guayabero rivers; *mico soldado* (Amazon piedmont in Cauca); *mico fraile, tití fraile, fraile* and *frailecito* in Putumayo and the area of Leticia (or Peruvian origin); *macaco de cheiro* or *saimiri* (Tupí *sai-mirim*, Leticia area); *chichico* in the Mirití-Paraná and the lower Caquetá river in the department of Amazonas; *mono ardilla* (animal trafficers); Brazil: *macaco de cheiro, macaquin, barizo* or *frailecillo*. German: *Todtenköpfchen, Todtenkopfaffe* or *Eichornaffe*. French: *saimiri*. Dutch: *aeckornaap* or *doodshoofdaapje*. Pidgin English in Guianas: *monki monki*.

Indian languages

Ararimá (Carijona); *pitipidi* (Curripaca, *fide* AMAZON, 1992); *jijiyo* (Cubeo); *titi* or *tsële* (Guahibo); *díyu* (Huitoto); *cayambero* or *yurac chichico* (Ingano of Aguas Claras-Cusumb, La Solita, Municipio Valpariso, Caquetá, *fide* J. V. Rodríguez-M.); *sawon* (Jébero); *süi* (Macú, *fide* Capitán Francisco Macú of the Yujup-Macú, lower Apaporis river); *nsuëma* (Macuna), *fide* Napoleón Miraña; *tiyi* (Muinane); *joinxo* (Ocaimo); *tití* or *tigtí* (Piaroa), *shaun, chiau* or *siaon* (Puinave); *bosisi* (Siona); *jëjëya* (Tanimuca); *maiñechá* (Ticuna); *sujingaininiami* (Tucano); *cuhuisú* (Yucuna).

Identification

These small primates are often quite variable in weight, the males being heavier than the females. Nine Colombian males weighed an average of 1082 g and five Colombian females weighed an average of 858 g (HERNÁNDEZ-C. & DEFLER, 1989). Two male *S. s. cassiquiarensis* from Vichada each weighed 1125 g in October, 1993. Head-body length in the above animals was about 385 mm with a tail length of about 470 mm. These specimens represent the upper end of the size scale for the species and may represent the fatted state.

The dorsum of *S. sciureus* is dark, while the ventrum is white or yellowish-white. The face is also white with the exception of a velvety black muzzle. The tail is the same color as the dorsum with the exception of a black tail-tip. The populations in the Llanos Orientales (*S. s. albigena*) and the piedmont and southward to the Caquetá River are rather uniform in aspect and differ

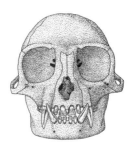
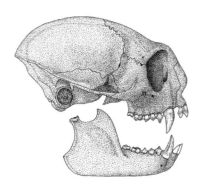
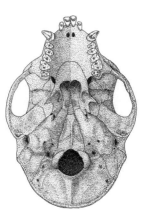
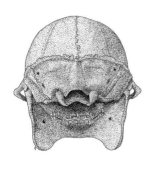

Figure 18. Views of skull of *Saimiri sciureus*.

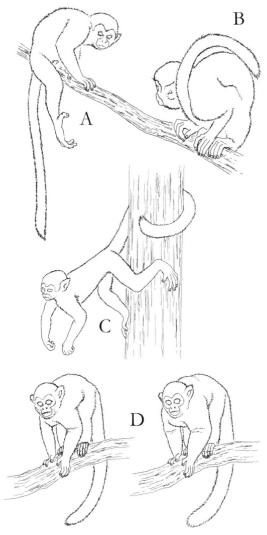

Figure 19. Various positions of *Saimiri*. a) Penis display. b) Resting. c) Use of tail by young. d) Standing. e) Aggressive display.

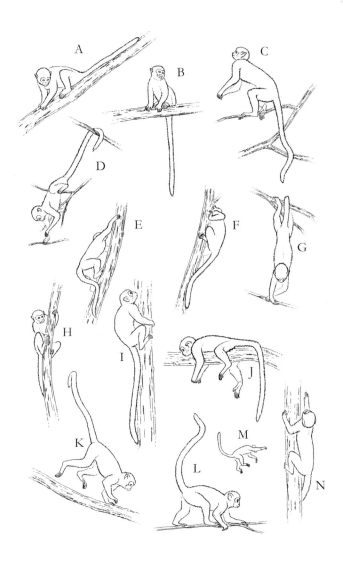

Figure 20. Various aspects of *Saimiri*. a-j) Resting positions. k-n) Locomotion.

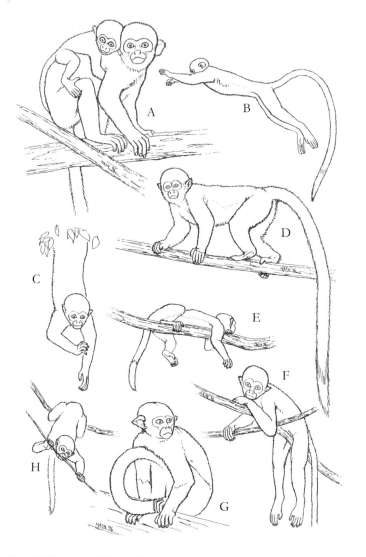

Figure 21. Diverse views of *Saimiri*. a) Mother resting with infant. b) Jumping. c) Hanging. d) Locomotion. e-h) Resting positions.

principally from those of the Colombian Amazon by coloring of the forearms and wrists, which are grizzled grey rather than yellowish. The subspecies are differentiated as follow:

• Forearms and wrists grizzled grey and finely scattered white with or without a tone of dull yellow, crown and neck pricipally grey olive ....*Saimiri sciureus albigena.*

• Forearms and wrists, feet and the major part of the inner and outer part of the forearms, yellowish or orange tones (typical of the Amazon)

• Above and with nuchal band (neck) between the shoulders that is paler the the crown and the shoulders......................*Saimiri sciureus cassiquiarensis.*

• Same color as *S. s. cassiquiarensis* but without the nuchal band ....*Saimiri sciureus macrodon.*

Geographic range

In Colombia *Saimiri* is found in the entire Colombian Amazon and in the eastern piedmont of the Cordillera Oriental of the Andes at least to the Metica River (near Villavicencio) and in a large part of the southern Llanos Orientales up to the Río Vichada. In the upper Magdalena River valley the species has a distribution that is similar to that of *Cebus apella;* its northernmost distribution in the Magdalena River Valley is not very well known, nor are its easternmost limits in Arauca or Casanare. The maximum altitude where the species has been observed is around 1500 m in Huila. The upper R. Magdalena and Llanos Orientales populations are referable to *S. s. albigena* while the forest to the north of the Apaporis River are *S. s. cassiquiarensis.* Populations below the R. Apaporis are *S. s. macrodon.*

Besides Colombia the species is found in Amazonian Perú to the Río Purus (except between the Marañon/Ucayali and Tapiche rivers where it is replaced by *S. boliviensis peruviensis*). Its distribution continues in southern Venezuela and the Guianas to the R. Amazonas and it is also found south of the R. Amazonas to the Xingú or the Iriri. Animals in Bolivia and in southern Perú which extends to the R. Purús and the R. Guaporé are referable to *S. boliviensis,* while the populations between the R. Purús and the Xingu are *S. ustus* (HERSHKOVITZ, 1984).

Saimiri are found in many different types of forest habitat; gallery forest, low forests of sclerophytic vegetation, forested slopes, palm forests, (particularly associations of *Mauritia flexuosa*) and rainforest that is both seasonally flooded and from unflooded (*terra firme*). *Saimiri* uses edge habitat frequently and readily survives in isolated forests which have been degraded by human activity. These animals are active in low to mid-levels of the forest and sometimes they go to ground level. (FLEAGLE *et al.*, 1981).

Natural history

This species has not been extensively studied in the field. THORINGTON (1967, 1967b) spent ten weeks observing squirrel monkeys in Meta Department in the Colombian Llanos Orientales. BALDWIN (1971, 1981) studied a semi-natural forest group maintained by DUMOND (1968) in southern Florida (the Monkey Jungle colony). Later, Baldwin and BALDWIN (1971, 1981) published notes and a synopsis of work on natural populations. While studying *Ateles belzebuth* in La Macarena National Park, KLEIN & KLEIN (1973) collected some information on *Saimiri sciureus* as well. MOYNIHAN (1976a) reported his own general observations in Colombia on the species. BAILEY *et al.* (1974) and SPONSEL *et al.* (1974) reported the results of their censuses on the Mike Tsalikis Isla de los Micos (Santa Sofia) colony near Leticia, although the island was an artificially maintained colony. Other more recent studies of the species in Colombia were accomplished by CARRETERO & AHUMADA (2002).

Some of the best field data for the genus come from TERBORGH´s (1983) study at Manu National Park in southern Peru for *S. boliviensis*. MITCHELL (1990, 1994) also studied *Saimiri* behavior at Manu. SOINI (1986a) studied *S. boliviensis* at the Pacaya-Samiria National Park as part of a synecological study of the entire community. *Saimiri sciureus* was part of a wider synecological study in Surinam (FLEAGLE & MITTERMEIER, 1980); FLEAGLE *et al.*, 1981). This species was also compared to field data for *S. oerstedi* (MITCHELL *et al.*, 1991) using MITCHELL´s (1990) study of *S. boliviensis*.

BOINSKI (1986, 1987a, 1987b, 1988a, 1998b, 1989, 1991, 1992a,b, 1993, 1994, 1966, 1999a,b,d, 2000), BOINSKI & MITCHELL (1992, 1995), BOINSKI & NEWMAN (1988), and BOINSKI & TIMM (1985), BOINSKI & CROPP (1999), BOINSKI *et al.* (2002) and BOINSKI & GERBER (2000) compared this species with the closely realted *Saimiri oerstedi* in Panama. A comparative study of various *Saimiri* was accomplished by COSTELLO *et al.* (1993). Another more recent study on diet comes from Brazil (LIMA & FERRARI, 2003 and LIMA *et al.*, 2000).

Because of the many *S. sciureus* that were exported to colonies in the United States and Europe the species has been studied extensively in captivity. These studies are an important source of information that have increased our understanding of the natural history of the species. Nevertheless, with the new revisions it is important to be cautious about the identification of the species, since before 1984 all *Saimiri* in South America were known as "*Saimiri sciureus*". Fortunately, since the majority of *Saimiri* were exported from Leticia and Iquitos it is probable that the majority were, indeed, *Saimiri sciureus sensu* HERSHKOVITZ (1984). Nevertheless, an example of one research project claims that the animals used were the "*Saimiri*, Brazilian type", forcing the readers to select which of the four species known in Brazil were in fact used.

Following I list various important research projects that were done with captive individuals and groups which help us to understand some aspects of the biology and behavior of *Saimiri sciureus*: **social behavior**: (BALDWIN, 1967, 1968, 1969, 1970, 1979, 1971, 1992; BALDWIN & BALDWIN, 1971, 1981; CASTELL *et al.,* 1969; CASTELL & MAURUS, 1967; CANDLAND *et al.,* 1980; COE & ROSENBLUM, 1974; FAIRBANKS, 1974; GREEN *et al.,* 1972; MASON, 1971, 1973b, 1974a,b, 11 1975, 1978; MASON & EPPLE (1969); PLOOG *et al.,* 1963, 1967; PLOOG & MCLEAN, 1963; PRUSCHA & MAURUS, 1976; ROBINSON & JANSON, 1986; ROSENBLUM, 1968; RUMBAUGH, 1968; SCOLLAY & JUDGE, 1981; SMITH *et al.,* 1982; VAITLE, 1969, 1977a, 1977b, 1978); **feeding:** (FRAGASZY , 1978); **reproduction**: (DUMOND, 1968; DUMOND & HUTCHINSON, 1967; HARRISON, 1973; LANG, 1967; LATTA *et al.,* 1967; ROSENBLUM, 1968a, 1968b; RUMBAUGH, 1965; SMITH *et al.,* 1983); conservation: (SOINI, 1972; COOPER & HERNÁNDEZ-C., 1975; KAVANAGH & BENNETT, 1984); **development** (BALDWIN, 1969; COE & ROSENBLUM, 1974; LONG & COOPER, 1968; WISWALL, 1965; ROSENBLUM, 1968); **vocalizations**: (BARCLAY *et al.,* 1991; BIBEN & SYMMES, 1986; WINTER, 1972; WINTER *et al.,* 1966; SCHOTT, 1975; SOLTIS *et al.,* 2002; KAPLAN *et al.,* 1978, SMITH *et al.,* 1982); **learning** (RUMBAUGH, 1968).

The sizes of *S. sciureus* groups vary usually around 25-45 individuals where they have been studied (KLEIN & KLEIN, 1973a), although the group size is reduced in the black-water area of the lower Apaporis to around 10-20 (DEFLER, pers. obs.). TOKUDA (1968) found an average of 18 individuals in 12 groups of animals on the R. Putumayo. Sizes reported for Santa Sofía Island of the Colombian Amazon River of 42-54 may be at the upper end of group sizes for this species, since the soils of Amazonian Islands are much more fertile than many other Amazonian sites and these animals received suplemental feeding. Lower ranges found near black-water sites like the lower Apaporis river, surely are due to the poverty of the food supply and Santa Sofía is not a natural situation.

Multi-male and multi-female *Saimiri* troops studed by BALDWIN (1985) included 65% infants and subadults, 29% adult females and 6% adult males. In the Pacaya-Samiria the *S. boliviensis* troops are multi-male and multi-female and only 30-37% of the troop was made up of adults.

In a Peruvian study one home range was about 40 ha for *S. boliviensis* (SOINI, 1986a). In La Macarena a home range for *S. sciureus* was calculated ab about 65-135 ha (KLEIN & KLEIN, 1975), while in Manu National Park home range for *S. boliviensis* equaled about 250 ha (TERBORGH, 1983). On Santa Sofía the day range for *S. sciureus* was around 1.5 km (ROBINSON & JANSON, 1986).

The density of individuals has been similar in the various studies, since about 50 individuals/km² have been reported by TERBORGH (1983) and by SOINI (1986a) (for *S. boliviensis*), 50-80/km² by KLEIN & KLEIN (1974) for *S. sciureus*. This may be contrasted by the much lower density of about 8-9 individuals/km² on the lower Apaporis River, 4 individuals/km² on the upper Río Puré and 8.5-11.5 individuals/km² at four sites on the upper R. Purité, all in eastern Colombian rainforest (DEFLER, 2003).

An activity budget calculated by TERBORGH (1983) for *S. boliviensis* showed 11% rest, 27% travel, 11% feeding on plants, 50% feeding on insects (61% total foraging). *Saimiri* uses primarily quadrupedal walking and running, although these animals leap very well. The tail is very versatile as a balance and a strut. This species is basically frugivorous and insectivorous. In Manu, *Saimiri boliviensis* spent 78% of its feeding time eating fruits 1cm in diameter or less, the majority of them being soft-skinned. The mean heights of foraging varied during four time periods from 18-32 m and averaged 27 m. The mean crown diameter used averaged 21 m, which was considerably larger than that used by *Saguinus imperator, Saguinus fuscicollis, Cebus apella* and *Cebus albifrons*. *Saimiri* concentrated on figs when they were available or they concentrated on pure insectivory.

In the same study *Saimiri* fed on 92 species of fruits which fell into 36 families. Of overwhelming importance was the Moraceae, which provided

22 species, made up of fully 16 species of *Ficus*, the most important genus used by *S. boliviensis*. Other families of descending importance in terms of number of species chosen were Annonaceae (8 spp.), Leguminosae (7 spp.), Sapindaceae (5 spp.), Flacourtiaceae (4 spp.), Myrtaceae (4 spp.), Ebenaceae (3 spp.), Mensipermaceae (3 spp.) with the rest of the species providing either 1 or 2 species. Of animal food eaten, frogs, lizards and birds were chosen but more often orthopterans and lepidopterans (especially larva and pupae) were eaten with a few hymenopterans and coleopterans and some other miscellaneous invertebrates thrown in. *Saimiri* searched for invertebrates overwhelmingly on the surface of leaves and vines and very little on branches (TERBORGH, 1983). These primates are consummate predators of invertebrates, and they surpassed (in the Manu study) the capture rates of *Saguinus fuscicollis*, *S. imperator*, *Cebus apella* and *C. albifrons*. One expects that *S. sciureus* has a very similar diet to the above *S. boliviensis* diet.

Males are sexually mature at 2.5-4 years and females breed for the first time at about 46.3 months (HARVEY *et al.*, 1987). Females have an estrus cycle of 18 days, although much variation has been reported by other studies (HARVEY *et al.*, 1987; BALDWIN & BALDWIN, 1981). They are promiscuous and copulation may be initiated by either sex. There is a restricted breeding season when males become fat (especially around the shoulders), excitable and reproductively active DUMOND & HUTCHINSON, (1967) – the fatted male syndrome. Gestation is 170 days. A birth season of 1-3 months is usual and this usually occurs during the wet season. In La Macarena the birth season is during the month of April (KLEIN & KLEIN, 1976) and the Leticia area has been variously placed in November through March, according to the author (KLEIN, 1972; COOPER, 1968; BALDWIN & BALDWIN, 1981). Singletons are born with no twins reported.

Neonates cling immediately to the mother, moving to her back. At this point the infant has a prehensile tail which aids in staying on the mother. There is no special neonate coloring. The first two weeks are spent in sleeping and nursing, but with increasing amounts of time looking around. From 2-5 weeks the other females take an interest in the baby and it becomes very active on the mother´s back as well as being carried at times by other females. At 5-7 weeks the baby begins to leave the mother, who sometimes retrieves it worriedly. She pays much attention to the infant at this time. Adult males usually show no attention to the infant, but if an adult male comes near, the baby is able to give an erect penile display, which the male ignores.

From eight weeks to four months the mother-infant interactions wane as the infant spends more time away. The majority of allogrooming in *Saimiri* occurs during the third and fourth month between the mother and her infant. The infants begin peer-interactions during this time, especially playing. At 4-6 months the adult males interact increasingly with the mothers due to the approaching mating season, and during these times the infants direct genital displays towards the males who usually ignore them. From the fifth to tenth month increasing independence results in less nursing and decreasing daytime contact. Usually the infant is weaned at about six months (BALDWIN & BALDWIN, 1981).

The core of social groups is of adult females, to which all other age-sex groups are attracted. Females travel together and prefer each other´s company (BOINSKI *et al.*, 2002). However, MITCHELL (1994) looked at alliances and coalitions among male *S. sciureus*. Adult males maintain a dominance hierarchy which becomes accentuated during the breeding season. The principal expression of dominance is the penile display, which is used as a type of self-assertion. Dominant males often obtain the most copulations although not always so. Allogrooming is not common between adults and is not often seen. BALDWIN & BALDWIN (1981) provide a list of vocalizations of *S. oerstedii* which are very similar to *S. sciureus*. However, no comparative study has been done to tell us how similar these two species are. Nevertheless, *S. sciureus* combines shrieks, chirps, squawks in an amazing variety of calls. Their chuck call was recently studied by de THOISY *et al.* (2002), SOLTIS *et al.* (2002) and BOINSKI & MITCHELL (1997). BOINSKI (1996a, 1996b) compared vocal coordination in *S. oerstedi* and *S. sciureus* and «caregiver» calls (BOINSKI & MITCHELL, 1992, 1995).

Displays include urine washing, in which a few drops of urine are applied to the palm of the hand, which then is rubbed on the sole of the foot. This is done during situations of high anxiety or excitement. Chest rubbing occurs as the sternal gland is rubbed against a substrate, although the basis for this is not well-understood. Pile up displays occur when adult males pile onto each others´ back in conjunction with penile displays or fights and is connected with dominance.

Saimiri sciureus is well-known to associate with *Cebus apella*. Usually *Saimiri* is seen in the front of these associations, although *Cebus* is more likely to lead movements of the two species. Some of these interspecific groups last for long periods of time. *Saimiri* seem to take the initiative in joining a *Cebus* group, which then results in leading the *Saimiri* to fruits. Sometimes *Saimiri* troops are seen associating in a similar manner with *C. albifrons* groups. It is common to see interspecific groups of *S. sciureus* associating with mixed groups of *C. apella* and *Lagothrix lagothricha*. *Saimiri sciureus* also associate with groups of *Cacajao melanocephalus*. The hawk *Harpagus bidentatus* commonly travels with this primate, just as it travels with most primates. Since *Saimiri* is a comparable small species, there are probably many predators which threaten it. This monkey gives alarm calls at large birds, snakes, tayras, dogs and cats (BALDWIN & BALDWIN, 1981).

Conservation state

Saimiri sciureus does well in secondary vegetation and it survives close to human beings. The species is not considered endangered in Colombia for the above reasons and it is classified LC in the IUCN classification for Colombia. Nevertheless, an active national market in this species as pets perniciously threatens many local populations with extinction and this should be discouraged. The greatest pressures because of the national pet trade are probably on *S. s. albigena* along the east side of the Eastern Cordillera and on populations in the upper Magdalena river valley.

Saimiri sciureus was the most numerous Colombian monkey exported from Colombia before 1974 when this trade was banned by Colombian law (Código de Recursos Naturales de 1974; INDERENA Resolución 0392). In 1970, for example, Colombia officially exported 5,563 squirrel monkeys mostly from Leticia (COOPER & HERNÁNDEZ, 1975), but this does not count the illegal exportations which took place. Most of these animals ended up in the pet trade in the United States, but because of their abundance and low cost, many captive colonies were also founded, fortunately resulting in a large number of captive studies of the species. It seems important to determine how different genetically the upper Magdalena population might be from the eastern plains populations of the same subspecies.

Where to see it

Saimiri is commonly observed in most of the Amazonian national parks as well as parks in the upper Magdalena River valley, and is often easy to find outside of the parks, especially if a local person is enlisted for information as to where to look for it. Along the front range of the Cordillera any forest slope is liable to have a few *Saimiri* in it. This species is often common around gardens if they are not shot at by the farmer.

Family **Aotidae**

This family includes at least nine or ten species, seven probably represented in Colombia, the most *Aotus*-rich country known: *Aotus zonalis, Aotus griseimembra, Aotus lemurinus, Aotus brumbacki, Aotus vociferans, Aotus* sp. nov. (karyotype described by TORRES *et al.*, 1998) and probably *Aotus trivirgatus* (unconfirmed but probable) (DEFLER *et al.*, 2000; DEFLER & BUENO, 2003). Of course it is possible that the *Aotus* karyotype described by TORRES *et al.* (1998) may actually be *Aotus trivirgatus*, although possibly animals around Manaus with a karyotype 2n=51, 52 as suggested by SANTOS MELO & THIAGO DE MELLO (1985) may actually be *Aotus trivirgatus* or another species altogether. These small, cat-sized primates (around 1 kg) are the only nocturnal neotropical primates and exhibit very high karyotypic variation and very little phenotypic variation, making it quite difficult to recognize species.

In agreement with GROVES (2001) this author places all *Aotus* into a family apart from the Cebidae, Atelidae and Pitheciidae, despite the genus having been associated in one study or another with each of these three families. The taxonomic confusion is due, no doubt, to the fact that the *Aotus* group has similar characteristics to each of the three clearly defined primate clades listed above, despite the group being very ancient, having been recognized from the mid-Miocene Villavieja Formation of the La Venta area of Colombia. Further morphological and molecular research should finally resolve the placement of this group with one of the better-defined clades, or the *Aotus* may continue to be classified in its own group.

Night Monkey

Plate 4 • Map 14

FAMILY **AOTIDAE**
Genus *Aotus* HUMBOLDT, 1812

Taxonomic comments

Previously this genus was described as containing one species (*Aotus trivirgatus* HUMBOLDT, 1812) throughout its entire range in Central and South America (HERZHKOVITZ, 1949; HERNÁNDEZ-C. & COOPER, 1976). HERSHKOVITZ' comments (1983) are paraphrased below. With the discovery of polymorphic karyotypes (2n=54, 2n=53, 2n=52) in *Aotus trivirgatus griseimembra* from northern Colombia (BRUMBACK *et al.,* 1971) and a distinct karyotype (2n-54) from differently colored Peruvian *Aotus,* an awareness grew that there was probably more than one species in the genus and that "*Aotus trivirgatus*" needed to be revised. BRUMBACK (1973) formally named the Colombian species "*Aotus griseimembra*" and the Peruvian species "*Aotus trivirgatus*", and later he announced the discovery of a third karyotype from a specimen supposedly from Paraguay, but now generally conceded to have come from the Villavicencio area in Colombia (BRUMBACK, 1974; HERSHKOVITZ, 1983).

MA (1981) and MA *et al.* (1976a, 1976b, 1977, 1978, 1980) discovered polymorphic chromosomes in the Panamanian *Aotus* and later discussed the karyotypes of two other species of *Aotus* which had previously been described by de BOER (1974). The methods used by Ma and collaborators for classifying the karyotype, phenotype and geographic information of different *Aotus* became standard for distinguishing *Aotus* used in biomedical research (HERSHKOVITZ, 1983).

HERSHKOVITZ (1983) revised the entire genus and recognized that there were at least nine allopatric species of *Aotus*, including at least three or four species for Colombia (*Aotus lemurinus, Aotus brumbacki, Aotus vociferans* and probably *Aotus trivirgatus*), although it was generally recognized (see below) that there were more species to be discovered within the country and elsewhere.

GALBREATH (1983) constructed a parsimonious cladogram of phylogenetic relationships of *Aotus* species *fide* HERSHKOVITZ (1983), using comparisons of banded karyograms of different forms. He hypothesized a reconstructed ancestral karyotype of 54 and a common origin for *A. vociferans* and *A. brumbacki* and that the sequence "*griseimembra-lemurinus-brumbacki-vociferans*" may constitute "both a geographic and a karyotypic ordering" (GALBREATH *op cit.*)(see Table I from GALBREATH).

FORD (1994) examined the cranial morphometrics (n=193) and pelage (n=105) coloration of adult males and females. The analysis strongly suggested that *Aotus lemurinus, A. brumbacki* and *A. vociferans* exhibited a clinal gradient in most characteristics while populations of *Aotus trivirgatus* had no overlap and exhibited strong differences between that species and the neighboring *A. vociferans*. This supports the view that *A. trivirgatus* is a separate species from *A. vociferans*, but the study is negative morphometric evidence for the other Colombian species as being valid species. FORD (1994) went on to suggest only two species for the grey-necked group rather than the minimum of four presently suggested by HERSHKOVITZ (1983).

Nevertheless, DEFLER *et al.* (2001) criticize her interpretation, based on the high karyotypic variability of such a *A. vociferans* grouping, the type of social grouping, and on ecological factors.

DEFLER *et al.* (2001, 2003) and DEFLER (2003) recognize on karyological evidence at least six Colombian *Aotus* species, one of which was underlined by TORRES *et al.* (1998) as an undescribed karyomorph, whose origin is not yet known. If *A. trivirgatus* is confirmed for eastern Colombia, Colombian species of *Aotus* could equal seven, exceeding the diversity of *Saguinus* for Colombia.

Colombian *Aotus* species belong to the recognized grey-necked species group of *Aotus* (hair on side of neck grayish-agouti to mainly brownish agouti and chromosome pairs 6 and 7 discrete), found mainly north of the Amazon River. They can be contrasted with the red-necked group (*A. nancymai, A. miconax, A. infulatus* and *A. azarae*), which has partly to entirely orange or yellowish hair on the sides of the neck, like the chest and chromosome pairs 6 and 7 with a translocation of the arms. The red-necked group is found principally south of the Amazon River.

The *Aotus* species are one of various sibling species complex known in neotropical primates (others being the *Callicebus* complex) and one of the few sibling species complexes known in the mammalian order, although in the last few years more such groups of mammals have been discovered. Other such species complexes exist in some rodents, in some shrews, and in some groups of more primitive primates. This type of species group is most common in animals in which chemical senses (olfactory) are more highly developed than the sense of vision (MAYR, 1969, 1971). In this respect it is interesting that *Aotus* is nocturnal and belongs to a group of primates (Platyrrhini) with highly developed scent glands, although we know very little about this type of communication. Assumedly the nocturnal habitat provides very little selection pressure for external morphological differences and visual displays. It will be interesting in the future to identify other characteristics which have been selected more strongly in this genus. Chemical signals are still rather difficult to study, but perhaps vocalizations will show good differences between the species. Sympatric populations of two species (*A. nancymai* and *A. vociferans*) have been observed in Peru (AQUINO & ENCARNACIÓN, 1994).

MARKS (1987) hypothesized that numerous karyotypic variations are due to genetic background and inhibition of gene flow and resulting inbreeding.

If a primate species is philopatric and lives in small social groups, this would tend to produce a high evolutionary rate in the karyotype. He uses *Hylobates* as an example, although *Aotus* is another example of the influences on karyotypic evolution of this type of social system. This theory of the origin of karyotypic variability has been discussed more fully by WILSON *et al*. (1975), BUSH (1975), BUSH *et al*. (1977) to suggest an explanation for high chromosome variability based on type of social structure with small deme size, although MARTIN (1990) lists various criticisms of these ideas.

At this time the best field study is of *Aotus boliviensis* from Manu National Park in southern Perú (WRIGHT, 1978, 1984, 1986a, 1986b, 1990, 1994a, 1994b), but these data are probably a good general description of the other species (WRIGHT, 1981, 1994b).

Being a sibling species complex, these small monkeys have generally the same aspect, and this has complicated the eventual recognition of the various species. Generally the adults weigh around .8-1 kg, although there seems to be recognizable slight variations in size, according to the species, and captive animals often may pass these weights.

Most noticeable, perhaps are the very large eyes, an adaptation to the animal's nocturnal habits, although, unlike many nocturnal mammals and reptiles, there is no tapetum lucidum present (JACOBS, 1977a, 1977b; HERSHKOVITZ, 1977). There are often three longitudinal stripes above and from the edges of the eyes, which either extend towards the back of the head, leaving the head a contrasting agouti or brown color or alternatively they may fuse extensively, rendering the head entirely dark.

The tail is short, relative to the body and ends in a brush-like tuft. At the ventral base of the tail, specialized hairs occur on a glandular field that is immediately distal to the perineum. This seems best-developed in the male, and at puberty this glandular field begins to generate an unpleasant and foul-smelling tar-like substance which stains the hairs black. The hairs of this glandular field are specialized and are course and thicker towards the distal end than at their roots, and they split toward the apices into two, then four, creating a terminal brush-like filae which interlock on their neighbors to give a mat-like effect. This mat of hairs holds the viscous and odiferous substance that is secreted from time-to-time, creating a strong, smelly effect which must be socially very effective at night (HALL, 1960; DIXSON *et al.,*

1980). The ventrum varies from white through buff to yellowish, according to the species. Some species possess prominent large white spots above the eyes, which may function as a startle tactic in the dark and poorly-lit forest

Based on the HERSHKOVITZ (1983) key to the Colombian species and subspecies a key to the Colombian *Aotus* species' phenotype is provided below DEFLER *et al.* (2001) and DEFLER & BUENO (2003) prefer to view former subspecies of *Aotus lemurinus* as species, due to their chromosome differences

Key to colombian *Aotus* species **(modified from HERSHKOVITZ, 1983)**

1. Entire side of neck including area behind and below ear, grayish agouti or brownish agouti like flank or outer side of arm; throat from entirely grayish or brownish agouti to entirely orange or buff. (*Aotus lemurinus, A. vociferans, A. brumbackii,* and *A. trivirgatus*).

1.1. *Aotus lemurinus, A. zonalis, A. griseimembra*

• interscapular whirl or crest absent

Inner side of limbs entirely grayish agouti like outer side or with orange or buffy chest and belly extending to or slightly beyond midarm or midleg, rarely to ankle; pelage of dorsum short and adpressed to long and lax; coloration of upper parts of body variable; middorsal band, if present, broad, blackish, brown or orange and not well defined, temporal stripes separate or united behind.

Aotus lemurinus (2n=58), hair of back and side long and soft

Aotus griseimembra (2n=54, 56), hands and feet clear agouti or light brown, hair short, adpressed

Aotus zonalis (2n=52, 53, 54) hands and feet blackish, hair short, adpressed

1.2. *Aotus vociferans* (2n=44,45,46)

• interscapular whorl or crest present

temporal stripes nearly always united behind; malar stripe well-defined to absent; pedal digital tufts thin, not extending beyond ungues; size smaller, tail length 340(308-363) mm.

1.3. *Aotus brumbackii* (2n=50)

• interscapular whorl or crest present with raised hairs directed backward and laterally; gular gland long (5 cm), thin, the surrounding hairs extending outward from sides.

Holotype: blackish temporal stripes extending from outer corner of eyes convergent but not united except by shadowy prolongations. Blackish mid-coronal stripe pointed on brow, expanding on crown to twice width of either temporal stripe; crown between blackish stripes buffy, the hairs with long blackish tips; malar stripe, an extension of temporal stripe, blackish and well defined; paired pale supraorbital patches creamy in front becoming buffy behind; buffy patches of each side not connected by pale band; side of head and neck below ears grayish agouti to buffy agouti, chin creamy; anterior border of throat drab, remainder of throat without skin.

1.4. *Aotus trivirgatus* (2n = 51.52 according to SANTOS MELLO & THIAGO DE MELLO (1986)

• interscapular whorl or crest absent

Temporal stripes well-defined and often parallel sided in front and usually separate in back (Caparú specimens had a very slight black wash between the ends of the stripes; inner side of limbs with orange or buff of chest and belly extending to wrist and ankle, terminal half of hairs unbanded; pelage of dorsum short, adpressed, upper parts dominantly grayish, sometimes buffy agouti with narrow, strongly upper parts dominantly grayish, sometimes buffy agouti with narow, strongly contrasting orange middorsal band.

2. Part or entire side of neck including area behind and below ear, and not less than medial portion and posterior half of throat orange or buff like chest (*Aotus nancymai* and *Aotus nigriceps*).

2.1. *Aotus nancymai* (information provided for possible confirmation in Leticia or trapezium)

interscapular whorl absent

throat with at least anterior portion of sides mostly or entirely grayish agouti.

upper surface of proximal portion of tail orange with blackish stripe or overlay; ventral surface, mostly blackish.

2.2. *(Aotus nigriceps* (information provided for possible confirmation in Leticia or trapezium)

interscapular whorl absent

throat entirely orange

Common names

Other English: Nocturnal Monkey, Douroucouli (The name douroucouli is an onomatopaic version of the vocalization of *Aotus* based on a report of JUSTIN GOUDOT, *fide* ALSTON, 1879, HERNÁNDEZ C. & COOPER, 1976).

Colombia: *marta*, *martica* and *martica* in northern Colombia; *marta* in the departments of Antioquia, Caldas, Quindio, Risaralda, Tolima and northern Valle; *mico dormilón* in Meta; *mico de noche* in the center of Colombia and Amazonia; *mono tigre* in Orinoquia (Venezuelan origin) and *mico cagao* in Santander. In general, with the exception of *maco cagao, mico dormilón* and *sorbehumo*, these names are also used for *Potos flavus* (Procyonidae), *Bassaricyon gabii* (Procyonidae), *Caluromys* sp. (Didelphidae) and *Cylclopes didactylus* (Myrmecophagidae). *Sorbehumo* also refers to hawks like *Buteo platypterus* and *Buteo swainsonii* due to their habit of hunting around fires. French: *Singe de nuit*; German: *Nachtaffe* ; Brazil: *macaco da noite*.

Indian languages

HERNÁNDEZ C. & COOPER, (1976); other sources, pronounced as in Spanish: *kuakuanama* or *watají* (Carijona); *un* (Indians of Chocó); *aspúlvur* (Cuna); *mocori* (Curripaca); *una* or *–ra* (Embará - Waura-na); *mucúali* (Guahibos of Vichada): *jimok+* (Huitoto); *tutamono* (Ingano); *kulok* (Jébero); *ocui* (Macú); *ucu* (Macuna); *teemu* (Miraña); *tohomimi* or *cacha* (Muinane); *cubaime* (Muzo, an extinct tribe of Boyacá, *fide* Simón (1882-1892); *ohótseyu, watsaña* or *wasayu* (Piaroa); *macuriya»* or *puípo* (Puinave in Vaupés); *tutamono* (Quechua, a dialect of the Inganos of Putumayo, also Kamsa and Indians of Sibundoy); *unamihu* (Siona); *mucutica* (Tanimuca); *jane* (Ticuna); *ñamisenimi* (Tucano); *moco'o* (Yucuna). Outside of Colombia *mirikina* (Guaraní, *fide* HILL, 1960).

Geographic range

The genus *Aotus* is known throughout Colombia, except in the northeast of Guajira, northern Vichada north of the Tomo River, at altitudes of more than 3200 m, and in some local forested areas of tree-savanna in Guainía, Vaupés and Guaviare in the geological zone known as the Guianean shield. *Aotus* has also not been collected in Nariño but has been reported there. With the exception of mangroves, *Aotus* lives in most forest types found in Colombia, including secondary growth and coffee plantations.

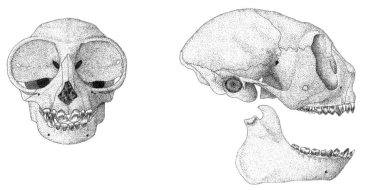

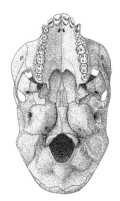

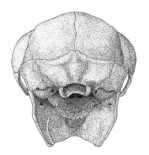

Figura 22. Views of the skull of *Aotus griseimembra.*

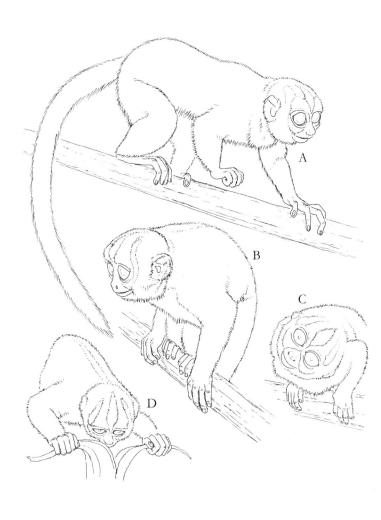

Figure 23. Different positions of *Aotus*. a) Moving, b) Resting on a branch, c) Alert position, d)Foraging

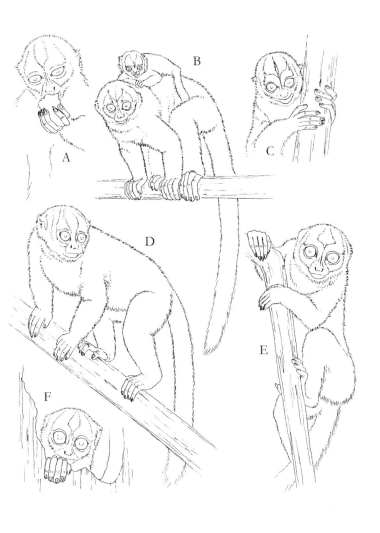

Figure 24. Different positions of *Aotus*. a) Feeding, b) Carrying an infant., c- e) Climbing, f) Taking a look from a tree hole.

Andean or Lemurine Night Monkey

Aotus lemurinus (I. GEOFFROY-ST. HILAIRE, 1843)

Chocoan Night Monkey
Aotus zonalis (GOLDMAN, 1914)

Grey-handed Night Monkey
Aotus griseimembra (ELLIOT, 1913)

Plate 4 • Map 1

Taxonomic comments

HERSHKOVITZ (1983) accepts *Aotus lemurinus* as a valid species, and divides it into two subspecies, *A. l. lemurinus* and *A. l. griseimembra*. The subspecies, according to HERSHKOVITZ (op. cit.) are distinguishable by karyotype only, *A. l. lemurinus* exhibiting 2N= 55, 56 and *A. l. griseimembra* exhibiting 2N= 52, 53, 54. Another karyotype 2N=58 (DEFLER *et al.*, 2001) was found by the young Colombian primatologist (now tragically deceased) JAIRO RAMÍREZ. RAMÍREZ who believed this *Aotus* was a new species, but he was unable to describe the species before his death. DEFLER *et al.* (2001) have carefully looked at the evidence for *Aotus hershkovitzi* and concluded that that this taxon represents the karyotype for *Aotus lemurinus lemurinus* and that karyotypes assigned by HERSHKOVITZ (1983:Table II) represent *Aotus l. zonalis* (*A. zonalis*).

HERNÁNDEZ-C. & COOPER (1976) called the population in the Pacific coastal lowlands of Colombia *Aotus trivirgatus zonalis*, which in the HERSHKOVITZ revision would be equivalent to *Aotus lemurinus zonalis*. HERNÁNDEZ-C. & COOPER (1976) used the name *A. t. lemurinus* for the populations in the upper altitudes of the three Andean cordilleras at altitudes from around 1000-1500 m upwards to the tree limit at about 3000-3200 m. DEFLER *et al.*, (2001) and DEFLER & BUENO (2003), having considered each karyomorph, conclude that all Panamanian and Chocoan lowland *Aotus* are referable to *Aotus zonalis*, while *Aotus* of the Cordilleras of the Andes, perhaps above about 1000-15000 m are *Aotus lemurinus* and include the specimens referable to *Aotus hershkovitzi*. This means that the karyotype of *Aotus lemurinus* is 2n=58. Lowland *Aotus* east of the Sinu river are referable to *Aotus griseimembra*, but the highland form of *Aotus* in the Serra Nevada de Santa Marta may actually be another taxon. Thus, this author separates at least three species from the previous subspecies of *Aotus lemurinus*.

Identification

According to HERNÁNDEZ-C. & COOPER (1976), *A. l. lemurinus* is a subspecies that is rather variable and which often has two color phases, which can be found in the same family group. One color phase is decidedly gray-brown and the other is more intensively reddish-brown on the upper parts. Nevertheless, one can find an intermediate color scale. The lower parts are always somewhat opaque yellow, indistinguishable from *A. griseimembra* and *A. zonalis*. The pelage is extremely long and soft and this is one of the most valuable characters for distinguishing *Aotus lemurinus*. The hands and the feet of this taxon are notable for their variation in color and by their absence of correlation with each other, even in one individual. The specimens examined from the Cordillera Occidental (around Cali) and the majority of the specimens from the Cordillera Central have hairs with black points on the hands and on the feet, and these reach at least to the distal carpals and tarsals. Some specimens from the Cordillera Occidental and some from the Cordillera Central seem to have a variable color of greyish (grizzled) in the regions of the metatarsals and the metacarpals, due to the reduced extension of the dark points, which permits the exposure of the hair bases which are less dark.

In the Cordillera Oriental of the Andes a complete extension of individual variation in the above characteristic is found, *i.e.* from the hair tips that are extensively black to hair tips that are very reduced in the black coloration, typical of *A. griseimembra*. Some variation of this character has also been observed between the hands and the feet of the same individuals.

Aotus griseimembra is characterized typically by its relatively short and somewhat tight pelage, with the lower parts having a brownish or yellow-brownish tone which varies from opaque to light yellow. The dorsal parts of the hands and feet are light coffee brown colored with darker hair tips, although not notably so (HERNÁNDEZ CAMACHO & COOPER 1976).

Aotus zonalis is distinguished as being completely homogeneous throughout its entire distribution in Colombia. The phenotype of this taxon corresponds to *A. griseimembra* in all respects except that the dorsal hair of the hands and feet is dark brown or black (HERNÁNDEZ-C. & COOPER, 1976).

Geographic distribution

Aotus lemurinus is found from about 1,000 -1,500 m upwards on the slopes of the Cordillera de los Andes, while *A. zonalis* is found in the lowlands from the Panamá border to the Ecuadorian border and in the north to about the Sinú River valley and possibly in the upper San Jorge valley to the region of Puerto Valdivia in northern Antioquia, although the exact contact with *A. griseimembra* is unknown (HERNÁNDEZ-C. & COOPER, 1976; DEFLER, 2003). *Aotus griseimembra* extends either from the Sinú River valley or more eastwardly to the Venezuelan border and includes the Magdalena River Valley and the uplands of the Sierra Nevada de Santa Marta.

The discovery of a new karyotype of *Aotus (A. hershkovitzi)* in uplands of the western slope of the Eastern Cordillera and the detection of other specimens of this taxon from the Cordillera Central of Colombia suggests that the *Aotus* of the cordilleras is *Aotus lemurinus* and that its karyotype is 2n=58. *Aotus hershkovitzi* is a junior synonym for *Aotus lemurinus*.

Aotus griseimembra is also found in Venezuela, to the west and south of Maracaibo and *A. zonalis* in Panama. *Aotus lemurinus, A. griseimembra* and *A. zonalis* are found in all types of forest including secondary forest and coffee plantations (although it should be noted that new strains of coffee are eliminating the necessity of the use of secondary forest vegetation as a shade for coffee) which is eliminating this habitat for *Aotus*). In a census done by HELTNE & MEJÍA (1978), however, *A. griseimembra* was found only in highly diverse forest, and this study found evidence that *Aotus* does not tolerate degraded forest well. Nevertheless, GREEN (1978) reported *Aotus* in secondary forest 5-15 years old that contained trees of 10-20 m.

Natural history

There have been very few field studies of these three species. THORINGTON *et al.* (1976) did a short study of the movements of one individual, probably *A. zonalis*, released with a transmitter on Barro Colorado. HELTNE & MEJÍA (1978) and HELTNE (1977) censuses *A. grisemembra* for the Pan American Health Organization (PHO). There are only scattered natural history notes to supplement the above. Moynihan (1964) studied individuals (*A. zonalis*) in cages on Barro Colorado for some of the first behavioral information.

DIXSON (1982, 1994) and DIXSON & FLEMMING (1981) studied reproductive behavior in *A. grisembembra*, while DIXSON *et al* (1980) also studied puberty in the male owl monkey. ERKERT & GROBERT (1986) examined in captivity the relationship of body temperature and light intensities in *A. lemurinus* (or more probably *A. griseimembra*). Umaña *et al.* (1984) describe the establishment of a breeding colony of this species.

HELTNE (1977) observed *Aotus griseimembra* groups of 2-4 animals in northern Colombia. GREEN (1978) calculated a density of 1.5 animals/km² of *A. griseimembra* for one area and HELTNE (1977) calculated a density of 150/km² in another small forest which probably was a refuge for individuals from nearby forest which had been cut down. Like all species of *Aotus*, these animals often spend the day in tree holes or in thick vegetation.

No time budget has been calculated for these species A study of *Aotus nigriceps* showed a time budget of 21% traveling, 53% feeding, 22% resting and 4% agonistic behavior (WRIGHT, 1981). *Aotus* is one of the most quadrupedal monkeys, although they are powerful leapers and runners (MOYNIHAN, 1964). Using stomach contents HLADIK *et al.* (1971) calculated a diet of 65% fruits, 30% foliage and 5% animal prey (including birds' eggs, insects and cocoons).

These monkeys are monogamous and generally give birth to one infant every year, although twins have been reported (WRIGHT, 1978, 1981, 1984, 1985, 1986a,b; DIXSON, 1982, 1994). The gestation period is 133 days for these species (HUNTER *et al.,* 1979). MERRIT (1993) studied reproductive parameters of what was probably *A. griseimembra*. After birth the infant passes to the father's care on the first or second day. He carries the infant at all times, passing it to the mother for nursing. DIXSON *et al.* (1980) and DIXON & FLEMMING (1981) describe the maturation of males of *A. griseimembra*, noting the development of the subcaudal scent-marking gland during puberty, which first becomes evident with the stiffening and discoloration of the overlying hairs at around 282-370 days. These changes are apparently androgen dependent, as treatment of a prepubertel male with testosterone caused the animal to develop its subcaudal gland. Development of this gland is apparently not retarded within a family group. This gland is found in other *Aotus* as well, recently being observed in two *A. trivirgatus* (*sensu* HERSHKOVITZ, 1983). The odor varies from week to week from extremely strong and offensive to human nostrils to practically nonexistent.

Some olfactory behaviors are utilized to convey social information, such as: (1) *social sniffing*, during an initial encounter, (2) *anal rubbing* on the substrate during slight hostile situations and (3) *urine washing*, moistening palms of hands and feet with urine, possible in slightly hostile situations. MOYNIHAN (1964) studied communication in this genus (probably *A. zonalis*) in captivity. He did not find many behavioral displays used to convey information. Agonistic *arched back displays* (somewhat like *Callicebus*) are used with head down, *pilo-erection*, *stiff legged jumping*, urination and defecation. Stiff-legged jumping, however, is also used in play behavior by *Aotus trivirgatus* (tame individuals in the Caparú Biological Station).

Vocalizations are quite variable and MOYNIHAN (1964) described nine distinct calls for *A. zonalis*: (1) *Rough grunt* – low tone produced with inflated gular sack and directed towards another group; (2) *Resonant grunt* – series of low notes repeated 10-17 times and usually accompanied with arch-back display with head down and followed by stiff legged jumping; hostile; (3). *Scream* - High-pitched and wavering and given when grabbed by humans; when siezed by predators? (4). *low trill* - Bubbling series of 3-12 short, low-pitched tones, rising in pitch with mouth closed and uttered to opposite sex or near a preferred food item. (5). *Moan* - Brief, soft note uttered with mouth closed and without throat inflation; hostile. (6). *Sneeze grunt* - Staccato, metallic clicks, 1-3 notes, sometimes accompanied by grunt in situation of perceived danger. (7). *Gulp* - Loud, liquid sound, 2-5 notes used during locomotion at dawn or dusk. (8). *Hoot* - Similar to owl and supposedly used as contact call. (9). *Infant squeaks* - High-pitched series when infants need attention (et-epimeletic). These three species undoubtedly have similar interspecific interactions to the other *Aotus* species.

Conservation state

These species are undoubtedly the most threatened *Aotus* and in Colombia all northern species are VU and *Aotus griseimembra* possibly EN, due to extensive destruction of habitat and to use in biomedical research (STRUHSAKER *et al.*, 1975), since the subspecies is highly susceptible to infection by the malarial *Plasmodium* and is the preferred research model for this deadly disease.

Where to see it

Aotus lemurinus, A. griseimembra and *A. zonalis* can be searched for in a remnant native forest of more that a few hectares (*A. lemurinus* above 1,500 m.). Particularly they should be evident during brightly-lit moonlight when the animals are most active and vocal.

Brumback's Night Monkey

Aotus brumbacki (HERSHKOVITZ, 1993)

Plate 4 • Map 14

Taxonomic comments

Aotus brumbackii has a diploid chromosome number of 50. The holotype was originally believed to be from Paraguay, southern Brasil and northern Argentina (BRUMBACK, 1974; BRUMBACK *et al.,* 1974). However, the specimen was matched via skins and karyotypes from animals around Villavicencio (YUNIS *et al.,* 1976) which provide the paratype (2N = 50 females; 49 males). In this species the Fundamental Number (NF) is equal to 58 (the number of chromosome arms). GROVES (2001) classification of this taxon as a subspecies of *A. lemurinus* is not karyotypically convincing and I do not accept it in this book.

Identification

The chest and belly of this species is pale orange, hair bases buff to whitish; inner side of arms and legs with pheomelanin of chest and belly extending to elbow and knee, while the temporal stripes may be separate or united behind (HERSHKOVITZ, 1983).

HERSHKOVITZ (1983) distinguished *Aotus brumbackii* phenotypically from *A. lemurinus* by the interscapular crest with raised hairs that are directed backwards and laterally and also by a large gular gland. *Aotus lemurinus* has neither an interscapular crest nor a gular gland. According to HERSHKOVITZ (*op cit.*) *Aotus vociferans* is distinguished from *A. brumbackii* by its more circular gular gland and by its interscapular whirl (instead of a crest), the whirl with centrifugal hairs.

Geographic range

The distribution of *Aotus brumbacki* is very imperfectly known, but extends at least from eastern Boyacá eastward into the forested parts of Meta. A paratype is from Agua Dulce, Restrepo, Meta at 1500 m. A second animal was collected from Tenza, Boyacá at 1543 m and other specimens were from "Villavicencio", a demarcation that is not very specific, since they were probably animals sold in the market or in the street. Other animals have been acquired from Cali, so that it is probable that the species has a wide distribution, perhaps in all of the Andean lowlands.

Aotus brumbackii may be a species endemic to Colombia. North of *A. brumbackii* a new karyotype (2N=58) has been discovered, which apparently is the real karyotype for *Aotus lemurinus,* the karyotypes published by HERSHKOVITZ (1983) as *A. lemurinus* being actually for *A. zonalis. Aotus vociferans* is distributed south of *A. brumbackii.*

It is still not clear which species of *Aotus* is found to the east, towards the Orinoco, although it does not appear to be *A. brumbacki.* Nevertheless, a female specimen of *Aotus* captured live near Maipures, on the banks of the Orinoco had a karyotype apparently of 2N = 50, as does *A. brumbackii,* but the Maipures chromosome arms were arranged differently and there were problems with the quality of the preparation which make it unreliable. The Maipures specimen (which is found in the collection at IvH did not agree morphologically with *A. brumbackii*). HERSHKOVITZ (com. pers.) identified this Vichada specimen as *A. trivirgatus,* although HERNÁNDEZ-C. (pers. com.) did not agree. This species is found in gallery forest and closed canopy forest.

Natural history

Although the species has not been confirmed, SOLANO (1995) observed what is probably this species for a total of about 345 hours of observation during six months on the right bank of the Duda River. The group observed by SOLANO (1995) contained one adult male, one adult female and a young animal which was said to be born in October. The home range was 17.5 ha during the six months of observation. The average day range was 837.3 m (n=53).

SOLANO (1995) calculated an activity pattern for these animals as follows: 32% moving, 33.2% resting, 15% eating, 2.8% social interactions, 16.4% vocalization. There was a clear increase in activity during the lunar light phase. These animals sleep in hollow trees and dense vegetation.

SOLANO (1995) calculated a generalized diet composed of 59% fruits, 13% flowers, 28% arthropods. Plant species consumed by *A. brumbackii* in Tiniguas National Park (*fide* SOLANO, 1995) are as listed: Arecaceae (*Jessenia bataua, Socratea exorrhiza* and *Iriartea deltoidea*); Bombaceae (*Bombacopsis quinata* and *Quararibea cf. asterolepis*; Burseraceae (*Protium cf. apiculatum, Protium sagotianum* and *Crepidospermum rhoifolium*); Cecropiaceae (*Cecropia membranaceae* and *Cecropia ficifolia;* Gutifferae (*Garcinia macrophylla*); Caricaceae (*Jacaratia digitata*); Leguminosae (*Inga* sp.); Melastomataceae (*Bellucia pentamera*); Moraceae (*Coussapoa orthoneura, Ficus insipida, Ficus* sp., *Perebea xanthochyma, Pourouma bicolor, Pseuldomedia laevis*; Nictaginaceae (*Guapira myrtiflora*); Passifloraceae (*Passiflora nitida*); Rubiaceae (*Alibertia etulis*); Papilionaceae (undetermined); Myrtaceae (*Eugenia cf. cowanii*); Undetermined; Undetermined.

Conservation status

This species is classified VU in Colombia, using the latest IUCN criteria (DEFLER *et al.*, 2003). The species' true distribution is not known, but where it has been confirmed around Villavicencio the forested areas have been destroyed. If the distribution proves to be restricted immediate actions will have to be taken for the species' protection. A priority for this species is to determine its geographic range which may be endemic to Colombia.

Where to see it

The only known places for observing this species are the remnants of forest around Villavicencio, Meta.

Amazonian Night Monkey

Aotus vociferans (SPIX, 1823)

Plate 4 • Map 14

Taxonomic comments

This species was originally described as having a diploid number of 46. Nevertheless, DESCAILLEAUX *et al.* (1990) have pointed out that the species exhibits at least three diploid numbers 46, 47, and 48, and that the karyotype 46 and 48 appear in the population with the same frequency (32/68; 35/68) with 47 being very rare. This karyotype is probably a balanced polymorphism like *Aotus griseimembra* and as probably is the case for *A. zonalis,* although more data on karyotypes are needed to confirm this for *A. vociferans* and *A. zonalis*

Common names

Other English: Nocturnal Monkey

Indian languages

Mocori (Curripaco); *mucúali* (Guahibos of Vichada); *jimok+* (Huitoto, *fide* TOWNSEND *et al.*, 1984); *tutamono* (Ingano); *kulok* (Jébero); *ocui* (Macú-yujup; *ucu* (Macuna); *teemü* (Mirañà); *tohomimi* or *cacha* (Muinane); *ohótseyu, watsañà, wasayu* (Piaroa); *macuriya, puípo* (Puinave in Vaupés); *tutamono* (Quechua, dialect of Ingano from Putumayo as well as Kamsá Indians of Sibundoy); *unamihué* (Siona); *mucuticá* (Tanimuca); *jane* (Ticuna); *ñàmisenimi* (Tucano); *moco'o* (Yucuna).

Identification

This small monkey has an average weight of 707.5 g for males and 698 g. for females, based on weights of twenty animals for each sex, captured in Peru (MONTOYA *et al.*, 1990). The short tail measures an average of 340 mm (range 308-471; n=36). The body usually is short-haired and dark, usually greyish with hands and feet clearer than the dorsum. According to HERSHKOVITZ (1983) this species possesses an interscapular whorl with centrifugal hairs. The temporal stripes nearly always are united behind while the malar stripe may not be present and the gular gland, which is more or less circular, has hairs which extend outwards from the center, although FORD (1994) considers that this character is not reliable as a diagnostic character. The stomach is a bright yellow, as compared to *Aotus* further to the north.

Geographic range

HERSHKOVITZ (1983) suggested that this species was found from the Tomo River in Vichada and perhaps from the upper Guyabero River southward, throughout the Colombian Amazon, although it is not clear to what altitude it reaches along the east side of the eastern Cordillera.

Two live specimens from the eastern Colombian Amazon and one from the lower Apaporis, Brazil do not fit the phenological description given by HERSHKOVITZ (1983) for *A. vociferans*, since neither do the malar stripes of these individuals unite in back nor do the animals possess an interscapular whorl. The phenotypic description of these three individuals best fits *Aotus trivirgatus* as defined by HERSHKOVITZ (1983). No karyotype has been done on this individuals. This species is found in moist forest of *terra firme* and in flooded forest along the edges of rivers and lakes.

The distribution of *Aotus vociferans* in the Brazilian Amazon is said to be between the Solimoes and Negro rivers, and extends south from Colombia to the Marañon and Amazonas rivers, including all of the Ecuadorian Amazon (HERSHKOVITZ, 1983).

Natural history

Very little natural history is available. AQUINO & ENCARNACIÓN (1986[a], 1986b, 1986c, 1989, 1994[a], 1994b) and AQUINO *et al.* (1990, 1993) gathered data on population structure, sleeping sites, distribution and densities of the species in Peru. Based on counts of 82 captured groups the average group size was 3.3 individuals. The groups usually were comprised of the breeding pair and one or two offspring. Groups of four were common between October and December, and sometimes they included adult male offspring. Sex ratio of the group was 1:1.1 (AQUINO *et al.*, 1990).

The group range is unknown, although a study of a group of four *A. nigriceps* in southern Peruvian rainforest showed a home range (during nine weeks) of 3.1 ha, which overlapped with other *Aotus* groups (WRIGHT, 1978, 1981, 1984, 1985, 1986a, 1986b, 1994a, 1994b). *Aotus vociferans* densities were censused in lowland forests at 33 individuals (10 groups)/km^2 and at 7.9 individuals (2.4 groups)/ km^2 in highland or upland forests (AQUINO & ENCARNACIÓN, 1986a). Another set of data showed 30-52 individuals/km^2 or 9-14 groups/km^2 (AQUINO *et al.*, 1990). The species is not commonly detected on the lower Apaporis River and seems to have a fairly low density. Day range for this species is unknown, although a study of a group of *A. nigriceps* averaged 252 m path length (range 60-450 m) (WRIGHT, 1978, 1994b).

No activity budgets have been calculated in the forest, although data from *A. nigriceps* and from laboratory studies indicate heightened activity during the early part of the evening with feeding bouts interspersed with frequent rest periods. Quadrupedal locomotion and leaping are particularly important for this monkey.

There are no data for natural habitat, but data on *A. nigriceps*, *A. brumbacki* and from the laboratory show an increase in activities during the earliest part of the night with periods of foraging alternated with frequent periods of rest. Moonlight strongly increases activity patterns in *A. nigriceps*, in *A. brumbacki* and in *A. vociferans* (WRIGHT, 1978; SOLANO, 1995; pers. obs.).

DEFLER (obs. pers.) observed two groups (presumably of this species but east of the Apaporis River) in hollow tree trunks in flooded igapó forest at the edge of the lake, facing the water and another group was seen at the top of a dead palm trunk at about 12 m. The groups at the lake's edge were discovered when the rising water was only about 1 m from the mouth of the sleeping nest. At lower water these nests would be higher at 6-10 m respectively. Another nest was observed in a dead palm tree with the opening via the broken off trunk at 12 m.

AQUINO & ENCARNACIÓN (1986a, 1994b) studied the sleeping sites of 35 distinct groups of *Aotus vociferans* and of 42 groups of *A. nancymai* in the northern Peruvian Amazon. *Aotus* used tree holes principally and for *A. vociferans* 11.4% of the nests were in the underbrush at about 10 m from the ground, 54.3% were found in the lower parts of the canopy (10-19 m), 20% were found in the middle parts of the canopy (19.1-25 m), while 8.6% were found in the high canopy. This study found that *A. vociferans* only used tree holes, while *A. nancymai* also slept in vine tangles and in dense vegetation. *Aotus* nests always have a second door for escape from danger. The two species sometimes share their nest with *Potos flavus, Bassaricyon gabbii* and *Coendu bicolor*.

Like all *Aotus*, this species has a diet made up of small fruits small invertebrates, perhaps complemented with leaves (see HLADIK *et al.*, 1971; HLADIK & HLADIK, 1969). Births may occur year-round with the majority from November-January (AQUINO *et al.*, 1990). One infant is born and is carried by the father, who passes the single infant to the female for nursing. The species has vocalizations similar to the other *Aotus* species, but interesting differences are expected when studied, since these animals, being nocturnal, depend on scent and vocalization rather than visual displays.

Conservation state

Aotus vociferans probably is not endangered but it is found in a habitat that still is very extensive and the animal is rarely hunted. The species has been frequently collected for a reasearch laboratory in Leticia, but no data are available regarding numbers taken from the wild. The species is classified LC by the IUCN internationally and now is classified LC by the Colombian national committee of IUCN classification of mammals.

Where to see it

This species can be observed south of the Guaviare/Guayabero River in lowland forest. It is particularly obvious on moonlit nights.

Aotus trivirgatus
(HUMBOLDT, 1812)

Plate 7d • Map 3

Taxonomic comments

According to HERSHKOVITZ (1983) and FORD (1994) the species is recognizeable by their parallel head bands (they are united in *A. vociferans*) and by their lack of any interscapular whirl or tuft. Parallel or united head bands was considered by FORD (1994) to be a highly reliable distinguishing characteristic between these two species, but her sample for *A. vociferans* was very small.

Geographic range

HERSHKOVITZ (1983) suggested the distribution of this species to be to the east of the Río Negro and Orinoco in Brazil and Venezuela, although probably extending westward to eastern Colombia, although there was no material to confirm this Colombian distribution. Nevertheless, several years back this author acquired (unpublished data) two live individuals, one from the middle Caquetá River above the mouth of the Mirití-Paraná and another individual from the lower Inírida River (the left-bank affluent Cañò Bocón), both of which fit HERSHKOVITZ' (1983) description of *A. trivirgatus*, and suggest that this species may be wide-spread in eastern Colombia and may be sympatric with *A. vociferans*. Additionally, HERSHKOVITZ (com. pers.) believed that a live animal collected in eastern Vichada, on the banks of the Orinoco near Maipures (no. x on map and in list of collections and observations; which appeared to have a karyotype 2N = 50 in poor preparations which are now lost) fitted the description of *Aotus trivirgatus* as well. Considerable more collection and observation will have to be done in order to understand the extent of the distribution of this species within Colombia.

An abstract for the X Congress written by SANTOS MELLO & THIAGO DE MELLO (1985) mentioned karyotypes of 2n=51 for males and 2n=52 for females from *Aotus* captured around Manaus. If *A. trivirgatus* turns out to have a karyotype of 2n=50, then the Manaus *Aotus* are likely to be another species entirely. It is vital to determine karyotypes for *A. trivirgatus* and for animals in eastern Colombian Llanos and Amazon regions to be able confirm the actual presence of this species in Colombia.

Aotus nancymai and *Aotus nigriceps* in Colombia?

Several years previously JORGE HERNÁNDEZ-CAMACHO and PHILIP HERSHKOVITZ visited the local primate lab in Leticia, Amazonas and observed individuals of *Aotus nancymai* and *Aotus nigriceps* in cages. These biologists were assured that the two species had been captured in Colombia along with the *A. vociferans* that were also observed in captivity there. Publications relying on material from this primate holding facility have also identified these three species as having been collected on the Colombian side of the Amazon River. *Aotus nancymai* is known from the opposite bank of the Amazon and *A. nigriceps* appears downriver (NINO-VÁSQUEZ *et al.*, 2000; DIAZ *et al.*, 2000; DAUBENBERGER *et al.*, 2001; MONTOYA *et al.*, 2002). Since there is no other published material confirming the presence of the two well-known right bank species also being from the left (Colombian) side of the Amazon, nor evidence of the collection sites and since *Aotus* are known to be brought from the south side of the Amazon to be sold to this laboratory, I remain skeptical that either *A. nancymai* or *A. nigriceps* exist in the Colombian part of Amazonia and do not include them here as part of the Colombian primate fauna.

Family **Pitheciidae**

This family is recognized in this book as including four genera: *Cacajao, Chiropotes* and *Pithecia* grouped as the Pitheciinae and *Callicebus* grouped as Callicebinae, according to the scheme of Schneider in SCHNEIDER & ROSENBERG (1996) and provisionally accepted by RYLANDS *et al.* (2000). Three genera from this family are found in Colombia, *Cacajao* with one species *C. melanocephalus, Pithecia* with one species *P. monachus* and *Callicebus* with two or (according to another taxonomic viewpoint) three or more species. The Pitheciinae are about the size of a domestic cat at around 2-3 kg, the *Cacajo* being slightly larger than the *Pithecia* and they have always been recognized as a coherent ecological group (ecological grade) of seed predators.

The Callicebinae are smaller at about 1-1.5 kg. with *C. torquatus* being the larger of the two species recognized in this book. It was not until recently that it was also recognized that seed predation (without the specialized dentary of the Pitheciinae) is also very important at least to *C. torquatus*, being perhaps a reflection of their phylogenetic connection with the specialized Pitheciinae.

Morphological and molecular studies confirm the relation of the *Callicebus* with the pithecine group through the work of ROSENBERGER (1981) and SCHNEIDER *et al.* (1996).

FAMILY **PITHECIIDAE**
Genus *Pithecia* DESMAREST, 1865

LANDAZABAL
2003

This genus contains five species (*Pithecia pithecia, Pithecia irrorata, P. aequatorialis, P. albicans* and *P. monachus*) (HERSHKOVITZ, 1987; RYLANDS *et al.*, 2000) of which one is found in Colombia, *Pithecia monachus* with two subspecies, *P. m. monachus* and *P. m. milleri*. The animals are cat-sized with a weight of about 1.5 – 3 kg. They have bushy tails, slightly longer than the combined head- body and long, shaggy pelage which they erect when alarmed, giving the fluffy effect most known by observers. But when calm the pelage is slicked back, revealing a much smaller body than apparent with the pilo-erection. The face is covered with very short hair, contrasting with the body hair. In Colombia *Pithecia* has not been seen north of the Caquetá except west of the Caguan River. It has not been registered east of the Caguan River, although it might be found there eventually.

Monk Saki

Pithecia monachus (E. GEOFFROY, 1812)

Taxonomic commentary

Plate 13 • Map 22

HERSHKOVITZ (1987b) recently revised this genus, dividing it into five species: *Pithecia pithecia, P. monachus, P. irrorata, P. aequitorialis* and *P. albicans*. Only *P. monachus* with two subspecies (*P. m. monachus* and *P. m. milleri*) are known for Colombia. This arrangement is recognized by RYLANDS *et al.* (2000). A previous arrangement by Hershkovitz (1979) had described a somewhat different scheme, which called *P. aequatorialis, P. monachus* and called the present *P. monachus, P. hirsuta*. The 1979 revision is entirely superceded by the 1987 version.

Common names

Saki from Tupí, *sagui*); Colombia: *volador* or *mico volador; huapo* near Leticia (of Peruvian origin); German: *zottige Bärenaffe, Zottelaffe* or *Mönschaffe*; French: *saki moine*.

Indian languages

Jitoje (Carijona, *fide* CAPITÁN ARMANDO PEREA Carijona of Marimanteca and Córdoba rapids, Caquetá River, 1985); *jidóbe* (Huitoto); *osomono* (Ingano);

compong (Macú-Yujup of the lower Río Apaporis, *fide* CAPITÁN FRANCISCO MACÚ, 1992); *ricu* (Macuna); *ópacua* (Miraña, *fide* NAPOLEON MIRAÑA, 1985); *juubaiga* or *juecha* (Muinane); *faagi* (Okaimo); *waʔo* or *suü* (Siona); *parahuacú* (Tanimuca); *puü* or *poh/wi* (Ticuna); *parahuacú* (Yukuna); *ukuenu* (Yuri); *Pithecia* is called *paraguaᵵu* in the Mato Grosso, suggesting that this name is a Guaraní or Tupí root that was adopted by the Yucunas and Tanimucas.

Identification

Head-body length is 370-480 mm with the tail varying about 404-500 mm. Both sexes of these small monkeys weigh between 2-3 kg (HERSHKOVITZ, 1987b) and in this particular species there is little sexual dichromatism, unlike that found in *P. pithecia* and *P. aequatorialis*, whose male coloration is much darker than the female, possessing contrastingly long, white facial hair, while the females are much more similar to the presumed basic stock and similar in coloration to both sexes of *P. monachus*. Nevertheless, in *P. monachus* the males are slightly larger than the females and the male canines are slightly more robust than the females. The chest and lower parts of the upper arms are mostly naked with V-shaped fringes of hair extending from the scapula to the sternum.

Basically *P. monachus* is a dark-bodied animal with course, long hair of more than 5 cm length, mixed throughout with buff-tipped hairs, giving a silvered appearance. The bushy tail is slightly longer than the combined head-body. The coronal hears are directed forward and form a hood that partially overlaps the short-haired frontal region. The facial skin is blackish but the naked throat and upper chest skin is more lightly colored, with the exception of a very black glandular field of thickened skin on the lower neck and upper chest. The lips possess short grayish hairs. In the males the testicles are black, contrasting with a vivid pinkish penis. The hands and feet are whitish or buff, contrasting with the darker limbs. The form of the body is long and gracile compared to many primates. Extra ability for radial movement in the torso allows extreme agility in its movements (Hershkovitz, 1987b).

• Male with chest the same color as the dorsum, although with much less hair, female the same..*Pithecia monachus monachus*.

• Male with chest colored brown and throat blackish orange, female colored the same as other subspecies...................................*Pithecia monachus milleri*

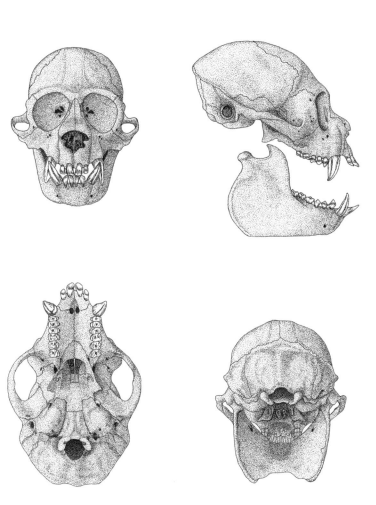

Figure 25. Aspects of skull of *Pithecia monachus*.

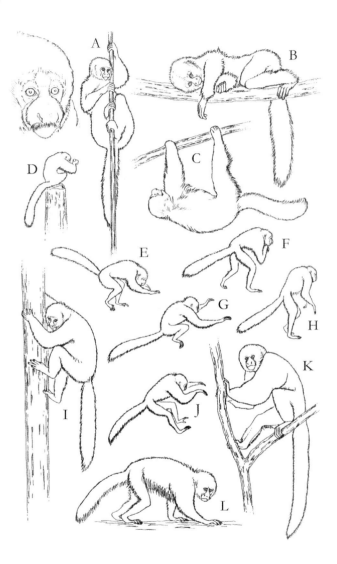

Figure 26. Various aspects of *Pithecia*.

(note: study of this subspecies in the Chicago Field Museum of Natural History prompts me to call this a very poorly defined subspecies, since I saw very little brown or orange-colored chest hair on any or the putative *P. m. milleri*. HERSHKOVITZ (1987), has himself characterized this as a very "weak" subspecies.)

Geographic range

In Colombia *P. m. monachus* is found south of the Caquetá River (including the Colombian trapezium) and east of Puerto Lequízamo and *P. m. milleri* is found west of the Caguan River and north and south of the Caquetá River, west of Puerto Leuízamo, the exact limit between the two subspecies undefined. Besides Colombia, the species is found throughout eastern Ecuador and Peru south to Ucayali and east into Brazil to the Juruá River in Amazonas.

The species is commonly found up to about 600 m in well-developed primary forest but it is also frequently seen in vegetation alongside watercourses where they usually frequent the middle and upper levels of the canopy (HAPPEL, 1981, 1982). They have been observed by this author sunning themselves in low flooded vegetation (2-3 m above the surface of the river) after rainfall. They can survive in isolated islands of trees if unhunted.

Natural history

Most of our knowledge of the natural history of this species is derived from a study by P. SOINI (1986a, 1987d, 1988b, 1995a, 1995b) who studied *P. monachus* on the Río Pacaya in the Samiria-Pacaya National Park in Perú. Some observations were made by TODUKA (1968) and by MOYNIHAN (1976) and by this author in Colombia.

These primates live in monogamous family groups of one adult male and female plus assorted young with group sizes varying from 2-8 and the average being 4 (n=94). The great majority (84%) of the groups contain 2-5 individuals, groups with 7 or 8 being relatively scarce. In several groups an adult female has been seen to be accompanied by two adult males or an adult male joined the original pair under observation. These supernumerary males may leave after a few years. Young animals tend to emigrate from the natal group at about 7-8 years of age. Solitary animals are common, moving throughout the territory of an established group or peripherally attached to a group (SOINI, 1988, 1995a).

Home ranges are unknown, but HAPPEL (1981, 1982) felt the animals were not territorial, since several groups were observed feeding together in the same tree. In the R. Pacaya study area the ecological density for this species was about 3.2 groups/km^2 or about 14.3 individuals/km^2, while HAPPEL (1981) recorded a density of 36 individuals/km^2 at her study site. On the Purité River in the Colombian trapezium densities have been calculated varying between 2 – 14.5 individuals/km^2 at four different sites in the same area (DEFLER, 2003). No day range or activity budget has been calculated for this species.

This species is primarily a quadrupedal walker (50%), runner (6%) and leaper (33%) (HAPPEL, 1982). Their long leaps when frightened have given origin to one of their common Spanish names *volador* (flyer), since they can appear to be flying with their long fur streaming out into the wind. Most locomotion in the HAPPEL (1982) study was only at 10-15 m above the ground while resting occurred at about 20-25 m. These animals stand on their hind legs when examining something at a distance and can also suspend themselves by their back feet when attempting to reach a morsel of food.

The diet of this species, based on frequency of consumption, consists principally of fruits (55%), seeds (38%), leaves (4%), and flowers (3%), plus some invertebrates, especially ants (HEYMAN & BARTECKI, 1990; IZAWA, 1975, 1976). Usually the seeds that are consumed are immature, while the fruits that are utilized are mature. In the SOINI (1987d) study this monkey ate 22 species of fruits, 30 species of trees and lianas for the seeds, 6 species of trees for the leaves and the flowers of 4 species of trees and the females seemed to have a stronger preference for ant eating than the males.

The most important species consumed (*fide* SOINI, 1987d) are listed as follow: (*Xylopia ligustifolia*, Annonaceae; *Brosimum rufescens*, Moraceae; *Maripa axilliflora*, Convolvulaceae; *Inga* spp., Leguminosae); most important seed species (*Eschweilera timbuchensis*, Lecythidaceae; *Rourea amazonica*, Connaraceae; *Alchornea latifolia*, Euphorbiaceae; *Roupala diolsii*, Proteaceae; *Pterocarpus amazonus*, Leguminosae); most important leaf species (*Pitocarpus amazonus*, Leguminosae); most important flower species (*Iriartea exorrhiza*, Palmae).

In Perú, the most popular fruit, *Xylopia ligustifolia*, made up 24% of all the vegetal food choices in the diet. The most popular seed source, *Eschweilera timbuchensis* made up 7% of all of the vegetal food choices while the most popular leaf source, *Pithecelobium multiflorum* made up 4% of all the vegetal choices. Only five species of flowers were consumed in this study with the inflorescence of the palm tree *Iriartea exorrhiza* being the most important (Soini, 1987d, 1995a). Only five species of flowers were consumed, with the inflorescence of the palm tree *Iriartea exorrhiza* being the most important (Soini, 1987d, 1995a).

The importance value of plant families in terms of species consumed by *Pithecia monachus* per family are listed as follow (*fide* Soini, 1987d): Leguminosae (8), Moraceae (5), Euphorbiaceae (3), Myrtaceae (3), Annonaceae (2), Araceae (2), Convolvulaceae (2), Sapotaceae (2), Dichapetalaceae (1), Malphighiaceae (1), Tiliaceae (1), Proteaceae (1), Flacourtiaceae (1), Hippocrateaceae (1), Melastomataceae (1), Palmae (1), Rosaceae (1), Bignoniaceae (1), Connaraceae (1), Combretaceae (1), Bromeliaceae (1), Olacaceae (1), Sapindaceae (1), Loranthaceae (1).

Happel (1982) saw *P. monachus* feeding on *Copaifera pubiflora* (Leguminosae), *Inga* sp. (Leguminosae), *Pithecellobium* sp. (Leguminiosae), *Brosimum* sp. (Moraceae), *Ficus* sp. (Moraceae), and *Pourouma* sp. (Moraceae). Happel (1981) also observed *P. monachus* drinking from salt licks which had high concentrations of the elements Na and Ca.

These animals form a long-lasting monogamous union. There is a definite birth season in the Río Pacaya (Perú) from September-February, but especially October-December (70% of births), based on 33 births. At Caparú a male *P. monachus* was maintained free-ranging, which was born around January, 1990 near La Pedrera, Amazonas in Colombia. This birth suggests that the northern part of the species range has a birth season similar to that of the Peruvian site. The gestation period is probably 5 1/2 months as is known for the closely related *P. pithecia*. In 69% of 16 groups studied the birth interval was two years, the range being 1-3. Males smell the anogenital region of the females, apparently to determine her sexual state.

In this monogamous species (unlike *Callicebus* and *Aotus*) the male does not help carry the infant, which usually rides the mother for the first five months. Occasionally a sibling will also help carry the baby. Of 29 infants whose development was followed in Peru, four died in the first six months of life. The first week or so the infant is carried ventrally and only in the second week transfers to the dorsum of the mother. Infants begin to leave their mothers´ back during the second month and are almost independent at five months. Danger from the ground usually causes the young independent animal to hide, while the rest of the group moves off a short distance, vocalizing and attempting to draw the predator after them.

The newborn has a dark, almost black, natal coat, which gradually becomes lighter than the adults after 2-8 weeks, to be transformed into the adult coloration by 2-3 years of age. Nursing occurs during the first eight months and weaning is probably not complete until the first year. Mothers often share their food with the youngsters, even opening tough pods into the third year (SOINI, 1987d).

Within the monogamous union, the adult male and female exhibit great affection for each other and for the young from several different years. However, neighboring groups are reacted to aggressively by means of vocalizations, but rarely physically. Home ranges seem to overlap extensively. No particular study has been done of vocalizations, but some prominent vocalizations are known, and HAPPEL (1982) distinguished eleven. An excited trill is used by individuals to convey danger, while aggressive and loud growls show aggressive intent. Other, softer trills and peeps are emitted to show interest in objects and pacific intent. A loud series of of *eh-eh-eh´s* expresses excitement in seeing a recognized conspecific at a distance. The species also uses *moans* and *wails* as well as a variety of other sounds.

Some interesting social signals have been observed in the free-ranging, tame male at Caparú. A sexual display is very prominent, with the male thrusting his chin and face forward, while closing his eyes and sticking out his tongue at the object of his desire. While making the facial display, the male broadens his shoulders by rolling them forward and by erecting the loose cape-like fur on them. Finally, during all of this, the pink penis stands erect, contrasting with the dark black scrotum. This display is liable to be affected either on two or four legs.

These animals are sometimes seen mixed with woolly monkeys or squirrel monkeys at the sides of rivers, although it may be that they were passive and the other species were passing by, since SOINI (1995b) does not believe that the species actively seeks associations. Both HAPPEL (1982) and SOINI (1995b) observed other similar seemingly passive associations with *Alouatta*, *Saimiri*, *Saguinus* and *Lagothrix*. SOINI (1995b) observed *C. apella* harassing *P. monachus* twice.

Conservation state

Pithecia monachus monachus seems to be a common species in many parts of its Colombian range and cannot be said to be endangered. The IUCN classification for this species is LC internationally and LC en Colombia, according to new IUCN (2001) criteria. Nevertheless, *P. m. milleri* is much more pressed than *P. m. monachus*, since the distribution of the former is in an area of southern Caquetá and western Putumayo which is very altered by colonization. *Pithecia m. milleri* is classified VU by the IUCN system. Theoretically the presence of the La Paya National Park should help to conserve *P. m. milleri*.

When to see it

Pithecia monachus is easily seen along waterways of small rivers south of the Caquetá River. The subspecies *P. m. monachus* is very common on the upper Cahuinarí River and can be sighted on the Cotuhué and Amacayacu Rivers of the Amacayacu National Park, while *P. m. milleri* can be observed inside the La Paya National Park.

Uacari or Uakari

FAMILY **PITHECIIDAE**
Genus *Cacajao* LESSON, 1840

This genus contains two species, *Cacajao calvus* and *Cacajao melanocephalus* (HERSHKOVITZ, 1987b; RYLANDS *et al.*, 2000), one of which *C. melanocephalus* is known from Colombia east and north of the Apaporis River. *Cacajao calvus rubicundus* may have been present in the Colombian trapezium during historic times and even up to the last century, but it now appears to be extinct there, due to hunting and a probable originally small population.

Cacajao is the largest of the Pitheciinae at around 2.6 – 3.5 kg with the males being larger and heavier than the females. The mandibles are also notably larger in males with longer canines. *Cacajao melanocephalus* are generally smaller than *C. calvus* and their tails average only about 35% of head-body length. *Cacajao* appear to be more extreme seed predators than are *Pithecia* and able to subsist on flowers and hard nuts of *Eschweilera* sp. The genus exhibits the interesting phenomenon that young animals are sexually cryptic up to about four years; careful examination is required to ascertain the true sex of such young animals (HERSHKOVITZ, 1993).

Black-headed Uacari

Cacajao melanocephalus (HUMBOLDT, 1812)

Taxonomic comments

This species has been recently revised by HERSHKOVITZ (1987a) who described (in accordance with a proposal of HERNÁNDEZ-C. & COOPER, 1976) two subspecies, *C. m. melanocephalus* Humboldt and *C. m. ouakary* SPIX. Only *C. m. oukary* is known for Colombia.

Karyotypically this species is very similar to *Cacajao calvus* (KOIFFMAN & SALDANHA, 1981; HERSHKOVITZ, 1987). The two species share the same low diploid numbers (2n=45, 2n=46) and only differ in two Robertsonian translocations (VIEGAS-PEQUINOT, KOIFFMANN & DURTILLAUX, 1985; AYRES, 1986). The diploid number (2n=45) for *C. melanocephalus* is the one published to date for this species (KOIFFMANN & SALDANHA, 1981). The number 2n=46 was, however, found by MJONSALVE *et al.* (unpublished) from a female *C. m. ouakary* from the lower Apaporis River in Colombia, and it is mentioned here for the first time, since the slides and the data from this work were subsequently lost and the data never published.

Common names

Colombia: *chucuto* (Guainía and northern Vaupés; origin Geral); *colimocho* or *ichacha* (lower Apaporis river and the lower Caquetá river in Colombia);

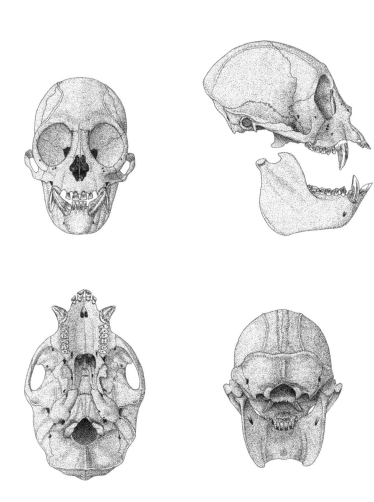

Figure 27. Views of skull of *Cacajao melanocephalus*. Male.

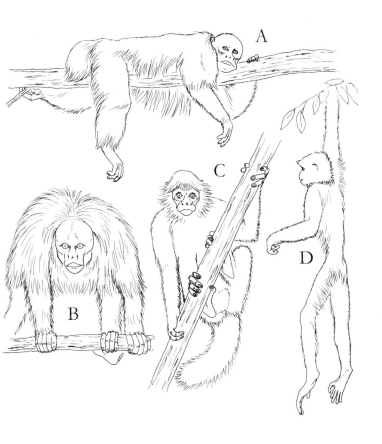

Figure 28. Various aspects of *Cacajao*. a) Resting on a branch. b) Aggression. c) Climbing. d) Hanging.

Figure 29. Various aspects of *Cacajao*. a-d) Positions of locomotion and hanging.

Brazil: *bicó* (Japurá River below mouth of Apaporis River); *chucuto, chucuzo, caruiri, cacajao*; German: *Schwarzkopf Uacari*; French: *ouacari..tete noire.*

Indian languages:

Karrubirri (Curripaco); *charutika* (Letuama); *ëh* (Macú-Yujup of the lower Apaporis River, *fide* Capitán FRANCISCO MACÚ); *nüestiamini* (Macuna); *cháu* (Puinave); *charurikaya* or *charuruka* (Tanimuka); *nüestiama* (Tukano); or *piconturo* fide Olalla from specimen tickets; *charú* (Yukuna); *puoghu* (Yuri).

Identification

These animals weigh about 2.5-3 kg with the males weighing more than the females. The head-body length is 300-500 mm with a mean for females (n=21) of 389 mm and for males (n=17) of 414 mm and the short tail varies around 153 mm for females and 173 mm for males. *Cacajao. melanocephalus ouakary* (the only subspecies known for Colombia), on the lower Apaporis river, is richly colored with a saddle and back of golden-yellow that contrasts sharply with the darker chestnut-red sides and underparts of the animals. The front limbs are dark-brown or blackish as are the lower parts of the hind limbs from the knee down. The hind limb flanks are chestnut red, this extending to the short tail as well. *C. m. melanocephalus* (which is found in s. Venezuela south of the upper Orinoco and east of the Casiquiari Canal to the Branco River in Brazil) has none of the golden-yellow on its back, being primarily blackish from the head to the mid-back and reddish brown or tawny at mid-back, not contrasting with the lower back or thighs.

The body of these animals is more gracile than robust, since they are experts at leaping from branch to branch; they appear like small cats as they move along. These monkeys are great leapers. The teeth are specialized for eating nuts and seeds. The canines are well-developed and splayed (*i.e.* they project outward) and are triangular in cross-section. These are utilized for opening fruits and (especially) seeds which are the main portion of the uacaris´ diet. Both the upper and lower incisors are long, narrow and procumbent, which allows seed endosperm to be scraped out of an opened seed. In general, this is the major ecological adaptation of the entire subfamily Pitheciinae and lends credence to the view of HERSHKOVTIZ (1977) and others that the pithecines can be seen as belonging to a specific ecological grade. This grade relates principally to the adaptation for seed predation.

Geographic range

Cacajao melanocephalus ouakary is found in Colombia between the Guayabero-Guaviare rivers and the Apaporis River with the most westerly record being near La Angostura, close to the southern tip of the Serranía La Macarena. A record of the American Museum of Natural History with no specimen was ascribed to a collection of CARLOS VELÁSQUEZ in 1942 for the northern tip of La Macarena, but this should be discounted until more concrete evidence is available, since the area is one of the better-known parts of La Macarena and no new material has come to light.

Cacajao m. ouakary is also found in Brazil, west of the Río Negro and north of the Japurá, as well as west of the Casiquiari Canal. *C. m. melanocephalus* is found east of the Casiquiare Canal in s. Venezuela and east of the Río Negro in northern Brazil, possibly to the Branco River. Its eastern limits are unknown. This species seems to prefer habitat along small to medium-sized black-water streams and lakes, including black-water flooded *várzea* (*igapó*) and the inland *terra firme* forests adjoining such *igapó*. These animals forage at all levels from the surface of water upwards to the canopy, readily descending to the ground to consume seedlings.

The two subspecies can be easily distinguished as follows (HERSHKOVITZ, 1987a):

• Bright colored with a golden-yellow mantle over the shoulders to mid-back which contrasts markedly with the dark chestnut-red of the sides and ventrum; arms are dark brownish black for their entire length as well as the lower parts (from the knees on down) of the legs and feet; upper parts of the feet are chestnut-red, a coloration with extends to the tail ....*Cacajao melanocephalus ouakary*.

• No golden-yellow mantle over shoulders to mid-back; instead the same area is basically black from the head to the mid-back and chestnut-red from mid-back, with no contrast with the lower parts ....*Cacajao melanocephalus melanocephalus*.

Natural history

Little research has been accomplished on this species until recently. DEFLER (1989c, 1991, 1999, 2001, 2003) studied the species on the lower Apaporis of Colombia for several year. BOUBLI (1993, 1994, 1997a, 1997b, 1998, 1999a,

1999b) and BOUBLI & DITCHFIELD (2000) studied it during 1994-1996 in the Cerro da Neblina National Park in northern Brazil, east of the Río Negro. LEHMAN & ROBERTSON (1994) censused this primate in southern Venezuela and planned a field study there, about which no information is available. BARNETT & DA CUNHA (1990, 1991) and DA CUNHA & BARNETT (1990), BARNETT & BRANDON-JONES (1995, 1996), BARNETT *et al.* (2000) and BARNETT & LEHMAN (2000) censused *C. melanocephalus* on the Río Negro, Brazil. FERNANDES (1993b) examined tail wagging in pithecines. FONTAINE (1981) reviewed what little was known at that time about the uacaris and RYLANDS (1994) wrote a description in the Brazil Red Book of endangered species.

These primates are sometimes seen in large groups of over 100 and perhaps approaching 200 animals. One good count by this author was 108 animals, and some were probably missed, since the count was only of animals moving along the edge of flooded forest (DEFLER, obs. pers.). But this is the temporary fusion of several smaller groups, since the species most often is seen in groups of about 20-30 on the lower Apaporis in Colombia, including multiple adult males and females, juveniles, and infants, apparently the basic social unit (DEFLER, 1989c, 1991, 2003). BOUBLI (1997a, b) studied a loosely-knit group of about 70 animals which did not break up into subunits, contrary to the animals in Colombia.

When only very poor feeding is available, groups of 20-30 also fission to small subunits and occasionally only 1 or 2 are seen moving separately from a larger group so that a numerical range of from 1 animal to over 100 animals has been observed (DEFLER, 1999a). The basic groups of 20-30 use a home range which seems to be greater than 500 ha on the lower Apaporis River. Nevertheless, occasionally such groups travel much more widely in the home ranges of adjacent groups when all groups travel together, so it is unclear how much total area any one group covers if these wider ranging movements are counted.

BOUBLI (1997a,b) estimated a range of over 1000 ha for his group of 70 animals, which apparently did not break up into subgroups. During certain times of the year when groups concentrate their activities in the flooded forest, this species seems very common and (ecological) population densities are very high. This concords with the season of greatest fruit abundance. However, calculated over a wide area the species is actually present in very low densities (crude densities) of less than 1 animal/km², while BOUBLI (1997b) suggested about 7 animals/km², based on his calculation of a home range for a 70 animal troop over more than 1000 ha. DEFLER (2001) calculated a crude density of 4.15 animals/km² and an ecological density of 12 animals/km² at Caparú. Day ranges at Caparú Biological Station in Colombia can vary from only about 50-100 m to over 5 km, but seem to average about 3 km. BOUBLI (1997b) found averages of 2,300 m (range 1,200-4,400 m).

An activity budget has not been calculated for the Caparú study site, but BOUBLI (1997b:153) calculated 22% rest, 27% travel, 20% feeding and 31% moving/foraging for his Brazilian site. These animals are predominantly quadrupedal walking, running and leaping with great frequency. Their leaps across gaps are spectacular and cat-like and some cover vertical spaces of 5-10 m and horizontal space of up to 5 m. While feeding and playing they often hang from their hind feet for lengthy periods of time. Resting, they sit for long periods, probably also at vigilance. A common sitting position is with one leg hanging free, perhaps for balance, somewhat as a long tail might be used.

Along the edges of Lake Taraira (lower Apaporis River) in Colombia this species sleeps in rather exposed positions over the flooded forest, sometimes only 10-15 m over the water and on a stable branch or on the tips of stiff vegetation exposed to the sky. In the *terra firme* forest or the unflooded igapó the animals sleep much higher and less-exposed at 25-30 m.

Cacajao melanocephalus is a specialized seed predator and the majority of its diet is made up of immature seeds, which the animals extract very easily with their sharp canines, opening up a fruit and splitting the seed using their tweezer-like procumbent incisors to scrape out the endosperm of each cotyledon. They also supplement their diet with some fruit pulp, leaves and arthropods. BOUBLI (1997b: 140) calculated a diet of 91% fruits, 3% leafy material (young leaves, mature leaves, petioles, bromeliads), 4% flowers and 2% insects. Of the fruits consumed, DEFLER (2003) observed that about 93% were consumed for the seeds and 7% for the pulp, judging from split seeds under each crop, whereas BOUBLI (1997b:143) calculated 77.8% of all fruits were consumed for their seeds, and 63.8% of feeding records in his study were of unripe seeds.

A few of the food species collected at the Estación Biológica Caparú are listed here (DEFLER, 2003): *Eschweilera* sp. (Lecythidaceae) (fruits, flowers, July, Oct., Dec.; seeds Jan, Feb.); *Brosimum utile* (Moraceae), fruits, seeds, Feb; *Minquartia guianensis* (Olacaceae), pulp, seeds, Sept.; *Minquartia punctata* (Olacaceae), pulp, seeds, Sept.; *Clusia* sp. (Clusiaceae), aril, seed; *Micrandra spruceana* (Euphorbiaceae) immature seed, Sept.; *Gomphilum gomphiifolia* (Sapotaceae), immature seed, June-July; *Pouteria laevigata* (Sapotaceae), immature seed, June-July; *Buchenavia* sp.; *Mabea* sp. (Euphorbiaceae); *Monopteryx uaucu* (Caesalpiniaceae), seed; *Aspidosperma oblongum* (Apocynaceae), seed, May; *Abuta* sp. (Menispermaceae), seed, May; *Inga* sp. (Leguminosae), aril, seed, May; *Macrolobium* sp. (Leguminosae), seed; bromeliad of igapó; *Amphizoma coriaceae* (Celastraceae), seed; *Mauritia flexuosa* (Palmae).

In the Pico da Neblina National Park, Brazil, BOUBLI (1997b:140-141) observed the following species preferences for his study group from 120 fruit species taken from 32 families: *Micrandra spruceana* (37%, Euphorbiaceae); *Eperua leucantha* (10%, Leguminosae); *Hevea* cf. *braziliensis* (7%, Euphorbiaceae); *Alexa imperatricis* (6%, Fabaceae); *Mauritia flexuosa* (4%, Palmae); *Eperua purpurea* (3%, Leguminosae); *Chromolucuma rubiflora* (2%, Sapotaceae); *Lisyostyles scandens* (2%, Convolvulaceae). Family prefences in the same study are as follow: Euphorbiaceae (23%), Caesalpinioideae (13%), Fabaceae (11%), Sapotaceae (6%), Lecythidaceae (6%), Palmae (5%), Chrysobalanaceae (4%), Convolvulaceae (4%), Moraceae (3%), Elaeocarpaceae (3%). The animals often lick rain-soaked leaves and descend trunks and branches over igapó to drink water directly from the lake´s surface. They also drink water from tree holes and bromeliads, dipping their hand into the water and drinking from the hand and fur.

Adult males display an erect penis of impressive length, which has scattered over most of the surface an array of fleshy barbs, apparently used to stimulate the female. HERSHKOVITZ (1993) suggests that these disproportionately long genitals may actually reach the uterus for direct insemination. There is a narrowly defined birth season around March-April on the lower Apaporis of Colombia at which time one infant is born. Females seem to give birth about once every two years. Some births apparently occur outside of this season, as evidenced by a full-term female which in early October, 1996 fell out of a tree near the Estación Biológica Caparú, breaking her back, while trying to give birth.

Infants and youngsters are sexually cryptic until about four years of age (HERSHKOVITZ, 1993). How this might be reflected in the social development of these primates is an interesting question yet to be studied. *Cacajao melanocephalus* are very social and easily form very large units in which they are sometimes found traveling. All members groom frequently and males are very tolerant of infants, which males guard assiduously from danger. Any play pair of young animals has an adult male sitting nearby on guard for any danger.

Tail wagging seems to be an important social signal, conveying well-being of the animal. Although FERNANDES (1993b) felt that tail wagging for two species of *Chiropotes* was a tension relief mechanism, I do not believe this to be the case in *C. melanocephalus*, since as soon as a situation becomes tense among the *Cacajao*, tail wagging does not increase, rather it stops altogether. An animal in a stressful situation, such as being punished or observing a close-by human, lowers its tail, sometimes almost between its hind legs, and the tail remains in this position until the animal again feels secure once again. A lowered tail conveys a signal to others that the situation is tense or insecure, and others are put on guard when they observe the tail in this position.

Some vocalizations of *Cacajao melanocephalus* are as follow (DEFLER, 2003): *purr* – cat-like sound when seeking contact with mother, being groomed or other affectionate behavior; *ki-ki-ki* – infrequent, low contact call when foraging; *harsh scream* – strident, harsh scream given by individuals seeking contact with others; *wai-wah-wah* – aroused vocalizations of comparatively pure tone; *chee-chee* – excited tone given for danger (all react to this sound be searching for danger or by moving away from area); *scream* – punished animal gives frantic, loud vocalization; *harsh ehh* – play sound given by adults and young when tussling with another.

Some displays given by this species follow (DEFLER, 2003): *penile display* – drops leg and shows erect penis (male agonistic ? or appeasement ?); *pilo-erection* – all fur becomes puffed up when the animal is excited, thus making the animal appear much larger; *tail wag in upright position* – when traveling calmly with others, upright tails wag back-and-forth incessantly perhaps conveying well-being; *tail-down* – when frightened or disturbed the tail is held in downward position and is not wagged; *posterior display* – flat-footed, tail in air, present rear view to observer (friendly intent?); *urine washing* – animal washes feet and vigorously scratch the urine on all parts of body from chest, belly and back.

Cacajao melanocephalus travels and forages with *Lagothrix lagothricha* during certain times of the year. *Saimiri sciureus* also commonly travels in association with *Cacajao* and *Alouatta seniculus* sometimes feeds within the sometimes numerous groups of *C. melanocephalus*, although this is probably more a passive association than the *Saimiri/Cacajao* groups. *Cebus albifrons* has also been seen with *Cacajao* groups in flooded forest. The greater ani (*Crotophaga major*) often travels in a small group through its home range with a group of *Cacajao*, along the edge of the lake where the bird often hunts insects in a social group of 7-10 members. The *Cacajao* often dislodge insects which the anis then capture. If the *Cacajao* feed long enough in one spot they attract many fish, which arrive to take advantage of the dropping and opened fruits and seeds. Such concentrations of fish then attract feeding dolphin (*Inia geoffrensis*) which then fish beneath the feeding monkeys. The ubiquitous small forest hawk *Harpagus bidentatus* often follows a *Cacajao* group, hawking disturbed insects.

Conservation state

This species is listed in Appendix I of CITES and is classified by the IUCN as LC internationally and NT (almost threatened) in Colombia. Because of its seasonal concentrations in flooded forests it is very vulnerable to hunting from a canoe and it would be fairly easy to hunt out all local populations for an area. Other species of primates hunted from a canoe still have other populations inland and would be much more difficult to extirpate than *Cacajao melanocephalus*.

Although this species is probably found in the two huge and recently established Natural Reserves in the Colombian Amazon (Nukak with 855,000 ha and Puinawai with 1,280,000 ha), the species is hunted in those regions and there is no confirmation concerning densities and the extent of the species´ presence, and densities are likely to be low.

Where to see it

The species can be observed fairly easily in Taraira or Caparú Lake in the indigenous reserve Yaigojé-Apaporis. Permission should be obtained from indigenous authorities and guides contracted from that reserve. There are seasons when the animals are more easily found along the lake edge than other times. Probably the best season to locate groups of this species is April-August.

Titi monkeys

FAMILY **PITHECIIDAE**
Genus ***Callicebus*** THOMAS, 1903

Taxonomic comments

Up until 1990 only three species of *Callicebus* were recognized in this genus (HERSHKOVITZ, 1963). HERSHKOVITZ (1990) revised his earlier view and recognized 13 species in four groups: *C. modestus* with only that species; *C. donacophilus* with three species (*C. donacophilus*, *C. olallae* and *C. oenanthe*); *C. moloch* group with eight species (*C. cinerascens*, *C. hoffmannsi*, *C. moloch*, *C. brunneus*, *C. cupreus*, *C. caligatus*, *C. dubius* and *C. personatus*); and the *C. torquatus* group with one species.

KOBAYASHI (1995) slightly modified the above scheme, creating five groups of species: *C. personatus* with its own group; he transferred *C. olallae* and *C. oenanthe* to the *C. modestus* group and established a group for *C. cupreus*, made up of *C. cupreus* and *C. caligulatus*, retained the rest of the *C. moloch* group and recognized the *C. modestus* group for a total of 15 species.

GROVES (2001) decided to recognize *C. ornatus*, *C. discolor* and *C. medemi* instead of leaving them as subspecies, which of course affects how we view *Callicebus* in Colombia and this deserves a comment. *Callicebus cupreus ornatus* (=*C. ornatus*) as an isolated taxon represents an argument for recognizing this type of isolated taxon as a separate species, since at a minimum, it now has an evolution independent of other populations of *Callicebus cupreus*. Nevertheless, it would be very valuable to carry out karyological, molecular and morphological research on *ornatus* to clarify just how different it has become from *C. c. discolor*. Phenotypically it is different, but that is not an adequate biological argument for distinguishing it as a separate species. There is a tendency in primatology today to recognize isolated populations as separate species, just as ornithologists did 100 years ago. However, *ipso facto* recognition of (naturally) isolated populations ignores possible karyological or molecular evidence which might show that *discolor* and *ornatus* may not be very different genetically. After all, probably there was connection between the *Callicebus cupreus* populations only about 12,000 years ago as the Ice Age came to an end and habitat was more to this species liking.

Callicebus c. ornatus is clearly related to *C. c. discolor*, but how genetically different they are must be determined, and this could be the object of an interesting research project in Colombia. Both taxa were reported to have the same karyotype of 2n=46 by BENIRSHKE (1975), but recently the Colombian geneticist M. BUENO (com. pers.) has found a new karyotype for two *C. c. ornatus* (2n=44).

Making the relationship between *C. c. discolor* and *C. c. ornatus* more interesting, apparently there is a population of another taxon (probably the *C. cupreus* group, without a white band on the forehead) in southern Caquetá, between the Orteguaza and Caquetá Rivers, which MOYNIHAN (1976) observed and informally described, but which was ignored by HERSHKOVITZ (1990). HERSHKOVITZ' (1968, 1977) theory of metachromism would suggest that in fact *C. c. ornatus* and *C. c. discolor* would have to derive from that southern Caquetá population, since according to the theory bleached fur derives from more darkly colored fur.

GROVES' (2001) decision to recognize *C. medemi* as a species on the grounds that "its range is isolated well to the north of other members of the group" is a non-starter on the grounds that *medemi* is not isolated but is found in continuous forest which harbors (to the east), *C. torquatus lucifer*. *Callicebus t. medemi* has no visible differences any greater than other subspecies from *C. t. lucifer,* and it doesn´t appear logical to name it a species, at least based on arguments so far presented. The nature of the contact zone between the two taxa, however, is completely unknown and could be the subject of another interesting research project.

VAN ROOSMALEN *et al.* (2002) would raise *ipso facto* all previous *Callicebus torquatus* subspecies to species, including (in Colombia) *lugens, lucifer* and *medemi*, on the grounds that "we increasingly find the concept of subspecies to be of minimal value in describing the diversity of neotropical primates", a rule of thumb that may not be the best method of recognizing species. This taxonomy would elevate the number of species in the genus to 28. Increasing support for the "phylogenetic species concept" should emphasize the necessity to bring together all possible characters for distinguishing the "smallest diagnosable cluster of individual organisms within which there is a parental pattern of ancestry and descent" (CRACRAFT (1983). It seems to this author that genetic information and particularly karyotypes and molecular characteristics would of necessity need to be added to the character cluster used to diagnose species. Sole use of phenotypic differences observed in pelage and the denial of variability as a species characteristic that can be identified by naming the population a "subspecies" will only take us back to a typological species era.

Actually the recent discovery of a diploid number 2n=16 (the lowest diploid number so far detected in primates) in populations of *Callicebus torquatus lugens* demonstrates that *Callicebus torquatus* is actually made up of at least two species, since previous work had discovered 2n=20 in other, unidentified populations (see below). It is possible, nevertheless, that more

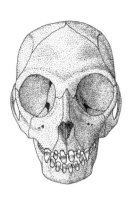

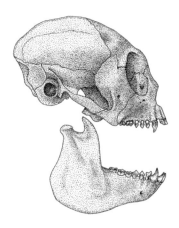

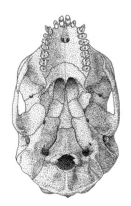

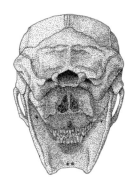

Figure 30. Views of skull of *Callicebus torquatus*.

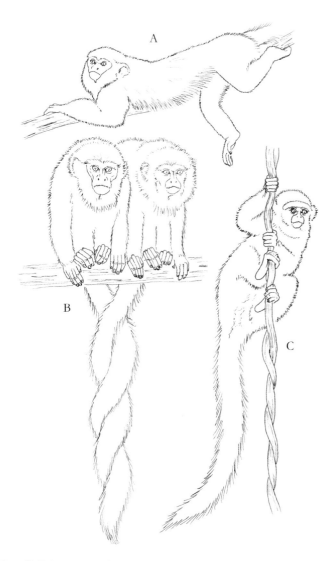

Figure 31. Various positions of *Callicebus*. a) Resting. b) Pair resting with entwined tail. c) Climbing.

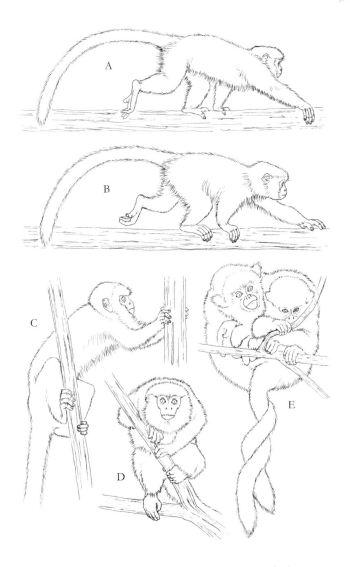

Figure 32. Various positions of *Callicebus*. a-d) Locomotion and climbing. e) Resting.

than one subspecies possess the number 2n=16. Naming subspecies as species will not clarify the systematics of any group, especially because a subspecies is a rubric for any population distinguishable phenotypically from other populations, no matter what the level of genetic difference involved. Subspecies are not necessarily evolutionary units, but they do call attention to variability which may represent many different levels of genetic difference, allowing us to then analyze with modern genetic tools.

In this book, I prefer to continue classifying *Callicebus* according to HERSHKOVITZ (1990). This may change in the future, but only based on identified karyotypic differences together with a suite of other evidence which would make it clear that we might be dealing with biological species. Contrary to the popular understanding of the nature of a biological species, all evidence must be brought to bear to test the hypothesis of reproductive isolation, karyology being one of the more persuasive lines of evidence (MAYR & ASHLOCK, 1991:107, 159-194).

Animals of this genus are usually the size of a cat or rabbit, weighing around 800-1500 g according to the species. The smallest species are perhaps *C. donacophilus* and *C. oenanthe* and the largest species is *C. torquatus*. They have long, non-prehensile tails and the sexes are usually the same size. Pelage color varies geographically from greyish agouti (considered to be the most primitive type of mammalian markings) to dark brown or black, which is completely saturated eumelanin with white bands (completely bleached) constrasting on the forehead, throat and hands, according to the species.

Common name

German: *Springaffe;* French: *callicèbe;* Brazil: *zogui-zogui.*

Indian languages

The epithet *tungue* preceded by a qualifying adjective, according to the species; *chutartunque =Callicebus cupreus; yanatungue=Callicebus torquatus*) is used by the Ingano; *cui* is used after a qualifying adjective by the Piaroa (*huat-cuí-huat-cuí,* which is onomatopoeic for the common territorial vocalization=*Callicebus cupreus; huecuí,* also onomatopoeic=*Callicebus torquatus*).

Etymology

Gr. *kali* ο κάλι for pretty + *cebus* the transliteration of the Gr. κήβος, used first by ARISTOTLE (Hist. Anim., ii. 81) to refer to primates with long tails, later altered to κῆπος , especially referring to *Cercopithecus* and later transferred by ERXLEBEN (1777) to a use referring to neotropical primates (HILL, 1960).

Dusky Titi

Callicebus cupreus (SPIX, 1823)

Plate 12 • Map 21

Taxonomic comments

Until recently this primate was clumped together with other species as *Callicebus molloch*, according to an earlier HERSHKOVITZ (1963) revision describing three species of *Callicebus* for South America. That was completely superceded by a new HERSHKOVITZ (1990) revision, describing thirteen species of *Callicebus*. *Callicebus moloch* is no longer appropriate for Colombia, another species on the south side of the Amazon River in Brazil having priority for the name. The first description of this species *C. cupreus* was published by SPIX in 1823. Two known subspecies are described by HERSHKOVITZ (1990) for Colombia (*C. c. discolor* and *C. c. ornatus*).

There is a recent tendency to raise subspecies of *Callicebus* to species level (GROVES, 2001; VAN ROOSMALEN *et al.*, 2002), which in some cases may be justified because of chromosome differences. Raising other subspecific populations to species on the basis of being isolated or on the basis of minor phenotypic differences seems more a philosophic decision than a decision based on biological criteria and such actions need to be evaluated on a case by case basis. This book will consider the genus from the viewpoint of HERSHKOVITZ´ (1990) recent revision.

MOYNIHAN (1976) described a population in the south of Caquetá department between the Orteguaza and Caquetá Rivers, but this population needs a more complete description to be included formally within an accepted taxonomy. A recent morphometric study of cranial characteristics seeks to organize the animals of this genus into a phylogeny (KOBAYASHI, 1995).

A new karyotype (2n=44) has recently been discovered in *Callicebus c. ornatus* which may support the tendency to classify this taxon as a species apart from *C. discolor* (BUENO, *et al* in press). However, the new karyotype could eventually be discovered in *C. c. discolor* and *C. c. cupreus* as well, which in this case would argue for the maintenance of the species with three subspecies, according to HERSHKOVITZ (1990). Until additional genetic information is available, the decision of some to raise these subspecies to species level must be considered a philosophical viewpoint or, at most, a new taxonomic hypothesis, which must be tested. There are advantages to leaving such taxa as trinomial subspecies, at least until such time as new information comes to light.

Common names

Colombia: *zocayo* or *zocay* in Meta (*C. c. ornatus*) and in the region of the upper Putumayo river (*C. c. discolor*), (probably of Quechua origin); *zogui-zogui* in Amazonia; *mico tocón* or *tocón* (*C. c. cupreus*) of Peruvian origin in the Leticia area; French: *callicébe roux;* German: *roter Springaffe; kopferfürbiger Springaffe;* Brazil: *zogui-zogui.*

Indian languages

A+k+ (Huitoto); *chuntartungue* (Ingano, *fide* RODRÍGUEZ-M., 1993); *gaai* (Muinane); *watcuí-watkui* (Piraroa); *duaré* (Ticuna in the Colombian trapezium); outside of Colombia: *uapo* (Pebas, *fide* GEOFFROY, 1951 in HILLE, 1962), which is also used by Spanish-speakers in Peru, although also used for *Pithecia.*

Identification

The head-body length is about 300-400 mm and the tail is usually 400-500 mm. The animals weigh around 1000 g when adult; 4 males weighed an average of 1,106 g. There is a slight trend towards increased body weight among males as compared to females. Basic body color is agouti with a reddish-orange ventrum and sideburns, throat, arms and legs of the same reddish-orange.

The three subspecies can be distinguished as follow:

• No white band over the eyes and only the end of the tail is whitish (not found in Colombia)..*Callicebus cupreus cupreus.*

• A band of white found over the eyes:

- above the white band another dark band (very contrasting); the wrists and the lower parts of the legs are reddish, as well as the feet and hands (not white)..*Callicebus cupreus discolor.*

- the white band over the eyes extends to the point of the ears; the feet and the hands are whitish; also the proximal part of the tail is reddish and the rest is whitish...*Callicebus cupreus ornatus*.

MOYNIHAN´s (1976) cursory description of an undescribed population in southern Caquetá mentions that this population had no white band over the eyes.

Geographic range

Callicebus c. discolor is found in Putumayo Department between the Guamués and Sucumbios rivers and may extend (according to some comments) to the Colombian trapezium south of the R. Putumayo (the latter needs to be confirmed and this author, despite efforts, has been unable to do so) (DEFLER, 1994b; DEFLER, 2003). *Callicebus c. ornatus* is found mostly north of the upper Guayabero from the Cordillera Oriental to the Ariari River, but there is an enclave population in the lowlands along the Guayabero River on the south bank, which may extend east of the mouth of the Atrato River down the Guaviare River, but this needs confirmation. The main population extends up the piedmont forest at least to the Upía River. Eastern limits of this subspecies are poorly defined, but except for the Guaviare River, the subspecies probably does not extend beyond the Metica River. None of the Colombian populations is large and each subspecies is disjunct from the other.

Outside of Colombia *Callicebus cupreus* is found to the west of the Purús and the Ituxi Rivers in Brazil to the Huallaga River in the south and north of the Marañon River to the piedmont of the Cordillera de los Andes. The northernmost limit for *C. c. discolor* appears to be the Guamués River in Colombia.

The habitat of this species seems to be low and poorly-developed forest, often along rivers growing on lower, poorly-drained ground. In the trapezium, it is said by some that *C. c. discolor* is found in the flood-plain of the Amazon (as is *C. torquatus*, although this species is found in well-developed *várzea* forest), where there is a vegetation of low trees and shrubs with much tangled vegetation, what one could call "poor forest". But I have been unable to confirm this.

Callicebus c. ornatus can be found around San Juan de Arama (Meta) in low gallery forests interspersed with savannas which have a canopy of 15-20 m, and which at times may be heavily altered by humans (MASON, 1968). MOYNIHAN (1976a) saw a group of the undescribed third subspecies in extremely bushy and dense vegetation in Caquetá, not more than 7 m high. He felt that this species could be almost thought of as a swamp monkey, since it seemed to prefer sites which were over wet or water-logged ground, but that designation seems somewhat exaggerated to me (DEFLER, 1994b, 2003). *Callicebus cupreus* enters secondary growth as well.

Natural history

MASON (1965, 1966a, 1966b, 1968, 1971) first studied *C. cupreus* in Meta Department near San Martín. MOYNIHAN (1966, 1976a) made observations of captives, and in Colombia he studied them in the field. KLEIN (1974) and KLEIN & KLEIN (1976) collected some distributional data on this species while studying principally *Ateles belzebuth*. ROBINSON (1977, 1979a, 1979b, 1981, 1982a,b) and ROBINSON *et al.* (1987) studied the vocalizations of this *C. c. ornatus* in low gallery forests east of La Sierra La Macarena, near San Juan de Arama (Meta). Later, POLANCO O. (1992), POLANCO AND CADENA (1993) and POLANCO *et al.* (1994) did her graduate thesis on *C. cupreus ornatus* in Tinigua National Park at the joint Los Andes/Japanese Centro de Investigaciones Primatológicas La Macarena (Meta) in the Duda River and PORRAS (2000) continued studying the same group.

In Brazil there have been recent projects published by BICCA-MARQUES (2000a, 2000b, 2003), BICCA-MARQUES & GARBER (2002), BICCA-MARQUES *et al.* (2002, 1998), SILVERA *et al.* (1998) and REY (1997), while in Ecuador YOULATOS & RIVERA (1999) and KNOGGE & HEYMANN (1995) have published work on *C. cupreus* in the field. WELKER *et al.* (1998). MÜLLER & ANZENBERGER (2002) and MÜLLER (1995) studied duetting while TIORADO & HEYMANN (2000) reported on diet differences between the sexes.

Some of the characteristics of the close family ties in this species have been elucidated in captive studies by MASON (1971, 1974a, 1974b, 1975, 1978); MASON & EPPLE (1969) and PHILLIPS & MASON (1976); CUBIOCOTTI *et al.* (1975); SASSENRATH *et al.* (1980) and FRAGASZY *et al.* (1982). KINZEY (1979) compared teeth in *C. cupreus* and *C. torquatus* and made some conclusions as to dietary differences, which HERSHKOVTIZ (1990) criticized as erroneous. KINZEY (1981) also wrote a review article for the genus. JONES & ANDERSON

(1978) published a good review article of what was before called "*Callicebus moloch*" and which reviews all of the information then available of *C. cupreus*.

The average group size of *C. cupreus* is usually 2-4, since the groups are made up of a monogamous pair and one or two off-spring. ROBINSON (1977) found an average of 2.4 in one gallery forest (n=10, including some yearly recounts) and an average of 3.87 (n=8) in another forest, but he did not speculate as to the cause of this difference. Occasionally groups of five are sighted.

MASON (1966b) found territories as small as 0.5 ha at his study site, but this isolated forest was probably protecting many refugees from other forests, since the density was exaggeratedly high. Robinson (1977) measured territories of 3.29 ha, 4.18 ha and 3.5 ha in gallery forest near San Juan de Arama, a bit south of San Martín, where Mason worked. POLANCO (1992) recently reported a territory of 14.2 ha for *C. cupreus* in Tinigua National Park at the joint Los Andes-Japanese study site (CIPLM) along the Duda River and in continuous forest. Later PORRAS (2000) confirmed the territory size of this group.

Densities reported vary from the exaggeratedly high density found by MASON (1966) of $400/km^2$ (130 groups/km^2) in a forest of 6.9 ha to the (perhaps) more natural situation of 5 individuals/km^2 found by ROBINSON (1977) in the gallery forest site where he worked. The Mason density is generally not accepted as "natural". The average day range of the MASON (1968) study group was 570 m, while that of POLANCO´s (1992) group was 615.5 m (range 268-1152, n=10).

An activity budget calculated for this species by POLANCO (1992) based on 245 hours of observation showed 48.4% rest, 20.2% locomotion, 20% feeding, 5.5% grooming, 4.9% vocalization and 1% direct interaction with other members of the group. Locomotion in these primates is basically quadrupedal, although they are also classified as "springers" by ERIKSON (1963). Occasionally individuals hang from their hind legs to reach a fruit, and their tails are often used as props or semi-anchors, by pressing it hard against a tree or branch while leaning outwards (DEFLER, 2003). Groups pass the night in very densely-leaved trees or vines where the animals sleep closely pressed together with the young attempting to be in the middle (ROBINSON, 1977; KINZEY, 1981).

All *Callicebus* are basically frugivores (KINZEY, 1981), however they consume a variety of other food-stuffs. In the POLANCO (1992) study the diet was 80.6% fruits, 9.4% insects, 7.6% leaves and 2.4% flowers. Some foods used by *C. cupreus* are listed as follow: Arecaceae (*Jessenia bataua, Iriartea exhorriza, Catoblastos* sp.); Bombacaceae (*Quaribea cf. asterolepsis*); Burseraceae (*Protium sagotianum, Protium robustum*); Cucurbitaceae (*cf. Guramia* sp.); Leguminosae (*Inga* sp.), Melastomataceae (*Miconia trinervie, Henrietella omiconia*); Meliaceae (*Trichilia stipitata, Trichilia mainasiana*); Moraceae (*Pourouma bicolor, Perebea santoxchyma, Trophis racemosa*); Myrtaceae (unidentified); Passifloraceae (*Passiflora* sp.); Simaroubaceae (*Picramnia latifolia*); Violaceae (*Leonia glicicarpa, Hybanthus prinifolius, Rinorea falcata*).

These monkeys form close monogamous and affectionate relationships which last for years and probably for the life of the partners. *Callicebus cupreus* are apparently seasonal breeders; The estrus period is said to be 17-20 days in *Callicebus cupreus ornatus* (SASSENRATH *et al.*, 1980). The gestation period is unknown, though *Callicebus brunneus* has a gestation of 167 days (WRIGHT, 1984 1985). Copulation during the breeding season is frequent and at times without preliminaries (MASON, 1966). The adult pair are very affectionate towards each other at all times, but they are aggressive towards neighboring groups, with the exception of occasional copulations which sometimes occur between neighboring males and females (MASON, 1966). Grooming is an important activity between male-female and male-young (KINZEY, 1981).

The strong affection between pair mates has been contrasted with *Saimiri sciureus* by MASON (1971, 1975) in laboratory studies. The affection of *Callicebus* group members, especially the pair, may be one of the mechanisms used to maintain the ties which bind such a group together, which apparently last until the death of one of the members.

The vocalizations of *C. cupreus ornatus* were considered by MASON (1966, 1976) to be one of the most complex vocalizations among the primates. They were the subject of research by ROBINSON (1977, 1979a, 1979b, 1981, 1982). He found that this species repeated calls to form phrases and that the animals combined these phrases into sequences. He defined six sequence types. Calls were of two classes: loud, generally low-pitched calls and high-pitched, quiet calls. The first are generally long-distance calls and intergroup calls, the second are short-distance calls with the group as listed here: A.

low-pitched and high-pitched quiet calls: (1). *Squeaks, whistles and trills* – pure tones of constant frequency, utilized to convey danger; (2). *Chirps* – low-pitched sound made towards an observer as the animals eat, a low-intensity alarm; (3). *Grunts* – short and noisy explosive sound produced after the duet between groups; (4). *Sneezes* – short explosive sound produced by the nose while being bothered by other or when fleeing the observer; (5). *Infant distress call* – faint buzz made by the infant after being rejected from the male´s shoulder. B. Strong sounds of low tones: (1). *Chirrups* – produced by inhalation and specific with respect to age and sex of the animal and utilized as an introduction to the duet; (2). *Moans* – long and low introduction of much modulation, given when disturbed by other animals and causing members of the pair to come closer together. Many times these are a prelude to the duet and when first awakening; (3). *Pants* – used in phrases as part of a longer sequence and distinguishable as to sex; (4) *Honks* – interposed with pants but the frequency is lower and always part of a sequence; (5) *Bellows* – a stronger vocalization of males and females and in sequences of duets different in the two sexes; (6). *Pump* – found in all of the sequences of rapid cadenza, ascending in tone with a short climax; (7). *Screams* – during violent fights.

Added to these can be listed *purr* (a low cat-like low frequency sound conveying affection and contentment), as observed in a free-ranging, male *Callicebus cupreus ornatus* at Caparú Biological Station (DEFLER, 2003). ROBINSON (1981) felt that the long-distance calls increased the probability of boundary encounters between neighboring groups and thus serve to define boundaries to the interacting groups. Boundary and territorial defense also included chases, usually of same sex individuals.

The vocalization most often heard is a type of duet of the pair, which both individuals learn over time to maintain and this may serve as a more imposing stimulus than an individual caller. The animals call after proceeding to the limits of the territory in the early morning. After arriving at the territorial limits they climb high up into a tree and the pair vocalizes in duet or only the males call. MÜLLER (1994, 1995) continued vocalization studies in captive groups.

Callicebus cupreus exhibit a variety of display patterns which have important communication value (MOYNIHAN, 1966, 1976a): *arched back display* (somewhat like a cat), conveys aggressive intent; *tail lashing* shows excitement and hostility as does pilo-erection; *chest rubbing* along a branch, possibly as a territorial statement; *tail twining* by members of the family group, obviously an affiliative behavior; *swaying body*, in response to alarm; *looking away*, in response to being disturbed; *head-down* postures performed by alarmed animal; *displacement scratching* on the chest when nervous; *lip protruding* when making certain vocalizations.

Conservation state

Callicebus cupreus populations in Colombia are not large and should be considered threatened, especially because of colonization. *Callicebus cupreus* subspecies are considered to be Vulnerable (VU) according to IUCN (2001) criteria, because of the high rate of colonization in their small areals. Since *C. c. ornatus* is endemic to Colombia, its conservation within the country is very important. This subspecies needs census work to evaluate its status in more detail. Fortunately it can survive in impoverished woodlots. *C. c. discolor* are also very few in Colombia and located only in the zone between the Guamués and Sucumbios Rivers in southern Putumayo. Censuses are also urgently needed. Finally, the undescribed subspecies observed by MOYNIHAN (1976) in southern Caquetá needs urgently to be evaluated. This population is likely to be VU or EN because of the enormous colonization pressures there and its seemingly small areal.

Where to see it

Callicebus cupreus ornatus can often be observed in western Meta on farms that have some small forests preserved. The species is easy to localize in the early morning when it loudly vocalizes its position. MOYNIHAN (1976a) observed *Callicebus c. discolor* on the banks of the Guamués River near Santa Rosa (Putumayo), where it still ought to be common, since there is still plenty of forest there. Any news of observations of *Callicebus cupreus* outside of the above two areas in Colombia would be very important to the author.

Widow Monkey

Callicebus torquatus (HOFFMANNSEGG, 1807)

Plate 11 • Map 2●

Taxonomic comments

The genus was recently revised from the HERSHKOVITZ (1963) concept of three species to thirteen neotropical species with three subspecies of *Callicebus torquatus* known for Colombia (HERSHKOVITZ, 1990, 1988a). GROVES (2001) elevates *C. t. medemi* to *Callicebus medemi* and VAN ROOSMALEN *et al.* (2002) elevate all of the subspecies to species. These last changes were made with few arguments to support the changes and were made apparently influenced by the increasing use of the phylogenetic species concept, which seeks to define species as the "smallest diagnosable cluster of individual organisms within which there is a parental pattern of ancestry and descent" (CRACRAFT 1983). This might work, given enough information, which is usually not the case. Distinguishing species based on phenotype alone is merely a repeat of typological practices. The recent discovery of a diploid number of 16 for *Callicebus torquatus lugens* in Brazil certainly suggests that (with the previously known 2n=20 of another, unidentified population of *C. torquatus*) there are at least two species in this complex (BONVICINO *et al.*, 2003). Nevertheless, in this treatment I shall continue to refer to the traditional subspecies of *C. torquatus* as defined by HERSHKOVITZ (1990) until the systematics of this species complex become clearer.

Common names

Other English: White-handed Tití; Colombia: *macaco* in the departments of Caquetá and Putumayo; *macaco caresebo* in the region of the Guayabero River; *viuda* or *viudita* in the Vaupés and Orinoco River regions; *huicoco* in the lower Caquetá River; *tocón* in the Amazon River; *foga-foga* close to Tarapacá in the Putumayo River; *ikay* en el medio R. Guaviare (*fide* POLITIS & RODRÍGUEZ, 1994); French: *callicebe…fraise;* German: *Wittvenaffe;* Brazil: *tití, titi.*

Indian languages

Podso (Andoke); *wikokó* (Carijona); *wao* or *huau* (Cubeo); *wawi* (Curripaco); *ojó-ojó* (Guahibo); *tunque* or *yanatunque* (Ingano); *huaúhua* (Letuama); *agu* (Nukak, *fide* Politis and Rodríguez, 1994); *cof* (Macuje-Yujup, *fide* Capitán Francisco of the lower Apaporis); *huaú* (Makuna); *cuaí* (Miraña); *gaahi* (Muinane); *agu* (Nukak); *aaxu* (Ocaima); *wecuí* (Piaroa); *tuí, tu* or *tup* (Puinave); *huaúhua* (Tanimuka); *tú* (Ticuna), *huaú* (Tukano); *huaácú* (Yucuna); *ääu* (Yuri); pronounce as in Spanish.

Identification

Five adults weighed an average of 1462 g (range 1410 – 1722 g) with a head-body length of around 290 – 390 mm and a tail length of about 350

– 400 mm (HERSHKOVITZ, 1990). The face has very little hair, being limited to sparse short white hairs over a black skin. There is no sexual dimorphism, although the male has canines a bit longer than the female. The species has the smallest karyotype known for primates a 2n=16 recently described by BONVICINO *et al.* (2003).

The species´ pelage is uniformly reddish brown or blackish brown, the tail is blackish mixed with some reddish hairs; hands and feed whitish or dark brown. This pelage contrasts in all of the subspecies with a band of white hair which extends upward from the chest and follows the neck, prolonging itself to the ears. This extension to the ears is weak in *Callicebus torquatus torquatus*, a subspecies not confirmed for Colombia and different from the other subspecies which have white extending to the base of the ears. The other subspecies are distinguished from the above subspecies as follows:

• The pelage is generally blackish mixed with dark brownish and some reddish brown hairs on the back and the flanks. Hands are white or yellowish yellowish..*Callicebus torquatus lugens*

• The pelage is basically blackish but intermixed with many hairs on the back (extending to the top of the crown) and flanks with many reddish brown hairs, giving the animal a definite reddish appearance in the sunlight...*Callicebus torquatus lucifer*

• The pelage is predominantly black and variable looking like *lugens* or like *lucifer* but with with the white chest and the hands are blackish instead of whitish...*Callicebus torquatus medemi*

Geographic range

Callicebus torquatus is found throughout lowland Colombian Amazonia up to about 500 m of altitude in Putumayo and probably about the same in Caquetá. The species has been observed on the left bank of the Guayabero River (OLIVARES, 1962), where it was collected in 1959 by JORGE HERNÁNDEZ CAMACHO, in La Macarena National Park, and recently it was observed by the Colombian biologist Rocío Palanco north of the Guayabero above La Cordillera de los Picachos National Park. The species is known in the Vichada selva between the Vichada and Guaviare Rivers and the northernmost Colombian population extends north of the Vichada River, reaching the middle Tomo River, where it probably extends to the upper Tomo, although this needs to be confirmed.

Callicebus torquatus is not found on the lower Tomo or lower Tuparro River nor is it found on the north bank of the lower Vichada River, contrary to the distribution map of Hershkovitz (1988, 1990). This error is due to the collection of a specimen by the English ornithologist Cherrie in about 1904 from Maipures, which evidently was a captive animal obtained in the then existing village of Maipures on the left bank. Extensive and concerted efforts have failed to identify the species for the entire area mentioned above (Defler, pers. obs.); nor is it known by locals for this region. The nearest *Callicebus* from Maipures in Colombia are found on the middle Tuparro River and south of the lower Vichada River (Defler, 2003).

Outside of Colombia this species extends from the Napo River northward throughout the Ecuadorian (Ulloa, 1986) and Peruvian Amazonia and throughout S. Venezuela to the Río Branco and the lower Río Negro in Brazil. South of the Amazon it is found west of the R. Purús to the Yavarí River (Hershkovitz, 1990).

Callicebus torquatus is seen most frequently in well-developed, tall forest with a closed canopy, usually over *terra firme*, but not exclusively so. The species also enters extensive *várzea* forest, especially if the forest is tall and well-developed (Defler, 1994a). Such *várzea* forest contrasts with the habitat needs of *Callicebus cupreus*, which also uses *várzea* forest and more commonly so. But *C. cupreus* survives in low, vine-covered, "poor" forest where *C. torquatus* is rarely found. *Callicebus torquatus* is also known from some gallery forest, north of the Vichada River, although this seems to be the exception for the species, which is much more extensively known from closed-canopy forest to the south. Contrary to Kinzey & Gentry (1978), the forest habitat of *C. torquatus* grows on a variety of substrates, varying from white sand through many types of clay soils (Defler, 1994a).

Natural history

This species was studied in Peru by Kinzey (1975, 1977a, 1977b, 1977c, 1978, 1981, Kinzey *et al.*, 1977; Kinzey & Rosenberger, 1975; Kinzey & Gentry, 1979, Kinzey & Wright, 1982, Kinzey & Robinson, 1981, 1983) and by Easley (1982), Easley & Kinzey (1986), Starin (1992), Milton & Nessimian (1984), Robinson *et al.* (1987)(a review article). In Colombia the species has been studied by Defler (1983b, 1994a, 2003), Forero (1985), Palacios & Rodríguez (1995), Palacios *et al.* (1997), Rodríguez, 1997 and Hernández-B. & Castillo-A. (2002).

Social groups are made up of a monogamous pair and one or two of its young. A count of ten groups in Vichada yielded an average of 3.5 per group (DEFLER, 1983). Occasionally groups of five are seen and unpaired individuals ("floaters") can also be detected from time to time. Second year youngsters usually leave the group, although they may remain into the third year before leaving. These young animals sometimes appear, moving peripherally to the group and then disappear again to move alone.

Measured home ranges have varied from about 15-25 ha. Appropriate habitat contains 4-5 groups/km² (14 + floaters), which may add another 8-10 individuals to the total ecological density/km². The average day range calculated by KINZEY (1977b) and KINZEY et al. (1977) was 819.4 m (n=22 days) for a research project in Perú and at the Estación Biológica Caparú the average was 807.2 m (range 513.7 – 1070 m, n=26)(PALACIOS & RODRÍGUEZ, 1995; PALACIOS et al., 1997).

EASLEY (1982) calculated a time budget based on 400 hours of observation as 62.7% rest, 16.5% moving, 16.1% feeding, 2.7% grooming, 1.6% playing and 0.3% vocalizing. PALACIOS & RODRÍGUEZ (1995) calculated 54.3% rest, 22.9% moving, 17.6% feeding, 4.07% grooming, 0.41% playing, and 0.42% vocalizing based on 240 hours of observation.

EASLEY (1982) analyzed the locomotive and positional behavior of the species, defining it as a generalized quadruped walking and running about 66.8% of the time. This species also engages in active jumping (23.9% of the time) and climbing (9.1% of the time). Sitting (62%3% is the most common posture, followed by lying (16.1%), walking (10.4%), jumping (4%), vertical clinging (3.1%), climbing (1.5%), running (0.8%), hanging suspended by the back legs (0.8%), horizontal clinging (0.7%) and standing (0.2%). If postures of locomotive behaviors are excluded from this analysis then the scores were sitting (74.8%), lying (19.3%), vertical clinging (3.7%), hanging suspended from the hind feet (0.9%), horizontal clinging (0.8%) and standing (0.2%). Previously KINZEY & ROSENBERGER (1975) pointed out that these animals fit into the "clinging and leaping" group of primates. Groups of C. torquatus sleep ontop of large branches of emergent trees, frequently a bit above the level of the main canopy (KINZEY, 1981).

Although fruits are the major portion of this primate´s diet, invertebrates and leaves are also consumed to a smaller degree. Lepidopteran larva, spiders and orthopterans are especially eaten with relish and probably occasional small lizards, judging by the hunting preferences of a tame, free-ranging adult female, which lived at the Caparú Biological Research Station on the lower Apaporis River (DEFLER, 2003). KINZEY (1977a) found the following range of dietary preference during his 135 hours study in Perú: 14% *Clarisia racemosa* (Moraceae); 13% unidentified (Guttifereae); 7% *Pithecellobium* sp. (Leguminoceae); 7% *Alchornaea* sp. (Euphorbiaceae); 6% *Maripa* sp. (Convolvulaceae); *Jessenia bataua* (Arecaceae); *Psychotria axillaris* (Rubiaceae); *Guatteria elata* (Annonaceae); *Virola* sp. (Myristicaceae).

EASLEY (1982) identified frequency of item choice in the diet of the same groups as above: 74.1% fruits, 15.8% insects, 8.8% leaves, 0.6% buds and flowers and 0.1% other. Of the 57 fruit species identified, the palm tree *Jessenia polycarpa* was the most commonly eaten in 22.7% of the feeding observations. The following lists the range of preference observed in this study: 22.7% *Jessenia polycarpa* (Arecaceae), 7.9% *Ocotea* no. 1 (Lauraceae); 6.6% *Tachigalia* sp. (Caesalpiniaceae); 5.9% *Beilschmiedia* sp. (Lauraceae); 5.8% *Ocotea* no. 2 (Lauraceae); 4.8% unidentified; 3.5% unidentified; 3.5% *Guatteria* sp. (Annonaceae); 3.4% *Annona* sp. (Annonaceae); 2.4% unidentified; 2.0% unidentified; *Guatteria* sp. (Annonaceae); 1.9% *Duguetia* sp. (Annonaceae).

PALACIOS & RODRÍGUEZ (1995) and PALACIOS *et al.* (1997) identified 62 species from 32 plant families in the diet of a study group of *C. torquatus lugens* in the Estación Biológica Caparú in eastern Colombia. The preference values of each family, according to species utilized, are as follow: Myristicaceae (25.02%); Euphorbiaceae (15.28%); Moraceae (14.37%); Arecaceae (8.68%); Caesalpiniaceae (7.85%) Rubiaceae (5.10%); Chrysobalanaceae (4.41%); Annonaceae (4.19%); Cecropiaceae (4.03%); Araceae (1.95%); Elaeocarpaceae (1.78%); Dilleniaceae (1.69%), Combretaceae (1.17%), Apocynaceae (1%); Aquifoliaceae (1%), Meliaceae (0.88%); Sapotaceae (0.85%); Burseraceae (0.81%); Apocynaceae (0.67%); Monimiaceae (0.23%); Piperaceae (0.22%); Melastomaceae (0.18%); Humiriaceae (0.13%) Celastraceae (0.11%); Myrtaceae (0.09%); Lecythidaceae (0.08%); Aquifoliaceae (0.07%); Sterculiaceae (0.07%); Solanaceae (0.05%); Clusiaceae (0.02%).

The most important species consumed during six months in this study are listed as follows: (PALACIOS *et al.*, 1997): 13.88% *Sandwithia heterocalyx* (Euphorbiaceae); 10% *Virola melinonii* (Myristicaceae); 8.35% *Iryanthera ulei* (Myristicaceae); 7.06 *Oenocarpus bataua* (Arecaceae); 6.53% *Heterostemon conjugatus* (Caesalpiniaceae); 5.10% Coussarea sp. (Rubiaceae); 5.02% *Ficu.* sp. (Moraceae); 4.53% *Iryanthera crassifolia* (Myristicaceae); 3.84% *Helicostylis tomentosa* (Moraceae); 3.39% *Brosimum rubescens* (Moraceae).

The estrus cycle seems to be about 16 days, based on observations of 14 cycles of a tame, free-ranging female living at the Estación Biológica Caparú (Vaupés, Colombia). During the period of receptivity (which lasts 2-3 days), the black labia and the clitoris became swollen and hard and behavior changes occurred. During the receptive period the female became much more affectionate towards its human "parents", purred loudly, somewhat like a cat and crouched in a lordic position when the base of the tail was stimulated. Contrarywise to her increased affection towards her perceived "family unit" (or two humans), she became much more aggressive than normal towards any "outsiders" (i.e. other human beings). During estrus the female tongue-flicked frequently, using this signal in two opposite contexts; she tongue-flicked as she attempted to approach her favorite humans while she also tongue-flicked as a preliminary to the attack of outsiders (especially male humans) (DEFLER, 2003).

One recognizable pair at the EBC in Vaupés had been observed together for 14 years and was said to be still together at least four years more after this author had left. During these 14 years the pair produced 10 young, all of which survived the first year. During four years no young were produced. In Vichada, 660 km to the north, young are usually produced in December or early January (DEFLER, 1983a). In Vichada this is a difficult season with sharply reduced fruit resources for many animals in this part of the country (which has an annual precipitation of about 2400 mm); a long dry season is just taking hold and January and February produce only a very few millimeters of precipitation for each month. A close analysis of the diet of *C. torquatus lugens* here would be interesting, in as much as it would serve to identify the resources which allow the species to have this birth pattern.

On the Guayabero River near La Macarena the birth season is apparently about the same time as in Vichada (HERNÁNDEZ-CAMACHO & COOPER, 1976; F. MEDEM, pers. com.). On the lower Apaporis River in Vaupés with about 3815 mm of precipitation throughout the year, the birth season is also centered around December, although some outlying births are known as early as the first of October (PALACIOS & RODRÍGUEZ, 1995). Nevertheless, the birth season is the same as the other two sites, despite the lack of a strong dry season. However, we know that fleshy fruits are beginning to increase from their yearly low during this time, so the question of resource use by the species remains very interesting (DEFLER, 2003).

ROBINSON et al. (1987) also report a birth season in Dec.-Jan. for the species in Peru at 4°S. Why this specific birth season should be characteristic for the species in such widely divergent places both north and south of the equator and with different phenological cycles must remain for the moment an open question. After birth, the newborn quickly acclimates to being carried by the male, and usually goes to the female for nursing only (ROBINSON et al., 1987).

Callicebus torquatus is very affectionate within the family unit, but the adult pair is aggressive towards neighboring pairs. The most common interaction with neighbors is counter-singing of the pairs, where one pair waits listening while the other pair vocalizes their duet, later the listeners answer, while the first vocalizers listen. There are instances when two pairs interchange vocalizations from very close together or from almost the same place in the forest. Sometimes these emotional interactions may finish in chases by the pair or an individual against the others. PALACIOS & RODRÍGUEZ (1995) found evidence of different types of agonistic interactions between different pairs. For example, a habituated group at EBC (perhaps for historic reasons) treats one pair of neighbors differently from another pair. The group studied at EBC interacts via counter-singing with the young pair which took territory left by the study pair, while the group both counter-sings and is aggressive in other ways towards a different neighboring group. Perhaps the young pair is made up of one of the offspring of the other group? Between group members a lot of affection is evident including much contact. The affection is probably one of the mechanisms which maintains the ties in these associations which apparently last until one of the members of the pair dies.

Vocalizations of this species are very complex, especially a long-call display, perhaps used to regulate spacing and to define territory (ROBINSON et al., 1987). Surprisingly, experimental playback of solo male calls caused the owners´ of a particular territory to move away from the recording, and recordings of duetting caused the territory owners to duet in return and to travel parallel to the speaker (KINZEY & ROBINSON, 1983). However, any proximate sound stimulus can cause duetting by territory owners, and many direct observations of duetting neighbors were observed to cause the territorial owners to move towards the calling, where the two couples sometimes confronted each other across a small space (DEFLER, 1983a). Lone individuals in the established territory of another pair do not normally vocalize, since they may be vigorously attacked if the "owners" of the territory find them (DEFLER, pers. obs.).

There is some evidence that *Callicebus* not only can determine sex from a long call but can identify duetting individuals, so it should perhaps not be surprising that a resident pair could distinguish a recording from a live monkey and move away from it. A human-raised and newly matured female *Callicebus torquatus lugens* on first shouting, attracted the resident forest group closer until they became accustomed to her presence, although they always answered her calling with their duetting, later neither coming closer nor moving away. The female´s vocalizatons sometimes attracted individual males in short order, which attempted to duet with the female. Since the female had been raised by humans, she did not show interest in duetting with the newly appeared males nor in establishing a relationship with them, and the males eventually desisted and left. The only exception to this was one male which attempted to establish a relationship during two years before giving up and leaving during an accidental 26-day absence of the female when she became lost in the forest (DEFLER, 2003).

Some vocalizations of *Callicebus torquatus lugens* are listed here (DEFLER 2003): (1). *morning duet* – the most commonly heard vocalization of the pair, singing in duet, complex and utilized to defend territory; it is interchanged with neighboring groups as counter-singing; (2) *danger peep* – various soft, high-pitched peeps but sometimes low intensity, advising of danger; very difficult to localize; (3) *purr* – sounds very much like a cat´s purr, used by all members of the group to show contentment, affection or request for food, grooming or contact; (4) *rough growl* – given by young animals when complaining of rain or when greeting adults; (5) *sharp scream* – when fighting to express extra disgust; (6) *play growl* – low, gargling growl used in play and changing in tone, terminating in interrogative tone; (7) *soft whine* – especially young animals but also adults when requesting something of another such as food or while grooming another; (8) *bark* – loud, sharp and sudden bark when molested by the unwelcome close presence of other larger primates such as *Lagothrix, Cebus, Ateles* or raptors.

Individuals of both sexes occasionally mark their chests with pungent wadded leaves, rubbing the leaf up onto the throat and chin to the mouth, where the wad is wetted and rubbed down again, repeated various times while looking up into the air. One wild male did this as he approached the tame estrus female, who was near a building, after this male had left the forest and while walking on the elevated poles which had been set up for monkey travel. Another foraging female marked herself in the presence of an observing human who was 20 m from her.

Displays are similar to *C. cupreus*, which were first described by MOYNIHAN (1966, 1967, 1976a). Some displays are listed here (DEFLER, 2003): (1) *piloerection* – agonistic; excited state when attacked or attacking; during danger; (2) *arched-back* – agonistic; before some attacks or when threatened; position held for several seconds; (3) *tail twinning* – when duetting or resting the pair often wind their tails around each others tail; (4) *tongue flicking* – in two contexts; aggressive just before attack or as space reducer towards mate and probably just before copulation (hand-raised female at EBC tongue flicks at human "parent", especially at height of estrus cycle; (5) *chest rubbing* – using a wadded leaf the individual rubs from throat to chest after first wetting the leaf with saliva; performed in presence of human observer; nervousness. KINZEY *et al.* (1977) observed play behavior only between the infant and male and between two juveniles. In Caparú the adult female occasionally exhibited play behavior towards her human "parent" with slight head shakes and sideways jumps on the ground (DEFLER, obs. pers.). Agonistic behavior is common between neighboring groups and can sometimes result in fights, although usually the aggression is limited to intergroup vocalization.

Callicebus torquatus usually attempts to move out of the path of passing troops of *Lagothrix lagothricha or Cebus apella,* although sometimes the small monkeys give a burst of loud and aggressive-sounding vocalization (*bark*) when they are approached closely by the larger species. *Callicebus* frequently hides and shows much caution towards raptors. Being frightened causes them to give alarm peeps, probably because they must be especially alert to predators. A *Felis weidii* was detected alongside a dead *Callicebus torquatus* during recent censuses on the Purité River in Colombia, although the monkey was not freshly killed. The local group was no longer observed after this.

Conservation state

This species is not considered to be endangered in Colombia, but where there are many colonists this primate tends to disappear, due to deforestation. The species is commonly hunted and eaten by Colombian indigenous peoples or used as bait for hunting larger carnivores or for fishing; however, where there is plenty of forest meat the species is found commonly close to Indian settlements. *Callicebus torquatus* is classified LC (formerly LR) in the national Colombian Red Book classifications. Actually *C. t. medemi* may be slightly endangered due to its presence in a zone of heavy colonization, but presently it is classified LC with the other subspecies. Fortunately *C. t. medemi* is found in La Paya National Park, but the Park itself is very vulnerable to surrounding colonization. *Callicebus torquatus lucifer* is protected in Amacayacu National Park and Cahuinarí National Park, while *Callicebus torquatus lugens* is protected in Chiribiquete National Park and El Tuparro National Park and in the two biological preserves Nukak and Puinawai.

Where to find it

Callicebus torquatus can be widely observed in its Colombian range. *Callicebus t. lugens* is easily located both in El Tuparro National Park (in the forests nearest Tapón on the west side) and at the Caparú Biological Station, but it is often found fairly close to many small Amazonian villages if they are not hunted. *C. t. lucifer* is easily seen along the rivers of Amacayacu National Park, especially the Amacayacu and Cotuhué Rivers, while *C. t. medemi* is a commonly observed species in La Paya National Park.

Family **Atelidae**

This family includes the genera *Alouatta*, *Ateles*, *Brachyteles*, *Lagothrix* and *Oreonax* (formerly part of *Lagothrix*). This is a clearly-defined clade of large, prehensile-tailed primates having ape-like postcranial adaptations such as "medially buttressed temporomandibular joints and a mandibular corpus whose lower border slopes distinctly ventrally, often giving the angle of the mandible an enlarged appearance" as well as dental similarities including a "reduced buccal cingulum on the cheek teeth and relatively enlarged hypocones on upper molars" (ROSENBERGER, 1981:23). The clade clearly includes the Brazilian Pleistocene fossils *Protopithecus brasiliensis* and *Caipora bambuiorum*, both rather large animals of 20-21 kg.

The molecular evidence also clearly clumps *Alouatta, Ateles, Brachyteles* and *Lagothrix* together, suggesting that *Brachyteles* and *Lagothrix* are most closely related (HARTWIG *et al.*, 1996; SCHNEIDER & ROSENBERGER, 1996; SCHNEIDER *et al.*, 1996; GOODMAN *et al.*, 1998). The family can be divided into two subfamilies: Alouatinae (*Alouatta*) and Atelinae (*Ateles, Brachyteles, Lagothrix,* and *Oreonax*). Until recently, most people included *Oreonax* as *Lagothrix flavicauda*, but HARTWIG *et al.* (1996) and GROVES (2001) have pointed out rather distinctive cranial characteristics for this taxon, while a cladistic analysis performed by GROVES (2001) clearly places it into a separate genus, previously named by THOMAS (1927) *Oreonax*.

Spider monkeys

Family **Atelidae**
Genus *Ateles* (**E. Geoffroy, 1806**)

Taxonomic comments

The spider monkeys are a variable group with many color variations (see KONSTANT *et al.*, 1985) whose systematics and taxonomy are confusing, especially due to the various taxonomies in use. There are still doubts as to how many species actually exist in the genus. Until recently there were two or three main points of view.

The first point of view is based on a revision of phenotype by KELLOGG & GOLDMAN (1944) who suggested there were four species: (1) *Ateles belzebuth* (three subspecies, *A. b. belzebuth* found east of the Andes in Colombia, Ecuador and Perú and in southern Venezuela; *A. b. hybridus* found to the east of Cauca/Magdalena Rivers in northern Colombia to the Cordillera de Mérida of Venezuela, although MONDOLFI & EISENBERG (1979) published other collection records from northern Venezuela; and *A. b. marginatus* found south of the lower Amazonas River and east of the Tapajós River; (2) *Ateles fusciceps* with two subspecies (*A. f. robustus* from northern Colombia and the Colombian Pacific coast; *A. f. fusciceps* of Ecuador and possible marginally in Colombia; (3) *Ateles geoffroyi* of Central America with nine subspecies, one of which – *A. g. grisecens* was said to be found on Colombia's Serranía del Baudó on the northern Pacific coast; (4) *Ateles paniscus* of the Guyanas and Perú, including *A. p. paniscus* of the Guyanas and *A. p. chamek* of Perú from the right bank of the Amazon/Mariñon River, extending into western Brazil and eastern Bolivia.

The above taxonomy was for many years the norm for *Ateles* (see HONACKI *et al.*, 1982) and was used by MITTERMEIER & COIMBRA-FILHO (1981), KONSTANT *et al.* (1985) and others as a conservative view, despite the difficulties of its antiquated system. The scheme has been criticized on the basis of much intergradation of pelage color between distinct species, hybridization between *A. geoffroyi* and *A. fusciceps* in Panamá (ROSSAN & BAERG, 1977) and (more important) the illogic of identifying subspecies of two species (*A. paniscus paniscus*, *A. p. chamek*, *A. belzebuth belzebuth*, *A. b. marginatus*) widely separated with the other species interdigitated between the subspecies of the other species. Nevertheless, chromosome variation between some taxa support the variation between various species (BENIRSCHKE, 1975; KUNKEL *et al.*, 1980).

The second taxonomic viewpoint proposed and sustained by HERSHKOVITZ (1969, 1972a, 1977), HERNÁNDEZ-C. & COOPER (1976) and WOLFHEIM (1983) is that *Ateles* is comprised of one variable species (*Ateles paniscus*). According to this viewpoint there are four (HERNÁNDEZ-C. & COOPER, 1976) or five subspecies (HERNÁNDEZ-C. & DEFLER, 1986, 1989) of *A. paniscus* in Colombia: *A. paniscus belzebuth* (Amazonia); *A. p. hybridus* (north of the Cordillera Occidental to the Magdalena), *A. p. fusciceps* (Panamanian frontier), *A. p. rufiventris* (=*A. p. robustus*) and *A. p. brunneus* (northern Central Cordillera). This viewpoint is supported by the natural hybridization that has been mentioned above (ROSSAN & BAERG, op. cit., 1977) and by the variations in coat color that have been observed.

A third viewpoint was published by GROVES (1989), who preferred to recognize six species of *Ateles* (*A. belzebuth*, *A. chamek*, *A. fusciceps*, *A. geoffroyi*, *A. marginatus* and *A. paniscus*). But taking new karyotypic (MEDEIROS *et al*, 1997) and morphometric (FROEHLICH *et al.*, 1991) information into account, GROVES (2001) recognized *A. paniscus*, *A. chamek*, *A. marginatus*, *A. fusciceps*, *A. geoffroyi* and *A. hybridus*.

FROEHLICH *et al.* (1991) analyzed *Ateles* morphometrically and came to a new conclusion about the systematics of this difficult genus. This latest viewpoint may more correctly reflect a natural *Ateles* phylogenetics than the first two views. Using a set of 50 measures of 284 specimens of skulls and skins, they constructed a dendrogram of taxonomic relationships and found at least three independent taxa. The taxa based on this dendrogram would be *Ateles paniscus* (of the Guyanas), *Ateles belzebuth* (including all the Amazon groups, *i.e. chamek* and *marginatus* as subspecies of this Amazonian species); and *Ateles geoffroyi* (which includes *hybridus*, which was previously included as a subspecies of *A. belzebuth* and *A. fusciceps*). This permits the recognition of *A. geoffroyi hybridus, A. g. rufiventris* (=*A. g. robustus*), *A. g. grisescens* and *A. g. brunneus* and *A. b. belzebuth* for Colombia.

FROEHLICH *et al.* (1991) take into account the known hybrids between *Ateles geoffroyi panamensis* and *Ateles fusciceps robustus*. This view maintains that there are no known areas of overlap between *hybridus* and *belzebuth*. This also solves the problems of the illogical placement of *A. paniscus chamek* and a subspecies of *Ateles belzebuth* (*marginatus*), so distant from the other populations of each taxon. *Ateles belzebuth* is thus seen as a clinal ring species. The viewpoint also unifies some variable and complex populations of central and northern South America in one species (*A. geoffroyi*), permitting finally the recognition of at least two species of *Ateles* for Colombia: *Ateles belzebuth* and *Ateles geoffroyi*.

The revision of *Ateles* using this third viewpoint of *Ateles* systematics has not yet been published. FROEHLICH (pers. com.) has recently commented to the author, however, that the *hybridus* group may be distinctive enough from the *A. geoffroyi* group to merit species status. COLLINS (1999) and COLLINS AND DUBACH'S (2000a, 2000b, 2001) discuss their research on mitochondrial and nuclear DNA of *Ateles* and point out the species status of *A. hybridus* as a monophyletic group "without clear ties to any other spider monkey clades". The species *Ateles hybridus* would include both *A. h. hybridus* (east of the Magdalena River into Venezuela) and *A. h. brunneus* in the northern parts of the Central Cordillera and endemic to Colombia. In this book and in agreement with FROEHLICH *et al.* (1991) and COLLINS & DUBACH'S (2000a, 2001) phylogenetic anlalysis of mitochondrial and nuclear DNA, the taxa *Ateles geoffroyi, A. hybridus* and *A. belzebuth* are recognized for Colombia.

Common names

Colombia: *mico araña*. Common Names Elsewhere: (adapted from HILL, 1966) *koaitá* (Tupi, cognate with Guaraní = *kaaita, fide* SIMPSON, 1941; *kawát* (MACUSHI, *fide* WILLIAMS, 1932, in SIMPSON, 1941; *kwata* (KALINA, DE GOEJE, 1909, in SIMPSON, 1941; *maquisapa* (Quechua, used general throughout Perú; *mono araña*. Brazil: *macacos-aranha, maquizapaas, coat*. French: *atéles, singes araignée*. German: *Klammeraffen*. Dutch: *slingeraapen*

Indian languages: **(see the description of the species)**

Identification

These are generally large monkeys weighing around 7-10 kg and possessing very long arms and legs. They usually don't have a thumb or it is vestigial. The tails are prehensile and the inside (ventral) surface is mostly naked and with a callosity which allows a good grip.

Ateles are found from southern Mexico, throughout Central America and extending through Colombia, Ecuador, Perú and Brazil to Bolivia. They are also found in the Guyanas (*A. paniscus*). In Colombia they are found on the northern and western slopes of the Sierra Nevada de Santa Marta, the upper valley of the Río Cauca, the lowlands of the Pacific and Atlantic coast to the border with Panama and in large parts of the Llanos Orientales and

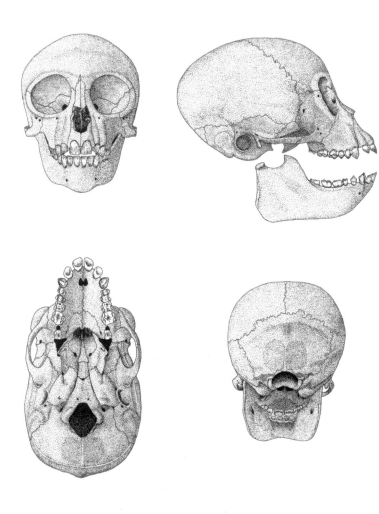

Figure 33. View of skull of *Ateles belzebuth*.

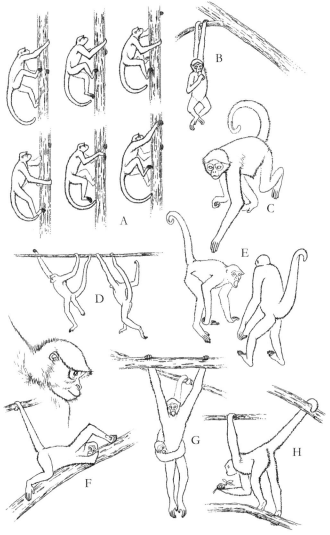

Figure 34. Various attitudes of *Ateles.* a) Climbing in tree. b) Resting hanging by tail and hand. c) Cuadrupedal locomotion. d) Brachiating. e) Agonistic behavior. f) Typical resting. g) Hanging with infant and vocalizing. h) Reaching branch to eat.

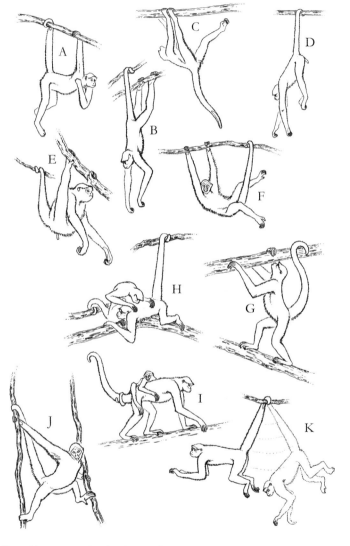

Figure 35. Various attitudes of *Ateles*. a-g) Different hanging postures. h) Grooming. i) Carrying infant. j-k) Play.

Amazonia, although much less than indicated by HERNÁNDEZ-C. & COOPER (1976).

Long-haired spider monkey

Ateles belzebuth **(E. GEOFFROYI, 1806)**

Taxonomic comments

Plate 8 • Map 15

Includes the subspecies *A. b. belzebuth* (=*A. paniscus belzebuth*), *A. b. chamek* (=*A. p. chamek*), and *A. b. marginatus* (=*A. p. marginatus*), although only *A. b. belzebuth* is found in Colombia (see comments on genus *Ateles*).

Common names

Other English: Yellow-bellied Spider Monkey; White-bellied Spider Monkey. Colombia: *marimonda* and *marimonda* in eastern Colombia; *marimba* and *braceadora* in the Llanos Orientales and Amazonia; *coatá* in the Leticia region (Tupí origin); *maquizapa* in the Putumayo and Leticia regions (origin Quechua); French: *coata à ventre blanc* or *atèle chuva* ; German : *Soldstirnaffe;* Dutch : *langhaarige slingeraap.*

Indian languages

Fiaguai (Andaquíes in southern Cauca and the Caquetá, CASTELLVÍ, 1938); *painaso* (Coreguajes of Caquetá, CASTELLVÍ, 1938); *cuverí* or *cuváiri* (Guahibo); *marimba* (Inganos, Aguas-Claras-Cusumbé, La Solita ; municipio Valparaiso, Caquetá, *fide* J. V. RODRÍGUEZ-M.) ; *cuata* (Macuna) ; *vana*, pl. *vanámu* (Miraña, *fide* NAPLOEÓN MIRAÑA, 1985) ; *méecu* (Muinane, WALTON, 1972, Summer Institute of Linguistics) ; *barata* (Piaroa); *chaira* (Puinave, Carera, unpublished) ; *painnazo* (Siona, CASTELLVÍ, 1938) ; *cuhuatacá* (Tanimuca) ; *juatá* (Yucuna) ; outside of Colombia, *marimonda, maquiçapa, urcu maquiçapa* (Inca, *fide* BARTLETT, 1871).

Identification

Body length is more or less 46-50 cm with a long prehensile tail which measures around 74-81 cm. The species in general has a naked face with black skin, at times a bit less pigmented around the orbits and mouth.

Usually there is a submalar stripe of white hair that is rather conspicuous and which terminates on the forehead at the base of the ear. There is often also a triangular frontal patch of white or yellowish-white and which may have some dark hairs. The dorsal parts of the body, including the hind feet, the hands and the majority of the tail are usually black. The point of the tail is invariably cream-yellow to black, although the ventral parts of the tail often are buffy-yellow to orange in color, the same as the ventral parts in general and usually including the majority of the hind feet. The feet and the parts of the ribs are always black, and this color can extend a variable distance to the anterio-lateral parts of the thigh and lower leg. Notable variations in many of these color characteristics are found locally and between individuals.

Geographic range

The distribution of this species is not well-known, despite numerous collection localities from the Colombian Amazon and the Llanos. KELLOGG & GOLDMAN (1944) erroneously included most of the Colombian Amazon and all of the Llanos Orientales in their map, while HERNÁNDEZ-CAMACHO & COOPER (1976) included piedmont in the Llanos and all of the Amazon. Nevertheless, there are many questions. It is fairly clear that the species is predominantly a closed-canopy forest species. Collection localities in the piedmont and Cordillera Oriental go no further north than the Upía River drainage in Boyacá department, although there are probably populations west in the Manacacías river (STEVENSON, pers.com.). Ample collection localities are known for the region of La Macarena and for eastern Caquetá and the Coehmani rapids in SE Caquetá department. West of the Yarí River it is apparently very sparse and almost always a surprise when observed.

Two females were observed on the upper Mesay River by Armando Perea (Carijona), although he had never seen the species in that region before (pers.com. to DEFLER, 1985). A female was shot by PEPE RODRÍGUEZ MAKUNA and CELESTINO YUKUNA on the right bank of the Apaporis River (pers.com. to DEFLER, 1988) and another female was shot in 1992 on the lower right bank of the Apaporis at a salt lick near the Estrella rapids. In the latter case, this author procured the skull and collected interesting notes from the hunter, JOSÉ RENALDO MUCA, including the observation that this female was living with a group of *Alouatta seniculus,* a little-known fact about lone

Ateles belzebuth, which has been reported by KLEIN (1972) and observed in the field by DEFLER in 1976.

The individuals described above all may have been long-distance immigrants, lost from a breeding population, since in each instance the Indian observers were greatly surprised to see this species of primate, and for some it was the first time ever that they had seen it, even though they lived in the area where the monkey appeared. Other surprise locations, distant from known breeding populatins were a collection of one example by the Christian Brother APOLINAR MARÍA in 1913 in the upper Magdalena River and an observation by Thomas Lemke in 1977 in La Cueva de los Guacharos National Park, also where the species was not known to occur, after much work by other biologists and naturalists.

To my knowledge no specimens are known to have been collected between the Caquetá and Putumayo Rivers, except for piedmont forest which nowadays is highly intervened by colonization, especially up to around 500-600 m. The species has been observed at altitudes of as much as 1300 m on the eastern slope of the Cordillera Oriental (HERNÁNDEZ-CAMACHO & COOPER, 1976). Recently a sight record was reported to this author for the uper Marandú River (Caquetá), where the species is said to be common (RAMÍREZ pers. com. to DEFLER, La Pedrera, 1995). There is still much work to be able to define adequately the Colombian range of this mammal.

For some years this author has been collecting possible locations where *A. belzebuth* may yet be found in the Amazon basin of Colombia, although in general this primate is not known in great sections of Amazonia. Closer to the mountains in Tinigua National Park, on the Peneya River and close to the Coehmani rapids (in Caquetá department) the species is sympatric with *Lagothrix lagothricha*, which has a similar frugivorous diet to that of *Ateles belzebuth*. It has been observed as well by IZAWA (pers. com.) on the west bank of the upper Guayabero River in Cerro de los Picachos National Park. It is probable that other sites are not fertile enough to permit *Ateles* to compete with the much more widely spread (and possible superior competitor) *Lagothrix lagothricha*.

Outside of Colombia this species extends to the Marañon River in Perú (*A. b. belzebuth*), is found east of the Marañon and throughout eastern Brazil (*A. b. chamek*) and is found in eastern brazilian Amazon (*A. b. marginatus*), but its scattered and patchy distribution is still poorly-known.

This species of spider monkey has been found in a variety of habitats, although most commonly in primary inland high rainforest. It is rarely seen in edge habitats but does enter flooded forest at certain times of the year (AHUMADA *et al*, 1998), when that habitat offers an abundance of fruits (KLEIN & KLEIN, 1976; DEFLER, pers. obs.). These primates usually chose the mid- to upper levels of the forest, including emergents (STEVENSON & QUIÑONES, 1993).

Natural history

The first study of this species in Colombia took place north of the Guyabero River during 1966-1967 (KLEIN, 1971, 1972, 1974; KLEIN & KLEIN (1971a, 1971b, 1973a, 1975, 1976, 1977). Other research in Colombia took place on the Peneya River (IZAWA, 1975, 1976, 1993; IZAWA *et al.*, 1979). Several shorter studies have taken place at the joint Japanese-Los Andes University research camp on the Duda River (in Tinigua National Park), 120 km away from the original site of the KLEINS (STEVENSON, 2001; STEVENSON *et al.*, 1992, 1998, 2000a,b, 2002; STEVENSON & QUIÑONES, 1993; AHUMADA, 1989, 1990, 1992); AHUMADA *et al.*, 1998 and IZAWA, 2002, 1990d, IZAWA (1993), INABA (2000, 2002), LINK (2002), MATSUSHITA & NISHIMURA (1999), NISHIMURA (2002b), SHIMOOKA (2000, 2002a-c, 2003), and DIDIER (1997).

In Ecuador the species has been studied by CANT *et al.* (1997, 2001, 2003), RUSSO *et al.* (2003), DEW (2002a, 2002b) and POZO (2001). In Brazil there are studies by IWANGA & FERRARI (2001), MENDES (1997), NUNES (1995a, 1995b, 1998) and in Venezuela (CASTELLANOS & CHANIN (1996).

An important study (*Ateles belzebuth chamek*) was recently completed in southern Perú in Manu National Park by MCFARLAND-SYMINGTON (1986, 1990; SYMINGTON (1987a, 1987b, 1988a-c, 1990) during 1982-1986. DURHAM (1971, 1975) made brief observations in southern Perú as well. WHITE (1986) published censuses and preliminary observations on this Peruvian population. Comparisons with work done on *A. paniscus* by VAN ROOSMALEN (1985) seems very similar as is pointed out in van ROOSMALEN & KLEIN (1988).

These monkeys usually associate in lowland forest with social groups which may vary from 16-40, and most often forage in subgroups averaging around 2.5-3.5 members per subgroup. Adult females tend to move in separate home ranges with several adult males moving throughout the home range of several females. All of the members of the larger group defend this territory from adjacent groups (or clans). This type of social system is rather unusual for primates; it is, however, reflected in the social system of chimpanzees (*Pan troglodytes*) and is seen as a strategy for utilizing fruits, since spider monkeys are obligate frugivores. SYMINGTON (1987a, 1988b) was able to demonstrate a correlation between foraging party size and the amount of fruit available, and her data are a good indication of the selective pressures which have probably led to this type of social system (MCFARLAND, 1986, 1990). However, this relationship seems to be weak for other populations of spider monkeys (AHUMADA *et al.*, 1989). DURHAM (1971) demonstrated a reduction of group size with altitude in southern Perú from a mean of about 16.55 individuals at 275 m, 10.2 individuals at 576 m, 5.7 at 889 m and 3.6 at 1424 m, which he correlated with the relative abundance of fruit trees.

Spider monkey clans in La Macarena used a home range of about 2.6-3.9 km² with about 20-30% overlap, according to KLEIN & KLEIN (1976), and at Manu two home ranges were somewhat less at 153 ha and 231 ha with about 16% overlap between groups (SYMINGTON, 1987a) and 150 ha in the Duda River in Tinigua National Park (STEVENSON *et al.,* 1991). IZAWA *et al.* (1979) studied grouping on the Duda River. Interestingly these three densities show a rather narrow range of densities at three distinct sites. Day ranges average around 1666 m at the Duda River site (STEVENSON *et al.,* 1991) to 1977 m at the Manu National Park site (SYMINGTON, 1987a,b). Population densities vary, according to the available information from 15-18 individuals/km² north of the Guayabero (KLEIN & KLEIN, 1976), while at Manu about 16-26 individuals/km². Calculations at the Duda River site in Tinigua National Park are about 17 individuals/km² (STEVENSON *et al.,* 1991).

An activity budget calculated by KLEIN & KLEIN (1977) for *A. belzebuth* showed that the study group spent 22.2% of their time feeding, 63% of their time resting and 14.8% of their time moving. Another time budget calculated for a nearby group in the Duda River showed the study animals feeding about 25% of the time, resting about 43% of the time, moving about 27% of the time and spending about 5% of the time in general social activities (STEVENSON *et al.,* 1991). The time budget for the southern Perú study showed 29% of the time feeding, 45% resting and 26% moving

(SYMINGTON, 1988b). This species is a keen brachiator, throwing itself recklessly through the vegetation.

These large primates are primarily frugivores and they consume a great variety of mostly ripe fruit. Their feeding behaviors are described by KLEIN & KLEIN (1977). SYMINGTON (1987a) found that her study groups consumed about 75% ripe fruit; the other 25% was made up of new leaves, flowers, seeds, honey, wood, caterpillars and termites. KLEIN & KLEIN (1977) observed their study animals, spending about 83% of their time consuming ripe fruits, with the remaining feeding time divided in 5% tree leaves and buds, 2% epiphytic leaves and stems and 10% decaying wood. STEVENSON (1992) saw their study group eating about 75% ripe fruit and a bit less than 25% young leaves, with the addition of a few insects and termite soil, fungus, flowers and tree bark.

KLEIN (1972) identified 45 species of plants from 20 families that were foods for *A. belzebuth*. The following list shows family values, calculated according to the number of species chosen per family: Moraceae (15), Palmae (5), Sapotaceae (3), Annacardiaceae (2), Guttiferae (2), Lauraceae (2), Leguminosae (2), Myristacaceae (2), Annonaceae (1), Burseraceae (1), Chrysobalanaceae (1), Connaraceae (1), Cucubitaceae (1), Cyclanthaceae (1), Euphorbiaceae (1), Flacourtiaceae (1), Marcgraveaceae (1), Oleaceae (1), Sterculiaceae (1) and Ulmaceae (1).

Many of the fruits consumed by *Ateles belzebuth* are consumed whole with the intact seeds, thus serving as a major disperser for many species of trees and vines used by this primate. Only a small percentage of seeds are masticated to be used as food. Smaller spider monkeys unable to swallow some of the seeds whole drop the seed directly after attempting to remove as much of the flesh as possible. Many of the fruits which are consumed by these monkeys possess a very hard epicarp or shell which smaller primates would not ordinarily be able to swallow. The strong canines of spider monkeys are used to bite into and to remove this outer shell, so that the pulp and seeds can be swallowed whole. This feeding adaptation allows all *Ateles* and *Lagothrix* to feed on the upper end of the spectrum of epicarp-protected fruits and to exploit a niche which is unavailable to many smaller animals with less powerful jaws.

This species of *Ateles* uses more species of fruit from the family Moraceae than any other family (ROOSMALEN & KLEIN, 1988; SYMINGTON, 1987a,b), a fact that is of interest inasmuch as this family of fruits is one of the most important families in number of species consumed by several other primates species, such as *Alouatta, Cebus apella, Cebus albifrons,Saimiri sciureus, Lagothrix lagothrix,* etc. (MILTON, 1980; IZAWA, 1979a, DEFLER, 1979a, DEFLER & DEFLER, 1996; PERES, 1994a). Members of this family do not show characteristics of hard-shelled fruits which are easily exploited by the feeding-adaptation of the atelines (as described above for *Ateles belzebuth* and for *Lagothrix lagothricha*). The fruits of this family are generally soft, poorly protected and available to a very wide number of species of animals, and so serve as the basic mainstay of many species at once in each habitat. It is the fruit of such families as the Myristicaceae (which until they open are unavailable to most animals), Sapotaceae (with hard or leathery epicarps that are difficult to open without a very hard bite), Sterculiaceae (which possess very hard shells) that are not so easily exploitable by smaller animals, where *Ateles* and *Lagothrix* has the advantage here with strong jaws and short canines.

Ateles is one of the few neotropical primates which eats soils from salt licks and also drinks the free-standing water found in these licks. They also eat the soil from termite nests found on tree trunks (IZAWA, 1975, 1993); IZAWA (1993) discusses the anlysis of minerals from these sources and finds the evidence inconclusive as to what minerals are being sought. *Ateles belzebuth* drinks water from tree holes by wetting the hands and drinking the drops falling from the fur. It also drinks directly from the water. EMMONS (1980) concluded that the macaws (*Ara* spp.) which frequent salt licks are after phosphorus, which is poorly represented in their diet. MUNNS (1984) proposed that the clays contribute to detoxification of animals which have eaten too much tanin or other mildly toxic substances. This species leaves the trees in order to drink from water in certain salt lick areas as well (IZAWA & MISUMO, 1990).

Females become adults at around 4-5 years of age when they begin their first estrus cycles. The complete sexual cycle lasts for about 26-27 days and according to VAN ROOSMALEN & KLEIN (1988) the period of receptivity lasts for 8-10 days. Copulation, as with other members of the atelines, lasts for a rather long time as primates go (8-25 minutes)(KLEIN, 1971). The form of copulation is unusual for primates as well, since it takes place ventrum to dorsum while the animals are sitting.

Gestation has been calculated in *Ateles fusciceps* and *Ateles geoffroyi* at 226-232 days (EISENBERG, 1973) and is probably similar to gestation in *A. belzebuth*. KLEIN (1971) did not believe that *A. belzeuth* showed seasonal births, but SYMINGTON (1987a) recorded a clear tendency for a birth peak during January-August with 59% of 46 births occurring during the first four months of the year, corresponding to the first part of the rainy season.

Of the subgroups observed by KLEIN (1971) and KLEIN & KLEIN (1973a) 63% contained at least one adult male and an adult female and half of these observations were of two or more adult males with the females. The size of the subgroup varied from 4-22 animals. Of those subgroups observed 34% contained only one sex and the majority of them were females with their babies.

It seems that the females do not defend their home range, which is a subsection of the males´ home range. The males are very friendly among the other males of the troop, but the same males are very aggressive towards males of different social groups. Vocalizations are very similar to other species of *Ateles*. A detailed analysis and comparison is yet to be accomplished.

Ateles belzebuth sometimes form associations with *Alouatta seniculus, Saimiri sciureus, Cebus apella* (KLEIN, 1972). This author observed a solitary female *A. belzebuth* living for various months with a troop of *A. seniculus*, and a recent shooting of another female *A. belzebuth* on the lower Apaporis also included the report from the hunter that the animal was living with a group of *A. seniculus*. JULLIOT (1994a) described a predation of *A. paniscus* by the crested eagle (*Morphnus guianensis*), indicating that other *Ateles* could as easily be attacted by these eagles.

Conservation state

Ateles belzebuth is the most endangered taxon in the Colombian Amazon (DEFLER, 1989, 2003; DEFLER *et al.*, 2003). It is listed in Appendix II of CITES and classified as VU, using the new IUCN (2000) classification system. Maps published on the geographic range of this species exagerate the total area (HERNÁNDEZ-CAMACHO & DEFLER, 1985, 1989).

In Colombia one of the previously most secure populations of *A. belzebuth* is being threatened by colonists within La Macarena National Park and Tinigua National Park. The only other large forest blocks which have extensive populations of this species are the Cordillera de los Picachos and the forest between the Caguan and the Yarí to the Caquetá River. The three national parks are surrounded by a colonization front and although these parks are in fairly good state at the moment, natural resources (especially timber) are being taken out by river and colonists continue to settle within the parks, while these parks haven´t sufficient park personel to protect them.

Other populations are on the slopes of the Cordillera Oriental but these are highly threatened by the many colonists in all of the piedmont of the region. Generally *A. belzebuth* has a distribution in Colombia that is congruent with much colonization.

Where to see it

The very easiest way to observe wild *A. belzebuth* is at the Ecological Research Center of La Macarena (CIEM), which is coordinated by the University of Los Andes in Bogotá, although lately it has become very complicated to travel there, due to problems of security. This population of *Ateles* is well-habituated to humans. Observing other populations requires more travel with less chance of success. I strongly recommend coordinating any such trip with professors CARLOS MEJÍA or PABLO STEVENSON of Los Andes University (Bogotá), since there are some particular problems involved in traveling to this site.

Geoffroy´s spider monkey

Ateles geoffroyi (KUHL, 1820)

Plate 8 • Map 14

Taxonomic comments

This species includes *Ateles geoffroyi grisescens* and *A. g. rufiventris* (=*A. g. robustus* is a junior synonym of *A. g. rufiventris*) in Colombia. The species includes all *Ateles* west of the Magdalena and Cauca Rivers and all Central American *Ateles*. In this book I treat *Ateles geoffroyi* as a different species from *Ateles hybridus*, and I include *Ateles fusciceps* with *Ateles geoffroyi* as suggested by FROEHLICH *et al.* (1991) and supported by his cladogram and by known hybridization (ROSSAN & BAERG, 1977) as well as comparative DNA research of COLLINS & DUBACH (2000a, 2001b, 2001). See the discussion for *Ateles* in this book.

Common names

Colombia: *marimonda* and *marimunda* in northern Colombia; *choiba* in the department of Antioquia in the middle Magdalena; *mono negro* in the department of Chocó and on the Pacific coast; *mica* in the department of Bolívar; *zamba* in the department of Antioquia.

Indian languages

Hierré, perré or *yerré* (Chocó Indians of *A. g. rufiventris*, *fide* VALLEJO, 1929) ; *jarachi* (Guak, CASTELLVÍ, 1938) ; *yarré* (Embara of the region of the Indian reserve of the upper Baudó; Coroboro Valley, upper Bojaya, upper Buey River; *citroa* (Tunebo, CASTELLVÍ, 1938).

Identification

The head-body length of *Ateles geoffroyi rufiventris* (=*A. g. robustus*) has a length of about 450-550 mm with a tail length of 700-850 mm and a weight that varies more or less between 6-9.5 kg. The face is black-skinned with essentially no hair and sometimes with some depigmentation around the orbits and the nasal orifices. There is no white submalar mark or it is represented by a few dispersed white hairs. The triangular frontal patch is usually absent, but apparently it is present in some local populations. The hair is completely black, at times with some light brown on the head and back. Often some dispersed yellowish hairs are found on the ventral surface of the body and on the interior surface of the thighs. One captive specimen which reportedly was from the Atrato River watershed was described as an animal with some reddish coloration on the belly, so the name *A. rufiventris* was given to it (SCLATER, 1872), a name which has priority to the often used *A. robustus*. Unfortunately the reddish coloration of the holotype is rarely seen in this subspecies of *Ateles geoffroyi*. They eye color is usually dark brown in *A. g. rufiventris*, but rarely the eye color can be lighter and even blue.

Ateles geoffroyi grisescens is characterized by a brown or rust-colored back (with black-tipped hairs) and by the head, limbs and tail colored completely black. It is possible to consider this subspecies as a transitional form between the light yellowish *A. g. geoffroyi* of Central America and the almost completely jet-black *A. g. rufiventris* of southeast Panama and adjacent parts of Colombia (HERNÁNDEZ-C. & COOPER, 1976).

Geographic range

Ateles g. rufiventris is found in all of the Pacific lowlands (except around Juradó in Chocó), the region of Urabá in Antioquia, the departments of Córdoba, Sucre and the north of Bolívar to the east to the lower Cauca River and in all of the western bank southward into Antioquia to around Concordia. In recent times the most northern limit was the southern bank of the Canal del Dique close to Cartegena. Nevertheless, it possibly may also extend further north to around Pendales, where there was a well-developed hygrotropophytic forest (HERNÁNDEZ-CAMACHO & COOPER, 1976). The most southerly example collected in Colombia was from Barbacoas in the department of Nariño.

Ateles g. griscesens is conjectured to be found in Colombia in the vicinity of Juradó, close to the Panamanian border on the Pacific coast (KELLOGG & GOLDMAN, 1944; HERNÁNDEZ-CAMACHO & COOPER, 1976). It may be found throughout the Serranía de Baudó perhaps as far as Cabo Corrientes, but its Colombian distribution needs to be studied, since, despite conjectures, there are no observations to confirm its presence in the country.

Ateles g. fusciceps (=*A. fusciceps fusciceps*) may be present in southern Colombia, south of the Mira River and would be continuous with known populations in neighboring Ecuador. The presence of this subspecies and *A. g. griscesens* needs to be confirmed for Colombia.

This species has been reported for more habitat types than other species of *Ateles*. FREESE *et al.* (1976a) found them in both deciduos and semi-deciduous evergreen forests. EISENBERG & KUEHN (1966) and ÁLVAREZ DEL TORO (1977) observed the species in mangrove swamps in Chiapas, Mexico. HERNÁNDEZ-CAMACHO & COOPER (1976) list them as occupying an ample range of habitats, including hygrotropophytic cloud forests. They reach altitudes of 2,000-2,500 m on the slopes of the Cordillera Occidental of the Andes in Colombia.

Natural history

This species has been studied more extensively in the field than have *Ateles belzebuth* or *A. paniscus*, although all three species have been the object of excellent research. One of the first investigations of neotropical primates was of *Ateles geoffroyi panamensis* in western Panamá (CARPENTER, 1935). DARE (1974a, 1974b, 1975), CAMPBELL (1994, 1999, 2001, 2002), CAMPBELL *et al.* (2003), HLADIK (1972a, 1972b, 1990), HLADIK & HLADIK (1969) and MITTERMEIER (1978) studied *A. geoffroyi* on Isla Barro Colorado, although it should be understood that this is an introduced population, which has some characteristics unique to it. MILTON (1993a, 1993b) studied diet and social organization in the same spider monkey population. A more natural population of *A. g. vellerosus* in Tikal National Park, Guatemala has been the object of several studies (COELHO, 1975; COELHO *et al.*, 1976a, 1976b, 1977), CANT (1978, 1986, 1990), BRAMBLETT *et al.* (1980), MUSKIN & FISCHGRUND (1981), FEDIGAN & BAXTER (1984), MILTON (1981a), PRUETZ & LEASOR (2000).

FREESE (1976a) censused *A. g. frontatus* in the Costa Rican dry forest of Santa Rosa National Park. Other projects in Costa Rica have been published by SORENSEN (1997, 1999), BERGESON (1998); SORENSON & FEDIGAN (2000), DEGAMA-BLANCHET & FEDIGAN (2003), PRUETZ & LEASOR (2000), ROSE *et al.* (2003), WEGHORST (2001). Recently CHAPMAN (1987, 1988a, 1988b, 1989a, 1989b, 1990a, 1990b), CHAPMAN & CHAPMAN, (1987, 1991, 1990a,b, 2000), CHAPMAN *et al.* (1989a, 1989b), CHAPMAN & LEFEVRE (1990a.b) carried out comparative studies of this species with *Alouatta palliata* and *Cebus capucinus* in Central America and ROBBINS *et al.* (1991) studied the nature of *A. geoffroyi*'s type of social grouping. YOUNG-OWL *et al.* (1992) censused the species in the Reserva Los Cedros in Ecuador.

In Mexico there has been research on the species by MOHENO (2002), PÉREZ-RUÍZ & MONDRAGON-CEBALLOS (1998, 2002), PÉREZ-RUIZ & MONDRAGÓN-CEBALLOS (2001), SILVA-LÓPEZ & JIMÉNEZ-HUERTA (2000), GONZÁLEZ-KIRCHNER (1999) and VICK & TAUB (1996a,b), RAMOS-FERNANDEZ AND AYALA-OROZCO (2003), SERIO-SILVA *et al.* (2003), ESTRADA-A. *et al.* (2002), MANDUJANO (2002), LARA & JORGENSON (1998), ESTRADA & COATES-ESTARDA-R. (1996), and MORALES-H. (2002) in El Salvador and DAHL *et al* (1996) on the status and taxonomy of spider monkeys in Belize. Unfortunately no research has been accomplished for *Ateles geoffroyi* in Colombia, a regrettable situation, since the species (like all Colombian *Ateles*) is endangered within the country, and we must have more information about it.

CARPENTER (1935) reported observing groups averaging about 8, although these groups form part of a larger social grouping or troop. Troops of 33 have been observed in Tikal National Park and sometimes up to 70-75 individuals (FEDIGAN & BAXTER, 1984). EISENBERG & KUEHN (1966) reported median sizes of 5 per group in Chiapas, Mexico. Other average sizes were 3.1 in Santa Rosalia National Park (FREESE, 1976a). CARPENTER (op cit.) made it clear that the groups which he observed formed a part of a larger grouping (troop), probably much like *A. belzebuth* and *A. paniscus* (ROOSMALEN & KLEIN, 1988), although later, CANT (1978) confirmed this in his own study. CARPENTER (1935) is the only one who suggests the larger grouping size was around 17 (at least at this site). FREESE (1976a) counted on occasion up to 20. Perhaps the actual total social unit is more like 25-30, as in the other two species of *Ateles*. VAN ROOSMALEN (1985) and VAN ROOSMALEN & KLEIN (1988) characterized groups of *A. paniscus* as being separated by agonistic interactions, particularly because the males defend the home range, while the females do not.

The only calculated home range for this species seems to be about 1-1.15/km² for Barro Colorado (DARE, 1974a,b) and 1.5 km² for the Santa Rosa National Park in Costa Rica (CHAPMAN, 1988). COEHLO *et al.* (1976a,b) and CANT (1978) calculate 26-45 individuals/km² in Tikal National Park. RICHARD (1970) estimated a maximum range of about 3000 m in Barro Colorado, be he did not state an average.

Ateles geoffroyi was one part of a comparative study of locomotion and positional behavior by MITTERMEIER (1978), which MITTERMEIER & FLEAGLE (1976) compared to the African primate *Colobus guereza*, although the population studied was the anomalous Barro Colorado population, whose origin was of animals from the Panamá City Market place (DARE, 1974a). According to the study, this species spent 22% in quadrupedal walking and running, 15.2% in brachiation, 10.5% in armswinging, 39.8% climbing, 11.4% in leaps, jumps and dropsand 0.8% in bipedalism during locomotion from tree to tree. These percentages varied somewhat during locomotion while feeding, especially in the great increase in climbing (horizontal climbing) to 47.2%. Half of their travel was over branches of about 2-10 cm with the rest distributed over other size ranges. While feeding, *A. geoffroyi* uses suspensory positons about 53% of the time, using one, two or three limbs with most of the weight supported by the tail. MITTERMEIER (1978) found that this population used the understory 34% of the time, the middle and lower part of the canopy 47.6% of the time, the upper part of the canopy 15.5% of the time and emergents 1% of the time.

CHAPMAN (1989a) listed various types of sites that were utilized for sleeping during 159 nights in Santa Rosa National Park, Cost Rica. He found that some sites were used various times and other sites were used only once. The size of the subgroup which sleeps together varies according to the quantity of trees that offered food at the moment. The trees used for sleeping were large with a diameter at breast height of 75.1 cm. Eight of the 11 trees that were repeatedly used were emergents situated at the base of valleys with very steep sides.

Ateles geoffroyi is an obligate frugivore and consumes a wide variety of fruits, supplementing its diets with some leaves. HLADIK & HLADIK (1969) and RICHARD (1970a,b) calculated a diet for *A. g. panamensis* of Barro Colorado at about 80% fruits and 20% leaves, as well as some bark-eating, flower buds and flowers. No insect eating was confirmed for this species in this study. CHAPMAN (1988a) calculated a diet composed of 77.7% fruits, 9.8% flowers, 1.2% mature leaves, 7.3% young leaves, 2.6% buds and 1.3% insects. A list of the families and number of different foods consumed in various studies in Panama follows: Moraceae (11 species); Anacardiaceae (6); Leguminosae (6); Guttiferae (4); Myrtaceae (3); Palmae (3); Bombaceae (3); Musaceae (2); Rutaceae (2); Burseraceae (2); Chrysobalanaceae (2); Dilleniaceae (2); Lauraceae (2); Annonaceae (2); Apocynaceae (1); Lecythidaceae (1); Leguminosae/Pap. (1); Meliaceae (1); Sapotaceae (1). The detail of diet which is available for *A. paniscus* and *A. belzebuth* lamentably is not available for this species. According to HERNÁNDEZ-C. & COOPER (1976), *A. geoffroyi* (and presumable *A. belzebuth*) is fond of the seje palm *Oenocarpus bataua* (syn: *Jessenia polycarpa*) and fattens noticeably on it when it is in season. These animals, like the other members of *Ateles*, are major seed dispersors and chose many large-seeded fruits with tough epicarps, which are not easily exploited by other, smaller primates. This does not preclude many softer and smaller-seeded fruits as well, and the family Moraceae seems to be the most important family in terms of numbers of species utilized. *Ficus* spp.,

Anacardium, and *Spondias* are especially popular fruits eaten by *A. geoffroyi* (HLADIK & HLADIK, 1969; HERNÁNDEZ-C. & COOPER, 1976).

Female *A. geoffroyi* become sexually mature at 4-5 years when they begin to cycle sexually, about every 26 days (HARVEY *et al.*, 1987), although generally a female doesn´t have her first birth until 7-8 years of age (CHAPMAN & CHAPMAN, 1990a). In Central America there seems to be a season when there are more copulations (MILTON, 1981a), which are long-lasting in duration and accompanied by a position unique to *Ateles*, with the male and female seated, male´s ventrum to female´s dorsum with the male´s legs hooked over the female´s thighs, holding her in place (KLEIN, 1971; EISENBERG, 1973). These sessions are unusually long as primate copulations go, usually lasting 5-10 minutes. Uterine bleeding is sometimes present after ovulation (GOODMAN & WISLOCKI, 1935). The sexual state of the female appears to be conveyed particularly by pheromones in the urine, since males are highly motivated to lick urine voided by females and the female clitoris may be specialized for the deposit of urine on nearby surfaces (KLEIN, 1971; VAN ROOSMALEN & KLEIN, 1988).

Gestation is 226-232 days (EISENBERG, 1973) and birth intervals are around three years, the longest in any neotropical primate (EISENBERG, 1976; DARE, 1974a; MILTON, 1981) and a characteristic which makes this species highly vulnerable, since long interbirth itnervals mean that any population perturbation (such as hunting or disease) will take a long time to be repaired by new births. After birth the infant clings to the mother´s ventrum and only at about 4-5 months does the young one start riding on mother´s back. The newborns have a different color from the adults. Juveniles move independently of the mother but close to her, although sometimes the male must make a bridge between trees to help the youngster pass over.

Adult males within the same group or clan are most often friendly to each other, but they are aggressive towards the males outside the group, and the friendly males cooperate in defending their territory, which includes also the separate territories of several females (FEDIGAN & BAXTER, 1984; VAN ROOSMALEN & KLEIN, 1988). The group males may travel and feed together, but they show more social cohesion and are more territorial than the females (FEDIGAN & BAXTER, 1984). The females, on the other hand, are usually friendly to all other females either inside their own group or outside it.

Males and females are always friendly and often forage in mixed groups. Although little time is dedicated to grooming in the wild, some researchers have hypothesized from captive studies that the amount of grooming is correlated with social rank (EISENBERG & KUEHN, 1966; EISENBERG, 1976; KLEIN & KLEIN, 1971).

All *Ateles* possess rich and varied vocalizations. *Ateles geoffroyi* has been studied and analyzed in captivity by EISENBERG (1976, 1977) as follows: (1) *whoop* – adult males, when excited at the joining of two subgroups or during confrontations with other troops; (2) *subvocalized whoops* – much softer than whoops, given 1-4 times with open mouth, given before alarm call or when separated from conspecifics; (3) *wails* – 1-3 notes with open mouth used to facilitate the unification of subgroups and to answer distant calls; (4) *ook-barking* – adult males and females, short notes made closing the mouth in an alert position; predators´ alert; (5) *tschook* – adult males and females; shorter than the ook-barking and used to establish contact; (6) *squawk* – 1-4 loud notes made with the open mouth, closing it slightly, made by juveniles after having made contact with adults; (7) *whoas* – juveniles; usually one note between a *squaqk* and a *wail* after losing contact; (8) *tsee-tee* – juveniles and adults; closed mouth to indicate position or to greet; (9) *light panting* – juveniles and adults during play; (10) *heavy panting* – juveniles and adults, lower in tone use in play; (11) *rhythmic panting* – infants; while nursing or ontop of ma; (12) *growling* – strong sound with mouth half open during brusque contact play when threatening others; (13) *whine* – flat, clear, high note given by adults when excited and before being chased or before shaking a branch; (14) *cough* – juveniles and adults; from a short whine to a stronger sound given before biting or pulling hair; (15) *squeals and twitters* – usually juvenile, sometimes adults, variable with short notes when approaching conspecifics or protecting an object; (16) *sobs* – adults; soft version of subvocalized whoops given before and after whoops by adults in the same situation.

Following is a list of displays identified by VON ROOSMALEN & KLEIN (1988): (1) *tail curling* – threat display; (2) *arched back* – during locomotion and during threats; (3) *turn away* – showing indifference to a conspecific; (4) *head-shake* – seeking contact and play; also a threat; (5) *piloerection* – strong agonistic display after being bit or when fighting, etc.; (6) *arm-scratching* – scratching arm from the shoulder when feeling threatened or uncomfortable;

(7) *branch-shaking* – threat towards a danger to the group; (8) *arm-swaying* – threat towards a danger to the group; (9) *charge* – rapid charge towards another animal; (10) *follow retreating animal* – play or agonistic behavior. There are no associations known. The predators of *A. geoffroyi* probably are the jaguar (*Panthera onca*), the ocelot (*Felis pardalis*), the margay (*Felis wiedii*) and the puma (*Puma concolor*), but there are no observations on this type of interaction (WOLFHEIM, 1983; DEFLER, 1994b, 1996c, 2003).

Conservation state

Ateles geoffroyi is one of the most endangered primates in Colombia because of its size and eatable flesh which makes it the target of much hunting. Additionally it is endangered because of its need for ample primary forests, all of which are becoming fragmented. Since this species has its geographic range in a part of Colombia with a high human population, and such anthropogenic factors exercise strong negative influence on this monkey. Censuses of this primate are vitally needed wherever they are found, so that the details can be known. *Ateles geoffroyi* is classified CR internationally in the IUCN system, but it has recently been classified EN for the Colombian populations, while internationally the subspecies are classified En for *A. g. griscescens*, VU for *A. g. rufiventris* and *A. g. fusciceps*, and EN within Colombia for any subspecies (DEFLER *et al.,* 2003; RODRÍGUEZ *et al.,* 2003).

Where to see it

This species is becoming very difficult to see in the wild. Perhaps Katios National Park is the best place where wild *Ateles g. rufiventris* can be observed. To observe *A. g. nigriscens* one would have to travel to the area around Juradó, Chocó and seek a guide. It would probably be easier to observe in nearby Panamá. No recent information is available about populations of this subspecies.

Brown Spider Monkey

Ateles hybridus (I. GEOFFROYI-ST. HILAIRE, 1829)

Plate 8 • Map 16

Taxonomic comments

Ateles hybridus hybridus and *A.h.brunneus* are recognized by this author, based on HERNÁNDEZ & COOPER (1976), the cladogram of FROEHLICH *et al*

(1991) and his comments to me (FROEHLICH, pers. com. to DEFLER, 1993) Morphometric analyses, as well as analyses of nuclear and mitochondrial DNA, underline that *A. hybridus* should be given specific status (FROECHLICH *et al.*, 1991; COLLINS, 1999; COLLINS & DUBACH, 2000a, 2000b, 2001).

Common names

Other English: Variegated spider monkey; marimonda. Colombia: *choiba* in the Department of Antioquia. French: *atèle métis* in French. German: *brauner Spinnenaffe*.

Indian languages

(Pronounced as in Spanish) *citroa*: (Tunebo Indians, *fide* CASTELLVÍ, 1938).

Identification

The body has a length of about 45-50 cm with a tail of abut 72-82 cm and weight varies more or less from 4-8 km or more. *Ateles hybridus hybridus* has a more or less naked face as *A. belzebuth*, but usually it doesn't have the conspicuous unpigmented areas as adults. The submalar strip and the frontal triangular patch are also present variably, but in general they are not as conspicuous as in *A. belzebuth*. The color of the dorsum fluctuates from light grayish-brown to a rich brown. The head, neck, front and hind limbs and dorsum of the tail are invariably darker than the back, and in some specimens the color approximates blackish-brown. The hind limbs usually are lighter and similar in color to the back with the exception of typically darker coloration over the knees. The ventrum of *A.h.hybridus* varies from dirty-white to buffy and does not contrast with the flanks in populations east of the Magdalena River. Iris color is usually light brown and sometimes grayish-blue (HERNÁNDEZ C. & COOPER, 1976).

Ateles h. brunneus is similarly marked to *A.h.hybridus* but all are darker specimens than *A.h.hybridus*. The entire lateral surface of the hind limb is dark brown and relatively uniform with the color of the back and feet. The dirty-white to buff ventrum contrasts with the brownish flanks, limbs, tail and head (HERNÁNDEZ C. & COOPER, 1976) or as follows:

• Hind limbs are usually lighter and similar in color to the shoulders with the exception of the typical dark tones over the ribs. The ventrum varies from white with brown tonalities to light yellow with no contrast with the sides of the animal..*Ateles hybridus hybridus*.

• The exterior sides of the hind limbs are dark brown and relatively the same color as the shoulders and the feet. The stomach varies from whitish to light yellow and contrasts with the brown on the sides of the animal as well as the brown on the extremities, the tail and the head ....*Ateles hybridus brunneus.*

Geographic range

Ateles hybridus hybridus is found from the right bank of the Magdalena River in the Departments of Magdalena, César (northward to the southern slopes of the Sierra Nevada de Santa Marta), the southwestern portions of Guajira in the northernmost parts of the Serranía de Perijá, and in the middle Magdalena River valley at least to the Departments of Caldas and Cundinamarca. There are also two populations of this subspecies on the slopes of the Cordillera Oriental of the Andes on the Venezuelan border; one population is found in the Catatumbo River watershed in the Department of Norte de Santander and the other population is found in the northeast piedmont forest of the Department of Arauca.

Ateles hybridus brunneus is found between the lower Cauca and Magdalena Rivers in the Department of Bolívar, Antioquia and Caldas. This population has sometimes been included with *A.h.hybridus,* but it is easy to distinguish the two.

Natural history

MONDOLFI & EISENBERG (1979) report *A.h.hybridus* as occurring in tall, evergreen forest, semi-deciduous tropical forests, riverine semi-deciduous tropical forest and semi-evergreen dense but not very tall tropical seasonal montane forest at elevations of 28-600 m above sea level.

GREEN (1978) censused this and sympatric primates at Cerro Bran near Puerto Rico (Bolívar) and made some observations while BERNSTEIN *et al* (1976a,b) studied the effects of forest degredation on the species near Ventura (Bolívar). Both of the above studies were done in the Sierra de San Lucas of southern Bolívar, not far from each other. MONDOLFI & EISENBERG (1979) reviewed old collection and observation sites to expand its known geographical distribution.

BERNSTEIN *et al.* (1976a) observed an average of about 4.5 animals per group in the Serranía de San Lucas in censuses, while GREEN (1978) observed an average of 3.3 animals per group (n=223). BERNSTEIN *et al.* (1976a) calculated a density of 9-14 individuals/km² in the San Lucas mountains. GREEN (1978) calculated densities of 8.2-9.6 groups/km² at his study site, which if multiplied by his average group size (3.3 individuals) seems to suggest higher densities at his Cerro Bran site when compared with the other study site. Day ranges and home ranges are unknown for this species, and little is known about the activity budget, but it is probably similar to other *Ateles*. GREEN (1978) found the highest activity levels between 6 A.M. and 8 A.M., similar to many other primate species. Locomotion and posture are similar to other *Ateles*. Brachiation with and without the tail was observed by BERNSTEIN *et al* (1976a). Leaps and falling jumps of up to 15 m were also observed.

One assumes that these *Ateles* use the same type of sleeping sites as reported by CHAPMAN (1989a,b) for *Ateles geoffroyi* in Costa Rica, which were usually large trees with a diameter at breast height of 75 cm. A specific study of the diet of these frugivorous monkeys has not been undertaken.

Behaviors are very similar to other species of *Ateles*, but there are no specific studies. Vocalizations and displays are probably very similar to other *Ateles*. BERNSTEIN *et al.* (1976a) observed branch-breaking displays directed at the human observers. BERNSTEIN *et al.* (1976a) observed *A. hybridus* within sight of woolly monkeys six times, with *Cebus albifrons* two times and with *Alouatta seniculus* three times during 77 encounters with the *Ateles*. *Ateles hybridus* is also known to associate with *Cebus nigrivittatus* (MONDOLFI & EISENBERG, 1978).

Conservation status

Ateles hybridus including both subspecies are considered CR in Colombia, using the IUCN (2000) criteria, especially due to the effect of forest fragmentation and hunting. They are undoubtedly more threatened than *A. belzebuth*. *Ateles hybridus brunneus* is probably the more threatened of the two subspecies because of its small geographic range and the increasing threat of colonists, summed with the long interbirth intervals of 3-4 years (COLLINS, 1999), although the presence of the guerilla en the Serranía San Lucas may have prevented more fragmentation (there is no solid information for this one way or the other). BERNSTEIN *et al.* (1976) showed the effect of forest disturbance, especially on *A.h.brunneus* and *Lagothrix l. lugens* and made a plea for the establishment of reserves for this and other endangered primate taxa (such as *Saguinus leucopus*). Fortunately the Serranía de San Lucas in southern Bolívar still contains extensive forest which has been identified as a possible national park site.

The establishment of a San Lucas National Park ought to have high priority in Colombia, since it would preserve many elements of the Nechí refugium, including *Saguinus leucopus* and *Lagothrix lagothricha lugens*, the northernmost known population of *Lagothrix lagothricha*. But the presence of political insurgents, the military and some mine fields made the region very difficult for work and for the presence of the government.

Although *A.h.hybridus* is found in Catatumbo-Barí, Tamá, and in El Cocuy National Parks, population densities are unknown and this primate is the object of hunting by the Bari Indians. Actual physical protection other than park boundaries is absent.

Where to see it

The above-mentioned areas would be where to observe these two taxa, although extreme caution and consultation with local authorities is advised for anybody attempting to observe primates in southern Bolívar. Lately some populations have been found in northern Antioquia. Lately it has been called to my attention that a population of *Ateles hybridus hybridus* in the Serranía de las Quinchas is observable and the region does not have security problems.

Ateles belzebuth

Woolly Monkey

FAMILY **ATELIDAE**

Genus *Lagothrix* E. GEOFFROY, 1812

Von Humboldt made the first observations of this monkey when he saw a young animal in San Fernando de Atabapo on the Orinoco River. The animal had actually come from the Río Guaviare at the mouth of the Río Mataven, although it is not recorded whether the specimen came from the north bank (Vichada) or the south bank (Guainía) of present-day Colombia (they are found on both sides of the river). Humboldt decided to name the animal *lagotriche* because of its fur (λάγως = hare + θρίξ, τρίχος = hair or fur). GEOFFROY (1812) saw an early copy of von Humoboldt's description and named the animals *Lagothrix*.

Many forms have been identified and usually several species have been accepted (GRAY, 1970; PELZELN, 1883; SCHLEGEL, 1876; FORBES, 1894; CA-

BRERA, 1897; ELLIOT, 1907, 1909; LÖNNBERG, 1945, 1940; HILL, 1962) until FOODEN (1963) united them all into two species (*Lagothrix lagothricha* with four subspecies and *Lagothrix flavicauda*). Recently workers have begun to split the genus again into several species and have begun to recognize *Lagothrix flavicauda* as a different genus, *Oreonax flavicauda* on the basis of skull and dental characteristics (RYLANDS *et al.*, 2000; GROVES, 2001). Actually no new evidence is available to split *Lagothrix* (except for analyses of *Oreonax* skull characteristics) and morphological, karyological and molecular studies are urgently needed to clarify this genus. In this book I recognize *Lagothrix lagothricha* as the only species with the subspecies *lugens* and *lagothricha* for Colombia, according to FOODEN (1963).

Woolly Monkey

Lagothrix lagothricha (HUMBOLDT, 1812)

Plate 9 • Map 17

Taxonomic comments

The specific name for this species has been spelled by many *lagotricha* as well as *lagothricha*, since von HUMBOLDT'S (1812) original description of the species used both spellings. The major and latest revisor FOODEN (1963) chose the spelling *lagothricha* as the correct spelling in accordance with Article 24 of the International Code of Zoological Nomenclature, although some primatologists demure that this form is an incorrect transliteration of the Greek *trichos* into the Latin. A recent review of the legality of the spelling, nevertheless, (verbatim, FOODEN, 1991; verbatim, HERSHKOVITZ, 1991) has just concluded that *lagothricha* is the correct original spelling according to the Code (DEFLER, 2003).

In FOODEN'S revision the species was divided into four subspecies: *L. l. cana, L. l. poeppigii, L. l. lagothricha* and *L. l. lugens*. However, some recent observations have found two sympatric "varieties" of this "species" in the Tefé river basin, each of which had somewhat different vocalizations (JOHNS, 1986). MOYNIHAN (1976) was of the opinion that the present species might be made up of two or more separate species. GROVES (2001) elevated all subspecies of the genus to species level on the basis that the skins that he had studied showed a marked distinction from each other, certainly not a strong argument for calling them separate species, especially considering some of the transition forms that I have seen between *lugens* and *lagothricha*. Ruiz-García & Álvarez (2003) found two different haplotypes in three enzymes of mitochondrial DNA when comparing *lugens* and *lagothricha*, which they offered as support for Groves (2001) decision to elevate the subspecies. However, Ruíz-García & Álvarez (2003) also found equivalent haplotype differences in *Saimiri sciureus albigena* and *Saimiri sciureus macrodon*. In this book I continue treating these taxa as subspecies until more evidence is accumulated for the relationship between *lagothricha* and *lugens*. It seems probable that more haplotypes of mitocondrial DNA will be found in other populations of *lugens*, given their complex biogeographical history.

Common names

Other English: Humboldt´s woolly monkey; Colombia: *mono choyo, choyo, choyo and choro* in the northern piedmont forest of Arauca, Boyacá and Meta; *churuco* or *chuluco* in the upper Magdalena valley and the departments of Caquetá, Vaupés and Amazonas; *mico cholo* (Llanos Orientales); *mico negro* or in Bolívar; barrigudo in Amazonas. French: *lagotriche mico churrusco* de Humboldt; German: *Wollaffe* or *Humboldtische Wollaffe*. Brazil: *macaco barrigudo*. Dutch: *volaap van Humboldt* or *volaap*.

Indian Languages (pronounced as in Spanish)

Ziyuh, seguay (Andaquí); *arimime* (Carijona); *caparro, kaapáro* (Curripaco, Amazon 1992); *caparro* (Cubeo), *caparo* (Geral), *capalu* (Guahibo), *jemo* (Huitoto); *mono, macoco* (Ingano); *zulo isolo* (Jeberó); *tseraca* (Letuama); *o* (Macú-Yujup of the lower Apaporis river, *fide* Capitain Franciso Macú); *seu* (Makuna); *cümü* (Miraña), *cümi* (generic), *jadicümi* (white color phase), *peecujeci* (black color phase) (Muinane); *pátchu* (Nukak); *j+mmo* (Ocaimo); *caparu* (Piaroa); *choicack* (Puinave); *guaó o yuwi nas* (Siona), *tseraca* (Tanimuca);, *omé, kap, mai/kú* (Ticuna); *seuniami* (Tukano); *savároma, «sowarama»* (Tunebo); *caparú* (Yucuna); *jemo* (Witoto); *ghoobi* (Yuri).

Identification

This is one of the largest of Colombian primates. An average of 13 adult weights gave 7 kg., although the range included one animal of 11.5 kg

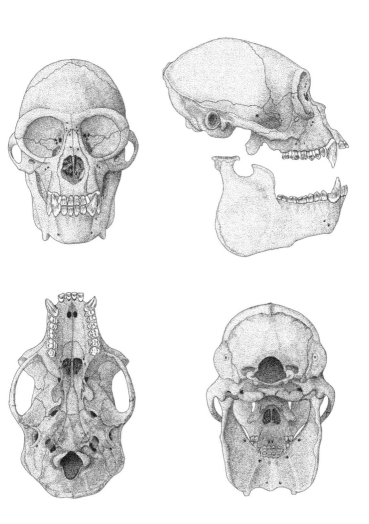

Figure 36. Views of skull of *Lagothrix lagothricha*.

Figure 37. Various positions of *Lagothrix*. a-f) Rest and hanging. g-h) Hanging infant. i-j) Bipedal and cuadrupedal locomotion.

(HERNÁNDEZ C. & DEFLER, 1985) and some animals in captivity have reportedly weighed up to 15 kg (WILLIAMS, 1967). Woolly head-body measurements are around 45 - 55 cm and tail measurements are about 60-65 cm.

The body of this animal is much more robust than the equally heavy spider monkey (*Ateles*), and the males are particulary more muscular and larger than the females. An adult male shows a bulging forehead, due to very muscular masseters and he is also very broad across the shoulders, possessing heavy arms, legs and tail as well. All of this is set off by long muttonchop whiskers on each side of the jaws. The animals are usually some dark variant of brown or grey with quite a lot of variation within any one troop.

Of the four subspecies described by FOODEN (1963) two (*L. l. lagothricha* and *L. l. lugens*) are known for Colombia. Fooden distinguishes *L. l. lagothricha* as "the trunk being pale brown and the crown of the head like the back or paler" while he describes *L. l. lugens* as "having a trunk silver-gray with the back dark blackish with or without contrasting dark crown patch and with a pale mid-sagittal coronal streak frequently present". These descriptions are widely seen as being inadequate inasmuch as there is great variation in each subspecies. HERNÁNDEZ-C. (pers. com.) believes that *lugens* is the most variable of the four subspecies, having as perhaps the most stable characteristic a longer and more luxureous coat than *lagothricha*.

This author finds that *L. l. lagothricha* also varies markedly. A mid-sagittal coronal strip is fairly common (but not mentioned in Fooden's review), and general body color varies from light blond to buff brown to almost black. At times patches of highly blond fur can be seen at the bases of tails and over shoulders. Grizzled forearms are common but not universal.

Geographic range

In Colombia *L. lagothricha* is known throughout the Amazonian lowlands, northward to the Uva River in Vichada, which is a left-bank affluent of the Guaviare River. The subspecies *L. l. lugens* is found to the north of the lower Guayabero River where it was always scarce (KLEIN & KLEIN, 1976), although it is more common in the Serranía de La Macarena and throughout the piedmont of the Uribe region (between La Macarena and the Cordillera Oriental). The range extends northward to the Venezuelan border in both piedmont and the eastern slopes of the Cordillera Oriental to about 3000 m, but it is highly endangered in this region. In central Colombia *L. l. lugens* extends from the upper Magdalena valley to at least southern Tolima west of the Magdalena river and historically at least to southern César department on the east bank on the west side of the eastern Cordillera and east of the Magdalena River the species is almost if not totaly extinct. There is an apparently isolated enclave in the Serranía de San Lucas in southeastern Bolívar and northern Antioquia at the northern end of the Central Cordillera. The San Lucas population (studied briefly by KAVENAUGH & DRESDALE, 1975) may or may not have been historically connected to the populations of the upper Magdalena Valley via the original primary forest, which is now almost completely wiped out. The boundary between the two subspecies is conjectural and clinal, located somewhere not far east of the Cordillera Oriental, according to HERNANDEZ-C. & COOPER, (1976) in Caquetá department, but this must be confirmed. Outside of Colombia *L. lagothricha* is found west of the r. Negro in Brazil north of the Amazonas. South of the Amazonas River it is found throughout the Ecuatorian and Peruvian Amazonian lowlands and extends into Brazil to the Tapajós Tiver.

Natural history

The major habitat of the woolly monkey is well-developed primary tropical moist forest (rainforest on inland, non-inundatable sites), and the largest groups and highest densities are found in seasonally specialized forest, such as flooded white-water forest at the beginning of the rainy season (DEFLER, pers. obs.). High densities in seasonally inundated forest are seasonal, while the fruit crops last. Later the woollies move inland to upland forests. The species also extends into cloud forests in Colombia, but sizes of groups diminish as they are found at increased altitude (DURHAM, 1975).

Preferred vertical use of the primary forest is the middle and upper canopy, although the animals descend to 10-12 m commonly, depending on the fruit, and occasionally go to the ground to pick up a favorite dropped item. In the Apaporis study the overwhelming height for most activities was at 20-30 m (DEFLER, 1990a).

Many of the studies of woollies have been accomplished in Colombia, first by NISHIMURA & IZAWA (1975) and K. IZAWA (1975, 1976) in the Peneya River, an affluent of the Caquetá River and later by NISHIMURA (1987, 1988ab, 1990a,b,c, 1992, 1994, 1996, 1997a,b, 1998a,b,c, 1999a,b,c, 2001, 2002a,b, 2003) and NISHIMURA *et al.* (1992a, 1992b), SAKURAI & NISHIMURA (1999) and YUMOTO *et al.* (1999) in the Duda river to the west of the Serranía de La Macarena. Recently STEVENSON (1992, 1997a,b, 1998, 2000, 2001, 2002) STEVENSON *et al.* (1992, 1994, 1998, 2000, 2002, 2003), STEVENSON & CASTE-LLANOS (2000), STEVENSON & QUIÑONES (1993) at the Duda river research site and DEFLER (1989a,b, 1990a, 1995, 1996a-c, in press), DEFLER & DEFLER (1996) studied the species near the lower Apaporis river in Vaupés. Studies of the spacing behavior of the same group of monkeys has continued with the thesis research of MUÑOZ (1991) followed by similar work by ARDILA & FLOREZ (1994).

Other work has been done in Perú by SOINI (1986a,c, 1990a) in the Pacaya-Samiria National Park on the Samiria River and some incidental observations by RAMÍREZ (1980, 1988) were accomplished in Manu National Park. In Brazil C. PERES (1987, 1988, 1990, 1991c, 1993a, 1994a, 1994b, 1994c, 1994d, 1995, 1998a,b) carried out some preliminary observations and censuses and studied the diet of this species. JOHNS (1986, 1991) included this species in his study of disturbed rainforest habitat. A new site in Yasuní National Park in eastern Ecuador has produced studies by DIFIORE (1997, 2001, 2002, 2003), DIFIORE & RODMAN (2001) and DEW (2002a,b). IWANAGA & FERRARI (2001, 2002) censused what they call *Lagothrix cana* in the SW Brazilian Amazon and studied party size and diet.

On the Apaporis River woolly monkey troops seem to comprise 20-24 individuals which are multi-male, multi-female and which separate into two temporary subunits, foraging nearby each other and coalescing again. One group in eastern Colombia contained five adult males, eight adult females and six juveniles and four infants (less than one year). At this same site the groups divide temporarily into two subunits which forage and eat closeby to each other, but again fuse into one unit. Probably the fission-fusion pattern is generated by a reduced food base. In Perú SOINI (1986) studied independent troops which were made up of 5 – 9 individuals. Some of these groups associated with each other, although they usually conserved a certain distance between themselves with the exception of a more intimate association at night. NISHIMURA (1990) and STEVENSON *et al.* (1998b) report troops of 12, 17 and 33 individuals on the Duda River. In other sites (probably more fertile) groups have been observed comprising around 40-50 (Izawa, 1976). Stevenson *et al.* (1998) showed that subgroup sizes correlate with fuit patch size.

These observations suggest that the species has a fission-fusion society in which there are clear differences in the size of the social unit, according to the food resources. The size of the groups is small with an increase in altitude and the size of a coherent unit increases in areas where soils are more fertile, permitting larger fruit crops, such as along the Peneya and Duda rivers (Durham, 1975; Hernández-C. & Cooper, 1976; Nishimura, 1990a-c; Izawa & Nishimura, 1988; Stevenson *et al.*, 1998). There is some evidence that especially the females migrate between troops rather than the males, which are more philopatric, but this needs to be examined much more closely (Stevenson *et al.*, 1994).

Very different home ranges have been reported: 124 ha and 108 ha for two groups (Difiore, 1997, 2003); 169 ha (Stevenson *et al.*, 1994); 350 ha and 450 ha (Nishimura, 1990b), 400 ha (Soini, 1990a) and 760 ha (Defler, 1989a,b, 1996a) and more than 800 ha (Peres, 1995). In the Apaporis study earlier density calculations for this species were 5.5 individuals/km^2. This same density was found by Soini (1986a, 1986b) at the black-water site in the Pacaya-Samiria National Park in northern Perú. A re-interpretation of the Caparú data suggest the data could actually be 13.5 individuals/km^2 (Defler, 2003). Other calculated densities of four sites in the upper Purité river in eastern Colombia were 4.8 – 13.7 individuals/km^2. Stevenson (1996) worked with a population showing a density of about 30 individuals/km^2 near the white-water Duda river in Tinigua National Park.

Defler (1995) calculated an activity budget for *Lagothrix lagothricha* of 29.9% rest, 38.8% move, 25.5% forage, and 5.8% "other activities" which includes social and nonsocial activities. For another group of woollies at the Duda river research site, Stevenson (1992) calculated 36% rest, 24% move, 36% forage and 4% in "other activities". Difiore (1997) calculated an activity budget of 17.1% rest, 35.5% move, 37.3% forage (includes 19.5% eat and 17.8% forage + attack), and 10.1% "other activities". These differences reflect interesting differences between three very different sites; at Caparú the *Lagothrix* are forced to travel more and rest less, due to the poorer forest habitat.

As the woollies travel and forage they use mostly a quadrupedal gait (41.8% total). Although they occasionally brachiate (1.7% total, with total suspensory locomotions equal to 8.6%, mostly using a simple armswing in contrast to *Ateles*), they do this much less than spider monkeys (*Ateles*) or muriquis (*Brachyteles arachnoides*). Woollies climb a total of 38.8% of their time, jump and leap about 10.8% (DEFLER, 1999b). Woollies use considerably less suspensory positions and locomotion than to the sometimes syntopic *Ateles sp.*, whose feeding adaptation is suspensory in contrast to *Lagothrix* DEFLER, 1999). CANT et al., 2001, 2003) compare suspensory locomotion and locomotor behavior to the syntopic *Ateles belzebuth*, whereas DEFLER (1999a) compared locomotory behavior to two other allopatric species of *Ateles* that had been previously studied by MITTERMEIER (1978). Woollies have an average day range of 2880 m on the Apaporis river (the highest so far registered) and an average of 1633 m on the Duda river, another demonstration of the comparative quality of habitat at the two sites (DEFLER, 1996a). DiFIORE (2003) observed day ranges for two groups of *L. l. poeppigii* at 1,792 m and 1,878 m.

Woollies are fairly deliberate in their movements and almost always utilize their strong prehensile tails for security as they move along, although the tail is probably more important in foraging, as it allows the heavy body to be supported by five points rather than four. Woollies can also hang completely supported by their tails, leaving the hands free to manipulate any food item. The feeding position most commonly is of an individual hanging from the tail with the hind feet in contact with some vertical substrate. Suspensory locomotion has recently also been studied in the species by STEVENSON (1999) and CANT et al. (2003).

Lagothrix usually chose rather high trees for sleeping (25-35 m). Often the group sleeps in several different trees, close enough to each other to communicate by vocalizations. It is common to sleep in the same tree or close to food, in order to begin eating immediately the next day.

Woolly monkeys are highly frugivorous primates, described as generalized opportunistic frugivores by STEVENSON (2004). In one study 83% of their food choices were fruit, and the woollies complemented their fruit diet with immature leaves (14% of their diet) and a few invertebrates and vertebrates (DEFLER & DEFLER, 1996). About 75% of the fruits selected by woollies have special characteristics which make the fruits fairly unavailable to most smaller primates. These characteristics are that the fruit has a hard outer shell which requires a strong bite to open and that there are one or only a few pits or seeds to which the endosperm of the fruits' flesh tightly adheres, making it necessary for the feeding animals to swallow the seeds whole in order to digest the flesh, later dispersing the seeds into other parts of the forest. Fruit hardness as a fruit choice has also been discussed by KINZEY & NORCONCK (1982) and fruit characteristics in the diet are discussed by STEVENSON (2004), who describes woolly preferences as "good-tasting" fruits from abundant plants that produce large quantities of fruits and large crops. The dispersed seeds insure the diversity of the forest, since they become the seedlings which will later become the future forest trees (STEVENSON, 2000, 2002a). Details of seed dispersal by woollies are described by STEVENSON (2000, 2002) and STEVENSON *et al.* (2003). Ecological overlap with other species is reduced by using different resources from other competing species during fruit bottlenecks (STEVENSON *et al*, 2000) and feeding group size is influenced by the food patch size in this species (STEVENSON *et al.*, 1998).

Thus, the feeding process of the woollies insures the maintenance of high diversity within the forest so that it can be said that the forest takes advantage of the woollies just as the woollies take advantage of the forest. Some general comments about primate seed dispersal can be found in HOWE (1980, 1993) and in JANSON (1983a). Pollination by primates may also be an important interaction (JANSON *et al.*, 1981).

Interestingly about 60% of these fruits mature to some variant of yellow or orange (10% green, 8% red, 7% brown). Studies of color perception in New World monkeys suggest that yellow-orange is the color which woolly monkeys (and most other Platyrrhines) best perceive against a background of green (FOBES & KING, 1982).

On the lower Apaporis river of Colombia the family Sapotaceae is very important in the woolly monkey diet and a very high percentage of food plants are vines. Other important families eaten are Moraceae and Leguminosae. In this study the ten most important species in order were in the following order of observed choices: (10.0%) *Chrysophyllum amazonicum* (Sapotaceae), (5.0%) *Manilkara amazonica* (Sapotaceae) (2.0%) *Protium* sp. (Burseraceae), (2.2%) *Iriartea ventricosa* (Palmae), (1.9%) *Buchenavia viridiflora* (Combretaceae), (1.8%) *Inga alba* (Leguminosae), (1.7%) *Micrandra spruceana* (Euphorbeaceae), (1.5%) *Cheiloclinium hippocrateoides* (Hippocrataceae), (1.5%) *Virola* sp. (Myristicaceae), and (1.3%) *Pouteria cuspidata* (Sapotaceae). Clumping all *Inga* species together equaled 13.5% of the total diet, while all species of

Pouteria (Sapotaceae) equalled 8% of the diet and all species of *Pourouma* (Cecropiaceae) equaled 5.4% of the diet.

In another diet study on the Duda River between Tinigua and La Macarena National Parks the Sapotaceae were of very minor importance, and the Moraceae were the most important family reported with 17 species chosen 28% of the time (STEVENSON *et al.*, 1994) as follow: (15.3%) *Gustavia hexapetala* (Lecythidaceae), (11.4%) *Spondias mombin* (Anacardiaceae), (10.4%) *Spondias venulosa* (Anacardiaceae), (3.5%) *Brosimum alicastrum* (Moraceae) and (2.9%) *Pouroma bicolor* (Moraceae).

PERES (1994a) studied a group of *L. l. cana* in the central Amazon (upper Urucu River, Amazonas, Brazil). The diet was made up of 73% fruits, 7% seeds, 14% young leaves, 6% exudates of pods and 3% flowers. He identified 225 species of food plants, and they represented 116 genera and 48 families. Arthropods were only 0.1% of the group's diet. The three most important families (Moraceae, Sapotaceae and Leguminosae) represented 43% of the species within the diet (PERES, 1994) as follow: (7.8%) *Chrysophyllum sanguinolenta* (Sapotaceae), (7.3%) *Parkia nitida* (Leguminosae), (6.4%) *Pouteria* sp. (Sapotaceae), (6.4%), *Ficus* sp. (Moraceae), (5.7%), *Pseudolmedia laevis* (Moraceae), (5.1%) *Clarissia racemosa* (Myricaceae), (3.1%), *Inga alba* (Leguminosae), (2.8%) *Couma macrocarpa* (Apocynaceae), (2.8%) *Ecclinusa ramiflora* (Sapotaceae), and (2.3%) *Brosimum utile* (Moraceae). *Lagothrix* frequently licks wet leaves and branches. Also they use tree holes filled with water.

Both HERSHKOVITZ (1977) and ROSENBERGER (1980) suggest that woolly monkeys belong to a grade of monkeys comprising all of the Atelinae (*Ateles, Brachyteles,* and *Lagothrix*) (*fide,* ROSENBERGER) and including *Cebus* according to HERSHKOVITZ (1977). The grade is said to be based on the animals' large bodies, their prehensile tails and molar features, which represents a particular ecological feeding adaptation. Certainly the Atelineae forage spreading their weight over five grasping organs rather than four and are able to open well-protected fruits. The *Cebus* may or may not fit into this scheme, although it does seem clear that the latter genus exhibits a sexual dimorphism in general mandibular strength, where the male animal may approximate the ateline abilities. But *Cebus* are only half the size of the other members of the three genera and the *Cebus* tail is not as facultative as these larger primates. *Brachyteles* depends much more on leaves than the other atelines as well.

Lagothrix seems to have a very similar ecological role to *Ateles*, since they chose fruits of similar size and taxonomic range, they both are important seed dispersers and they both depend highly on their prehensile tails for their foraging behavior. *Lagothrix* and *Ateles*, have long been suspected of exerting a sort of competitive interaction on each other, and the patchiness of one or the other's distribution, where it is sympatric with the other, has been taken as evidence of this (EISENBERG, 1979; HERNÁNDEZ C. & COOPER, 1976). Competative interactions deserve more analysis, and such research is presently possible at the joint Japanese/University of Andes research site on the Duda river in Tinigua National Park in southern Meta and in Yasuní National Park in eastern Ecuador where the two species are sympatric and study groups are habituated. In seven sites in eastern Colombia this species is the dominant species with about half the total biomass being made up of *Lagothrix lagothricha*.

Most of the known spots in Colombia where *Lagothrix* is known to be sympatric with *Ateles* are near or on the Cordillera Oriental. Further out from the Cordillera *Ateles* seems to disappear, probably due to the empovrishment of the soils and subsequent diminishment of available fruit crops.

Lagothrix become adults between 5-7 years of age and nuliperous females emigrate to other groups at about 6 years of age and on the average have their first baby at around nine years (NISHIMURA, 2003). The females seem to have a sexual cycle of 25 days and the dominant male controls the copulations, peripheralizing the other males of the group. When the male approaches (or if he doesn't approach) the estrus female rapidly opens and closes her mouth, showing her teeth in a *tooth chatter display* (EISENBERG, 1976). The male mounts from behind and copulates an average of about 6 minutes (NISHIMURA, 1988c, 1990a).

The females give birth usually to a single infant every two-three years. NISHIMURA (2003) calculates an average of 36.7 months on the Duda river, although twins have been observed and mean age at first parturition is 9 years and average interbirth interval about 36.7 mo (n=13)(NISHIMURA, *op cit*). NISHIMURA (op cit.) reports a birth season on the Duda river of July-December, although other sites show more scattered births. During the first weeks the mother and baby are closely attended by an adult male who gradually relaxes his vigilance as time goes by. The mother may nurse the infant up to about six months, but infants begin trying out solid foods during their first weeks of life, since all young primates are close to the mothers' feeding activities and have the constant option of being able to try out whatever she is eating.

Lagothrix females don't seem very social in comparison with *Macaca* or other very social primates. There is little grooming and what there is usually occurs between the mother and young, which also show the closest proximal spacing of all dyads in the group (STEVENSON, 1998). Females tend to move from place to place with other females, while males often remain close to each other (STEVENSON, 1998). The males may remain in their natal group, as opposed to the females, which seem to pass from group to group. All members of the group protect infants, and adult males gladly carry infants from place to place and play with them.

These primates have a variety of musical and bird-like vocalizations and many of their common names in Spanish and Indian dialects are onomotopaic borrowings from the sounds that they make. Particularly impressive is a strong "long-call" which individuals and the group make to contact other groups in the area. From the tops of emergent trees this call must travel hundreds of meters. The sound is best-described as a high-pitched warble. There are many other softer sounds which are used among nearby animals, but an ear-splitting shriek given by one animal punished or threatened by another may serve in its extreme intensity to actually deter further agression, since the volume is exceedingly painful at a human's close range. A few vocalizations are listed as follow: *e-olk* - excited, prior to feeding; *slow eolk* - content, while feeding; *uh-uh-uh* - repeated rapidly during play and used like a laugh; often accompanied with a head-shake; *long call* - high intensity warble-like call to other groups or individuals out of sight of caller; *scream* - when punished by bite or slap or when frightened by another; *whine* - interrogative pleading sound when begging from another; *wah* - given by female to placate infant and invite youngster onto her back. A concerned female often uses this vocalization, *tuf-tuf* - puffing-sound given behind hand or into armpit of another as placation, friendliness, etc. and *chuck* - softly repeated sound given by spread out group while foraging; contact call. In the population at Tinigua, females emit more contact calls than other age/sex classes, and except for adult females and individuals with less individuals in proximity emit more contact vocalizations than individuals at the center of the group (STEVENSON, 1997).

Some particularly outstanding behavioral displays are listed here: *chest rub* - performed by all members of group from 1 year old on up; perhaps male most frequent in out-of-the-way place (like under a branch) actor wets bark with saliva then rubs chest upwards across saliva repeatedly until chest hair is completely wet with saliva; *head shake* - any age shakes head rapidly side to side as friendly space reducer and during play; often accompanied by *uh-uh-uh* vocalization; *hand tuff* - *tuff-tuff* sound made into hand as friendly display after tension or after any brief absence from each other; *armpit tuff* - alternate to *Hand Tuff* where actors raise arm and other animal blows *Tuff-tuff* into the arm-pit; this variant to the hand tuff is used exclusively by some animals while others use *Hand Tuff* instead; *Branch shake Display* - shaking branches in tree when excited and desirous of threatening perceived danger; *Teeth chatter* - female in estrus chatters teeth while partially protruding tongue in order to advertise her condition and *Anal drag* - sitting animal pulls anus over substrate.

Several authors have suggested that, because of their size, woolly monkeys have very little predation pressure. But woolly monkeys begin as small animals, not as large adults and the disappearance of non-adults is common. Carlos LEHMAN (1959) reported a predation on woollies of the Black-and-Chestnut Eagle (*Oroaetus isidori*), and HERNÁNDEZ & COOPER (1976) report local names in Putumayo for *Harpia harpyia* and *Morphnus gianensis* as "aguilas churruqueras", suggesting predation, at least occasionally. Woolly monkeys give few warning calls because of flying birds, but they have a developed repertoire of warning calls and branch breaking behavior which they frequently utilize towards humans and other potential predators on the ground. Tame free-ranging *Lagothrix* in the Caparú Biological Station were attacked by puma and by margay. It seems probable that cats such as ocelots (*Felis pardalis*), margays (*Felis wiedii*), puma (*Felis concolor*) and the jaguar (*Panthera onca* = *Leo onca*) could take these primates, especially at night when the monkeys are not so able to detect and escape the danger. NISHIMURA (2003) found that it was rare in his group for an individual to reach 30 years, although one exception survived to an estimated 32 years of age.

Besides potential predators it is common to see this species travel through the forest with the primates *Saimiri sciureus* and *Cebus apella*. Less commonly seen are associations with *Cebus albifrons* and *Cacajao melanocephalus*. The small kite *Harpagus bidentatus* and the white hawk *Leucopternis albicollis* commonly trail after foraging bands of woolly monkeys, and the feeding activities and falling fruits also attract the birds *Psophia crepitans* and *Crax alector* (and other species of *Crax*) as well as the mammals *Mazama americana* (the red brocket deer), *Mazama gouazoubira* (gray brocket deer), *Dasyprocta fuliginosa* (the agouti), *Tayassu tajacu* (the collared peccary) and *Tayassu peccari* (the white-lipped peccary), all of whom hope to profit by the "messy" eating of these primates.

Conservation status

The species is included in Appendix II of the CITES convention and each subspecies (viewed by the IUCN specialist committee as different species) are now classified as LC internationally. The Colombian subspecies have been classified VU for *L. l. lugens* and NT for *L. l. lagothricha* (DEFLER, 1996c). Woolly monkeys are certainly among the most threatened wherever they occur, due to their being a frequent target of meat hunting and the fact that the species is adapted to primary forest (see MITTERMEIER, 1987a; DEFLER, 1989c; PERES, 1987, 1990a, 1991c).

The local pet trade in Colombia is still a threat to this species, since it is not uncommon for hunters to kill a mother carrying an infant, with the hope of procuring the infant for the hunter's children. Sadly such a pet is usually neglected after a time and in some cases the naturally curious primate baby is cruelly punished (sometimes with boiling water) when it begins to make predations into the kitchen. Another fate of these primate pets is accidental strangulation, when the animals are kept tied around the waist as a preventitive to their robberies and destruction in the house. Perhaps saddest of all is the neglected and half-grown pet, left in a small cage to eventually die of some infection, complicated by its abnormally lonely existence. It is fortunate that in Colombia the international trade as pets in these and other primates is prohibited, since international money would fuel a great increase in this type of activity.

The subspecies *L. l. lugens* is undoubtedly the most threatened of the four subspecies and since it is found almost entirely within the boundaries of Colombia, it must be the responsability of Colombia to insure that it is protected. Although this subspecies is nominally protected in about 12-13 conservation units totaling around 2.479.249 ha. (DEFLER, 1994), many of those populations are at risk, due to illegal hunting and habitat destruction. Also, many of these parks were established primarily for other types of habitats, and they protect little of the habitat required for this primate. For example, Puracé National Park contains a population of woollies, but the park was established to protect paramo as well as high montane forest. So these conservation units actually represent much less habitat for the woolly than the total hectares indicate.

A future for this and for *L. l. lagothricha* in Colombia is wholly dependent on increasing physical protection for the populations already legally protected by law and in increasing the number of conservation units which include these primates. Nevertheless, concentration on the first of these objectives would go a long way towards protecting *L. l .lugens*, but increased management costs money which the government does not have available.

Fortunately both subspecies are being studied by biologists in Colombia, but census work is needed to increase our knowledge of the presence or absence of these animals throughout their presumed range, including whether or not the species is present or absent in certain units of the national park system. It seems especially important to identify populations of *L. l. lugens* in the Cordillera Central, which could potentially be conserved and which would increase our knowledge of the species' distribution, since there may be relictual populations in that Cordillera which are unknown. Additionally there are spots on the west slopes of the Cordillera Oriental which may still harbor small populations of this primate and these isolated forests must be identified with the possibility of protecting them.

Where to see it

To observe *Lagothrix lagothricha* it is necessary to travel to remote parts of the Amazon where there are still good populations. The tamest group of *L. l. lagothricha* is undoubtedly at Caparú Biological Research Station in southern Vaupés. Nevertheless, it can still be located in Amacayacu, La Paya, and Cahuinarí National Parks, although it may take a few days of looking. The best place to observe *L. l. lugens* is in the Ecological Research Center of La Macarena in Tinigua National Park. Any populations of this highly endangered subspecies that are observed by anyone in the Cordillera de los Andes should be reported to this author so we can determine if some protection of the population is possible.

Howler Monkeys

Family **Atelidae**
Genus ***Alouatta*** Lacépède, 1799

This genus contains at least six species and probably more which will only be recognized when more karyological and molecular studies are carried out (Rylands *et al.*, 2000). Groves (2001) lists ten species. There are only two taxa recognized for Colombia presently, *Alouatta palliata aequatorialis* and *Alouatta seniculus seniculus* although molecular and chromosome analyses may change this in the future. This genus is proving to be exceedingly complex in terms of phylogeny based on karyotype, DNA or morphology and at this time it has the highest chromosome variability known within the platyrrhines with some drastic karyotypic differences between some species (DE OLIVEIRA *et al.* 2002). What has been traditionally considered to be *Alouatta seniculus* seems to be made up of a complex of populations with somewhat different chromosomes. So far, the southern red howler *A. sara* has been distinguished from *A. seniculus* and it is possible that in Colombia at least two species might be distinguished in the future from northern and southern populations.

Mantled Howler

Alouatta palliata (J. A. GRAY, 1849)   **Plate 10 • Map 18**

Taxonomic comments

The Colombian population may be referred to as *A. p. aequatoriales* (FESTA, 1903). Earlier it was common to use *A. villosa* as a synonym for this species, a name which has priority over *A. palliata*. However, because of the recognition that *A. villosa* is also a synonym for *A. pigra*, which seems to be a separate species from *A. palliata*, *A. palliata* stands as the corect name. FROEHLICH & FROEHLICH (1986) used dermoglyphics as a taxonomic technique.

Common names

Other English : Black Howling Monkey . Colombia : *mono zambo, mono, aullador negro, mono negro, mono cotudo* along the Colombian Carribean coast ; *mono negro* on the Colombian Pacific coastal area (sometimes applied to *Ateles paniscus fusciceps*) ; *mono chongo* and *chongón* in the region of the lower Pacific coast adjacent to Ecuador ; *güeviblanco* (Chocó). French: *hurleur…manteau.* German: *Mantelbrüllaffe*

Indian languages

Kotudú (Noaham in the Chocó, ORTÍZ, 1940); *cuara* (Cho+có); *uu* (Cuna).

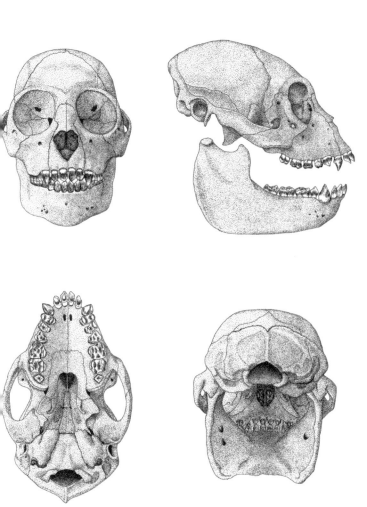

Figure 38. Views of skull of *Alouatta palliata*.

Figure 39. Diverse views of *Alouatta*. a-h) Hanging locomotion and tail use.

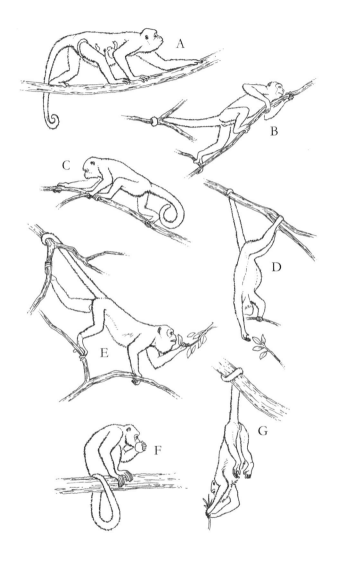

Figure 40. Various positions of *Alouatta*. a-g) Locomotion with infant and while tail hanging.

Figure 41. Various positions of *Alouatta*. a-c) Bridging so infant can pass c) quadrupedal locomotion. d-g) Rest. h) Observation and curiosity.

Figure 42. Various attitudes of *Alouatta*. a-e) Various positions. f-g) Hanging and climbing.

Identification

The head-body length of this species is about 481 - 675 mm with the males being on average longer at around 561 mm and the females at about 520 mm. The tail is around 545 – 655 mm and the body weight around 4.5 – 9.8 kg for males and 3.1 – 7.6 kg for females. *Alouatta palliata* is basically a black animal, although it has lighter brown or blond hair on the back, and light guard hairs along the sides and shoulders give the animal an effect like a mantle or saddle. This species of *Alouatta* is only about half as sexually dimorphic as *Alouatta seniculus*. The highly developed hyoid bone is present, just as in *A. seniculus*, and acts as a resonator to enable the animals to make their famous calls or howels.

Geographic range

The distribution of *A. palliata* in Colombia includes all of the Pacific lowlands, west of the Western Cordillera. The species is not found in mangrove swamps along the coast and probably it is also not found in the forests adjacent to *nato* (a swampy forest type named for the principle tree, the nato or *Dimorphandra oleifera* - Leguminosae). This primate is found in the majority of non-inundatable forest types. The altitudinal limit in the zone of piedmont forest is unknown. In the northern regions between the Pacific and Atlantic Oceans the species is found in the drainage of the Atrato River and in the valley of the Sinú River, where it is sympatric with *A. seniculus*. A specimen at Turbaco, collected by CARRIKER during the first years of the XXth century and information of DUGAND from the decade of the 1950´s, about the presence of the species in Los Pendales (the two locations are in the region of Cartagena), suggests that the most northeasterly extension of *A. palliata* could be more extensive than indicated on most distribution maps. Nevertheless, recent investigations in the Cerros of María in the Department of Sucre and in the north of Bolívar have not detected *A. palliata*, although *A. seniculus* was observed. Based on comments related above, it seems probable that historical populations of *A. palliata* included a very low population density of the species sympatric with the more common *A. seniculus* between the Atrato and Magdalena rivers. Besides the Pacific coast of Ecuador, *A. palliata* extends throughout Central America into southern Mexico, although it is replaced by *A. pigra*, in Belize and parts of s. Mexico.

In the drainage of the Atrato and Sinú the species is found from rainforest to semi-deciduous (semi-verde) forests. From studies in Central America we know that this monkey is found in a wide variety of forests, especially lowland evergreen forest but also mangrove forest, dry forests, deciduous forest, riparian forests as well as secondary and subxeric forests. These primates prefer the mid-to upper-level of the canopy but, similar to *A. seniculus*, they have no problem on the ground or even swimming across rivers.

Natural history

Alouatta palliata is the best-studied neotropical species, taking into account the number of studies accomplished. The species has been studied in Mexico by the following: BENITEZ-RODRÍGUEZ *et al.* (1993); CANALES-ESPINOSA (1994); COATES-ESTRADA & ESTRADA (1986); ESTRADA (1982, 1983, 1984); ESTRADA & COATES-ESTRADA (1983, 1984a, 1984b, 1985, 1986a, 1986b, 1991, 1993a, 1993b); ESTRADA *et al.* (1977, 1984, 1993ª, 1993b. 1999ª, 1999b, 2001); ESTRADA & TREJO (1978); GARCÍA *et al.* (2001); GÓMEZ *et al.* (2001); MUÑOZ *et al.* (2002); ORTIZ-MARTÍNEZ *et al.* (1999); RODRÍGUEZ-TOLDEDO *et al.* (2003); SERIO-SILVA & RICO-GRAY (2002, 2003); SERIO-SILVA *et al* (1999, 2002); SOLA-NO *et al.* (1999).

In Costa Rica the following studies can be listed: CHAPMAN (1996a, 1987, 1988ª, 1988b, 1990ª); CHAPMAN *et al.* (1989a); CHAPMAN & CHAPMAN (1986, 1990b); CLARKE (1981, 1982, 1983, 1990); CLARKE & GLANDER (1981, 1984, 2001, 2002); CLARKE & ZUCKER (1994); CLARKE *et al.* (1998, 1999, 2001a,b, 2002a, 2002b); FEDIGAN (2003); FEDIGAN & JACK (2001); FEDIGAN *et al.* (1995, 1996, 1998) GLANDER (1974, 1975a, 1975b, 1977, 1978a, 1978b, 1979, 1980, 1981, 1982, 1983, 1992a. 1992b); GLANDER *et al.* (1991); FREESE (1976a); JONES (1978, 1979, 1980a, 1980b, 1981, 1982, 1983, 1985, 1994, 1995, 1996, 2000, 2002a, 2002b,); ROSE *et al.* (2003); SORENSON (1999); SOUTHWICK (1955, 1962, 1969); STONER (1994, 1996a,b); and ZUCKER & CLARKE (1992, 1998, 2003).

Also, *A. palliata* has been studied extensively in Panamá, especially at Barro Colorado Island in Lake Gatún of the Panama Canal, where researchers of the Smithsonian Institute have accomplished many neotropical primate projects. Early studies begin with CARPENTER (1934, 1953). Populations on Barro Colorado Island have been monitored since the 1930´s, first only occasionally but in the recent past more regularly and intensively, so that we have a historical series of data of increases and decreases of population levels. The research on Barro Colorado Island permits hypotheses based on abundant data (CARPENTER, 1934, 1953, 1960, 1962, 1964, 1965; CHIVERS 1969; COLLIAS & SOUTHWICK, 1952; SOUTHWICK, 1955, 1962, 1969; ALTMANN 1959, 1966; HLADIK, 1967, 1972a, 1978; HLADIK & HLADIK, 1969; HLADIK *et al.* 1971; SMITH, 1977; RICHARD, 1970a, 1970b; MITTERMEIER, 1973; MILTON, 1975, 1977, 1978, 1980, 1981b, 1982, 2001; MILTON & MITTERMEIER, 1977; MILTON *et al.* 1979, 1980; NAGY & MILTON, 1979a, 1979b; CHAMBERLAIN *et al.*, 1993; EISENBERG & THORINGTON, 1973; SCOTT *et al,* 1976a; HELTNE *et al.* 1976; OTIS *et al.*, 1981; FROEHLICH *et al.,* 1981; FROEHLICH & THORINGTON, 1982a, 1982b; MENDEL, 1975, 1976; YOUNG, 1981a, 1981b, 1982a, 1982b, 1983; LEIGHTON & LEIGHTON, 1982; YOUNG, 1981a, 1981b, 1982a, 1982b, 1983). WANG & MILTON 2003; WHITEHEAD, 1985, 1986a, 1986b, 1987, 1989, 1992, 1994, 1995 studied long-distance vocalization and spacing in *A. palliata.* Another research site in Panamá at the Hacienda Baqueta has yielded another data set (BALDWIN & BALDWIN, 1972, 1973a, 1974, 1976, 1978) as well as GALINDO & SRIHONSE (1967).

Census data between Costa Rica and Panama was compared by HELTNE *et al.* (1976) and MILTON *et al.* (1979) studied their basal metabolism while grouping behavior and sexual ratio are discussed by SCOTT *et al.* (1978). TEAFORD & GLANDER (1991) discuss dental microwear in the species. In Nicaragua BENZANSON *et al.* (2002) studied social and group structure, McCANN *et al.* (2003) groups in coffee plantations, MITCHELL *et al.* (2002) studied strategies of resource use and WILLIAMS-GUILLEN & McCANN (2002), WINKLER (2000), GARBER *et al.* (1999a, 1999b) studied ecology and conservation of the species. In Ecuador CHARLAT *et al.* (2000) surveyed the species in the NE of the country while WINKLER (2000) studied social organization.

In Colombia the first observations are attributed to HERSHKOVITZ (1949) of the sympatry between *A. palliata* and *A. seniculus* in a small area west of the Atrato River (reported again in HERNÁNDEZ-CAMACHO & COOPER, 1976), *A. palliata* being found on the higher land and *A. seniculus* in the lower land. Since then only one project has been reported, that of RAMÍREZ-ORJUELA & SÁNCHEZ-DUEÑAS (2002), who estudied the diet of the species in the area around Cabo Corrientes (Chocó).

These monkeys are found in social groups which vary from 6-23 and usually are somewhat larger than the average *A. seniculus* groups that have been studied (CROCKETT & EISENBERG, 1987). MILTON (1982) found mean sizes of as much as 20.8 when reviewing data for Barro Colorado Island, whereas the highest mean for *A. seniculus* found by CROCKETT (1984a) for only one year at Hato Masaguaral in Venezuela was 10.5. In groups of *A. palliata* there are usually two or three adult males, unlike the tendency in *A. seniculus* to have only one male per group. Even in one zone there is large group size variance (GAULIN *et al.*, 1980).

Usually besides 2-3 males there are 4-6 females and sometimes as many as 7-10 females. LEIGHTON & LEIGHTON (1982) and CHAPMAN (1990a) found that these primates formed groups with predictable sizes, densities and spatial use based on the types of food available.

The groups maintain home ranges which vary from 10-60 ha, depending on the habitat (CARPENTER, 1934; CHIVERS, 1969). Nevertheless, home ranges of 3-7 ha were registered in a Panamanian forest with a high density of *Alouatta* which were refuges from other cutover forests of nearby areas (BALDWIN & BALDWIN, 1976). Such small home ranges are probably not natural and are due to crowding.

A density of 1050 individuals/km² was registered in a forest of the Panamanian coast where the population had concentrated after deforestation in the surrounding area. Perhaps more normal densities are between 16-90 individuals/km² on the Island of Barro Colorado in Panama; 23 individuals/km² in Mexico (ESTRADA, 1982, 1984) or 90 individuals/km² (probably representing a concentration of refuges) in Costa Rica (HELTNER *et al.*, 1976; GLANDER, 1978a; CLARKE, 1983; MASSEY, 1987). These monkeys have an average day range of 123 m (range 11-503 m), 443 m (range 104-792 m) and 596 m (range 207-1261 m) according to various studies (ESTRADA, 1982, 1984; HELTER *et al.*, 1976; MILTON, 1980; COLLIAS & SOUTHWICK, 1952; CROCKETT & EISENBERG, 1987). In seasonal forests of Costa Rica the species exhibits a marked difference in its range, although behavioral patterns were the same as past years.

MILTON (1980) calculated an activity budget for two groups on Barro Colorado: 65.54% rest, 10.23% move, and 16.24% eat. *Alouatta palliata* use quadrupedal locomotion during about 70% of their movements. They jump very little and their postures during eating are frequently suspensory. The most common activity is rest while sitting or reclining. One research project showed 47% quadrupedalism, 37% suspensory locomotion, and 10% "bridging". The postures utilized by these animals include 53% sitting,

20% standing, 12% reclining and 11% suspensory via hind feet and tail (GEBO, 1992; CANT, 1986). These primates sleep on a high horizontal branch of an average-sized tree close to a place where they last ate the day before.

In general terms, *Alouatta palliata* consume leaves and fruits in roughly equal quantities, but flowers are also sought out. Usually only a small percentage of the species consumed make up the most important species in terms of feeding time.

MILTON (1980) studied this species' feeding strategies. She found the diet to be 48.2% leaves, 42.1% fruits and 17.9% flowers. The diet included 103 species of foods from 41 families. In terms of the number of species chosen per family, family preferences are indicated as follow: Leguminosae (16 species); Moraceae (15); Bignoniaceae (8); Bombaceae (5); Anacardiaceae (4); (3 species) Annonaceae, Apocynaceae, Burseraceae, Dilleniaceae, Euphorbaceae, Myrtaceae, (2 species) Guttiferae, Malpighiaceae, Marcgraviaceae, Melastomataceae, and 1 species for the following families, Araceae, Boraginaceae, Chrysobalanceae, Combretaceae, Connaraceae, Convolvulaceae, Flacourtiaceae, Gnetaceae, Hippocrateaceae, Lauraceae, Lecythidaceae, Meliaceae, Mnisperaceae, Palmae, Piperaceae, Ulmaceae, Violaceae, Polygonaceae, Rubiaceae, Rutaceae, Simaroubaceae, Verbenaceae. In the same study the ten most preferred foods in terms of the percentage of time eaten by *A. palliata* are as follows: *Ficus yoponensis* (Moraceae-20.95%); *Ficus insipida* (Moraceae-14.89%); *Brosimum alicastrum* (Moraceae-6.08%); *Platypodium elegans* (Leguminosae-5.65%); *Inga fagifolia* (Leguminosae-3.86%); *Poulsenia armata* (Moraceae-3.63%); *Spondias mombin* (Anacardiaceae-2.63%); *Cecropia insignis* (Moraceae-2.24%); *Hyeronima laxiflora* (Euphorbiaceae-1.99%); and *Lacmellea panamensis* (Apocynaceae-.67%).

The percentage of time eating fruits from each family listed above shows that 47.79% was devoted to eating fruits from the Moraceae, 9.5% was spent eating from the Leguminosae, 2.62% from the Anacardiaceae, 1.99% from the Euphorbiaceae and 1.67% from the Apocynaceae. Clearly the Moraceae is the most important family in the diet of this species of primate and especially *Ficus*, which makes up 35.93% of feeding time. MILTON (1980) demostrated these monkeys' preference for new leaves, which have higher levels of protein than old leaves. ROCKWOOD & GLANDER (1979) compared foraging in both howler monkeys and leaf-cutting ants. TOMBLIN & CRANFORD (1994) compared ecological niche differences between *A. palliata* and *C. capucinus*.

A year-long study at Los Tuxtlas, Mexico identified 27 species of food plants and 89% of the feeding time was devoted to only 8 species. In this research, the most important family in terms of time spent feeding was also the Moraceae (58.4%) from which 5-6 species were chosen in the following importance order: *Ficus* spp., *Poulsenia armata, Brosimum alicastrum, Cecropia obtusifolia* and *Pseudomedia oxyphyllaria*. After Moraceae in importance value was the Lauraceae (22.6% of the time consumed) and four species of the Leguminosae were consumed 4.9% of the time.

A study in Costa Rica (GLANDER, 1981) based on 1071 hours of observation concluded that the animals had a diet based on the following time percentages of use: 19.5% mature leaves, 44.2% new leaves, 12.5% fruits, 18.2% flowers and 5.7% flowers. This study identified 62 food species from 27 families, but in this case the Leguminosae with 15 species was the preferred family. Following in importance were the Moraceae and the Anacardiaceae, each with 5 species chosen. The principal food species are listed as follow: *Andira inermis* (Leguminosae/Pap. – 12.15%); *Pithecellobium saman* (Leguminosae – 10.04%); *Pithecellobium longifolium* (Leguminosae – 7.92%); *Anacardium excelsum* (Anacardiaceae – 7.23%); *Licania arborea* (Chrysobalanaceae – 7.06%); *Manilkara achras* (Sapotaceae – 6.19%); *Astronium graveolens* (Anacardiaceae – 5.46%); *Pterocarpus hayseii* (Leguminosae – 4.71%); *Muntingia calabura* (Elaeocarpaceae – 3.77%); *Ficus glabrata* (Moraceae – 3.55%).

CHAPMAN (1987) and CHAPMAN & CHAPMAN (1990b) pointed out the flexibility, adaptability and variability of the diets of this and other species of primates, considering that each month there may be a completely different diet for the same group and even the type of food may change. Such a pattern makes it probable that competition between species would be a strong selective pressure determining diets. CHAPMAN (1987) found that the howlers in the National Park of Santa Rosa in Costa Rica spent 49% of their time eating leaves, 28% of their time eating fruits, and 22.5% eating flowers. These monkeys use up many of the patches of fruits eaten, but there are always large patches which they do not use up. An important ecological role of this species is seed dispersal (CHAPMAN, 1989b). These monkeys role as foragers on leaves is compared to leaf cutter ants (ROCKWOOD & GLANDER, 1979). *Alouatta palliata* drinks water from tree holes and probably from bromeliads.

The dominant male of the group copulates with the females. The males are sexually mature at 42 months, while the females mature in 36. Females show a sexual cycle of 16.3 days (GLANDER, 1980; CLARKE & GLANDER, 1984). It is probably that such factors as chemical pheromones are important during the sexual cycle, since the males nose the genitals and taste the urine of the females. Sexual displays of both sexes includes *tongue flicking*.

Gestation lasts for 186 days (CROCKETT & EISENBERG, 1987). In this species there is no seasonality of births. JONES (1985) found that the females showed an increase in the synchronization of estrus cycles in habitats that were more seasonal and pairing behavior between males and females were mutually coordinated and controlled. Some birth-related behaviors are described by MORENO *et al.* (1991). Some reproductive behaviors are discussed by JONES (1995).

Usually one infant is born, and it is relatively altricial and without a functional tail. The tail does not begin to function for about two months. Infants abandon their ventral position on the mother after 2-3 weeks, thereafter riding dorsally on the mothers' back. The young begin to become independent from the mother at about 15 days of age, but they don't go far. Maternal care lasts up to 1 ½ years until the infants are finally weaned. It is possible to predict the grade of sociability of the males from a very early age (CLARKE, 1990).

Mothers are relatively passive although they wait for their infants at difficult crossings in the canopy so that they may carry them across. At times male adults and subadults carry the infants and help pass the infants from one tree to another. Other females try to gain contact with new infants, although the mothers try to prevent such contact. Infants spend a lot of time playing with juveniles.

Intragroup behavior is very peaceful, although intergroup behavior can be quite violent. Solitary males from outside the group or coalitions try to take over established groups by banishing the group male and killing the infants, which induces estrus in the females so that they will copulate with the new male. The most recognized vocalization is of course the howl, a vocalization given by the adult males when they are bothered by another group or by some environmental factor such as thunder or airplanes. This vocalization is accompanied by the grunts and growls of the female and juvenile members of the group. Various vocalizations are listed (*fide* NEVILLE *et al.* 1988): *roar* or *howl* – deep and loud vocalizations given by adult males when disturbed by other group or environmental factors such as thunder or airplanes; *incipient roar* – short, popping roar or adult males when disturbed as above; *high roar coda* – high tone at end of common roar of adult males when finishing; *roar accompaniment* – high-pitched wailing of females and juveniles when accompanying male roar; *male woof* or *bark* – loud, deep barking given in clusters of 1-4 repitions by adult males when disturbed; *female woof* or *bark* – higher-pitched barking sound of adult females when disturbed; *incipient male woof* or *bark* – muffled male woof of adult males when slightly disturbed; *incipient female woof* or bark – muffled female woof or higher pitch than male woof when slightly disturbed; *oodle* – rhythmically repeated in-out pulses of air of adult males when disturbed and aggressive; *whimper speak* – high-sounding whimper by infants, juveniles and adult females when frustrated; *eh* – soft, expirant repeated every few seconds by infants to maintain contact; *cackle* – high, repeated cackling laugh by infants, juveniles and adult females when threatened; *caws* – series of three crying notes by infants when lost or separated from mother; *wrah-ha* – 2-3 syllable of mother separated from her infant; *yelp* – like yelp of dog from infants, juvenile adult females when highly frightened; *screech* – loud EEEeee by infants, juveniles and adult females when highly frightened; *infant bark* – cat-like purr given by infant when in close body contact with mother.

Various displays are common in this species and are similar to those in *A. seniculus*. Some are listed as follow: *urine rubbing* – rubbing urine on soles of feet, palms of hand, ventral surface of tail and on throat by adults and subadults when in distressed social situation; *lingual gesture* – lip smacking or rhythmic in and out and up and down tongue movements used by adult females as a copulatory invitation; *clitoral display* – rear present of erect clitoris of subordinate subadults or adults to dominants in tense situation; *vulval display* – rear present with tail up of adult and subadults in agonistic situation; *scrotal display* – rear present with tail up and testicles descended of males to males in an agonistic situation to prevent escalation; *shaking branch* – shaking branches of all members of group is mostly play behavior. There are few observations of predation on *A. palliata*. CARPENTER (1934) observed an ocelote (*Felis weidii*) attempting to catch some *A. palliata*. YOUNG (1982b) observed aggressive interactions between howlers and turkey vultures. Competition with *Ateles* in harvesting *Bagassa* fruits was described by SIMMEN (1992).

Conservation state

Alouatta palliata is listed in CITES in Appendix I and by the IUCN as LC. In Colombia the species is considered to be VU. Little is known about the species in Colombia, since it has been only studied in one project, but casual observations and interviews with local inhabitants suggest that the pressure of hunting by both indigenous and African-Colombian communities and the loss of habitat have affected the populations negatively, in many cases producing the "empty forest" syndrome. For example, a recent long-term study of prey species of the local Indian communities in Utria National Park showed there were very few *A. palliata* and *Ateles geoffroyi* within the national park where indigenous communities reside (H. RUBIO, pers. com.).

Where to see it

Alouatta palliata is probably best observed along the west bank of the Atrato River within the Katios National Park, where *A. palliata* is sympatric with *A. seniculus. Alouatta palliata* is found on the higher land and *A. seniculus* is found in the lower, often flooded land.

Red Howler Monkey

Alouatta seniculus (LINNAEUS, 1766)

Plate 10 • Map 19

Taxonomic comments

This species has traditionally been included with its allopatric conspecifics in the subfamily Allouatinae in the family Cebidae (for example, HERSHKOVITZ, 1977). Systematists now prefer to include the six or more described species of the genus in the family Atelidae, subfamily Alouattinae along with the subfamily Atelinae *Ateles, Brachyteles, Oreonax* and *Lagothrix*, based on recent biomolecular research which has demostrated a relationship of *Alouatta* with the other three genera mentioned above (FLEAGLE, 1988, 1999; ROSENBERGER, 1981). Currently all populations of this species in Colombia are considered to be one subspecies, *Alouatta seniculus seniculus*, although a taxonomic analysis of Colombian populations could change this. SAMPAIO *et al.* (1996) present a taxonomy based on biochemical and chromosome data. HIRSCH *et al.* (1991) studied the *Alouatta* species using morphological characteristics of crania showing species and sex differences and demonstrating relationships of species.

Common names

Other English: red howler. Colombia: *mono, mono colorado, mono cotudo, cotudo* in northern and central Colombia: *roncador* in the center of Colombia; *araguata, araguato, bramador,* in Arauca and other parts of the Llanos; *bonso, mono berreador, berreador,* and *bonso* in Tolima and the Llanos Orientales; *mono, mono cotudo* and *cotumono* in Amazonia; *guariba vermelho* occasionally in the Leticia region (of Brazilian origin). Brazilian Portuguese: *guariba ruiva; guariba vermelha.* French: *hurleur alouate, hurleur roux, singe rouge.* German: *roter Bröllaffe.* Surinamese: *baboen.*

Indian languages

Vowels as in Spanish: *kemuime* (Bora); *arawata, garabata* (Carijona); i*xoj-hoxi-mai* (Colima); *emu* (Cubeo); *uma* (Cuna); *itsi, litsi* (Curripaca, *fide* Amazon 1992); *tsyi-púR* (Emberá Chami); *nëjë* (Guahibo); *coto* (Ingano, *fide* J. V. RODRÍGUEZ M., 1993); *arabata* (Guak,); *mongón* (Kwaiker); *tum* (Makú of lower Apaporis, river, *fide* Captain Francisco MAKÚ, 1992); *ug* (Makuna); *inomé* (Miraña *fide* CASTELLVÍ, 1938), *tome, name* or *ihva cümü* (Miraña fide NAPOLEON MIRAÑA, 1985); *yu, illumi* (Muinane); *juuju* (Ocaimo); *imü* (Piaroa); *caa, ka, tzoigar* (Puinave); *koto, kotu, kotomonu* (Quechua, *fide* CASTELLVÍ, 1938); *chifurrú* (Quimbaya-extinct tribe of the departments of Quindío and Risaralda, *fide* BASTIAN, 1878); *emú, ?emó* (Siona, *fide* CASTELLVÍ, 1938); *ijiya* (Tanimuca); *nëe* (Ticuna), *seunimi* (Tucano); *bíbara* (Tunebo); *ju-chi* (Wayú); *íu* (Witoto); *arishav* (Yukpa); *jimo* (Yukuna, *fide* various Yucunas of the Mirití-Paraná river); *lólo* (Yuri), many approximate Spanish sounds (RODRÍGUEZ M. *et al.,* 1995; and other sources).

Identification

Along with *Ateles* and *Lagothrix, Alouatta seniculus* is one of the largest Colombian primates, reaching head-body lengths of 44 – 69 cm, tails of 54-79 cm with male weights averaging 7.5 kg (n=8) and females about 6.3 kg (n=9) (HERNÁNDEZ & DEFLER, 1985). Of these data one male weighed

12.5 kg and one female weighed 10 kg. A more homogeneous population in Venezuela at Hato Masaguaral had a male average of 6.5 km (n=10) and females 4.5 kg (n=7) (THORINGTON *et al.*, 1979). The overall color is a deep mahogany red with some individuals exhibiting a reddish gold saddle, especially when in the sunlight. Occasionally these individuals show the same yellow-gold at the distal half of their tail. The head is large and the face naked and colored blackish with very long hairs in a beard.

The most striking feature of this monkey is the loud vocalization, especially in the male. This vocalization is produced by a specialized and enlarged hyoid bone and attendant larynx complex, allowing an exceedingly powerful sound to be produced, comparable to some of the great sounds of nature, like the roar of lions and the screams of macaws. These primates have a very enlarged lower mandibular as well, probably part of the evolutionary process of the development of their oversized hyoid and larynx. The monkeys also possess a strong prehensile tail which is capable of supporting the entire body while feeding.

Geographic range

Alouatta seniculus is absent on the Pacific coast of Colombia and the desert of the Guajira Peninsula, and the species has not yet been reported from the Department of Nariño. Otherwise it is present throughout the country, except in non-forested areas and in mountainous regions above the cloud forest belt. Its upper altitudinal limits are known to be as high as 3,200 m locally in the central Andes. It is poorly known in the Atrato basin, but specimens have been collected on the west bank of the Atrato river near Unquía as well as on the east bank near Suatatá. Its southern limits in this region are not known.

Outside of Colombia *A. seniculus* is found in the greater part of Venezuela, on the island of Trinidad and north of the Amazon River in the Guyanas and northern Brazil. It is also found in Amazonian Perú, Ecuador, and Bolivia as well as in western Brazil.

Natural history

The habitat of this monkey is extremely varied and includes mangrove swamps of the Caribbean coast, gallery forests of the eastern plains and other relatively dry regions (including oak forests), and extremely small, isolated forest patches and second growth as well as a variety of types of closed canopy lowland forest (HERNÁNDEZ C. & COOPER, 1976).

Compared to other neotropical primates, *Alouatta seniculus* is one of the best-studied species. The conspecific *A. palliata* is the most-studied of the neotropical primates so far, but *A. seniculus* has been studied at more sites than *A. palliata*. Also, *A. seniculus* (unlike *A. palliata*) has been studied at 9 different sites in Colombia, La Macarena (along the Guayabero river) (KLEIN & KLEIN, 1976) as a study incidental to one of *Ateles belzebuth*, and along the Duda river, just west of the Sierra La Macarena (IZAWA, 1975, 1988a, 1989a, 1990c, 1991, 1992, 1993, 1997b, 1997c, 1999; IZAWA & LOZANO, 1989, 1990a,b, 1991, 1994; KIMURA, 1995, 1997, 1999; PULIDO, 1997; STEVENSON *et al.,* 2002, 2000, 1998; TOKUDA, 1988; KOBAYASHI & IZAWA, 1992; YUMOTO *et al.,* 1999; KOBAYASHI & IZAWA, 1992; CEBALLOS, 1989). Studies have also taken place on the Peneya river (Caquetá), an affluent of the Caquetá river (IZAWA, 1975, 1976), at Finca Merenberg in cloud forest of the central Andes of the upper Magdalena River (GAULIN, 1977; GAULIN & GAULIN, 1982) as well as cloud forest of the northern Cordillera Central of the Andes in the Department of Risaralda (CABRERA, 1994), in the Colombian llanos in El Tuparro National Park (DEFLER, 1981), in dry forest at Colosó, Sucre (BARBOSA, 1988; CUERVO-DIAZ *et al.,* 1986), in the premontaine forest of Risaralda (MORALES-JIMÉNEZ, 2002) and the Amazonian site of Caparú (PALACIOS AND RODRÍGUEZ, 2001).

Several major studies have been done in Venezuela, especially at the most extensively utilized llanos research site in Hato Masaguaral (Guárico State) (CROCKETT, 1984a, 1984b. 1985, 1987a, 1987b, 1996; CROCKETT & EISENBERG, 1987; CROCKETT & JANSON (2000); CROCKETT & SEKULIC, 1982, 1984; CROCKETT & POPE (1988, 1993); CROCKETT & RUDRAN, 1987a, 1987b; POPE (1990, 1992, 1996, 1998); RUDRAN, 1979; RUDRAN & FERNÁNDEZ (2003); SEKULIC, 1981, 1982a, 1982b, 1982c, 1982d, 1982e,1983a, 1983b); SEKULIC & EISENBERG, 1983; and AGORAMOORTHY (1992, 1994, 1995, 1997, 2001) and AGORAMOORTHY & RUDRAN (1992, 1993, 1994, 1995) and AGORAMOORTHY & HSU (1995). Other studies at another llaneran site in Venezuela were by BRAZA (1978, 1980), BRAZA *et al.* (1980, 1981, 1983).

At least two studies (probably more) have taken place in Perú (SOINI, 1986a) and the species has been studied as well in Trinidad (NEVILLE, 1972a, 1972b, 1976b, 1983; NEVILLE *et al.,* 1988), PHILLIPS (1998b) and PHILLIPS *et al.*

(1998), as well as general census studies of FREESE (1977b) and FREESE *et al.* (1982). Additionally the species diet was studied in French Guiana (GUILLOTIN *et al.,* 1994; JULLIOT, 1994a, 1994b, 1996a, 1996b, 1996c, 1997; JULLIOT & SABATIER, 1993; KESSLER, 1998; RICHARD-HANSEN & VIE, 1996; RICHARD-HANSEN *et al.,* 1998, 2000; SIMMEN *et al.,* 2001; SIMMEN & SABATIER, 1996; de THOISY & PARC, 1999; de THOISY & RICHARD-HANSEN, 1997; FEER, 1999; YOULATOS, 1998 and VERCAUTEREN DRUBBEL & GAUTIER, 1993).

The species was also included in sympatric primate studies in Surinam (MITTERMEIER & VAN ROOSMALEN, 1981) while NEVES & RYLANDS (1991) studied the species in isolated patches of forest in central Amazonia, Brazil. Other Brazilian studies are by GILBERT (1995), IWANAGA & FERRARI (2002), PINTO & SETZ (2000), SANTAMARÍA-GOMEZ (2000) and in Ecuador (DE LA TORRE *et al.,* 1999. and in Perú (SOINI, 1986a). Despite the many sites and studies, there still remain many details of this primate's distribution, ecology and behavior to be clarified

This species generally is found in social groups varying from 2 or 3 up to 16, although the mean troop size for any area is usually between 6-9 animals per troop (CROCKETT & EISENBERG, 1987). Most commonly the typical troop is made up of 1-2 adult males and 2-3 adult females with attendent young, thus the makeup and size of the groups is smaller than that of *A. palliata*.

The largest groups have been observed in the Hato Masaguaral, Venezuela (5.9-16). In two sites in El Tuparro National Park (Colombia) the group average was 5.8-6.8 individuals (DEFLER, 1981), in Surinam 4.3 individuals (MITTERMEIER in WOLFHEIM, 1983) in Perú 6 (TERBORGH, 1983), and in Bolivia 7.4 (FREESE *et al.,* 1982). Knowledge of population densities and group sizes allows us to identify preferred habitat conditions for the species. Densities and group sizes do not correlate with any simple factor like annual precipitation, rather probably with the type of available food.

The troops of *A. seniculus* occupy ranges that vary from 4-182 ha, assumedly according to the resources available to them; calculated densities have varied from 15 to as high as 118 individuals/km^2 (CROCKETT & EISENBERG, 1987). PALACIOS & RODRÍGUEZ (2001) measured the largest known home range of 180 ha located in the rainforest of Caparú.

From the present data it appears that the highest densities and largest groups (two aspects that are positively correlated in the species) are attained in the central Llanos of Venezuela at the Hato Masaguaral in Guárico State at a nature preserve owned by the Blohm family. The opposite extreme in density can be seen on the lower Apaporis River in extreme eastern Colombia where a low ecological density is attained only along the edges of Taraira lake made up of groups whose average density is 15 animals/km^2 and whose crude density would be much less; the species disappears altogether inland from the lake, perhaps a result in part of competition with *Lagothrix lagothricha* of the area (PALACIOS & RODRÍGUEZ, 2001).

Very low densities are found at many *terra firme* sites in the Colombian Amazon with higher densities along lakes and rivers (DEFLER, 2003) Available data suggest that densities (and covarying group sizes) become reduced in more humid habitats in close-canopy forest. Average precipitation at Masaguaral is about 1,462 mm with a strong dry season of about four months, while precipitation at the lower Apaporis site in Colombia is ~4000 mm (n=7 years) with no strong dry season. It seems that this species finds its ideal habitat in the central llanos of Venezuela.

Other density data from the Colombian llanos at sites with more precipitation and more species of sympatric primates also show densities intermediate to the other two sites (DEFLER, 1981). At other sites in the Venezuelan llanos 25-54 individuals/km^2 were calculated (BRAZA *et al.*, 1981). Some of the lowest densities estimated were present in Colombia in a cloud forest (15 individuals/km^2), although other unpublished estimates of E. PALACIOS (pers. com.) in lowland rainforest in a blackwater watershed are much lower.

Alouatta seniculus, like other members of the genus, do not travel far in their foraging, since about half of their diet is made up of leaves. Average day ranges calculated for different studies are 340 m (SEKULIC, 1982a); 445 m (RUDRAN, 1979); 540 m (NEVILLE, 1972a); 542 m (CROCKETT, 1984a,b); and 706 m (GAULIN & GAULIN, 1982). The lower ranges correlate with groups on dry llaneran sites in Venezuela, while the highest average is of a group in cloud forest in a Colombian study. Data collected by BRAZA *et al.* (1981) for 16 days (8 during the wet season and 8 during the dry season) showed the following time budget: 18%/15% motion, 20%/24% feeding, 38%/43% sleeping, 24%/18% other activities and resting.

Alouatta seniculus is principally a quadruped. They invest 80% of their time walking and only 4% of their time leaping (FLEAGLE & MITTERMEIER, 1980; SCHÖN- YBARRA, 1984, 1986, 1987, 1988). Bridging behavior across gaps in the canopy was studied by YOULATOS (1993) and head-first descent by YOULATOS & GASC (1994). These animals sleep alone or in small groups over horizontal branches in large and medium sized trees.

One study in cloud forest in Huila calculated a diet for this species made up of 7.5% mature leaves, 44.5% new leaves, 42.3% fruits, 5.4% flowers and 0.1% petioles, based on 340 hours of focal study during 10 months. The main species eaten were from the genera *Ficus, Cecropia, Morus,* and *Quercus* (GAULIN, 1977; GAULIN & GAULIN, 1982).

In another research project in cloud forest located in the Regional Natural Park Ucumari in the department of Risaralda in the Central Cordillera, *Alouatta* ate fruits of the following species: *Ficus insipida, Ficus gigantosyce, Eupatorium* sp., *Sapium cuatrecasasii, Quercus humboldtii, Nectandra* spp. *Souroubea* sp, *Guarea* sp., *Morus insignis, Cecropia tasmanii, Prunus integrifolia, Solanum* sp., *Billia columbiana, Oreopanax* sp., 3 species from the Lauraceae and 2 species from the Leguminosae (CABRERA, 1994). The use of acorns (*Quercus*) in the diet is particularly interesting, inasmuch as the genus is often thought of as a North American, though it actually reaches southern Colombia.

At Hato Masaguaral in Venezuela the diet depended mainly on a few key resources: *Ficus, Copernicia tectorum* (a palm), and *Albizia cf. caribea* (SEKULIK, 1982a; NEVILLE *et al.*, 1988). In La Macarena a short study showed that *Ficus* was the most important fruit consumed, while the major part of the diet was the consumption of new and old leaves (KLEIN & KLEIN, 1975). In a study in French Guiana in lowland primary forest, the diet was made up of 195 species of plants from 47 families, which consisted of 54% young leaves, 21.5% mature fruits, 12.6% flowers and occasional old leaves, immature fruits, termitaria soil, bark and moss and perhaps rarely small vertebrates (DE THOISY & PARC, 1999). The most important families in number of species utilized were: Moraceae (30 spp., includes Cecropiaceae), Sapotacea (28 spp.), Fabaceae (16 spp), and Mimosaceae (10 spp.) (JULLIOT, 1994b, 1994c; JULLIOT & SABATIER, 1993). The most important fruits consumed are listed in the following (*fide*, JULLIOT, 1994b): *Dryptes variabilis* (Euphorbiaceae) 3.99%; *Solanum* sp. (Solanaceae) 5.08%; ? 6.46%; *Goupia glabra (Celastraceae)* 8.39%; *Vouacapoua americana* (Caesalpiniaceae) 8.43%; (unidentified Sapotaceae)9.4%; *Bagassa guianensis* (Moraceae)10.2%.

The most important leaf species eaten are listed as follow (*fide* JULLIOT, 1994b): *Pithecelobium jupunba*-Mimosaceae (7.94%); *Dicorynia guianensis*-Caesalpiniaceae (4.55%); *Bocoa prouaensis*- Fabaceae (4.51%); *Tetragastris altissima*-Burseraceae (4.4%); *Tabebuia serratifolia*-Bignoniacea (4.18%); *Pourouma minor*-Moraceae (3.47%); *Inga bourgoni*-Mimosaceae (3.13%); *Inga* spp. - Mimosaceae (3.13%); *Philodendron linnaei* - Araceae (3.3%); *Neea* sp.2 - Nyctaginaceae (2.69%); *Neea* sp.1 (2.61%); *Swartzia panacoco* - Fabaceae (2.54%); *Eperua falcata* - Caesalpiniaceae (2.5%); *Pouteria filipes* Sapotacea (2.42%). The most important species eaten for the flowers are listed as follow (*fide* Julliot, 1994b):

Micropholis cayennensis - Sapotacea (26.55%); *Eperua falcata* (Caesalpiniaceae) -13.45%; *Odontadenia* sp. (Apocynaceae) - 13.19%. Julliot (1996a) discusses the importance of *A. seniculus* as a seed disperser, while STEVENSON *et al.* (2003) evaluate germination rates of dispersed seeds.

In Tinigua National Park the most important fruit species eaten by howlers are as follow (STEVENSON *et al.*, 1991): *Brosimum alicastrum* – Moraceae (20%); *Ficus andicola* – Moraceae (9%); *Pseudolmedia hirta* – Moraceae (7%); *Ficus americana* – Moraceae (5%); *Ficus sphenophylla* – Moraceae (5%); *Castilla ulei* – Moraceae (5%); *Brosimum utile* – Moraceae (5%). Overall, 30% of the time spent feeding on fruit is spent feeding on Moraceae, followed by Mimosaceae and Burseraceae (4% and 3% respectively).

The interesting habit of eating earth at places in the lowland forest called "salados" or "salt licks" was reported by IZAWA (1975) and soils from a salado were analyzed by YOSHIHIKO & IZAWA (1990) and discussed by IZAWA (1993). EMMONS (1984) previously reported an analysis of salado soils from other sites, since many mammals and birds eat such soils. Chemical analyses from the EMMONS study showed that Na is probably the element most sought after, although analyses by YOSHIKO & IZAWA (1990) and IZAWA (1993) were inconclusive. Indians of the Mirití-Paraná (Paraná=river in Geral) in the Colombian Amazon report that the density of *Alouatta* is higher around salt licks on the Mirití. This primate drinks water from tree holes and from bromeliads.

This species is sexually dimorphic and usually age scaled when the male adult is the only one to breed. Males reach maturity around 58-66 months and females are mature at about 43-54 months. Female sex cycles average about 26 days, including 2-3 days of estrus. Males smell the females' urine, sticking their nose in the puddles of urine or in the stream to do so. Sexual solicitation includes tongue oscillation (*tongue flicking*). Males mount from behind with their hind feet on the branch.

Gestation is about 191 days (CROCKETT & SEKULIK, 1984) after which time a single infant is born. There does not seem to be a particular birth season, except that there were fewer births during the early part of the wet season in one analysis at Hato Masaguaral in Venezuela (CROCKETT & RUDRAN, 1987a, 1987b; CROCKETT & SEKULIC, 1982, 1984. SEKULIC (1982d) described the birth of two infants, which were assisted by an adult male, guarding nearby.

When the infant is first born, its tail seems useless, and the tail does not become sufficiently coordinated to be of use until about 6-8 weeks. During this time there is a gradual change from being carried ventral-ventral to the position dorsal-ventral, and the baby monkey can then grab ahold of the mother's tail with its own. Infants begin to eat solid food during the second month, but they probably nurse until the tenth month (CEBALLOS, 1989). Other females show interest in a young infant and they try to touch it. Young animals are interested in playing with other members of the group (FIGUEROA, 1989). MACK (1979) also describes the growth and development of red howler infants in the wild.

Adult males pay little attention to infants. Male take-over of groups are known, when the new male sometimes in coalition with another, drives the group male off and kills the infants, causing the group females to become sexually receptive again, and ready to mate and bear the infants of the new group male (SEKULIC, 1981; CROCKETT & SEKULIC, 1984). POPE (1990) showed that a coalition of more than one male exhibited more success in dominating and taking over the groups of other males.

Howler societies are generally considered to be age-graded societies or containing an older and larger male, who may be joined by one or two younger and smaller males. Both supernumerary males and supernumerary females disperse from their natal group (SEKULIC, 1982a, 1982c; CROCKETT, 1984a, CROCKETT & SEKULIC, 1984) to join other social units. Solitaries are frequently detected. DEFLER (pers. obs.) observed a solitary subadult male traveling on the ground, several kilometers from the nearest forest. D. PINTOR (Parques Nacionales/Colombia), the head of El Tuparro National

Park picked up a solitary male caught in a whirlpool where the Tuparro river enters into the Orinoco, enabling the animal to disembark onto the large fluvial island between the two divisions of the Maipures rapids. A Makuna hunter known to the author, killed a lone, swimming male in the Caquetá River in 1980 and gave the skull to the author as a gift (pers. obs., DEFLER). SEKULIC (1982a) describes the efforts of a solitary female to become integrated in an established group.

The small social groups are usually quite tranquil and pacific with the young playing contentedly as the group spends often more than half the day resting, high in the trees. Any animal which uses foliage as a major part of its diet is obligated for physiological reasons to spend long periods of its time digesting the large bulk of foliage. The red howlers often chose high sun-exposed sites for this resting behavior, probably to increase the speed of the digestive process. Usually the most violent behavior between groups are prolonged screams and howls, but occasionally a group interaction leads to physical contact. DEFLER (obs. pers.) watched a group fight and chase until one of the males fell from the tree as he tried to avoid the males of the other group.

Grooming is most frequent between adult and subadult females, although all members of the group except male juveniles take part in such interactions (NEVILLE, 1972b), which are related to social structure (SÁNCHEZ-V. *et al.*, 1998). Howling behavior is the one characteristic which defines this genus and it was long assumed that this impressive vocalization serves to apprise other groups of individual presence, facilitating mutual avoidance and perhaps defense of resources (SCHÖN-YBARRA, 1986, 1988). However, SEKULIC (1982b, 1982c, 1982d, 1983a) carried out a detailed study of the behavior and decided that howling is also used to assess opponents in male (who are the loudest howlers) or female competition. Sekulic also supposed that the behavior probably strengthens the pair bond. Howling is often precipitated not only by the nearness of another group or the cries of nearby howlers, but it is commonly triggered as well by thunder, rain, wind, airplanes and other animal noises.

Aside from howling, this species is surprisingly quiet in its natural habitat. A description of various vocalizations are presented as follow: *incipient roar* - a series of short and incipient roars emited by the dominant male during general disturbance, aggressive; *roar accompaniment* - elevated tone produced by juveniles and young adults, alarm, aggressive; *male woof* - adult male, repeated in a series of 1-4 barks; when disturbed; *female woof or bark* - higher tone than male and given in accompanyment with male when disturbed; *incipient woof* - higher tone in females than in males when slightly disturbed; *oodel* - adult males, rhythmically repeated in pulsations of air, intense disturbance during intergroup confrontation, aggressive; *whimper* - infants, juveniles and adult females when frustrated by general disturbance; *eh* – infants, rapid vocalizaton of breathy expiration, when exploring and playing; *squeeking hinge* - infants when nervous and unable to keep up; *infant bark* - infants when nervous; *purr* - infants when in contact with the mother and content.

Other stereotyped social signals include tongue flicking in sexual contexts, back-arching in aggressive contexts, chin-throat-chest rubbing, anogenital rubbing and back-rubbing. Coordinated defecation and urination is common and may be used at times as a defense against ground predators. Branch shaking is common when playing or in alarm. The males lip purse and genital display during agonistic interactions. Adults crouch when in submission and yawn when alarmed. They lip smack in a sexual context, besides many other displays.

Known associations with other animals are limited to temporary *laissons* with *Ateles belzebuth* (KLEIN & KLEIN, 1976) and *Cebus albifrons* (DEFLER, 1979a). DEFLER observed a lone female *Ateles belzebuth* who might have been the last of the population in that area, traveling and feeding with a group of *Alouatta seniculus* on several occasions in La Macarena. In 1992, José Reinaldo Muca, an Indian of the lower Caquetá river area, killed a female *Ateles belzebuth* who was associated with a group of *Alouatta seniculus*. KLEIN & KLEIN (1976) observed *A. belzebuth* in association with *A. seniculus* in La Macarena on various occasions.

One observation of predation was the frequent use of the species as food by the eagle *Harpia harpijya* (RETTIG, 1978; SHERMAN, 1991). Another was the hunting of several howler monkeys by jaguars in a disturbed site in Venezuela (PEETZ *et al.*, 1992). It seems probable that other predators pursue this large primate, especially since they leave the trees to cross savannas and rivers.

Conservation

This species is not considered endangered by CITES, the IUCN LC or by Colombia. Nevertheless, it is difficult to generalize about the conservation state of *A. seniculus* in the country. In the majority of the forested lowlands it is possible to hear the howls of this monkey, and it seems well-protected in the Colombian national parks (DEFLER, 1994). Forest destruction is the principal threat to this and most primates, and due to the dispersal abilities of the species, its continued presence is evident after the disappearance of the majority of other primates.

The hunting of this species for food is variable depending on the region, *e.g.* this monkey is eaten very seldom in northern Colombia, but in the Llanos Orientales it is eaten by Indians and by white immigrants from the interior of the country. In the upper valley of the Magdalena river and the Cauca river this species has been eaten for a long time, and it is very difficult to find the species in that region. In the same region the hyoid bone frequently is used by the peasants as a cup for drinking, as they believe it to be an effective therapy for curing goiter (HERNÁNDEZ C. & COOPER, 1976).

In the Colombian Amazon the state of this primate is difficult to gauge, since it doesn't seem to be common around human populations. The crude densities of the species in many parts of Amazonia are much lower than in the Llanos Orientales, since in much Amazonia habitat the species lives only on the edges of large and medium-sized lakes and rivers and not inside the forest away from these bodies of water. This preference is probably related to the quality of the forest and the competitors present, since some forests which seem more fertile also contain *A. seniculus* in their interiors. This aspect of *A. seniculus* populations needs to be evaluated since it makes the species vulnerable in many parts of the Amazon where the problem has not been recognized.

Where to see it

This is not a difficult species to observe, since it advises its presence by loud vocalizations. Especially along waterways in the morning the howl of these animals can be used to locate them. They are easily heard in most of the Colombian national parks where they are present.

Plate 1.

a. *Callimico goeldii.*
   Map 1 • Page 126

b. *Cebuella pygmaea.*
   Map 2 • Page 135

A

B

Plate 2.

a. Saguinus geoffroyi.
Map 3 • Page 163

b. Saguinus inustus.
Map 4 • Page 169

c. Saguinus leucopus.
Map 5 • Page 173

d. Saguinus oedipus.
Map 6 • Page 188

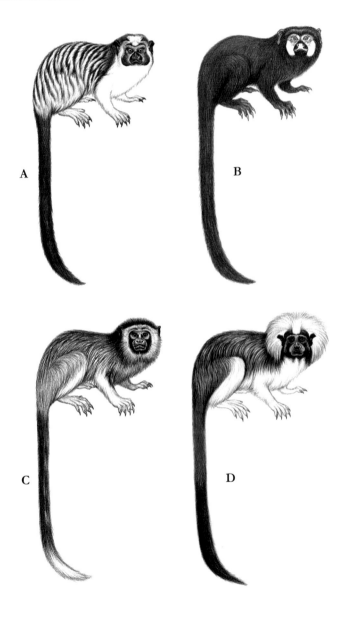

A

B

C

D

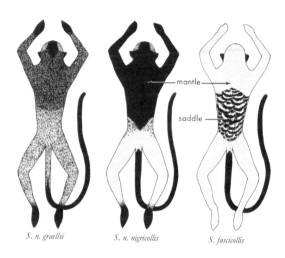

S. n. graellsi   S. n. nigricollis   S. fuscicollis

Plate 3.

a. *Saguinus fuscicollis fuscus.*
   Map 7 • Page 150

b. *Saguinus nigricollis nigricollis.*
   Map 8 • Page 179

c. *Saguinus nigricollis graellsi.*
   Map 8 • Page 179

d. *Saguinus nigricollis hernandezi.*
   Map 8 • Page 179

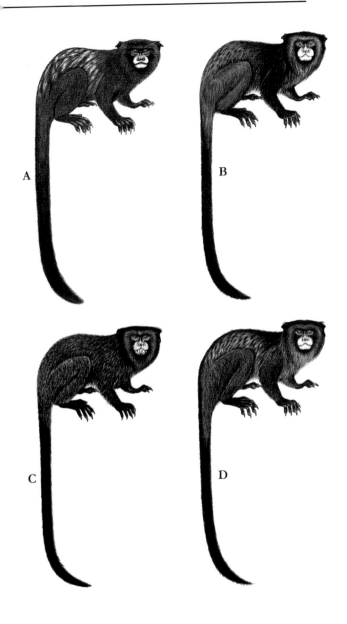

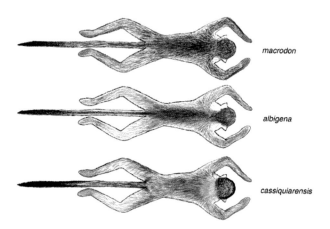

macrodon

albigena

cassiquiarensis

Plate 4.

a. *Saimiri sciureus albigena.*
   Map 9 • Page 237

b. *Saimiri sciureus cassiquiarensis.*
   Map 9 • Page 237

c. *Saimiri sciureus macrodon.*
   Map 9 • Page 237

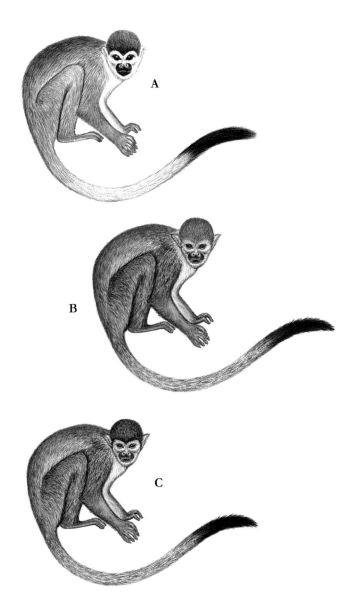

Plate 5.

a. Cebus albifrons.
   Map 10 • Page 207

b. Cebus capucinus.
   Map 11 • Page. 227

c. Cebus apella.
   Map 12 • Page 216

Plate 6.

a. *Cebus albifrons albifrons.*
b. *Cebus albifrons cesarae.*
c. *Cebus albifrons malitiosus.*
d. *Cebus albifrons versicolor.*
e. *Cebus albifrons cuscinus.*

Plate 7.

a. *Aotus brumbacki.*
   Map 13 • Page 267

b. *Aotus griseimembra.*
   Map 13 • Page 262

c. *Aotus lemurinus.*
   Map 13• Page 262

d. *Aotus trivirgatus.*
   Map 13• Page 273

e. *Aotus vociferans.*
   Map 13 • Page 269

f. *Aotus zonalis.*
   Map 13 • Page 262

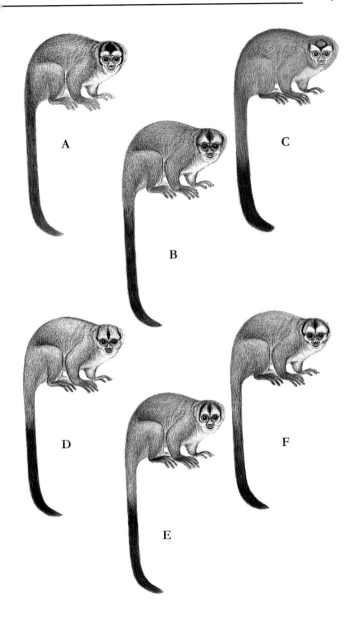

Plate 8.

a-b. Ateles belzebuth.
    Map 14 • Page 331

c-d. Ateles hybridus.
    Map 15 • Page 347

e. Ateles geoffroyi.
    Map 16 • Page 339

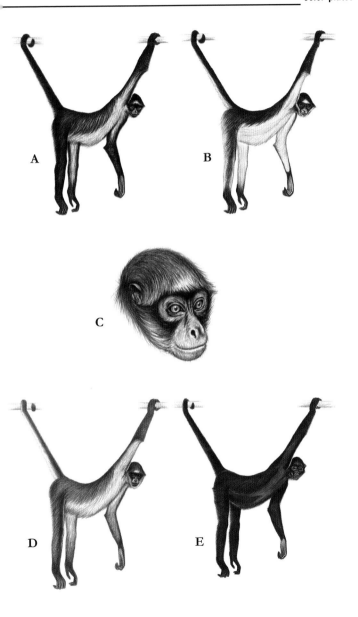

Plate 9.

a. Lagothrix lagothricha lagotricha.
    Map 17 • Page 357

b. Lagothrix lagothricha lugens.
    Map 17• Page 357

A

B

Plate 10.

a. *Alouatta palliata.*
Map 18 • Page 370

b. *Alouatta seniculus.*
Map 19 • Page 384

A

B

Plate 11.

a. *Callicebus torquatus lucifer.*
   Map 20 • Page 314

b. *Callicebus torquatus lugens.*
   Map 20 • Page 314

c. *Callicebus torquatus medemi.*
   Map 20 • Page 314

A

B

C

Plate 12.

a. *Callicebus cupreus ornatus.*
Map 21 • Page 306

b. *Callicebus cupreus discolor.*
Map 21 • Page 306

A

B

Plate 13.

a. Pithecia monachus milleri.
   Map 22 • Page 278

b. Pithecia monachus monachus.
   Map 22 • Page 278

A

B

Plate 14.

Cacajao melanocephalus ouakari
Map 23 • Page 287

a. Vista frontal.
b. Vista dorsal.

A

B

Plate 15.

a. Cebus albifrons albifrons.
Map 10 • Page 207

b. Aotus lemurinus.
Map 13• Page 262

A

B

Maps

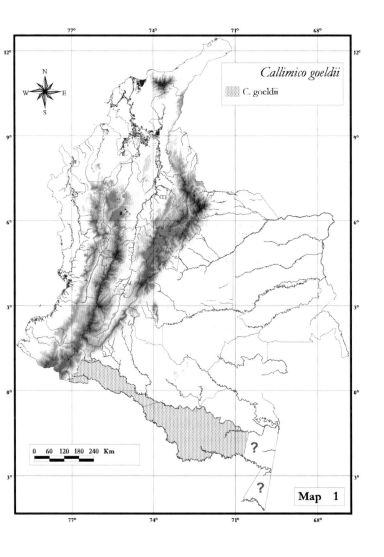

Callimico goeldii

C. goeldii

N
W E
S

0 60 120 180 240 Km

?

?

Map 1

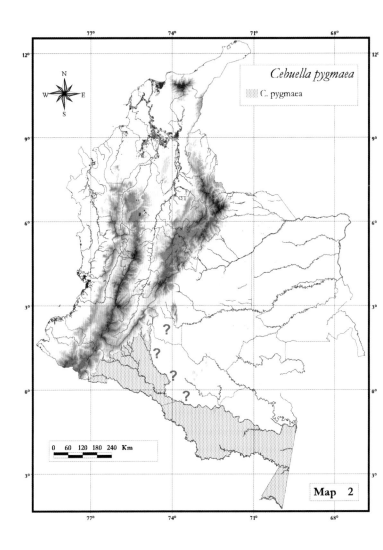

Cebuella pygmaea

C. pygmaea

Map 2

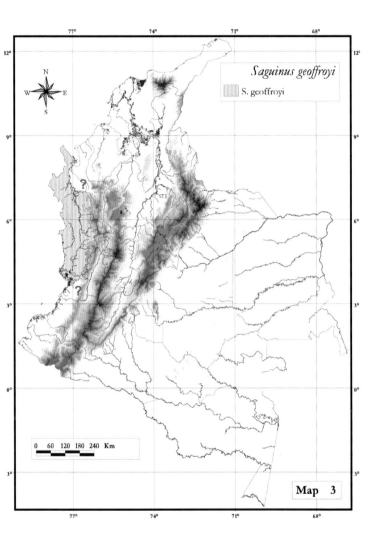

Saguinus geoffroyi

S. geoffroyi

Map 3

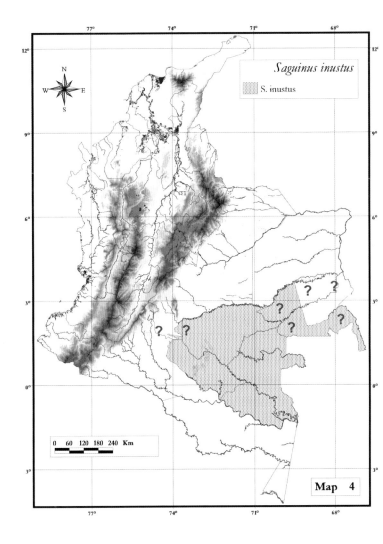

Map 4

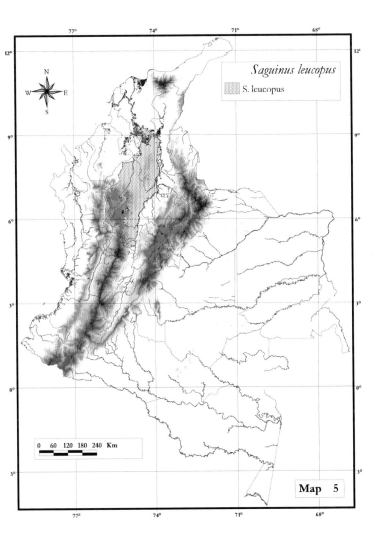

Saguinus leucopus

S. leucopus

Map 5

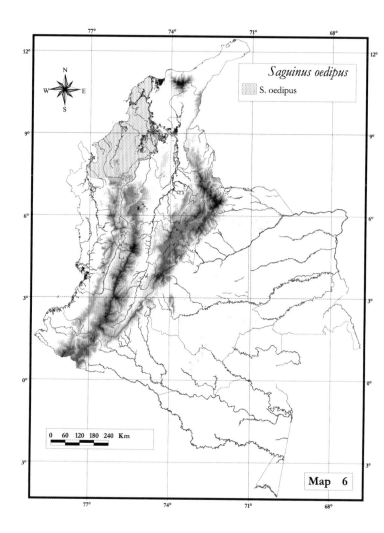

Saguinus oedipus

S. oedipus

Map 6

0 60 120 180 240 Km

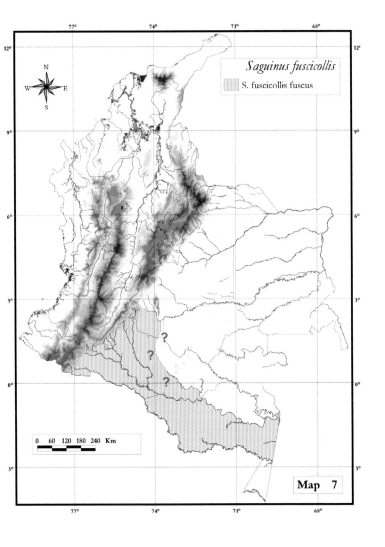

Saguinus fuscicollis

S. fuscicollis fuscus

Map 7

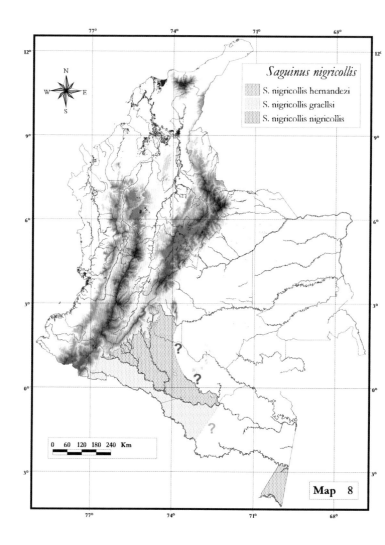

Saguinus nigricollis

S. nigricollis hernandezi
S. nigricollis graellsi
S. nigricollis nigricollis

0   60 120 180 240 Km

Map 8

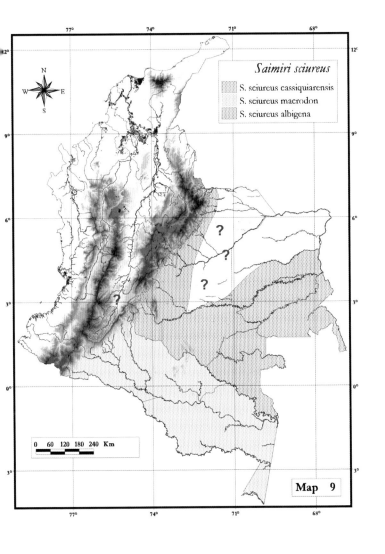

Saimiri sciureus

S. sciureus cassiquiarensis
S. sciureus macrodon
S. sciureus albigena

Map 9

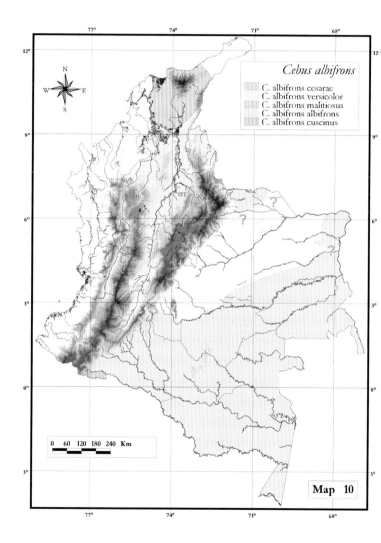

Cebus albifrons

C. albifrons cesarae
C. albifrons versicolor
C. albifrons malitiosus
C. albifrons albifrons
C. albifrons cuscinus

0 60 120 180 240 Km

Map 10

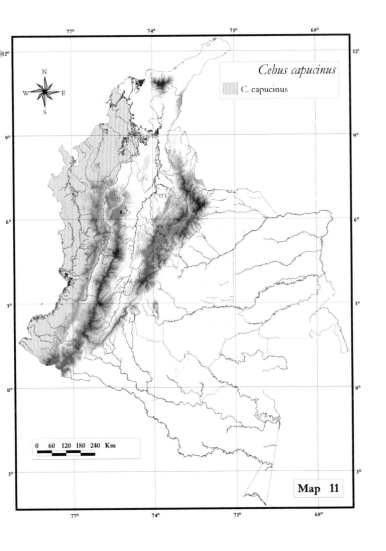

Cebus capucinus

C. capucinus

Map 11

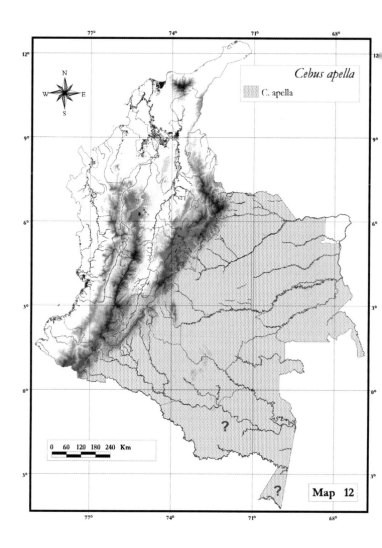

Cebus apella

C. apella

Map 12

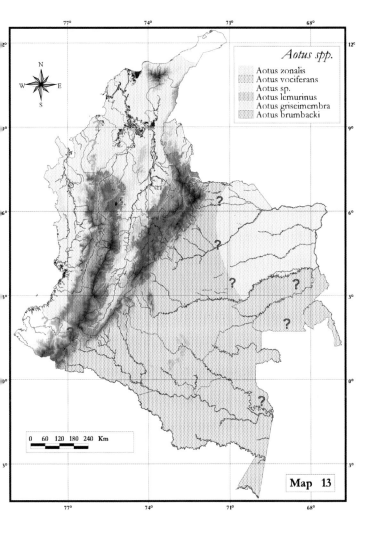

Aotus spp.

Aotus zonalis
Aotus vociferans
Aotus sp.
Aotus lemurinus
Aotus griseimembra
Aotus brumbacki

0 60 120 180 240 Km

Map 13

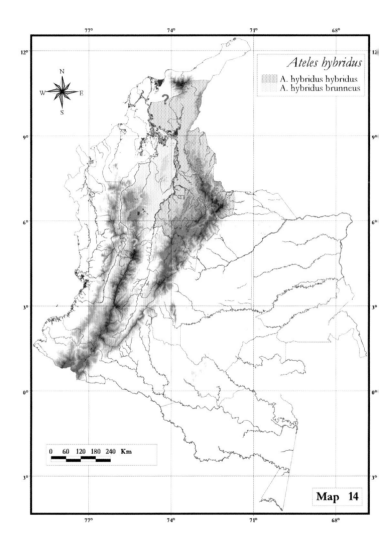

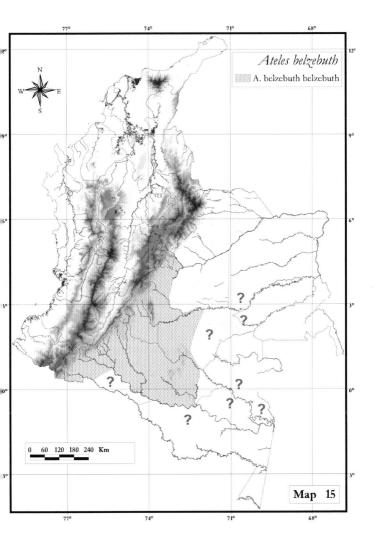

Ateles belzebuth

A. belzebuth belzebuth

Map 15

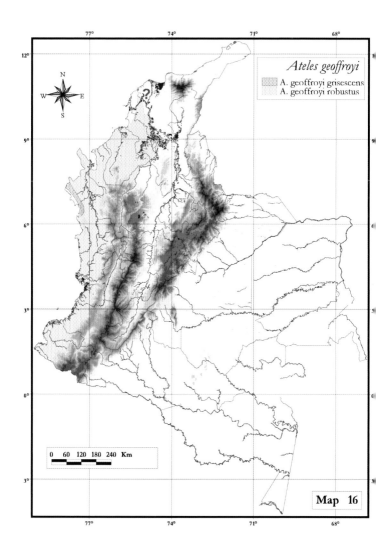

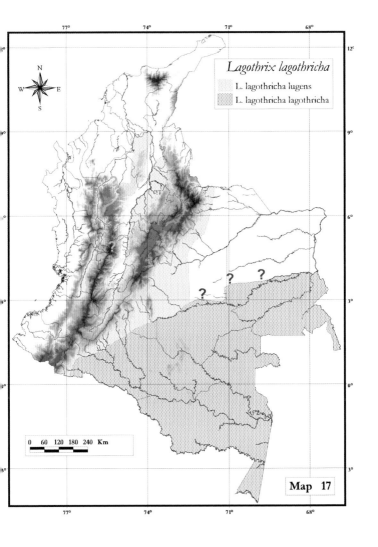

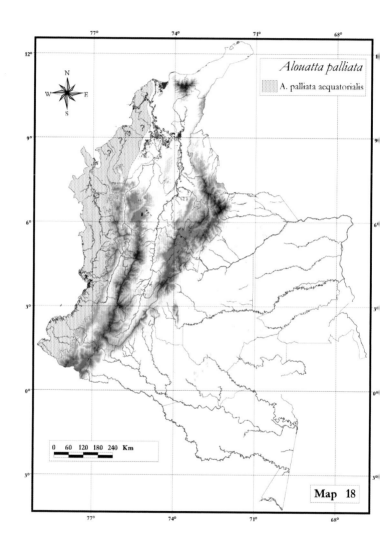

Alouatta palliata

A. palliata aequatorialis

0 60 120 180 240 Km

Map 18

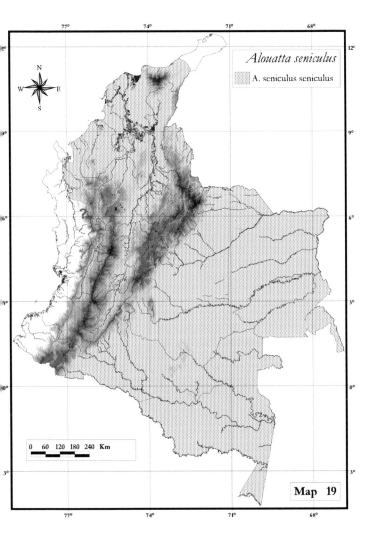

Alouatta seniculus

A. seniculus seniculus

0 60 120 180 240 Km

Map 19

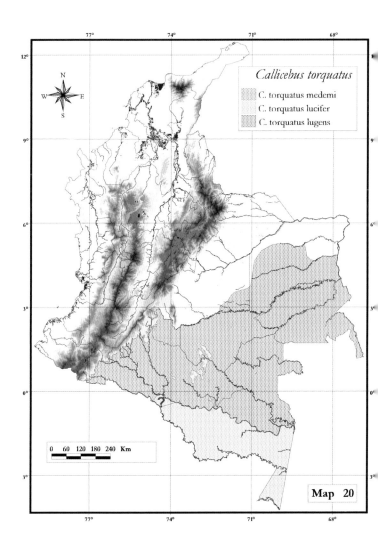

Callicebus torquatus

C. torquatus medemi
C. torquatus lucifer
C. torquatus lugens

0 60 120 180 240 Km

Map 20

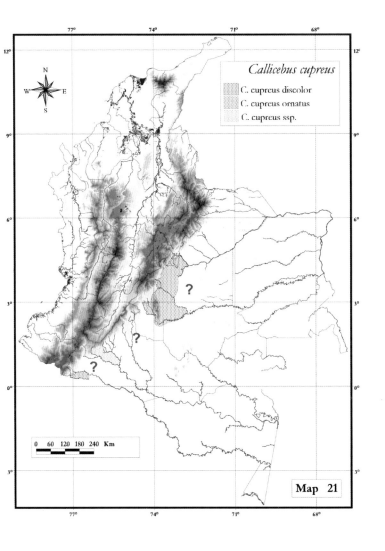

Callicebus cupreus

C. cupreus discolor
C. cupreus ornatus
C. cupreus ssp.

Map 21

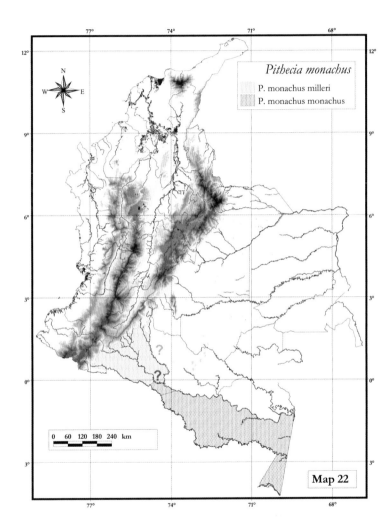

Pithecia monachus

P. monachus milleri
P. monachus monachus

Map 22

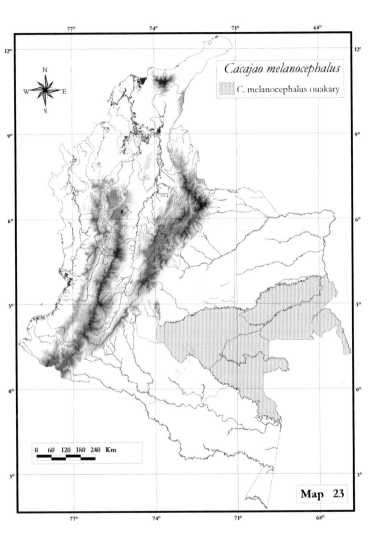

Cacajao melanocephalus

C. melanocephalus ouakary

Map 23

0 60 120 180 240 Km

Glossary

A

Abdomen – the ventral portion of the body where the viscera are located.

Activity budget - a classification of the way an animal or a group of animals distribute their available time among the categories of activity important for survival and reproduction (syn. time budget).

Adapid (adapoid) primates - the earliest relatives of the lemurform primates; these primates appeared in the early Eocene in both North America (*Notharctus*) and Europe.

Adult – individual that has reached sexual maturity and maximum growth (applied here to primates).

Advanced - a character of an organism which has changed through evolution from how it appeared in the ancestors of the organism.

Agonistic – behavior between individuals which includes both aggression and threat as well as appeasement and evasion.

Aligned sequences – two homologous nucleotide sequences with no extra insertions or deletions.

Allele – any of various forms of a gene or given characteristic (such as eye color).

Allogrooming - one individual grooming another individual.

Allopatric – (allopatry) or located in separate places, as in two populations of an organism that are found in two separate locations isolated from each other.

Allopatric speciation – mechanism thought to be most important in forming two species, whereby the population of one species becomes isolated from the main population, allowing changes to occur in the reproductive isolating mechanisms so that it evolves into a new species.

Altricial – infants which require extensive parenting before they become independent.

Alveolus - tooth socket in the mandibles.

Amazonia - a geographic term which is often used in different ways. in Colombia the term has referred to (1) the watershed of the Amazon River; (2) the forest land east of the Andes up to the Guaviare/Guayabero River; or (3) the totality of the forested land from the Vichada River south, including (according to this author) the national parks La Macarena, Tiniguas and Picachos as well as the eastward-facing forested slopes of the Cordillera Oriental de los Andes.

Amazon caatinga - (term proposed and discussed by Anderson, 1981) a well-recognized vegetation type which grows over white-sand soils and sometimes confusingly called caatinga but best distinguished from caatinga of northeastern Brazil, which is a special arid vegetation type. Amazon caatinga grows in a climate that is suitable for forest, but the limiting factors of the white sand cause stress factors which permit a vegetation type ranging from a completely open type of vegetation (sny. savana 1 or white sand savanna) to closed vegetation (syn. savana 5 or white sand savanna forest). certain genera of plants are characteristic of Amazon caatingas. Amazon caatingas have a xeromorphic aspect with thick leaves and thick bark, abundant lichens and mosses on branches (Pires and Prance, 1985). (syn. campinarana, sabana 1 or white sand savanna to sabana 5 or white sand savanna forest).

Androgen – hormones which stimulate the development of male sexual organs and male secondary sexual characteristics.

Androgyny – expression of both feminine and masculine characteristics.

Anogenital – region between the anus and genitals.

Anthropoidea - the higher primates, represented by New World (Ceboidea) and old world primates (monkeys, apes and humans).

Antiphonal – alternate or responsive singing (vocalizations) by two vocalizers (singers).

Appropriate development - the concept of the careful use of technologies and development that is in accordance of the needs of the people and the land with the philosophy of searching for a balance that does not destroy integral ecosystems and cultural values.

Arch posture - often aggressive display given by sharply bending the back like a house cat.

Areal – referring to the geographic distribution of a species.

Aril – an accessory covering or appendage of certain seeds, arising from the placenta and sometimes evolved to attract animals to swallow it (and the seed) for posterior dispersal.

Arthropod – biological division of the animal kingdom consisting of those animals with a hard exoskeleton at some stage of their life and paired, jointed legs; corresponding to insects, spiders, etc.

Atelid – one of the members of the family Atelidae (*Ateles, Lagothrix, Brachyteles, Alouatta*)

Ateline – one of the member of the subfamily atelinae (as above).

Auditory bulla - a bony chamber which houses the bones of the inner ear.

Auditory meatus - the oppen passage of the ear leading from the outside to the tympanum.

Austral – to or of the south.

B

Bajos – low lying lands, often frequently flooded Amazon in tropical forests.

Balanced polymorphism – three karyotypes in a population which maintain the same frecuencies within the population.

Basal metabolism – energy consumed at rest.

Bifurcate - to divide or fork into two branches.

Biogeographic Province – a major biogeographic division.

Biome - a particular region or collection of regions which possess the same physical and climatic characteristics and which support a fauna and flora which show adaptation to these conditions.

Bilophodont - molar teeth in which the mesial and distal pairs of cusps form ridges or lophs.

Black water - sediment free water colored by tannins and other organic acids aquired via their origen over white sand soils and providing no nutrients to aquatic or adjacent flooded forest ecosystems.

Boreal – to or of the north, referring to any distribution to the north.

B.P. – before present time.

Bridging – behavior of adults which form a bridge between two trees by holding onto both so that the young may pass over their bodies to the next tree.

Bunodont - teeth that have low and rounded cusps.

C

Campinas – a synonynm in Brazil for Amazon caatingas.

Campinerana - a synonym in Brazil for Amazon caatingas, the special type of vegetation which grows over white sands. this vegetation has been confusingly termed "sabana" in Colombia, both by local inhabitants and in the litterature (Carvajal *et al.*, 1979).

Canines - one of the four sharp-pointed teeth between the incisors and bicuspids; also called cuspid or, in the upper jaw, eyeteeth.

Canopy – the highest part of the tree.

Caparú – Estación Biológica Caparú, biological research station located in the eastern Colombian Amazon on the lower Apaporis River.

Carpal – bone(s) of the wrist between bones of the hand and the radius.

Catarrhini – Old World monkeys and apes.

Caudal – of or pertaining to the tail.

Caducifolious (caduceus) – deciduous (vegetation which periodically loses its leaves and grows new ones).

Caviomorph rodent - Neotropical hystricomorph rodents, sometimes classified in their own suborder Caviomorpha.

Centrale bone - wrist bone posterior to trapezoid or carpal 3.

Cheirogaleid – the smalles and most primitive of the Malagasy primate families and comprising mouse lemurs, dwarf lemurs, fork-marked lemurs and others, classified into 5-6 genera (*Microcebus*, *Allocebus*, *Mirza*, *Cheirogaleus*, and *Phaner*).

Chromosome – filamentous carrier of genes, found in the cell nucleus and most visible during mitosis and meiosis.

Cingulum - a u-shaped enamel ridge at the base of the posterior surface of the crown of a tooth.

Cis-Andean - a species restricted to one side of the Andes due to the efficacy of the mountains as an effective bio-geographic barrier to that species' dispersal.

CITES – Convention on International Trade in Endangered Species of Wild Fauna and Flora is an international agreement between governments. Its aim is to ensure that international trade in specimens of wild animals and plants does not threaten their survival.

Clade – a natural group with the same evolutionary lineage.

Cladistic – a type of analysis of evolutionary lineages or organisms using the precepts defined by Hennig.

Clinal – characteristics which over several populations change gradually from one state to another.

Clitoris – an erectile organ of the female of most vertebrates, at the anterior part of the vulva; the homolog of the penis.

Coleopterans – a large order of insects which includes the beetles.

Colhuehuapian – South American land mammal age of the lower Miocene Epoch of about 19-20 million years B.P.

Congeneric (**cogener**) – of the same genus.

Conspecific – of the same species.

Continental plates - tectonic plates or subdivisions of earth's surface on which continents move about with reference to each other.

Convergence - to move to a common point from different directions, like the fingers of a hand closeing.

Core area - the area whithin the home range which is most commonly utilized or frequented.

Coronal – of the crown of the head.

Cotyledon – primary or rudimentary leaf of the embryo of seed plants.

Counter singing - the habit of many birds and some primates (like *Callicebus* spp.) to vocalize (or sing) in a manner that is coordinated with each other.

Cranial sutures – the meeting or contact place of the various cranial bones;

Crude density – density of an organism calculated on an area which includes both habitat including the particular organism and habitat which does not include the organism, since most organisms are not uniformly spread out.

Cusp - one of the elevations on the chewing surface of a tooth.

D

Day range - the average distance which an individual or a group moves during one day from one sleeping site to the next.

Dermatoglyphics – of or pertaining to prints of fingers, hands, feet or callosity of tail.

Dimorphic sexes – two sexes, each of a different average size.

Diploid – chromosome numbers which include the complement from the mother and the complement from the father.

Distal - in anatomy, farthest from the center or the point of attachment or origin.

Divergence - to go in different direction from a common point, like the fingers of a hand opening.

Diversity centers – biogeographic areas from which it seems a group of organisms have spread out.

Dorsal (dorsum) – the superor or back part of an organism, located over the vertebral column.

Doubentonioid - doubentonioid-like; the daubentoniid or aye-aye is a 3 kg black animal with course, shaggy hair, huge ears and a large bushy tail, possessing more extreme morphological specializations than any other living primate. it is related to the lemurs, is nocturnal and has the third digit of each hand extremely elongated for probing for insects in holes.

E

EBC – the Estación Biológica Caparú, biological research station located on the lower Apaporis River in the eastern colombian amazon.

Ecogeographic – taking into account both ecology and geography.

Ecological density - population density of ecologically appropriate habitat were the organism is present .

Ecological tolerance – the capacity of an organism to adapt itself to its habitat or surroundings (=resilience).

Ecophysiological – pyhsiological adaptacion.

Edaphic - of soils.

Endemic – an organism found only in a defined area and not outside it.

Endocarp – internal part of a fruit where the seeds are located.

Endosperm - the nourishment for the embryo, which surrounds the embryo in the seed of the plant.

Entotympanic bone - bone at the base of the cranium at the root of the nose, containing numerous perforations for the filaments of the optic nerve.

Eocene - a geological epoch within the Tertiary Period (ca. 54-38 million years B.P.).

Et-epimeletic – calling and signaling for care attention.

Epithet – a pharase or word used adjectively to express some characteristic attribute or quality

Estrous (estral) – pertaining to the sexual cycling of females.

Ethmoid - the perforated bone or bones at the front part of the base of the skull, forming part of the septum and walls of the nasal cavity and through which the olfactory nerves pass.

Ethology – study of animal behavior, especially in their natural enivronment, with a naturalist viewpoint, rather than a behavioristic one.

Etymology – study of the origin and transformation of words.

Eumelanin – common form of melanin which is typical of the darker colors such as browns and blacks.

Eurothermic – tolerant of a broad range of temperatures.

Euthemorphic - shaped or like a placentic mammal.

Exudate feeding - use of tree sap as a food source.

F

Facultative polyandry – able to live with one or several male mates.

Fatted – peculiar state of male *Saimiri sciureus* which gain weight before the breeding season.

Fide – according to.

Fila - thread-like.

Flagship species - a species that is especially emphasized or underlined in conservation, often because of its appealing characteristics so as to appeal to humans to join a wider conservation effort.

Flank – the sides of an animal.

Flexion - the act of bending a limb; the position which a limb assumes when it is bent.

Foraging – active searching for food.

Frontal bone – the cranial plate or bone in the front, above the orbits.

Frugivore – an animal which specializes in eating fruit.

Fundamental number – the total number of chromosome arms in the karyotype.

G

Galagids – a diverse group of 3-4 African genera (*Galago*, *Otolemur*, *Euoticus* and *Galagoides*) and about 10 species generally known as bushbabies.

Gallery forest – riverside forest; forest growing along rivers, sometimes flanked laterally by extensive savannas.

Gene pool – the totality of all the genetic variation (all alleles) in a population.

Genetic drift – changes of genetic frequencies due to sampling errors.

Genetic flow – movement of alleles through populations.

Genetic heterogeneity – much variability in genetic constitution.

Genital display - use of genitals usually by prominently displaying them to convey a social message.

Genus – a taxonomic category under which species are classified.

Glandular field - an area where many glands are concentrated, such as in the pubic area of *Saguinus* or the chest area in *Pithecia*, *Callicebus*, etc.

Glenoid fossae – where the mandibular condyle articulates with the cranium (syn. mandibular fossae).

Gracile – graceful.

Gular gland –throat gland.

Gular sack – inflatable sack on the throat.

H

Habitat - the locality, site and particular type of local environment occupied by an organism.

Hallux - the thumb.

Haplorhini – suborder of Primates including the tarsioids, catarrhine and platyrrhines.

Haplotype – allelic composition for several different loci on a chromosome, e.g. $A_1B_3C_2$.

Head-body length – usual measurement of a specimen that is to be prepared and included in a museum collection.

Hemiephiphyte - a plant that spends only part of its life cycle as an epiphyte, producing both aerial and subterranean roots at different times.

High caatinga – tall forest growing on white sand.

Holotype - in taxonomy, the single specimen designated or indicated as the type specimen of a nominal species by the original author at the time of publication or the single specimen when no type was specified but only one specimen was present.

Home range - the area utilized by an animal or group of animals for normal travel and feeding throughout the year.

Homiothermic – maintain essentially the same body temperature.

Homolateral – on one side.

Hybrid (hybridization) – the offspring of two plants or animals from two different species, races, varities, subspecies or genera.

Hygrophytic forest – a forest such as mangroves, which does well in salty conditions.

Hymenopterans – insects of the Order Hymenoptera, including ants, wasps, bees and others.

Hypertrophy – exaggerated development of some characteristic or plant/body part.

Hypocone - a molar cusp that is distal to the protocone.

I

Igapó – generally forests which become flooded by **blackwater** rivers; as opposed to **várzeas**, which are forests which are flooded by **whitewater** rivers.

Immigration – the migrations of animals from other, more distant populations or places to the immediate population.

Incisors - any of the front teeth between the canines in either jaw.

Indriids – group of Madagascar primates of 3 genera (*Propithecus*, *Indri* and *Avahi*) known as sifaka, indris and woolly lemurs and specialized as leapers.

Interscapular crest – a crest of hair between the two shoulders.

Interscapular whorl – a whorl of hair between the two shoulders.

Interspecific – between species as in competition.

Interorbital foramen - holes connecting the two orbits, found in some primates.

Intertropical convergence zone - a zone or belt of unstable air where the northern and southern tradewinds converge, causing precipitation.

Intraspecific – within the species as in competition.

Inungulate - without claws or nails.

Isothermic values - a line on which every point represents the same temperature.

IUCN – International Union for the Conservation of Nature

K

Karyology – or or pertaining to chromosomes.

Karyotype – the form, structure and number of chromosomes of any individual or species.

Key species - a species in a community that is highly important for the maintenance of other species.

L

Lacrimal bones - small, thin bony forming the front part of the inner wall of each orbit.

La Venta – fosiliferous deposits of upper Magdalena, Huila which have yielded about half the total fossils of Neotropical primates known, limited to the Middle-Miocene of 12-13 Ma.

La Victoria – name of geological formation proposed by Guerrero (1997) for the lower unit of the Honda group in the upper Magdalena valley

(Colombia) of the so-called La Venta fossil beds and dated 13.5-12.9 Ma.

Lemurid – lemurs of the family Lemuridae comprised of 5 genera (*Lemur, Eulemur, Varecia, Pachylemur* and *Hapalemur*).

Lemurioid - like lemurs.

Lemurs - small primates, especially of Madagascar but some in Africa and Asia; the Lemuridae of Madagascar are usually diurnal and group-living but others are solitary and nocturnal.

Lepimurids – primates belonging to the family Lepimuridae of two genera (*Lepimur* and *Megaladapis*) and about ten species, known as the sportive or weasel lemurs.

Leptidopterans – insects of the Order Leptidoptera and including butterflies and months among others.

Lingually – the side nearest the tongue.

Long calls - a type of call used by most neotropical primates for communication over great distance. the call is usually strident and carries far.

Lordic position - peculiar position taken by some sexually receptive female mammals (e.g. *Callicebus*) with rear end raised and tail raised and to one side prepared for copulation.

Lorisoid - loris-like.

Loris - either of two kinds of small, slow-moving Asiatic lemurs that live in trees and are nocturnal.

Lunate - the second bone from the thumb side of the proximal row of bones of the carpus.

M

Malar - of the cheek or cheekbone.

Mammae - the glands for secreting milk, present in all female mammals.

Mandibular condyle – a rounded protuberance on the mandible which articulates with the mandibular fossae.

Mangrove – salt-loving (halophytic) plants found along many coasts.

Mantle – longer hair (fur) over shoulders and upper chest and back.

Maxillary – or or pertaining to the jaw

Medial plate - part of pterygoid process and usually smaller than the lateral plate.

Melanin – pigment that is dark and which colors skin and hair.

Meridional (meridian) - of the south.

Meristem – meristematic, undifferentiated tissue at base of developing palm leaf or other structures.

Mesic – very moist.

Metacarpal – a bone between the wrist (carpal) and the fingers.

Metatarsal – a bone between the ankle (tarsus) and the toes.

Mesocarp – the middle layer of pericarp, as in the fleshy part of certain fruits.

Mid-saggital crest – a protuberance of the skull along the mid-line of some male primates where strong masseter muscles are inserted.

Miocene - a geological epoch within the Tertiary period (ca. 26-5 million years B.P.).

Molar - teeth adapted for grinding.

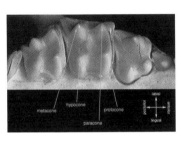

Montane – forest between approximately 1000-3400 m in the Cordillea de los Andes.

Monogamy (monogamous) – having only one mate at a time.

Monophyletic – various species with the same ancestor.

Monotypic genus – a genus which contains only one species.

Morphology – the form and structure of plants and animals.

Morphometrics – the measurement and statistical manipulation of the various morphological parts of an organism.

N

Nares – the openings of the nose through which air is taken.

Neotropical – corresponding to the American (North, Central and South) tropical region.

Neotype – the establishment of an official description for an organism, when the original description is for some reason inadequate.

Niche – the position or function of an organism in a community of plants and animals.

Nominotypic – the subspecies has the same name as the second name of the species (ex. *Cebus albifrons albifrons*).

Nuchal band - contrasting band across the nape (=back of the neck).

Nuclear area – area of home range where most activitities take place.

O

Obligate frugivore - adapted for feeding primarily on fruits.

Occlusal surface - the part of the teeth where the upper and lower teeth come into contact.

Occlusion – the act of closing the mouth and bringing the teeth from both mandibles together.

Ochraeous – a tone of yellow.

Oligocene - a geological epoch within the Tertiary period (ca. 38-26 million years B.P.).

Omnivore – an animal which has a very broad diet, including many plants and animals.

Onomatopoeic – names which mimic sounds made by the referent.

Orbits (orbitals) - the bony cavities containing the eyes.

Orthopterans – insects from the Order Orthoptera, comprising grasshoppers, katydids, crickets, etc.

Ottic bones – the bones of the inner ear.

P

Palmigrade - walking on the palm of the hand.

Palinology – the study of ancient deposits of fossil pollen to determine ancient floras.

Paracone – one of the main cusps of the upper molar, located on the cheek (buccal) side of the tooth.

Paraconule - small molar cusps adjacent to the protocone cusp.

Parapatric - two species or populations which exist side by side with little or no overlap.

Parapithecid - earliest and most primitive higher primates from Oligocene Epoch El Fayum deposits of Egypt and showing a dental formula 2/2, 1/1, 3/3, 3/3 and having several other similarities to New World primates.

Paratype - in taxonomy, any specimen from the type series remaining after the designation of the holotype or isotype.

Parietal bone – either of a pair of membrane bones forming, by their union at the sagittal suture part of the sides and top of the skull.

Pedal digits - toes.

Pentadactyl - five fingers and toes.

Penile display - the use of the erect and visible penis as a social signal, as in *Saimiri, Pithecia, Cebus, Lagothrix* and many other Neotropical primates.

Penis bone - small bone in penis of most primates, although secondarily lost in some primates like humans and woolly monkeys (syn. baculum).

Perarid - in the Holdrige system, tropical and subtropical ecosystems exhibiting extremely low precipitation characteristics of some extreme deserts.

Pericarp – the walls of a ripened ovary or fruit sometimes consisting of three layers (epicarp, mesocarp and endocarp).

Perinium - area between the genitalia and the anus.

Petrosal bone - of or pertaining to the petrous portion of the temporal bone; petrous portion is the hard dense portion of the temporal bone containing internal auditory organs.

Phenology – study of the cycles of plants and animals.

Phenotype - the physical appearance of an individual, such as the color, size, shape, etc.

Pheomelanin – type of melanin which produces reds and yellows in skin and pellage.

Pheromone – volatile chemical substance often produced by glands, used for intraspecific communication.

Phylogenetic – evolutionary connectiveness.

Philopatric – an animal which remains in a particular area and especially to the area where it was born, without emmigrating from it.

Phylogenetic position - evolutionary relationship of one organism to other organisms.

Phylogeny - the evolutionary or geneological relationships among a group of organisms.

Piedmont – lying along or near the foot of a mountain range.

Pilo-erection - the erection or fluffing out of hair, usually when an animal is upset, frightened or angry, conveying an impression of a larger animal.

Pioneer species – species which first colonize an area which has been drastically altered or destroyed.

Pisiform bone - the pealike bone on the ulnar surface of the carpus.

Placenta - a vascular organ within the uterus, connected to the fetus by the umbilical cord and serving as a structure through which the fetus receives nourishment from and eliminates waste matter into the circulatory system of the mother.

Plantigrade - walking on the entire sole of the foot..

Platyrrhines - primates belonging to Platyrrhini, comprising the Neotropic monkeys and having a broad, flat nose.

Plesiadapiform - primates of the order Plesiadapiform from North American Paleocene/Eocene; a species-rich radiation that probably did not give rise to South American primates.

Pollux – thumb.

Polyandry - social system with two or more sexually active males and a single sexually active female.

Polygamy – mating with more than one female.

Poliploidy – the development of chromosome numbers by duplication of the basic haploid number.

Politypic – having several types.

Population crash - a sharp decrease in population from high numbers to low numbers.

Postorbital process - an elongation of a bony process behind the eye.

Prehensile – adapted for seizing, grasping or talking hold of as the tails of some Platyrrhine monkeys.

Primary forest – climax vegetation or vegetation at its most mature phase.

Primitive - character (s) that are more similar to characters of the earlier precursors or ancestors of an organism.

Procumbent - inclined forward, protruding as in some primate incisors such as *Pithecia* and *Cacajao*.

Prosimians - includes all primates that are not Anthropoids, except for tarsiers. a polyphyletic term generally falling into disuse.

Proximal - in anatomy, situated nearest the center of the body or the point of attachment of a limb, etc.

Pterygoid process - a process each side of the sphenoid bone, consisting of two plates (the lateral and the medial) separated by a notch.

R

Reintroduction – process of re-adapting animals to their natural habitat after they have been in captivity.

Relict – a remnant.

Robersonian translocation - the exchange of genetic material between two chromosomes.

S

Saddle - area over the lower back.

Saltatorial - characteristic or adapted for leaping.

Salt lick – area where animals go to consume concentrated soil chemicals such as sodium and phosphorus which naturally are in short supply.

Sampling error – change in frequency (for example of alleles) due to chance.

Satelite adults – males which remain in the periphery of the group, which is focused on females and dominant males.

Scaphoid - bone at the radial end of the proximal row of the bones of the carpus; bone in front of the talus or ankle (syn. navicular).

Sclerophyl – typical type of vegetation in dry environments, small leaves with waxy coverings, spines, fleshy stems, adapted to little precipitation.

Secondary vegetation – vegetation that is recuperating after having been altered in some way such as being cut.

Sensu – according to.

Sexual dimorphism – the phenomenon of different sizes for each sex; many times the male is larger than the female, although sometimes the female is larger than the male.

Sibling species - morphologically similar or identical populations that are reproductively isolated.

Spathe – bract or pair of bracts, aften large and colored and enclosing a flower cluster.

Species - related organisms which interbreed and produce fertile offspring or are of the same evolutionary lineage.

Sphenioid - pertaining to the compound bone of the base of the skull, at the roof of the pharynx.

Sphenopalatine suture - line of junction between the sphenoid and the palate.

Staccato – musical notes or vocalizations which are played or vocalized in a way that each note or vocalization is disconnected from those the proceed and follow it.

Stenothermic – existing in or preferring a narrow range of temperatures.

Sternal - of or near the sternum, a thin, flat structure of bone and cartilage to which most of the ribs are attached in the front of the chest in most vertebrates.

Strepsirhini – suborder of the more primitive primates, consisting of lemurids, cheirogaleids, doubentonioids, lepilemurids, indriids, lorisids and galagids.

Sublingua - under the tongue.

Submalar stripe – stripe on the side of the head under the malar.

Subspecies - race or division of species based on any readily recognized characteristics which, taken together, distinguish one population from another population of the same species.

Subxerophytic – conditions that are semi-arid with the evapotranspiration of the plants exceeding the total precipitation.

Superhumid - in the Holdridge system, an ecosystem with extremely high precipitation.

Supraorbital region - above the orbits or eye sockets.

Suprapubic - above the pubic area.

Sympatric speciation – posited speciation of one population from another while occupying the same geographic area.

Sympatry - two species, populations or taxa occur together in the same geographic area; the populations may occupy the same habitat (biotic sympatry) or different habitats (neighboring sympatry).

Symphysis - the growing together of bones that were originally separate.

Synecological – pertaining to an ecological grouping of several species.

Synonym – a second taxonomic designation for the same species (another name designating the same animal).

Systematics - the ordering or classifying of organisms into a system that reflects their relatedness and their evolution.

T

Talar – of the talus.

Talus - the anklebone or the entire ankle.

Tapetum lucidum - reflective structure in choroid at back of eye which improves night vision by reflecting light back to the retina; found in all nocturnal primates except *Aotus*.

Tarsal – bone of the foot.

Tarsioid - tarsier-like.

Tarsier - small prosimians from East Indies and Philipines, related to lemurs; nocturnal and insect-eating.

Taxon – designates a taxonomic category at any level.

Taxonomic - of taxonomy or the science of describing and naming organisms using a formalized and internationally accepted system.

Tectonics - the forces or conditions within the earth that cause movements and changes in the earth's crust.

Tegula(e) – claw of callitrichines.

Temporal stripe – stripe along the sides of the head.

Terra firme - uplands, not inundated as opposed to **varzea**, **igapó**, etc.

Territory - a home range that is defended from other conspecifics.

Thermic tolerance - the degree to which an organism can tolerate a range (highs or, in the tropics, esp. lows) of temperatures.

Time budget - syn. activity budget

Tongue display - (syn. tongue flick in some cases); tongue may be protruded straight out, waved and wiggled or may be bent over the lower lip and jaw (*Cebus albifrons*). often such a display is in the context of aggression or friendliness (sexual).

Tongue flick - rapid movement of tongue as a display, which may be thrust out of the mouth as in *Callicebus* or moved up-and-down in the mouth as in *Lagothrix*.

Tooth chatter display - rapid opening and closing of teeth with lips withdrawn to show teeth, often with tongue display as well.

Trans-Andean - a species found on opposite side of the Andes, having been able to disperse past the biogeographic barrier of the mountains from the ancestor species.

Triquetrum – carpal bone on ulnar side of wrist.

Tympanic ring – (syn. ectotympanic) bony element which supports the tympanic membrane.

Type location – place where the holotype of an organism was found and which is the included as part of the official taxonomic description.

U

Ultrasonics – frequencies above the audio-frequency range.

Ungues - Hails of hands and feet.

Urine kick washing - wetting feet with urine and scratching side of body as *Saimiri* when nervous.

Urine washing - animal moistens palms with urine then wipes it on his foot soles with rapid swipes.

V

Varzea - seasonally flooded forest with whitewater.

Vertical clinging and leaping - a type of locomotion and posture where the animals cling to vertical support and leap between vertical supports.

Vicariance – the influence of topographic barriers on the evolution and distribution of organisms.

Vicariant effects – effects due to topographic barriers.

Villavieja – a geological Formation of La Venta fossil beds in Colombia, overlying the La Victoria Formations and dated 12.9-11.5 Ma.

Volcaniclastic – deposits from active volcanos of ashes, lava, etc.

Vulva – external part of female genitals.

W

White sand – podzols or extremely leached soils consisting in the main of white sands with very little fertility and supporting very species poor (although sometimes highly endemic) communities of plants and animals.

Whitewater - sediment-loaded water flowing from Andean Cordillera and with a light chocolate or whitish color and extremely turbid; provides some nutrients to aquatic and adjacent flooded forest and is considered much more fertile than blackwater.

X

Xeric – dry.

Z

Zygomatic arch – the bony arch below the orbit of the skull that is formed by the union of the zygomatic process of the zygomatic bone.

Zoogeography – study of the distributions of animals and how they got that way.

Bibliography

Abbott, D. H., J. Barrett and L. M. George 1993. Comparative aspects of the social supression of reproduction in female marmosets and tamarins. Pp. 152-163 in: A. B. Rylands (ed), *Marmosets and Tamarins: Systematics, Behaviour, and Ecology*. Oxford University Press, Oxford.

Ab'saber, A. N. 1977. Espaços ocupados pela expansãos dos climas secas na américas do sul, por ocasião dos períodos glacais quaternários. *Paleoclimas (Inst. Geogr. Univ. Sao Paulo)* 3:1-19.

Ab'saber, A. N. 1982. The paleoclimate and paleocology of Brazilian Amazonia. Pp. 41-59 in: G. T. Prance (ed.), *Biological Diversification in the Tropics*. Columbia University Press, New York.

Addessi, E. and E. Visalberghi. 2002. Acceptance of novel foods in tufted capuchin monkeys: The role of social influences. [Abstract] *Caring for Primates. Abstracts of the XIX Congress*. The International Primatological Society. Beijing: Mammalogical Society of China pp. 154-155.

Agoramoorthy, G. 1992. Infanticide by adult and subadult males in free-ranging red howler monkeys of Venezuela. presentation at the Nato/Advanced Studies Institute's Conference on the Ethological Roots of Culture, Cortona, Italy, 21 June - 3 July 1992.

Agoramoorthy, G. 1994. An update on the long-term field research on red howler monkeys, *Alouatta seniculus*, at Hato Masaguaral, Venezuela. *Neotropical Primates* 2(3):7-9.

Agoramoorthy, G. 1995. Red howling monkey (*Alouatta seniculus*) reintroduction in a gallery forest of Hato Flores Moradas, Venezuela. *Neotropical Primates* 3(1):9-10.

Agoramoorthy, G. 1997. Apparent feeding associations between *Alouatta seniculus* and *Odocoileus virginianus* in Venezuela. *Mammalia* 61(2):271-2723.

Agoramoorthy, G. 2001. Strategies and counterstrategies of infanticide in red howler monkeys. [Abstract]. *Advances in Ethology* 36:109.

Agoramoorthy, G. and M. J. Hsu 1995. Population status and conservation of red howling monkeys and white-fronted capuchin monkeys in Trinidad. *Folia Primatologica* 64(3):158-162.

Agoramoorthy, G. and R. Rudran 1992. Adoption in free-ranging red howler monkeys, *Alouatta seniculus* of Venezuela. *Primates* 33(4):551-555.

Agoramoorthy, G. and R. Rudran 1993. Male dispersal among free-ranging red howler monkeys in Venezuela. *Folia Primatologica* 61(2):92-96.

Agoramoorthy, G. and R. Rudran 1994. Field application of Telazol (tiletamine hydrochloride and zolazepam hydrochloride) to immobilize wild red howler monkeys (*Alouatta seniculus*) in Venezuela. *Journal of Wildlife Diseases* 30(3):417-420.

Agoramoorthy, G. and R. Rudran 1995. Infanticide by adult and subadult males in free-ranging red howler monkeys, *Alouatta senciculus*, in Venezuela. *Ethology* 99(1):75-88.

Ahumada, J. A. 1989. Behavior and social structure of free-ranging spider monkeys (*Ateles belzebuth*) in La Macarena. *Field Studies of New World Monkeys, La Macarena, Colombia* 2:7-31.

Ahumada, J. A. 1990. Changes in size and composition in a group of spider monkeys at La Macarena (Colombia). *Field Studies of New World Monkeys, La Macarena, Colombia* 4:57-60.

Ahumada, J. A. 1992. Grooming behavior of spider monkeys (*Ateles geoffroyi*) on Barro Colorado Island, Panama. *International Journal of Primatology* 13(1):33-49.

Ahumada, J. A., P. R. Stevenson and M. J. Quiñones 1998. Ecological response of spider monkeys to temporal variation in fruit abundance: The importance of flooded forest as a keystone habitat. *Primate Conservation* 18:10-14.

Alberico, M., A. Cadena, J. Hernández-C. and Y. Muñoz-S. 2000. Mamíferos (Synapsida: Theria) de Colombia/ Mammals of Colombia. *Biota Colombiana* 1(1):43-75.

Alderman, C. I. 1989. A general introduction to primate conservation in Colombia. *Primate Conservation* (10):44-51.

Allen, J. A. 1914. New South American monkeys. *Bulletin of the American Museum of Natural History* 33:647-655.

Alston, E. R. 1879. Mammalia. In: *Biología Centrali-Americana*, vol. 1. with field notes by Leo E. Miller. *Bulletin of the American Museum of Natural History* 35:559-610.

Altmann, S. A. 1959. Field observations on a howling monkey society. *Mammalia* 40:317-330.

Altmann, S. A. 1966. Vocal communication in howling monkeys. 7.5 i.p.s. Tape Library of Natural History Sounds, Laboratory of Ornithology, Cornell University.

Alvarez del Toro, M. 1977. Los mamíferos de Chiapas. Universidad Autonoma de Chiapas, Tuxtla Gutierrez, Chiapas, Mexico.

Ameghino, F. 1891. Nuevos restos de mamíferos fósiles descubiertos por Carlos Ameghino en el Eoceno Inferior de la Patagonia austral. *Revista Argentina Historia Natural* 1:289-328.

Ameghino, R. 1910. *Montaneia anthropomorpha*. Un género de monos hoy extinguido de la Isla de Cuba. Nota Preliminar. *Anales Museu Nacional de Buenos Aires*, s. 33: 317-318.

Andrew, R. J. 1963. The origin and evolution of the calls and facial expressions of the primates. *Behaviour* 20:1-108.

Aquino, R. 1991a. Captura y evaluación de *Aotus vociferans* en el valle del Río Nanay. Anexo 1. Pp. ? in: J. Moro et al. (eds.), *Project on the Reproduction and Conservation of Nonhuman Primates, Iquitos, Peru - Annual Report 1991*. PAHO Veterinary Public Health Program, Iquitos.

Aquino, R. 1991b. Post-captura y evaluación de *Aotus nancymai* en el bosque inundado del Río Tahuayo 4/17-5/1/91]. Anexo 4 in: J. Moro et al. (eds.), *Project on the Reproductin and Conservation of Nonhuman Primates, iquitos, Peru – Annual Report 1991*. PAHO Veterinary Public Health Program, Iquitos.

Aquino, R. and F. Encarnación 1986a. Characteristics and use of sleeping sites in *Aotus* (Cebidae: Primates) in the Amazon lowlands of Peru. *American Journal of Primatology* 11(4):319-331.

Aquino, R. and F. Encarnación 1986b. Population structure of *Aotus nancymai* (Cebidae: Primates) in the Amazon lowlands of Peru. *American Journal of Primatology* 11(1):1-7.

Aquino, R. and F. Encarnación 1986c. Population densities and geographic distribution of night monkeys (*Aotus nancymai* and *Aotus voiferans*) (Cebidae: Primates) in northeastern Peru. *American Journal of Primatology* 14:375-381.

Aquino, R. and F. Encarnación 1988. Population densities and geographic distribution of night monkeys (*Aotus nancymai* and *Aotus vociferans*) (Cebidae: Primtes) in northeastern Peru. *American Journal of Primatology* 14(4):375-381.

Aquino, R. and F. Encarnación 1994a. Primates of Peru. *Primate Report* 40:1-127.

Aquino, R. and F. Encarnación 1994b. Owl monkey populations in Latin America: Field work and conservation. Pp. 59-95 in: J. F. Baer, R. E. Weller and I. Kakoma (eds.), *Aotus: The Owl Monkey*. Academic Press, San Diego.

Ardila C., J. O. and A. N. Florez 1994. Aspectos de la ecología de un grupo silvestre de *Lagothrix lagothricha lagothricha* (Humboldt, 1812), Primate-Atelidae. Graduation Thesis in Biology, Universidad Nacional de Colombia, Bogotá.

Aquino, R., P. Puertas and F. Encarnacion 1990. Supplemental notes on population parameters of northeastern Peruvian night monkeys, genus Aotus (Cebidae). *American Journal of Primatology* 21:215-221.

Aquino, R., P. Puertas and F. Encarnacion 1993. Effects of cropping on the *Aotus nancymae* population in a forest of Peruvian Amazonia. *Primate Report* 37:31-40.

Bibliography

Arredondo, O. and L. S. Varona 1983. La validez de *Montaneia anthropomorpha* Ameghino, 1910 (Primates Cebidae) *Poeyana, Instituto de Zoologia Academia de Ciencias de Cuba* 255:1–21.

Aureli, R., L. Rebecchini, C. Ser and S. Fiori 2003. Effects of subgroup composition and home-range use on time budgets of wild spider monkeys in the Yucatan Peninsula. [Abstract] *American Journal of Primatology* 60(Suppl 1):141.

Ayres, J. M. 1985. On a new species of squirrel monkey, genus *Saimiri*, from the Brazilian Amazonia (Primates Cebidae). *Papeis Avulsos Zoologia Sau Paulo* 36(14):147-164.

Ayres, J. M. 1986. Uakaris and Amazonian flooded forest. Unpublished Ph.D. Thesis, University of Cambridge, Cambridge, England.

Ayres, J. M. C. and T. H. Clutton-Brock 1992. River boundaries and species range size in Amazonian primates. *American Naturalist* 140(3):531-537.

Bailey, R. C., R. S. Baker, D. S. Brown, P. von Hildebrandt, R. A. Mittermeier, L. E. Sponsel and K. E. Wolf 1974. Progress of a breeding project for non-human primates in Colombia. *Nature* 248:453-455.

Baker, M. 1996. Fur rubbing: use of medicinal plants by capuchin monkeys (*Cebus capucinus*). *American Journal of Primatology* 38(3):263-270.

Baker, M. 1999. Fur rubbing as evidence for medicinal plant use by capuchin monkeys (*Cebus capucinus*): Ecological, social, and cognitive aspects of the behavior. Unpublished Ph.D. thesis, Dissertation Abstracts International A59(10):3870.

Baldwin, J. D. 1967. A study of the social behavior of a semifree-ranging colony of squirrel monkeys. Unpublished Ph.D. thesis, Jowns Hopkins University.

Baldwin, J. D. 1968. The social behavior of adult male squirrel monkeys (*Saimiri sciureus*) in a seminatural environment. *Folia Primatologica* 9:281-314.

Baldwin, J. D. 1969. The ontogeny of social behavior of squirrel monkeys (*Saimiri sciureus*) in a seminatural environment. *Folia Primatologica* 11:35-79.

Baldwin, J. D. 1970. Reproductive synchronization in squirrel monkeys (*Saimiri*). *Primates* 11:317-326.

Baldwin, J. D. 1971. The social organization of a semifree-ranging troop of squirrel monkeys (*Saimiri sciureus*). *Folia Primatologica* 14:23-50.

Baldwin, J. D. 1985. The behavior of squirrel monkeys (*Saimiri*) in natural environments. Pp. 35-53 in: L. A. Rosenblum and C. L. Coe (eds.), *Handbook of Squirrel Monkey Research*. Plenum Press, New York.

Baldwin, J. D. 1992. Determinants of aggression in squirrel monkeys (*Saimiri*). Pp. 72-99 in: J. Silverberg and J. P. Gray (eds.), *Aggression and Peacefulness in Humans and Other Primates*. Oxford University Press, New York.

Baldwin, J. D. and J. I. Baldwin 1971. Squirrel monkeys (*Saimiri*) in natural habitats in Panama, Colombia, Brazil and Peru. *Primates* 12:45-61.

Baldwin, J. D. and J. I. Baldwin 1972. Population density and use of space in howling monkeys (*Alouatta villosa*) in southwestern Panama. *Primates* 13:371-379.

Baldwin, J. D. and J. I. Baldwin 1973a. Interactions between adult female and infant howling monkeys (*Alouatta palliata*). *Folia Primatologica* 20:27-71.

Baldwin, J. D. and J. I. Baldwin 1973b. The role of play in social organizations: Comparative observations on squirrel monkeys (*Saimiri*). *Primates* 14:369-81.

Baldwin, J. D. and J. I. Baldwin 1974. Warum brüllen brülaffen? *Umschau* 22:712-713.

Baldwin, J. D. and J. I. Baldwin 1976. Vocalizations of howler monkeys (*Alouatta palliata*) in southwestern Panama. *Folia Primatologica* 26:81-108.

Baldwin, J. D. and J. I. Baldwin 1978. Exploration and play in howler monkeys (*Alouatta palliata*). *Primates* 19:411-422.

aldwin, J. D. and J. I. Baldwin 1981. The squirrel monkeys, genus *Saimiri*. Pp. 277-330 in: A. F. Coimbra-Filho and R. A. Mittermeier (eds.). *Ecology & Behavior of Neotropical Primates*, vol. 1. Academia Brasileira de Ciências, Rio de Janeiro.

angs, O. 1905. The vertebrata of Gorgona Island, Colombia. I, introduction; III, Mammalia. *Bulletin Museum of comparative Zoology, Harvard* 46:85-91.

arbosa, C. 1988. Observaciones sobre el comportamiento de una manada del mono *Alouatta seniculus* (Linnaeus, 1766) (Mammalia: Primates) en el Arroyo Colosó, Sucre, Colombia. *Trianea: (Acta Científica y Tecnológica Inderena)* 1:123-130.

arbosa, C., A. Fajardo P. and H. Giraldo 1988. Evaluación del hábitat y status del mono tití de cabeza blanca, *Saguinus oedipus* Linnaeus, 1758, en Colombia. Final Report of the Status of the Cotton-top Tamarin Project, *Saguinus oedipus* en Colombia, 39 pp.

arclay, D., M. Maurus and E. Wiesner 1991. Mutual dependencies between vocal and visual signals of squirrel monkeys. *Primates* 32(3):307-320.

arnett, A. and A. C. da Cunha 1990. A preliminary study of the golden-backed uakari (*C. m. ouakary*). *Primate Eye* (41):13.

arnett, A. and A. C. da Cunha 1991. The golden-backed uacari on the upper Río Negro, Brazil. *Oryx* 25(2):80-88.

arnett, A. A. and D. Brandon-Jones 1995. The ecology and biogeography of *Cacajo*. [Abstract] *Primate Eye* 57:10-11.

arnett, A. and D. Brandon-Jones 1996. Ecological knowledge of *Cacajao*. [Abstract] *Folia Primatologica* 67(2):104.

arnett, A. A. and S. M. Lehman 2000. Ecological and historical correlates to the biogeography of *Cacajao*. [Abstract] American *Journal of Physical Anthropology* (Suppl 30):103.

arnett, A. A., S. Borges and C. V. Castilho 2000. Golden-backed uakari, *Cacajao melanocephalus ouakary*, in Jau National Park, Amazonas, Brazil. *Primate Eye* 70:33-37.

arros, R. M. S., J. C. Pieczarka, M. C. O. Brigido, J. A. P. C. Muniz, L. R. R. Rodrigues and C. Y. Nagamachi 2000. A new karyotype in *Callicebus torquatus* (Cebidae, Primates). *Hereditas* 133(1):55-58.

arroso, C. M. 1995. Molecular phylogeny of the Callitrichinae. *Neotropical Primates* 3(4):186.

arroso, C. M. L., H. Schneider, M. P. C. Schneider, I. Sampaio, M. L. Harada, J. Czelusniak and M. Goodman 1997. Update on the phylogenetic systematics of New World monkeys: Further DNA evidence for placing the pygmy marmoset (*Cebuella*) within the genus *Callithrix. International Journal of Primatology* 18(4):651-674.

artecki, U. and E. W. Heymann 1987. Field observation of snake-mobbing in a group of saddle-back tamarins, *Sguinus fuscicollis nigrifrons*. *Folia Primatologica* 48(3-4):199-202.

artecki, U. and E. W. Heymann 1990. Field observations on scent-marking behaviour in saddle-back tamarins, *Saguinus fuscicollis* (Callitrichidae, Primates). *Journal of Zoology* 220(1):87-99.

astian, ? 1878. *Die Kulturländer des alten Amerikas*. T. i. Berlin, p. 243, note 1.

ender, M. A. and L. E. Mettler 1958. Chromosome studies of primates. *Science* 128:186-190.

enirschke, K. 1975. Biomedical research. Pp. 3-11 in: Institute of Laboratory Animal Resources (ed.) *Research in Zoos and Aquariums*. National Academy of Sciences, Washington, D.C..

enirschke, K. and M. H. Bogart 1976. Chromosomes of the tan-handed titi (*Callicebus torquatus*, Hoffmannsegg, 1807). *Folia Primatologica* 25:25-34.

enirschke, K., L. E. M. de Boer and M. Bogart 1976. The karyotypes of two uakari species *Cacajao calvus* and *C. rubicundus* (Primates: Platyrrhini). *Genen en Phaenen* 19:1-6.

Bibliography

Benitez-Rodriguez, J., J. Jimenez-Huerta, G. Silva-Lopez, A. G. Christen and V. M. Castillo1993. (Studie of *Ateles* and *Alouatta* in the center for biological investigations at Veracruz University). *Estudio Primatológicas en Mexico* 1:119-127 (in Spanish).

Bennett, C. L., S. Leonard and S. Carter 2001. Abundance, diversity, and patterns of distribution o primates on the Tapiche River in Amazonian Peru. *American Journal of Primatology* 54(2):119-126.

Bennett-Defler, S. and T. R. Defler 1998. Los primates de Amazonas. *Cambio* (oct. 31, 1998):18-21.

Benson, W. W. 1982. Alternative models for infrageneric diversification in the humid tropics: Tests with passion flower vine butterflies. Pp. 608-640 in: G. T. Prance (ed.), *Biological Diversification in the Tropics* Columbia University Press, New York.

Benzanson, M. P. A. Garber and J. Rutherford 2002. Patterns of subgrouping, social affiliation and socia networks in Nicaraguan mantled howler monkeys (*Alouatta palliata*). *American Journal of Physical Anthropolog* Suppl. 34:54.

Bergeson, D. J. 1996. The ecological role of the platyrrhine prehensile tail. [Abstract] *American Journal o Physical Anthropology* (Suppl. 20):64-65.

Bergeson, D. J. 1997. The positional behavior and prehensile tail use of *Alouatta palliata*, *Ateles geoffroyi*, anc *Cebus capucinus*. Unpublished Ph.D. thesis. *Dissertation Abstracts International* A57(9):4011.

Bergeson, D.J. 1998. Patterns of suspensory feeding in *Alouatta palliata*, *Ateles geoffroyi*, and *Cebus capucinus* Pp. 45-60 in: E. Strasser, J. Fleagle, A. Rosenberger and H. McHenry (eds.), *Primate Locomotion: Recen Advances*. Plenum Press, New York.

Bernstein, I. S. 1964. A field study of the activities of howler monkeys. *Animal Behavior* 12:92-97.

Bernstein, I. S., P. Balcaen , L. Dresdale, H. Gouzoules, M. Kavanagh, T. Patterson and P. Newman-Warner 1976a. Differential effects of forest degradation on primate populations. *Primates* 17:401-411.

Bernstein, I. S., P. Balcaen, L. Dresdale, H. Gouzoules, M. Kavanagh, T. Petterson and P. Newman-Warner 1976b. An appeal for the preservation of habitats in the interests of primate conservation. *Primate.* 17(3):413-415.

Biben, M. and D. Symmes 1986. Play vocalizations of squirrel monkeys (*Saimiri sciureus*). *Folia Primatologice* 46:173-182.

Bicca-Marques, J. C. 2000a. Cognitive aspects of within-patch decisions in wild *Saguinus imperator*, *Saguinu fuscicollis*, *Callicebus cupreus*, and *Aotus nigriceps*. *Neotropical Primates* 8(1):52.

Bicca-Marques, J. C. 2000b. Cognitive aspects of within-patch foraging decisions in wild diurnal anc nocturnal New World monkeys (*Saguinus imperator*, *Saguinus fuscicollis*, *Callicebus cupreus*, *Aotus nigriceps* Brazil). Unpublished Ph.D. thesis. *Dissertation Abstracts International* A61(1):249.

Bicca-Marques, J.C. 2003a. How do howler monkeys cope with habitat fragmentation? Pp. 283-303 in: L K. Marsh (ed.), *Primates in Fragments: Ecology and Conservation*. New York: Kluwer Academic/Plenum Publ

Bicca-Marques, J. C. 2003b. The win-stay rule in whithin-patch foraging decisions in free-tanging tit monkeys (*Callicebus cupreus cupreus*) and tamarins (*Saguinus imperator imperator* and *S. fuscicollis weddelli*) [Abstract] *American Journal of Physical Anthropology* Suppl 36:66.

Bicca-Marques, J. C. and P. A. Garber 2002. The use of visual, olfactory, and spatial information during foraging in wild nocturnal and diurnal anthropoids: A comparison among *Aotus*, *Callicebus*, and *Saguinus* [Abstract] *American Journal of Physical Anthropology* (Suppl 34):45.

Bicca-Marques, J. C. and P. A. Garber 2003. Experimental field study of the relative costs and benefits tc wild tamarins (*Saguinus imperator* and *S. fuscicollis*) of exploiting contestable food patches as single- anc mixed-species troops. *American Journal of Primatology* 60(4):139-153.

Bicca-Marques, J. C., C. A. Nunes and K. Schacht 1998. Preliminary observation on handedness in wild tamarins (*Saguinus* spp.) and titi monkeys (*Callicebus cupreus*). *Neotropical Primates* 6(3):88-90.

Bicca-Marques, J. C., P. A. Garber and M. A. O. Azevedo-Lopex 2002. Evidence of three resident adult male group members in a species of monogamous primate, the red titi monkey (*Callicebus cupreus*). *Mammalia* 66(1):138-142.

Blumer, E. S. and G. Epple unpublished manuscript. *Saguinus leucopus*: notes on its behavior and vocal repertoire. 29 pp.

Bluntschli, H. 1931. *Homunculus patagonicus* und die ihm zugereibten Fossil funde an den Santa-Cruz-schichten Patagoniens: eine morphologische Revision an Hand der original Stücke in der Sammlung Ameghino zu la Plata. *Gegenbaurs Morpho. Jahrbuch* 67(2):811-892.

Boher-Benttii, S. and G. A. Cordero-Rodriguez 2000. Distribution of brown capuchin monkeys (*Cebus apella*) in Venezuela: A piece of the puzzle. *Neotropical Primates* 8(4):152-153.

Boinski, S. 1986. The ecology of squirrel monkeys in Costa Rica. Unpublished Ph.D. thesis. University of Texas.

Boinski, S. 1987a. Mating patterns in squirrel monkeys (*Saimiri oerstedi*): Implications for seasonal sexual dimorphism. *Behavioral Ecology and Sociobiology* 21(1):12-21.

Boinski, S. 1987b. Birth synchrony in squirrel monkeys (*Saimiri oerstedi*): A strategy to reduce neonatal predation. *Behavioral Ecology and Sociobiology* 21:6:383-400.

Boinski, S. 1988a. Sex differences in the foraging behavior of squirrel monkeys in a seasonal habitat. *Behavioral Ecology and Sociobiology* 23:177-186.

Boinski, S. 1988b. Use of a club by a wild white-faced capuchin (*Cebus capucinus*) to attack a venomous snake (*Bothrops asper*). *American Journal of Primatology* 14(2):177-179.

Boinski, S. 1989. The positional behavior and substrate use of squirrel monkeys: Ecological implications. *Journal of Human Evolution* 18:659-677.

Boinski, S. 1991. The coordination of spatial position: A field study of the vocal behaviour of adult female squirrel monkeys. *Animal Behaviour* 41(1):89-102.

Boinski, S. 1992a. Monkeys with inflated sex. *Natural History* 101(7):42-49.

Boinski, S. 1992b. Olfactory communication among Costa Rican squirrel monkeys: A field study. *Folia Primatologica* 59(3):127-136.

Boinski, S. 1993. Vocal coordination of troop movement among white-faced capuchin monkeys, *Cebus capucinus*. *American Journal of Primatology* 30(2):85-100.

Boinski, S. 1994. Affiliative patterns among male Costa Rican squirrel monkeys. *Behaviour* 130(3-4):191-209.

Boinski, S. 1996a. The huh vocalization of white-faced capuchins: A spacing call disguised as a food call? *Ethology* 102(10):826-840.

Boinski, S. 1996b. Vocal coordination of troop movement in squirrel monkeys (*Saimiri oerstedi* and *S. sciureus*) and white-faced capuchins (*Cebus capucinus*). Pp. 251-269 & 541-542 in: M. A. Norconk, A. L. Rosenberger and P. A. Garber (eds.), *Adaptive Radiations of Neotropical Primates*. Plenum Press, New York.

Boinski, S. 1999a. The social behavior of wild *Saimiri sciureus sciureus* in Suriname. [Abstract] *American Journal of Physical Anthropology* Suppl 28:94.

Boinski, S. 1999b. Geographic variation in behavior of a primate taxon: Stress responses as a proximate mechanism in the evolution of social behavior. Pp. 95-120 in: S. A. Foster and J. A. Endler (eds.), *Geographic Variation in Behavior: Perspectives on Evolutionary Mechanisms*. Oxford University Press, New York.

Boinski, S. 1999c. The social organizations of squirrel monkeys: Implications for ecological models of social evolution. *Evolutionary Anthropology* 8(3):101-112.

Boinski, S. 1999d. Comparison of the life history patterns of three squirrel monkey species: Ecological bases and phylogenetic constraints. [Abstract] *Primate Report* (Special issue 54-1):13-14.

Boinski, S. 2000. Social manipulation within and between troops mediates primate group movement. Pp. 421-469 in: S. Boinski and P. A. Garber (eds.), *On the Move: How and Why Animals Travel in Groups*. University of Chicago Press, Chicago.

Bibliography

Boinski, S. and R. M. Timm 1985. Predation by squirrel monkeys and double-toothed kites on tent-making bats. *American Journal of Primatology* 9(2):121-127.

Boinski, S. and J. D. Newman 1988. Preliminary observations on squirrel monkey (*Saimiri oerstedi*) vocalization in Costa Rica. *American Journal of Primatology* 14(4):329-343.

Boinski, S. and C. L. Mitchell 1992. Ecological and social factors affecting the vocal behavior of adult female squirrel monkeys. *Ethology* 92(4):316-330.

Boinski, S. and A. F. Campbell 1995. Use of trill vocalizations to coordinate troop movement among white faced capuchins: A second field test. *Behaviour* 132(11-12):875-901.

Boinski, S. and C. L. Mitchell 1995. Wild squirrel monkey (*Saimiri sciureus*) caregiver calls: contexts and acoustic structure. *American Journal of Primatology* 35(2):129-137.

Boinski, S. and A. F. Campbell 1996. The huh vocalization of white-faced capuchins: A spacing cal disguised as a food call? *Ethology* 102(10):826-840.

Boinski, S. and C. L. Mitchell 1997. Chuck vocalizations of wild female squirrel monkeys (*Saimiri sciureus*) contain information on caller identity and foraging activity. *International Journal of Primatology* 18(6):975-993.

Boinski, S. and S. J. Cropp 1999. Disparate data sets resolve squirrel monkey (*Saimiri*) taxonomy: Implications for behavioral ecology and biomedical usage. *International Journal of Primatology* 20(2):237-256.

Boinski, S. and P. A. Garber (eds.) 2000. *On the Move: How and Why Animals Travel in Groups.* The University of Chicago Press, Chicago, 811 pp.

Boinski, S., R. P. Quatrone and H. Swartz 2000. Substrate and tool use by brown capuchins in Suriname: Ecological contexts and cognitive bases. *American Anthropologist* 102(4):741-761.

Boinski, S., K. Sughrue, L. Selvaggi, R. Quatrone, M. Henry and S. Cropp 2002. An expanded test of the ecological model of primate social evolution: Competitive regimes and female bonding in three species of squirrel monkeys (*Saimiri oerstedi, S. boliviensis,* and *S. sciureus*). *Behaviour* 139(2-3):227-261.

Boinski, S., R. P. Quatrone, K. Sughrue, L. Selvaggi, M. Henry, C. M. Stickler and L. M. Rose 2003. Do brown capuchins socially learn foraging skills? Pp. 365-390, in: D. M. Fragaszy and S. Perry (eds.), *The Biology of Traditions: Models and Evidence.* Cambridge University Press, New York.

Bolen, R. A. 1999. The use of olfactory cues, spatial memory, and social information in foraging by capuchin monkeys (*Cebus apella*) and owl monkeys (*Aotus nancymae*). Unpublished Ph.D. thesis. *Dissertation Abstracts International* B59(12):6165.

Bonvicino, C. R., M. E. B. Fernández and H. N. Seuanez 1995. Morphological analyses of *Alouatta seniculus* species group (Primates, Cebidae). A comparison with biochemical and karyological data. *Human Evolution* 10(2):169-176.

Bonvicino, C. R., V. Penna-Firme, F. F. do Nascimento, B. Lemos, R. Stanyon and H. H. Seuánez 2003. The lowest diploid number (2n=16) yet found in any primate: *Callicebus lugens* (Humboldt, 1811). *Folia Primatologica* 74:141-149.

Boubli, J. P. 1993. Southern expansion of the geographical distribution of *Cacajao melanocephalus melanocephalus. International Journal of Primatology* 14(6):933-937.

Boubli, J. P. 1994. The black uakari monkey in the Pico da Neblina National Park. *Neotropical Primates* 2(3):11-12.

Boubli, J. P. 1997a. A study of the black uakari, *Cacajao melanocephalus melanocephalus*, in the Pico da Neblina National Park, Brazil. *Neotropical Primates* 5(4):113-115.

Boubli, J. P. 1997b. Ecology of the black uakari monkey *Cacajao melanocephalus melanocephalus* in Pico da Neblina National Park, Brazil. Unpublished Ph.D. thesis, University of California, Berkeley.

Boubli, J. P. 1998. Dietary ecology of Humboldt's black uakari, *Cacajao melanocephalus melanocephalus,* in Pico da Neblina National Park, Brazil. [Abstract] *Congress of the International Primatological Society, Abstracs* 17:391.

Bibliography

Boubli, J. P. 1999a. Feeding ecology of one group of Humboldt's black uakari (*Cacajao melanocephalus melanocephalus*) in a forest on white-sand soils of Pico da Neblina National Park, Brazil. *International Journal of Primatology* 20(5):719-749.

Boubli, J. P. and A. D. Ditchfield 2000. The time of divergence between the two species of uacari monkeys: *Cacajao calvus* and *Cacajao melanocephalus*. *Folia Primatologica* 71(6):387-391.

Bramblett, C. A., S. S. Bramblett, A. M. Coelho and L. B. Quick 1980. Party composition in spider monkeys of Tikal, Guatamala: A comparison of stationary vs. moving observers. *Primates* 21:123-127.

Brand, H. M. 1981a. Urinary oestrogen excretion in the female cotton-topped tamarin (*Saguinus oedipus oedipus*). *Journal of Reproductive Fertility* 62:467-473.

Brand, H. M. 1981b. Husbandry and breeding of a newly-established colony of cotton-topped tamarins (*Saguinus oedipus oedipus*). *Laboratory Animal Newsletter* 15:7-11.

Braza, F. 1978. El araguato rojo (*Alouatta seniculus*). Unpublished Ph.D. thesis, Universidad de Seville, Seville.

Braza, F. 1980. El araguato rojo (*Alouatta seniculus*). *Donana Acta Vertebrata Número Especial* 7(5):1-175.

Braza, F., T. Azcarate and F. Alvarez 1980. Parámetros del tamaño del grupo en el araguato rojo (*Alouatta seniculus*) en los llanos de Venezuela. I Reunión Iberoamericana de Zoólogos de Vertebrados, la Rabida (Huelva), Estación Biológica de Donana, España, pp. 315-323.

Braza, F., F. Alvarez and T. Azcarete 1981. Behavior of the red howler monkey (*Alouatta seniculus*) in the llanos of Venezuela. *Primates* 22:459-73.

Braza, F., F. Alvarez and T. Azcarete 1983. Feeding habits of the red howler monkeys (*Alouatta seniculus*) in the llanos of Venezuela. *Mammalia* 47:205-214.

Brown, K. S. 1977a. Geographical patterns of evolution in neotropical lepidoptera. systematics and derivation of known and new Heliconiini (Nymphalidae: Nymphalinae). *Journal of Entomology* (b) 44:(3):201-242.

Brown, K. S. 1977b. Centros de evolução, refúgios quaternários e conservação de patrimônios genéticos na região neotropical: padrões de diferenciação em Ithomiinae (Lepidoptera: Nymphalinae). *Acta Amazônica* 7(1):75-137.

Brown, K. S. 1979. Ecología geográfica e evolução nas florestas neotropicas. Parte vi na serie padrões geográficas de evolução em lepidópteros neotropicais. Unpublished Ph.D. thesis, Universidade Estadual de Campinas, Brazil.

Brown, K. S. 1982. Paleoecology and regional patterns of evolution in neotropical forest butterflies. Pp. 255-308 in: G. T. Prance (ed.), *Biological Diversification in the Tropics.* Columbia University Press, New York.

Brown, K. S. 1987a. Biogeography and evolution of neotropical butterflies. Pp. 66-104 in: T. C. Whitmore and G. T. Prance (eds.), *Biogeography and Quaternary History in Tropical America.* Clarendon Press, Oxford.

Brown, K. S. 1987b. Conclusions, synthesis, and alternative hypothesis. Pp. 175-196 in: T. C. Whitmore and G. T. Prance, *Biogeography and Quaternary History in Tropical America.* Clarendon Press, Oxford,.

Brown, A. D. 1989. Distribución y conservación de *Cebus apella* (Cebidae: Primates) en el noroeste argentino. Pp. 159-166 in: C. J. Saavedra, R. A. Mittermeier and I. Bastos-santos (eds.), *La Primatología en Latinoamérica.* WWF-U.S., Washington, D.C.

Brown, K. S. and Ab'saber, A. N. 1979. Ice-age forest refuges and evolution in the neotropics: Correlation of paleoclimatological, geomorphological and pedological data with modern biological endemism. *Paleoclimas (São Paulo)* no. 5.

Brown, A. D. and O. J. Colillas 1984. Ecología de *Cebus apella*. Pp. 301-312 in: M. Thiago de Mello (ed.), *A Primatologia no Brasil.* Sociedade Brasileira de Primatologia, Brasilia.

Brown, A. D. and G. E. Zunino 1990. Dietary evidence in *Cebus apella* in extreme habitats: evidence for adaptability. *Folia Primatologica* 54:187-195

Bibliography

Brown, K. S., P. M. Sheppard and J. R. G. Turner 1974. Quaternary refugia in tropical America: evidence from race formation in *Heliconius* butterflies. *Processes of the Royal Society of London* B 197:369-378.

Brown, A. D., S. C. Chalukian, L. M. Malmierca and O. J. Colillas 1986. Habitat structure and feeding behavior of *Cebus apella* (Cebidae) in el Rey National Park, Argentina. . Pp. 137-152 in: D. M. Taub and F. A. King (eds.), *Current Perspectives in Primate Social Dynamics*. Van Nostr& Reinhold Co., New York,.

Brumback, R. A. 1973. Two distinctive types of owl monkeys (*Aotus*). *Journal of Medical Primatology* 2:284-289.

Brumback, R. A. 1974. A third species of the owl monkey (*Aotus*). *Journal of Heredity* 65:321-323.

Brumback, R. A., R. D. Staton, S. A. Benjamin and C. M. Lang 1971. The chromosomes of *Aotus trivirgatus* Humboldt, 1812. *Folia Primatologica* 15:264-273.

Buchanan, D. B., Mittermeier, R. A. and van Roosmalen, M. G. M. 1981. The saki monkeys, genus *Pithecia*. Pp. 392-417 in: A. F. Coimbra-Filho and R. A. Mittermeier *eds.), *Ecology and Behavior of Neotropical Primates*. Academia Brasileira de Ciências, Rio de Janeiro.

Buchanan-Smith, H. 1990. Polyspecific association of two tamarin species, *Saguinus labiatus* and *Saguinus fuscicollis*, in Bolivia. *American Journal of Primatology* 22(3):205-214.

Buchanan-Smith, H. 1991. Encounters between neighboring mixed species groups of tamarins in northern Bolivia. *Primate Report* (31):95-99.

Buchanan-Smith, H. M. 1999. Tamarin polyspecific associations: Forest utilization and stability of mixed-species groups. *Primates* 40(1):233-247.

Buchanan-Smith, H. M., S. M. Hardie, C. Caceres and M. J. Prescott 2000. Distribution and forest utilization of *Saguinus* and other primates of the Pando Department, northern Bolivia. *International Journal of Primatology* 21(3):353-379.

Buckley, J. S. 1983. The feeding behavior, social behavior, and ecology of the white-faced monkey, *Cebus capucinus*, at Trujillo, northern Honduras, Central America. Unpublished Ph.D. thesis, University of Texas, Austin.

Burger, J. 2001. Visibility, group size, vigilance, and drinking behavior in coati (*Nasua narica*) and white-faced capuchins (*Cebus capucinus*): Experimental evidence. *Acta Ethologica* 3(2):111-119.

Bush, G. L. 1975. Modes of animal speciation. *Ann. Rev. Ecol. Syst.* 6:339-364.

Bush, G. L., S. M. Case, A. C. Wilson and J. L. Patton 1977. Rapid speciation and chromosomal evolution in mammals. *Process of the National Academy of Sciences U.S.A.* 74:3942-3946.

Cabrera, A. 1957. *Catalogo de los mamíferos de America del Sur*. Editorial Coni, Buenos Aires, i-xxii, 1-308.

Cabrera, J. A. 1994. Ecología y demografía del mono aullador (*Alouatta seniculus*) en un bosque andino bajo, en el Parque Regional Natural Ucumari. Pp. 399-419 in: O. Rangel (ed.), *Ucumarí: Un Caso Típico de la Diversidad Biótica Andino*. Corporación Autónoma Regional de Risaralda.

Cabrera, J. A. 1997. [Changes in play activity in infant and juvenile howler monkeys (*Alouatta seniculus*). *Neotropical Primates* 5(4):108-111 (in Spanish).

Calegaro-Marques, C. , J. C. Bizca-Marques and M. A. de O. Azebedo 1995. Two breeding females in a *Saguinus fuscicollis weddelli* group. *Neotropical Primates* 3(4):183.

Calle, Z. 1992a. A field observation of infant development and social interactions of a wild black-capped capuchin (*Cebus apella*) female and infant at La Macarena (Colombia). *Field Studies of New World Monkeys, La Macarena, Colombia* 4:1-8.

Calle, Z. 1992b. A field study of the social interactions of one year old black-capped capuchins (*Cebus apella*) at La Macarena (Colombia). *Field Studies of New World Monkeys, La Macarena, Colombia* 4:9-26.

Calle, Z. 1992c. Informe de actividades y resultados: censo preliminar y recomendaciones para el manejo de una población natural de *Saguinus leucopus* en la zona de influencia del Proyecto Hidroeléctrico La Miel ii. Unpublished manuscript.

Calouro, A. M., P. A. Garber, A. Stone and W. de Aquino Chaves 2000. Censusing a primate community in Brazil: A multimethod approach. [Abstract] *American Journal of Physical Anthropology* (Suppl 30), p. 117.

Campbell, A. F. 1994. Patterns of home range use by *Ateles geoffroyi* and *Cebus capucinus* at La Selva Biological Station, Northeast Costa Rica. [Abstract] *American Journal of Primatology* 33(3):199-200.

Campbell, C. J. 1999. Fur rubbing behavior in free-ranging black handed spider monkeys (*Ateles geoffroyi*) in Panama. [Abstract] *American Journal of Primatology* 49(1):40.

Campbell, C. J. 2001. The reproductive biology of black-handed spider monkeys (*Ateles geoffroyi*): Integrating behavior and endocrinology. Unpublished Ph.D. thesis, *Dissertation Abstracts International*. A62(1), 229 pp.

Campbell, C. 2002. The influence of a large home range on the social structure of free ranging spider monkeys (*Ateles geoffroyi*) on Barro Colorado Island, Panama. [Abstract] *American Journal of Physical Anthropology* (Suppl 34):51-52.

Campbell, C. M. and D. Frailey 1984. Holocene flooding and species diversity in southwestern Amazonia. *Quat. Res.* 21:369-375.

Campbell, C. J., F. Aureli, C. A. Chapman, G. Ramos-Fernandez G., K. Matthews, S. E. Russo, S. Suarez and L. Vick 2003. Terrestrial behavior of spider monkeys (*Ateles* spp.): A comparative study. [Abstract] *American Journal of Physical Anthropology* (Suppl 36): 74.

Canales-Espinosa, D. 1994. Howler monkeys (*Alouatta palliata*): Clinical evaluation of two groups captured in fragmented habitats.). *Ciencia y el Hombre* 18:71-87.

Canavez, F. C., M. A. Moreira, F. Simon, P. Parham and H. N. Seuanez 1999. Phylogenetic relationships of the Callitrichinae (Platyrrhini, Primates) based on beta2-microglobulin DNA sequences. *American Journal of Primatology* 48(3):225-236.

Candland, D. K., E. S. Blumer and M. D. Mumford. 1980. Urine as a communicator in a New World primate, *Saimiri sciureus. Animal Learning Behavior* 8:468-80.

Cant, J. G. H. 1978. Population surveys of the spider monkey, *Ateles geoffroyi*, at Tikal, Guatamala. *Primates* 19:525-535.

Cant, J. G. H. 1986. Locomotion and feeding postures of spider and howling monkeys: Field study and evolutionary interpretation. *Folia Primatologica* 46(1):1-14.

Cant, J. G. H. 1990. Feeding ecology of spider monkeys (*Ateles geoffroyi*) at Tikal, Guatemala. *Human Evolution* 5(3):269-281.

Cant, J. G. H., D. Youlatos and M. D. Rose 1997. Postural behavior of *Lagothrix lagothricha* and *Ateles belzebuth* in Amazonia Ecuador. [Abstract] *American Journal of Physical Anthropology* (Suppl 24):87-88.

Cant, J. G. H., D. Youlatos and M. D. Rose 2001. Locomotor behavior of *Lagothrix lagothricha* and *Ateles belzebuth* in Yasuni National Park, Ecuador: General patterns and nonsuspensory modes. *Journal of Human Evolution* 41(2):141-166.

Cant, J. G. H., D. Youlatos and M. D. Rose 2003. Suspensory locomotion of *Lagothrix lagothricha* and *Ateles belzebuth* in Yasuni National Park, Ecuador. *Journal of Human Evolution* 44(6):685-699.

Carosi, M. and E. Visalberghi 2002. Analysis of tufted capuchin (*Cebus apella*) courtship and sexual behavior repertoire: Changes throughout the female cycle and female interindividual differences. *American Journal of Physical Anthropology* 118(1):11-24.

Carosi, M., A. E. Ulland and S. J. Suomi 2001. Urine washing behaviour in tufted capuchin monkeys (*Cebus apella*): Testing a few hypotheses. [Abstract] *Folia Primatologica* 72(3):130.

Carpenter, C. R. 1934. A field study of the behavior and social relations of howling monkeys. *Comparative Psychology Monographs* 102(2):1-168.

Carpenter, C. R. 1935. Behavior of red spider monkeys (*Ateles geoffroyi*) in Panama. *Journal of Mammalogy* 16:171-180.

Bibliography

Carpenter, C. R. 1953. Grouping behavior of howler monkeys. *Extrait Arch. Neerlandaises Zool.* 10(supp 2):45-50.

Carpenter, C. R. 1960. Howler monkeys of Barro Colorado Island. 16 mm film, black and white, Pennsylvani State University library, Pennsylvania.

Carpenter, C. R. 1962. Field studies of a primate population. Pp. 286-394 i n: E. Bliss (ed.), *Roots of Behavio* Harper & Rowe, New York.

Carpenter, C. R. 1964. *Naturalistic Behavior of Nonhuman Primates.* Pennsylvania State University Press, Universit Park, pp. 454.

Carpenter, C. R. 1965. The howlers of Barro Colorado Island. Pp. 250-291 in: I. Devore (ed.), *Prima Behavior.* Holt, Rinehart and Winston, New York,.

Carretero, P. X. and J. A. Ahumada 2002. *Saimiri sciureus-Cebus apella* asociación en La Macarena, Colombia [Abstract] *Caring for Primates. Abstracts of the XIXth Congress.* The International Primatological Society Beijing: Mammalogical Society of China. p. 322.

Cartelle, C. and W. C. Hartwig 1996. A new extinct primate among the Pleistocene megafauna of Bahia Brazil. *Processes of the National Academy of Sciences, USA* 93(13):6405-6409.

Casamitjana, J. 2002. The vocal repertoire of the woolly monkey *Lagothrix lagothricha. Bioacoustics* 13(1):1-19

Castaño U., C. 1990. *Guía del Sistema de Parques Nacionales de Colombia.* INDERENA reference of long-haired spider monkeys (*Ateles belzebuth*). Pp. 451-466 in: M. A. Norconk, A. L. Rosenberger and P. A. Rosenberge (eds.), *Adaptive Radiations of Neotropical Primates.* Plenum Press, New York,.

Castaño U., C. 1993. *Situación General de la Conservación de la Biodiversidad en la Región Amazónica: Evaluación d las Areas Protegidas Propuestas y Estratégias.* UICN, Quito, 111 pp.

Castell, R. and M. Maurus 1967. Über das sogenannte Urinmarkieren von Totenkopfaffen (*Saimiri sciureus* in Abhëngigkeit von Umwelbedingten und emotionalen Factoren. *Folia Primatologica* 6:170-176.

Castell, R., H. Krohn and D. Ploog 1969. Rückenwalzen bei Totenkopfaffen (*Saimiri sciureus*) Korperpflege und soziale Funktion. *Zeitschrift für Tierpsychologie* 26:488-497.

Castellanos, H. G and P Chanin. 1996. Seasonal differences in food choice and patch preference of long-haired spider monkeys (*Ateles belzebuth*). Pp. 451-466 & 459 in: M. A. Norconk, A. L. Rosenberger and P A. Garber (eds.), *Adaptive Radiations of Neotropical Primates.* Plenum Press, New York.

Castellanos, M. C., S. Escobar and P. R. Stevenson 1999. Dung beetles (Scarabaeidae: Scarabaeinae) attracted to woolly monkey (*Lagothrix lagothricha* Humboldt) dung at Tinigua National Park, Colombia. *Coleopterist. Bulletin* , Bogotá, 198 pp.

Castellvi, M. 1938. Materiales para estudios glotológicos. *Boletin de. Estudios Históricos* 11(84):368-372.

Castillo-García, J. 2001. Comportamiento social de *Cebus apella* con énfasis en acicalamiento, Estación Biológica Caparú (Amazonas) Colombia. Graduation thesis in Biology, Universidad Inca de Colombia, Bogotá.

Castro Rodriguez, N. E. 1990. *La Primatología en el Perú: Investigaciones Primatológicas (1973-1985).* Proyecto Peruano de Primatología, Lima.

Castro, N. and P. Soini 1977. Field studies on *Saguinus mystax* and other callitrichids in Amazonian Peru. pp 73-78 in: D. V. Kleiman (ed.), *The Biology and Conservation of the Callitrichidae.* Smithsonian Institution Press, Washington, D. C.

Castro, N., J. Revilla and M. K. Neville 1990. Carne de monte como una fuente de proteínas en Iquitos, con referencia especial a monos. Pp. 17-35 in: N. E. Castro-Rodríguez (ed.), *La Primatología en el Perú.* Proyecto Peruano de Primatología «Manuel Moro Sommo», Lima.

Causey, O. R., H. W. Laemmert Jr. and G. S. Gayes 1948. the home range of Brazilian *Cebus* monkeys in a region of small residual forests. *American Journal of Hygene* 47:304-314.

Ceballos, C. 1989. Food change of a fourth month adopted infant in a wild group of *Alouatta seniculus*. *Field Studies of New World Monkeys, La Macarena, Colombia*. 2:41-43.

Cebul, M. S. and G. Epple 1984. Father-offspring relationships in laboratory families of saddle-back tamarins (*Saguinus fuscicollis*). Pp in: D. M. Taub (ed.), *Primate Paternalism,*. Van Nostrand Reinhold, New York.

Chalukian, S. C. 1985. Comportamiento alimenticio de *Cebus apella paraguayanus* en el Parque Nacional el Rey, Salta (Argentina). *Boletin de Primatologia Argentino* 3(1):15-26.

Chamberlain, J., G. Nelson and K. Milton 1993. Fatty acid profiles of major food sources of howler monkeys (*Alouatta palliata*) in the Neotropics. *Experientia* 49(9):820-824.

Chapman, C. A. 1986a. Ecological constraints on group size for three species of Neotropical primates. [Abstract] *American Journal of Primatology* 10(4):394.

Chapman, C. A. 1986b. Boa constrictor predation and group response in white-faced monkeys. *Biotropica* 18(2):171-172.

Chapman, C. A. 1987. Flexibility in diets of three species of Costa Rican primates. *Folia Primatologica* 49:90-105.

Chapman, C. A. 1988a. Patch use and patch depletion by the spider and howling monkeys of Santa Rosa National Park, Costa Rica. *Behaviour* 105:99-116.

Chapman, C. A. 1988b. Patterns of foraging and range use by three species of Neotropical primates. *Primates* 29(2):177-194.

Chapman, C. A. 1989a. Spider monkey sleeping sites: Use and availability. *American Journal of Primatology* 18:53-60.

Chapman, C. A. 1989b. Primate seed dispersal: the fate of dispersed seeds. *Biotropica* 21(2):148-154.

Chapman, C. A. 1990a. Ecological constraints on group size in three species of neotropical primates. *Folia Primatologica* 55:1-9.

Chapman, C. A. 1990b. Association patterns of spider monkeys: The influence of ecology and sex on social organization. *Behavioral Ecology and Sociobiology* 26:409-414.

Chapman, C. A. and L. J. Chapman 1986. Behavioural development of howling monkey twins (*Alouatta palliata*) in Santa Rosa National Park, Costa Rica. *Primates* 27(3):377-381.

Chapman, C. A. and L. J. Chapman 1987. Social responses to the traumatic injury of a juvenile spider monkey (*Ateles geoffroyi*). *Primates* 28(2):271-275.

Chapman, C. A. and L. J. Chapman 1990a. Reproductive biology of captive and free-ranging spider monkeys. *Zoo Biology* 9:1-9.

Chapman, C. A. and L. J. Chapman 1990b. Dietary variability in primate populations. *Primates* 31(1):121-128.

Chapman, C. A. and L. J. Chapman 1991. The foraging itinerary of spider monkeys: When to eat leaves? *Folia Primatology* 56:162-166.

Chapman, C. A. and L. J. Chapman 2000. Determinants of group size in primates: The importance of travel costs. Pp. 24-42 in: S. Boinski and P. A. Garber (eds.), *On the Move: How and Why Animals Travel in Groups*. University of Chicago Press, Chicago.

Chapman, C. A. and L. Lefevre 1990. Manipulating foraging group size: Spider monkey food calls at fruiting trees. *Animal Behaviour* 39:891-896.

Chapman, C. .A. and L. M. Fedigan 1990. Dietary differences between neighboring *Cebus capucinus* groups: Local traditions, food availability, or response to food profitability? *Folia Primatology* 54:177-186.

Chapman, C. A., L. J. Chapman and K. E. Glander 1989a. Primate populations in northwestern Costa Rica: Potential for recovery. *Primate Conservation* 10:37-44.

Chapman, C. A., L. J. Chapman and L. Lefebvre 1989b. Spider monkey alarm calls: Honest advertisement or warning kin? *Animal Behaviour* 39:197-198.

Bibliography

Chase, J. E. and R. W. Cooper 1969. *Saguinus nigricollis* - physical growth and dental eruption in a small population of captive-born individuals. *American Journal of Physical Anthropology* 30:111-116.

Chavez, R., I. Sampaio, M. P. C. Schneider, H. Schneider, S. L. Page and M. Goodman 1999. The place of *Callimico goeldii* in the callitrichine phylogenetic tree: Evidence from von Willebrand Factor Gene Intron II sequences. *Molecular Phylogenetics and Evolution* 13. 392-404.

Chevalier-Skolnikoff, S. 1989. Spontaneous tool use and sensorimotor intelligence in *Cebus* compared with other monkeys and apes. *Behavioral and Brain Sciences* 12(3):561-627.

Chevalier-Skolnikoff, S. 1990. Tool use by wild *Cebus* monkeys at Santa Rosa National Park, Costa Rica. *Primates* 31(3):375-383.

Cheverud, J. M. and A. J. Moore 1990. Subspecific morphological variation in the saddle-back tamarin (*Saguinus fuscicollis*). *American Journal of Primatology* 21(1):1-15.

Chiarello, A. G. 2003. Primates of the Brazilian Atlantic forest: The influence of forest fragmentation on survival. Pp. 99-121 in: L. K. Marsh (ed.), *Primates in Fragments: Ecology and Conservation*. Kluwer Academic Plenum Publ., New York.

Chivers, D. J. 1969. On the daily behavior and spacing of howling monkey groups. *Folia Primatologica* 10:48-102.

Christen, A. 1968. Haltung und Brutbiologie von *Cebuella*. *Folia Primatologica* 8:41-49.

Christen, A. 1974. Fortpflanzungsbiologie und Verhalten bei *Cebuella pygmaea* und *Tamarin tamarin* (Primates, Platyrrhini, Callitrichidae). Fortschritte der Verhaltensforschung. *Zeitschrift der Tierpsychologie* Supplement 14, pp. 1-79.

Christen, A. 1994. Goeldi's monkey, *Callimico goeldii*, in northern Bolivia. Pp. 73-78 in: B. Thierry, J. R. Anderson, J. J. Roeder and N. Herrenschmidt (eds.), *Current Primatology, vol. I.: Ecology and Evolution*. University Louis Pasteur, Strasbourg.

Christen, A. 1998. The most enigmatic monkey in the Bolivian rain forest – *Callimico goeldii*. *Neotropical Primates* 6(2):35-37.

Christen, A. 1999. Survey of Goeldi's monkeys (*Callimico goeldii*) in northern Bolivia. *Folia Primatologica* 70(2):107-111.

Christen, A. 2000. *Callimico goeldii* (Goeldi´s monkey), the most enigmatic South American monkey. [Abstract] *American Journal of Physical Anthropology* (Suppl 30):155.

Christen, A. and T. Geissmann 1994. A primate survey in northern Bolivia, with special reference to Goeldi's monkey, *Callimico goeldii*. *International Journal of Primatology* 15(2):239-274.

Christen, A. and L. M. Porter 1999. Field surveys of Goeldi's monkey in northern Bolivia. [Abstract] *Primatology and Anthropology into the Millenium*. Centenary Congress of the Anthropological Institue and Museum in Zurich, 1899-1999: Programme and Abstracts. University of Zürich, Zürich, p. 11.

Ciochon, R. L. and R. S. Corruccini 1975. Morphometric analysis of platyrrhine femora with taxonomic implications and notes on two fossil forms. *Journal of Human Evolution* 4:193-217.

Ciochon, R. L. and A. B. Chiarelli 1980. Paleobiographic perspectives on the origin of the Platyrrhini. Pp. 459-493 in: R. L. Ciochon and A. B. Chiarelli (eds.), *Evolutionary Biology of the New World Monkeys & Continental Drift*. Plenum Press, New York.

Clarke, M. R. 1981. Aspects of male behavior in the mantled howlers (*Alouatta palliata* gray) in Costa Rica. *American Journal of Primatology* 3:1-22.

Clarke, M. R. 1982. Socialization, infant mortality, and infant-nonmother interactions in howling monkeys (*Alouatta palliata*) in Costa Rica. Unpublished Ph.D. thesis, University of California, Davis.

Clarke, M. R. 1983. Infant-killing and infant disappearance following male takeovers in a group of free-ranging howling monkeys (*Alouatta palliata*) in Costa Rica. *American Journal of Primatology* 5:241-247.

Clarke, M. R. 1990. Behavioral development and socialization of infants in a free-ranging group of howling monkeys (*Alouatta palliata*). *Folia Primatologica* 54:1-15.

Clarke, M. R. and K. E. Glander 1981. Adoption of infant howling monkeys (*Alouatta palliata*). *American Journal of Primatology* 1:469-472.

Clarke, M. R. and K. E. Glander. 1984. Female reproductive success in a group of free-ranging howling monkeys (*Alouatta palliata*) in Costa Rica. Pp.111-126 in: ed. M. Small (ed.), *Female Primates: Studies by Women Primatologists*. Alan R. Liss, New York.

Clarke, M. R. and K. E. Glander 2001. Presence of mother and juvenile dispersal in free-ranging howling monkeys (*Alouatta palliata*) in the tropical dry forest of Costa Rica. *American Journal of Physical Anthropology* Suppl 32:50.

Clarke, M. R. and K. E. Glander. 2002. Female immigration patterns in mantled howling monkeys (*Alouatta palliata*) on La Pacifica, Guanacaste, Costa Rica. *American Journal of Physical Anthropology* Suppl. 34:112.

Clarke, M. R. and E. L. Zucker 1994. Survey of the howling monkey population at La Pacifica: A seven-year follow-up. *International Journal of Primatology* 15(1):61-73.

Clarke, M. R., Glander, K. E. and E. L. Zucker 1998. Infant-nonmother interactions of free-ranging mantled howlers (*Alouatta palliata*) in Costa Rica. *International Journal of Primatology* 19(3):451-472.

Clarke, M. R., E. L. Zucker, and C. M. Crockett 1999. Assessment of the howling monkey (*Alouatta palliata*) population on Hacienda La Pacifica, Guanacaste, Costa Rica. *American Journal of Anthropology* Suppl. 28:108-109.

Clarke, M. R., C. M. Crockett and E. L. Zucker 2001a. A comparison of methods used to census mantled howler in the dry tropical forest of Costa Rica. *Laboratory Primate Newsletter* 40(4):4-6.

Clarke, M., D. Arden, D. Epstein and M. Gilbert 2001b. Activity patterns of adult male howling monkeys (*Alouatta palliate*) in the dry forest of Costa Rica: Comparison by age, habiat and social group. [Abstract] *American Journal of Primatology* 54(Suppl 1):28.

Clarke, M. R., D. A. Collins and E. L. Zucker 2002a. Responses to deforestation in a group of mantled howlers (*Alouatta palliata*) in the tropical dry forest of Costa Rica. [Abstract] *International Journal of Primatology* 23(2):365-381.

Clarke, M. R., C. M. Crockett, D. A. Collins and E. L. Zucker 2002b. Mantled howler population of Hacienda La Pacifica, Costa Rica, between 1991 & 1998: Effects of deforestation. *American Journal of Primatology* 56(3):155-163.

Cleveland, J. and C. T. Snowdon 1982. The complex vocal repertoire of the adult cotton-top tamarin, *Saguinus oedipus oedipus*. *Zeitschrift für Tierpsychologie* 58:231-270.

Cleveland, J. and C. T. Snowdon 1984. Social development during the first twenty weeks in the cotton-top tamarin (*Saguinus oedipus oedipus*). *Animal Behaviour* 32:432-444.

Cleveland, A., A. R. Rocca, E. L. Wendt and G. C. Westergaard 2003. Throwing behavior and mass distribution of stone selection in tufted capuchin monkeys (*Cebus apella*). *American Journalof Primatology* 61(4):159-172.

Coates-Estrada, R. and A. Estrada 1986. Fruiting and frugivores at a strangler fig in the tropical rain forest of los Tuxtlas, Mexico. *Journal of Tropical Ecology* 2(4):349-357.

Coe, C. L. and L. A. Rosenblum 1974. Sexual segregation and its ontogeny in squirrel monkey social structure. *Journal of Human Evolution* 3:551-561.

Coehlo, A. 1975. Social organization and resource availability in Guateman howler and spider monkeys: A sociobioenergetic analysis. Paper presented at 44th annual meeting of American Association of Physical Anthropologist, Denver.

Coelho, A. M. Jr., L. S. Coelho, C. A. Bramblett, S. S. Bramblett and L. B. Quick 1976a. Ecology, population characteritics, and sympatric association in primates: a socio-bionergetic analysis of howler and spider monkeys in Tikal, Guatemala. *Yearbook of Physical Anthropology* 20:96-135.

Bibliography

Coelho, A. M . Jr., L. S. Coelho, C. A. Bramblett, S. S. Bramblett and L. B. Quick 1976b. Resource availability and population density in primates: a socio-bioenergetic analysis of the energy budgets of Guatemalan howler and spider monkeys. *Primates* 17:63-80.

Coelho, A. M. Jr., C. A. Bramblett and L. B. Quick 1977. Social organization and food resource availability in primates: a socio-bioenergetic analysis of diet and disease hypotheses. *American Journal of Physical Anthropology* 46:253-264.

Coimbra-Filho, A. F. and R. A. Mittermer (eds.) 1981. *Ecology and Behavior of Neotropical Primates*, Vol. 1. Academia Brasileira de Ciências, Rio de Janeiro, 496 pp.

Colinvaux, P. A. 1987. Amazon diversity in the light of the paleoecological record. *Quarterly Science Review* 6:93-114.

Colinvaux, P. A. 1996. Quaternary environmental history and forest diversity in the neotopics. Pp. 359-406 in: J. B. C. Jackson, A. F. Budd and A. G. Coates (eds.), *Evolution and Environment in Tropical America*. The University of Chicago Press, Chicago.

Colinvaux, P. A., K. Olson and K.-b. Liu 1988. Late-glacial and Holocene pollen diagrams from two endorheic lakes of the Inter-Andean plateau of Ecuador. *Rev. Palaeobot. Palynol.* 55:83-100.

Collias, N. and C. W. Southwick 1952. A field study of population density and social organization in howling monkeys. *Processes of the American Philosophical Society* 96:143-156

Collins, A. 1999. Species status of the Colombian spider monkey, *Ateles belzebuth hybridus*. *Neotropical Primates* 7(2):39-41.

Collins, A. C. 2001. The importance of sampling for reliable assessment of phylogenetics and conservation among Neotropical primates: A case study in spider monkeys (*Ateles*). *Primate Report* 61:9-30.

Collins, A. and J. Dubach 2000a. Phylogenetic relationships among spider monkeys (*Ateles*) haplotypes based on mitochondrial DNA variation. *International Journal of Primatology* 21(3): 381-420.

Collins, A. and J. Dubach. 2000b. Biogeographic and evolutionary forces responsible for speciation in *Ateles*. *International Journal of Primatology* 21(3):421-444.

Collins, A. C. and J. M. Dubach 2001. Nuclear DNA variation in spider monkeys (*Ateles*). *Molecular Phylogenetics and Evolution* 19(1):67-75.

Converse, L. J., A. A. Carlson, T. E. Ziegler and C. T. Snowdon 1995. Communication of ovulatory state to mates by female pygmy marmosets, *Cebuella pygmaea*. *Animal Behavior* 49(3):615-621.

Cooper, M. A., I. S. Bernstein, D. M. Fragaszy and F. B. M. de Waal 2001. Integration of new males into four groups of tufted capuchins (*Cebus apella*). *International Journal of Primatology* 22(4):663-683.

Cooper, R. W. 1968. squirrel monkey taxonomy and supply. Pp. 1-29 in: L. S. Rosenblum and R. W. Cooper (eds.), *The Squirrel Monkey*. Academic Press, New York.

Cooper, R. W. and J. I. Hernández-C. 1975. A current appraisal of Colombia's primate resources. Pp. 37-66 in: G. Bermant and D. G. Lindburg (eds.), *Primate Utilization & Conservation*. John Wiley & Sons, New York.

Cortes-Ortiz, L., E. Bermingham, C. Rico, E. Rodríguez-Luna, I. Sampaio, and M. Ruiz-Garcia 2003. Molecular systematics and biogeography of the Neotropical monkey genus, *Alouatta*. *Molecular Phylogenetics and Evolution* 26(1):64-81.

Costello, R. K., C. Dickenson, A. L. Rosenbergher, S. Boinski and F. S. Szalay. 1993. Squirrel monkey (genus *Saimiri*) taxonomy: A multidisciplinary study of the biology of species. Pp. 177-210 in: W. H. Kimbel and l. B. Martin (eds.), *Species, Species Concepts, and Primate Evolution*. Plenum Press, New York.

Cracraft, J. 1983. Species concepts and speciation analysis. Pp. 159-187 in: R. F. Johnston (ed.), *Current Ornithology*. Plenum Press, New York.

Crandlemire-Sacco, J. 1987. Habitat differences and ecological distinctions between two populations of *Saguinus fuscicollis*. [Abstract] *American Journal of Physical Anthyropology* 72(2):190.

Crandlemire-Sacco, J. 1988. An ecological comparison of two sympatric primates: *Saguinus fuscicollis* and *Callicebus moloch* of Amazonian Peru. *Primates* 29(4):465-475.

Crockett, C. M. 1984a. Emigration by female red howler monkeys and the case for female competition. Pp. 159-173 in: M. F. Small (ed.), *Female Primates: Studies by Women Primatologists.* Alan R. Liss, New York.

Crockett, C. M. 1984b. Family feuds. *Natural History* 93(8):54-63.

Crockett, C. M. 1985. Population studies of red howler monkeys (*Alouatta seniculus*). *National Geographic Research* 1(2):264-273.

Crockett, C. M. 1987a. Diet, dimorphism and demography: Perspectives from howlers to hominids. Pp. 115-135 in: W. G. (ed.), *The Evolution of Human Behavior: Primate Models.* State University of New York Press, New York,.

Crockett, C. M. 1987b. Infanticidia en mamíferos: Teorías y evidencia. *Boletin de Primatología Argentina* 5(1-2):13-27.

Crockett, C. M. 1996. The relation between red howler monkey (*Alouatta seniuclus*) troop size and population growth in two habitats. Pp. 489-510 & 550-551 in: M. A. Norconk, A. L. Rosenberger and P. A. Garber (eds.), *Adaptive Radiations of Neotropical Primates.* Plenum Press, New York.

Crockett, C. M. and R. Sekulic 1982. Gestation length in red howler monkeys. *American Journal of Primatology* 3:291-294.

Crockett, C. M. and R. Sekulic 1984. Infanticide in red howler monkeys. Pp. 173-191 in: G. Hausfater and S. Blaffer Hrdy (eds.), *Comparative & Evolutionary Perspectives,* Aldine Publishing Company, New York.

Crockett, C. M. and J. F. Eisenberg 1987. Howlers: variations in group size and demography. Pp. 54-68 in: B. B. Smuts, R. M. Seyfarth, R. W. Wrangham and T. T. Struhsaker, *Primate Societies.* University of Chicago Press, Chicago.

Crockett, C. M. and R. Rudran 1987a. Red howler monkey birth data I: Seasonal variation. *American Journal of Primatology* 13:347-368.

Crockett, C. M. and R. Rudran 1987b. Red howler monkey birth data II: Interannual, habitat, and sex comparisons. *American Journal Primatology* 13:369-384.

Crockett, C. M. and T. Pope 1988. Inferring patterns of aggression from red howler monkey injuries. *American Journal of Primatology* 15(4):289-308.

Crockett, C. M. and T. R. Pope 1993. Consequences of sex differences in dispersal for juvenile red howler monkeys. Pp. 104-118 in: L. A. Fairbanks (ed.), *Juvenile Primates: Life History, Development, & Behavior.* Oxford University Press, New York.

Crockett, C. M. and C. H. Janson 2000. Infanticide in red howlers: Female group size, male membership, and a possible link to folivory. Pp. 75-98, in: C. P. van Schaik and C. H. Janson (eds.), *Infanticide by Males and Its Implications.* Cambridge University Press, Cambridge.

Cropp, S. and S. Boinski 1999.Phylogenetic relationships among *Saimiri* spieces based on nuclear and mitochrondrial DNA evidence. [Abstract] *American Journal of Physical Anthropology* (Suppl 28):112.

Cropp, S. J., A. Larson and J. M. Cheverud 1999. Historical biogeography of tamarins, genus *Saguinus*: The molecular phylogenetic evidence. *American Journal of Physical Anthropology* 108(1):65-89.

Cuartas-Calle, C. A. 2001. [Partial distribution of tamarins (*Saguinus leucopus*, Callitrichidae) in Departmento de Antioquia, Colombia.] *Neotropical Primates* 9(3):107-111.

Cubiciotti, D. D., III and W. A. Mason 1975. Comparative studies of social behavior in *Callicebus* & *Saimiri*: Male-female emotional attachments. *Behavioral Biology* 16:185-97.

Cuervo Diaz, A., C. E. Barbosa and J. de la Ossa 1986. [Ecological and ethological aspects of primates with emphasis on *Alouatta seniculus* (Cebidae), from the Coloso region, San Jacinto Mountains (Sucre), on the northern coast of Colombia.]. *Caldasia* 24(68-70):709-741.

Bibliography

Cummins-Sebree, S. E. and D. M. Fragaszy 2001. The right stuff: Capuchin monkeys perceive affordances of tools, Pp. 98-92 in: G. A. Burton, R. C. Schmidt (eds.), *Studies in Perception and Action VI*. Lawrence Erlbaum, Mahwah, NJ.

Da Cunha, A. C. and A. A. Barnett 1990. Sightings of the golden-backed uacari, *Cacajao melanocephalus ouakary*, on the upper Rio Negro, Amazonas, Brazil. *Primate Conservation* 11:8-11.

Dahl, J. F., K. N. Karas and P. S. Dunham 1996. Spider monkeys of Belize: Taxonomy and status. [Abstract] *IPS/ASP Congress* 1996:728.

Dare, R. 1974a. The social behavior and ecology of spider monkeys, *Ateles geoffroyi* on Barro Colorado Island. Unpublished Ph.D. thesis, University of Oregon.

Dare, R. 1974b. Food-sharing in free-ranging *Ateles geoffroyi* (red spider monkeys). *Laboratory Primate Newsletter* 13:19-21.

Dare, R. 1975. The effects of fruit abundance on movement patterns of free-ranging spider monkeys, *Ateles geoffroyi*. [Abstract] *American Journal of Physical Anthropology* 42:29.

Davis, L. C. 1987. Morphological evidence of positional behavior in the hindlimb of *Cebupithecia sarmientoi* (Primates: Platyrrhini). Master's thesis, Arizona State University, Tempe.

Davis, L. C. 1988. Morphological evidence of locomotor behavior in a fossil platyrrhine. [Abstract] *American Journal of Physical Anthropology* 75(2):202.

Davis, L. C. 1994. Locomotor and postural adaptations of an unusual platyrrine, *Callimico goeldii*. [Abstract] *American Journal of Physical Anthropology* (suppl. 18):84.

Dawson, G. A. 1976. Behavioral ecology of the Panamanian tamarin, *Saguinus oedipus* (Callitrichidae, Primates). Unpublished Ph.D. Dissertation Abstracts International B37:645-645.

Dawson, G. A. 1977. Composition and stability of social groups of the tamarin, *Saguinus oedipus geoffroyi*, in Panama: Ecological and behavioral implications. Pp. 23-27 in: D. G. Kleiman (ed.), *The Biology and Conservation of the Callitrichidae*. Smithsonian Institution Press, Washington, D. C.

Dawson, G. A. 1978. Composition and stability of social groups of the tamarin, *Saguinus oedipus geoffroyi* in Panama: Ecological and behavioral implications. Pp. 23-37 in: D. G. Kleiman (ed.), *The Biology and Conservation of the Callitrichidae*. Smithsonian Institution Press, Washington, D. C.

Dawson, G. A. 1979. the use of time and space by the Panamanian tamarin, *Saguinus oedipus*. *Folia Primatologica* 31:253-284.

Dawson, J. D. 1976. Behavioral ecology of the Panamanian tamarin, *Saguinus oedipus*. Unpublished Ph.D. thesis, Michigan State University, East Lansing.

Dawson, G. A. and W. R. Dukelow 1976. Reproductive characteristics of free-ranging tamarins (*Saguinus oedipus geoffroyi*). *Journal of Medical Primatology* 5:266-275.

De Boer, L. D. M. 1974. Cytotaxonomy of the Platyrrhini (Primates). *Genen en Phaenen* 17(1-2):1-115.

Defler, T. R. 1979a. On the ecology and behavior of *Cebus albifrons* in eastern Colombia: I. Ecology. *Primates* 20:475-490.

Defler, T. R. 1979b. On the ecology and behavior of *Cebus albifrons* in eastern Colombia: II. Behavior. *Primates* 20:491-502.

Defler, T. R. 1980. Notes on interactions between the tayra (*Eira barbara*) and the white-fronted capuchin (*Cebus albifrons*). *Journal of Mammalogy* 61:156.

Defler, T. R. 1981. The density of *Alouatta seniculus* in the Llanos Orientales of Colombia. *Primates* 22:564-569.

Defler, T. R. 1982. A comparison of intergroup behavior in *Cebus albifrons* and *Cebus apella*. *Primates* 23:385-392.

Defler, T. R. 1983a. Some population characteristics of *Callicebus torquatus* in eastern Colombia. *Lozania (Bogotá)* 38:1-8.

Defler, T. R. 1983b. Observaciones sobre los primates del bajo Mirití-Paraná, Amazonas, Colombia. *Lozania* (Bogotá):46:1-13.

Defler, T. R. 1983c. A remote park in Colombia. *Oryx* 17(1):15-17.

Defler, T. R. 1985a. Contiguous distribution of two species of *Cebus* monkeys in El Tuparro National Park, Colombia. *American Journal Primatology* 8:101-112.

Defler, T. R. 1985b. Those crafty, capricious, clever capuchins. *Animal Kingdom* 89:16-21.

Defler, T. R. 1987. Ranging and use of space in a group of woolly monkeys (*Lagothrix lagotricha*) in the NW Amazon of Colombia. *International Journal Primatology* 8:420.

Defler, T. R. 1989a. Recorrido y uso del espacio en un grupo de *Lagothrix lagotricha* (Primates: Cebidae) mono lanudo o churuco en la Amazonia Colombiana. *Trianea (Acta científica y tecnológica, Inderena)* 3:183-205.

Defler, T. R. 1989b. Wild and Woolly. *Animal Kingdom* September/October:???.

Defler, T. R. 1989c. The status and some ecology of primates in the Colombian Amazon. *Primate Conservation* (10):51-56.

Defler, T. R. 1990a. Salvajes y lanudos. translation of article "Wild and woolly" *Ecológica* no. 4. ????

Defler, T. R. 1990b. Primates and the Colombian Amazon. *International Primate Protection League* 17(2):8-12.

Defler, T. R. 1991. *Cacajao melanocephalus* (Humboldt, 1811) Primates, Cebidae. *Trianea (Acta científica y tecnológica INDERENA)* 4:557-558.

Defler, T. R. 1994a. *Callicebus torquatus* is not a white-sand specialist. *American Journal of Primatology* 33(2):149-154.

Defler, T. R. 1994b. La conservación de los primates colombianos. *Trianea (Acta científica y tecnológica INDERENA)* 5:255-287.

Defler, T. R. 1994c. Reconocimiento biológico en los bajos Ríos Uva y Guaviare y en la región entre los Ríos Inírida y Guainía con especial enfasis en primates (12 de marzo-8 de abril, 1994). report to INDERENA.

Defler, T. R., 1994d. Biodiversidad en la Amazonia Colombiana: Reconocimientos Biológicos en tres Áreas con Énfasis en los Primates. unpublished report to INDERENA.

Defler, T. R. 1995. The time budget of a group of wild woolly monkeys (*Lagothrix lagothricha*). *International Journal of Primatology*. 16(1):107-120.

Defler, T. R. 1996a. Aspects of the ranging pattern in a group of wild woolly monkeys (*Lagothrix lagothricha*). *American Journal of Primatology* 38:289-302.

Defler, T. R. 1996b. An IUCN classification of primates for Colombia. *Neotropical Primates* 4(3):77-78..

Defler, T. R. 1996c. The IUCN conservation status of *Lagothrix lagothricha lugens* Elliot, 1907. *Neotropical Primates* 4(3):78-80.

Defler, T. R. 1999a. Fission-fusion behavior in *Cacajao melanocephalus ouakary*. *Neotropical Primates* 7(1):5-8

Defler, T. R. 1999b. Locomotion and posture in *Lagothrix lagothricha*. *Folia Primatologica* 70(6):313-327.

Defler, T. R. 2001. *Cacajao melanocephalus ouakary* densities on the lower Apaporis River, Colombian Amazon. *Primate Report* 61(November, 2001):31-36.

Defler, T. R. 2003a. Densidad de especies y organización espacial de una comunidad de primates: Estación Biológica Caparú, Departamento del Vaupés, Colombia. Pp. 21-37 in: F. Nassar & V. Pereira (eds.), *Primatología del Nuevo Mundo.* Fundación Araguatos, Bogotá.

Defler, T. R. 2003b. *Primates de Colombia.* Conservación Internacional Colombia, Bogotá.

Defler, T. R. unpublished manuscript. An analysis of the state of the forest cover in the Colombian Amazon: A study of the extent and pattern of forest conversion. Conservation International.

Defler, T. R. and D. Pintor 1985. Censusing primates by transect in a forest of known primate density. *International Journal of Primatology* 6(3):243-259.

Bibliography

Defler, T. R.. and S. Defler 1996. The diet of a group of *Lagothrix lagothricha* in the NW Amazon. *International Journal of Primatology* 17(2):161-190.

Defler, T. R. and J. I. Hernández-Camacho 2002. The true identity and characteristics of *Simia albifrons* Humboldt, 1812. *Neotropical Primates* 10(2):49-64.

Defler, T. R. and M. L. Bueno 2003. Karyological guidelines for *Aotus* taxonomy. [Abstract]

Twenty Sixth Annual American Society of Primatology Conference, Calgary, Canada, American Journal of Primatology.

Defler, T. R. and J. V. Rodríguez (2003). A Reassessment of the Present EN - IUCN Classification for *Saguinus oedipus*. Unpublished position paper delivered to IUCN specialist group chairman.

Defler, T. R., M. L. Bueno and J. I. Hernández-Camacho 2001. Taxonomic status of *Aotus hershkovitzi*: Its relationship to *Aotus lemurinus lemurinus*. *Neotropical Primates* 9(2):37-52.

Defler, T. R., J. V. Rodriguez and J. I. Hernández-Camacho 2003. Conservation priorities for Colombian primates. *Primate Conservation* 19:10-18.

DeGAma-Blanchet, H. and L. M. Fedigan 2003. The effects of forest fragment size and isolation on monkey density in a Costa Rican tropical dry forest. [abstract] *American Journal of Primatology* 60(Supp 1):57-58.

De Goeje, C. H. 1909. *Verh. Adad. Wetens., Ámsterdam,* Afd. Letterk. N.r. x, No. 3.

De la Ossa, J., J. G. Moreno and C. Segura 1988. Anotaciones sobre el comportamiento agresivo en la conformación de una colonia semicautiva de *Saguinus oedipus* (Linnaeus, 1758)(Mammalia: Primates) *Trianea* 1:131-139.

De la Torre, S. 1994. Feeding habits of *Saguinus nigricollis graellsi* in northeastern Ecuador. [Abstract] *Congress of the International Primatological Society* 15:82.

De la Torre, S. 1996. Notes on the distributions of the Ecuadorian callitrichids. *Neotropical Primates* 4(3):88

De la Torre, S. A. 2000. Environmental correlates of vocal communication of wild pygmy marmosets *Cebuella pygmaea*. Unpublished Ph.D. thesis, Dissertation Abstracts International B61(1):141.

De la Torre, S. and C. T. Snowdon 2002. Environmental correlates of vocal communication of wild pygmy marmosets, *Cebuella pygmaea*. *Animal Behaviour* 63(5):847-856.

De la Torre, S., F. Campos and T. De Vries 1992. Seasonal reduction in the home ranges and birth peak bimodality of *Saguinus nigricollis graellsi* (Primates: Callitrichidae) in Amazonian Ecuador. [Abstract] *XIV th Congress of the International Primatological Society*, Strasbourg, 1992:316.

De la Torre, S., F. Campos and T. de Vries 1995. Home range and birth seasonality of *Saguinus nigricollis graellsi* in Ecadorian Amazonia. *American Journal of Primatology* 37(1):39-56.

De la Torre, S., C. T. Snowdon and M. Bejarano 1999. Preliminary study of the effects of ecotourism and human traffic on the howling behavior of red howler monkeys, *Alouatta seniculus,* in Ecuadorian Amazonia. *Neotropical Primates* 7(3):84-86.

De la Torre, S., C. T. Snowdon and M. Bejarano 2000. Effects of human activities on wild pygmy marmosets in Ecuadorian Amazonia. *Biological Conservation* 94(2):153-163.

Delson, E. and A. L. Rosenberger 1984. Are there any anthropoid primate living fossils? Pp. 50–61 in: N Eldredge and S. M. Stanley (eds.), *Living Fossils.* Springer Verlag, New York.

De Oliveira, E. H. C. 1996. Cytogenetic and phylogenetic studies of *Alouatta* from Brazil and Argentina *Neotropical Primates* 4(4):156-157.

De Oliveira, E. H. C., M. Meusser, W. B. Figueiredo, C. Nagamachi, J. C. Pieczarka, I. J. Sbalqueiro, J Wienberg and S. Müller 2002. The phylogeny of howler monkeys (*Alouatta,* Platyrrhini): Reconstruction by multicolor cross-species chromosome painting. *Chromosome Research* 10:669-683.

Descailleaux, J., A. M. Garcia, L. Rodriguez, R. Aquino et al. 1986. [A new chromosome complement for the genus *Aotus*.]. *Primatologigia no Brasil* 2:387. (in Portuguese)

Descailleaux, J., A. M. Garcia, L. Rodriguez, R. Aquino et al. 1987. Chromosomal rearrangementes in South American primates (Platyrrhini). [Abstract] *Internacional Journal of Primatology* 8(5):557.

Descailleaux, J. R. Fujita, L. A. Rodríguez, R. Aquino and F. Encarnación 1990. Rearreglos cromosómicos y variabilidad cariotípica del género *Aotus* (Cebidae: Platyrrhini). Pp. 572-578 in: R. Castro (ed.), *La Primatología en Perú*. Proyecto Peruano de Primatología, Lima.

de Thoisy, B. and C. Richard-Hansen 1997. Diet and social behaviour changes in a red howler monkey (*Alouatta seniculus*) troop in a highly degraded rain forest. *Folia Primatologica* 68(6):357-361.

de Thoisy, B. and T. Parc 1999. Predatory behaviour by a red howler monkey (*Alouatta seniculus*) on green iguanas (*Iguana iguana*). *Neotropical Primates* 7(2):46-47.

de Thoisy, B. and C. Richard-Hansen 1997. Diet and social behaviour changes in a red howler monkey (*Alouatta seniculus*) troop in a highly degraded rain forest. *Folia Primatologica* 68(6):357-361.

de Thoisy, B., O. Louguet, F. Bayart and H. Contamin 2002. Behavior of squirrel monkeys (*Saimiri sciureus*) – 16 years on an island in French Guiana. *Neotropical Primates* 10(2):73-76.

Dew, J. L. 2002a. Synecology and seed dispersal in woolly monkeys (*Lagothrix lagotricha poeppigii*) and spider monkeys (*Ateles belzebuth belzebuth*) in Parque Nacional Yasuní, Ecuador). Unpublished Ph.D. thesis, Dissertation, Univeristy of California, Davis.

Dew, J. L. 2002b. How specialized are ripe-fruit specialists? Dietary selection in the face of sympatric competitors and shifting fruit abundance. [Abstract] *American Journal of Physical Anthropology* Suppl 34:63.

De Waal, F. B. M., L. M. Luttrell and M. E. Canfield 1993. Preliminary data on voluntary food sharing in brown capuchin monkeys. *American Journal of Primatology* 29:73-78.

De Waal, F. B. M. and J. M. Davis 2003. Capuchin cognitive ecology: Cooperation based on projected returns. *Neuropsychologia* 41(2):221-228.

Diamond, J. 1984. Historic extinctions: A Rosetta Stone for understanding prehistoric extinctions. Pp. 824-862 in: P. S. Martin and R. G. Klein (eds.), *Quaternary Extinctions: A Prehistoric Revolution*. The University of Arizona Press, Tucson.

Diamond, J. 1992. *The Third Chimpanzee*. Harper Collins, New York.

Diaz, D., M. Naegeli, R. Rodríguez, J. J. Nino-Vasquez, A. Moreno, M. E. Patarroyo, G. Pluschke and C. A. Daubenberger 2000. Sequence and diversity of MHC DQA and DQB genes of the owl monkey *Aotus nancymaaae*. *Immunogenetics* 51:528-537.

Di Bitetti, M. S. 2001. Home-range use by the tufted capuchin monkey (*Cebus apella nigritus*) in a subtropical rainforest of Argentina. *Journal of Zoology* 253(1):33-45.

Di Bitetti, M. 2002. Food-associated calls in the tufted capuchin monkey (*Cebus apella*). *Dissertation Abstracts International* B62(9):3883.

Di Bitetti, M. 2003. Food-associated calls of tufted capuchin monkeys (Cebus apella nigritus) are functionally referential signals. *Behaviour* 140(5):565-592.

Di Bitetti, M. S. and C. H. Janson 2001. Social foraging and the finder's share in capuchin monkeys, *Cebus apella*. *Animal Behaviour* 62(1):47-56.

Di Bitetti, M. S., E. M. L. Vidal, M. C. Baldovino and V. Benesovsky 2000. Sleeping site preferences in tufted capuchin monkeys (*Cebus apella nigritus*). *American Journal of Primatology* 50(4):257-274.

Didier, L. G. 1997. Leading behavior in a free ranging group of spider monkeys (*Ateles belzebuth*) in La Macarena, Colombia. *Field Studies of Fauna and Flora, La Macarena, Colombia*. 10:29-31.

Di Fiore, A. F. 1997. Ecology and behavior of lowland woolly monkeys (*Lagothrix lagotricha poeppigii*, Atelinae) in eastern Ecuador. Unpublished Ph.D. thesis, University of California at Davis.

Di Fiore, A. 2001. Ranging behavior and foraging ecology of lowland woolly monkeys (*Lagothrix lagotricha*). [Abstract] *American Journal of Physical Anthropology* (Suppl 32), p. 59.

Bibliography

Di Fiore, A. 2002. Molecular perspectives on dispersal in lowland woolly monkeys (*Lagothrix lagothricha poeppigii*). [resumen] *American Journal of Physical Anthropology* Suppl 34:63.

Di Fiore, A. 2003. Ranging behavior and foraging ecology of lowland woolly monkeys (*Lagothrix lagothricha poeppigii*) in Yasuni National Park, Ecuador. *American Journal of Primatology* 59(2):47-66.

Di Fiore, A. and P. S. Rodman 2001. Time allocation patterns of lowland woolly monkeys (*Lagothrix lagotricha poeppigii*) in a neotropical terra firma forest. *International Journal of Primatology* 22(3):449-480.

Dixson, A. F. 1980. Androgens and aggressive behavior in primates: A review. *Aggressive Behavior* 6:37-67.

Dixson, A. F. 1982. Some observations on the reproductive physiology and behavior of the owl monkey. *International Zoo Yearbook*. 22:115-19.

Dixson, A. F. 1994. Reproductive biology of the owl monkey. Pp. 113-132 in: J. F. Boer, R. W. Welker and I. Kakoma (eds.), *Aotus: the Owl Monkey*. Academic Press, New York.

Dixson, A. F. and D. Fleming 1981. Parental behaviour and infant development in owl monkeys (*Aotus trivirgatus griseimembra*). *Journal of Zoology, London* 194:25-39.

Dixson, A. F., J. S. Gardner and R. C. Bonnery 1980. Puberty in the male owl monkey (*Aotus trivirgatus griseimembra*): a study of physical & hormonal development. *International Journal of Primatology* 1(2):129-139.

Dobroruka, L. J. 1972. Social communication in the brown capuchin, *Cebus apella. International Zoo Yearbook* 12:43-45.

Dominy, N. J., P. A. Garber, J. C. Bicca-Marques and M. A. O. Azevedo-Lopes 2003. Do female tamarins use visual cues to detect fruit rewards more successfully than do males? *Animal Behaviour* 66(5):829-837.

Drapier, M., E. Addessi and E. Visalberghi 2003. Response of *Cebus apella* to foods flavored with familiar or novel odor. *International Journal of Primatology* 24(2):295-315.

Dumond, F. V. 1968. The squirrel monkey in a semi-natural environment. Pp. 87-145 in: I. Rosenblum and R. W. Cooper (eds.), *The Squirrel Monkey*. Academic Press, New York.

Dumond, F. V. and T. C. Hutchinson 1967. Squirrel monkey reproduction: the "fatted" male phenomenon and seasonal spermatogenesis. *Science* 158:1067-70.

Durham, N. M. 1971. Effects of altitude differences on group organization of wild black spider monkeys (*Aeles paniscus*). *Processes of the 3rd International Congress of Primatology, Zurich*, vol. III. Karger, Basel, pp. 32-40.

Durham, N. M. 1975. Some ecological, distributional, and group behavioral features of atelinae in southern Peru, with comments on interspecific relations. Pp. 87-103 in: R. H. Tuttle (ed.), *Socioecology and Psychology of Primates*. Mouton, The Hague.

Dutrillaux, B., M. Lombard, J. B. Carroll and R. D. Martin 1988. Chromosomal affinities of *Callimico goeldii* (Platyrrhini) and characterization of a Y-autosome translocation in the male. *Folia Primatologica* 50:230-236.

Easley, S. P. 1982. Ecology and behavior of *Callicebus torquatus*, Cebidae, primates. Unpublished Ph.D. thesis, Washington University, St. Louis.

Easley, S. P. and W. G. Kinzey 1986. Territorial shift in the yellow-handed titi monkey *(Callicebus torquatus). American Journal of Primatology* 11(4):307-318.

Egler, S. G. 1991. Double-toothed kites following tamarins. *Wilson Bulletin* 103(3):510-512.

Egozcue, J. 1969. Primates. Pp. 357-389 in: K. Benirshke (ed.), *Comparative Mammalian Cytogenetics*. Springer-Verlag, New York.

Eisenberg, J. F. 1973. Reproduction in two species of spider monkeys, *Ateles fusciceps* and *A. geoffroyi. Journal of Mammalogy* 54:955-957.

Eisenberg, J. F. 1976. Communication mechanisms and social integration in the black spider monkey, (*Ateles fusciceps robustus*) and related species. *Smithsonian Contributions to Zoology* 213:1-108.

Eisenberg, J. F. 1977. Comparative ecology and reproduction of New World monkeys. Pp. 13-22 in: D. G. Kleiman (ed.), *The Biology and Conservation of the Callitrichidae*. Smithsonian Instituion Press, Washington, D. C.

Eisenberg, J. F. 1979. Habitat, economy, and society: some correlations and hypotheses for the Neotropical primates. Pp. 215-262 in: I. S. Bernstein and E. O. Smith (ed.), *Primate Ecology and Human Origins*. Garland Press, New York.

Eisenberg, J. F. 1983. *Ateles geofroyi* (mono araña, mono colorado, spider monkey). Pp. 451-453 in: D. H. Janzen (ed.), *Costa Rican Natural History*. University of Chicago Press, Chicago.

Eisenberg, J. F. 1989. *Mammals of the Neotropics: The Northern Neotropics*, vol. 1. The University of Chicago Press, Chicago.

Eisenberg, J. F. and R. E. Kuehn 1966. The behavior of *Ateles geoffroyi* and related species. *Smithsonian Miscellaneous Collections* 151(8):1-iv, 1-63..

Eisenberg, J. F. and R. W. Thorington, Jr. 1973. A preliminary analysis of neotropical mammal fauna. *Biotropica* 5:150-161.

Elliot, D. G. 1909a. Descriptions of apparently new species and subspecies of *Cebus,* with remarks on the nomenclature of Linnaeus's *Simia apella* and *Simia capuucina. Bulletin American Museum of Natural History* 26:227-231.

Elliot, D. G. 1909b. Descriptions of apparently new species and subspecies of monkeys of the genera *Callicebus, Lagothrix, Papio, Pithecus, Cercopithecus, Erythrocebus* and *Presbytis. Annals of the Magazine of Natural History* 31(8):244-274.

Elliot, D. G. 1913. *A review of the primates.* Vols. I and II. American Museum Natural History Monographs, New York.

Emmons, L. H. 1984. Geographic variation in densities and diversities of non-flying mammals in Amazonia. *Biotropica* 16(3):210-222.

Emmons, L. H. 1997. *Neotropical Rainforest Mammals: A Field Guide.* 2ND Edition. The University of Chicago Press, Chicago.

Endler, J. A. 1977. Geographic variation, speciation and clines. *Monographs of Population Biology* 10.

Endler, J. A. 1982. Pleistocene forest in refuges: Fact or fancy? Pp. 641-657 in: G. T. Prance (ed.), *Biological Diversification in the Tropics*. Columbia University Press, New York.

Epple, G. 1967. Vergleichende Untersuchungen über Sexual - und Social - Verhalten der Krallenaffen (Hapalidae). *Folia Primatologica* 7:37-65.

Epple, G. 1968. Comparative studies on vocalization in marmoset monkeys (Hapalidae). *Folia Primatologica* 8:1-40.

Epple, G. 1970. Maintenance, breeding and development of marmoset monkeys (Callitrichidae) in captivity. *Folia Primatologica* 12:56-76.

Epple, G. 1971. Discrimination of the odor of males and females by the marmoset, *Saguinus fuscicollis* ssp. *Processes of the 3rd International Congress of Primatology, Zürich,* 3:166-171.

Epple, G. 1972. Social behavior of laboratory groups of *Saguinus fuscicollis*. Pp. 50-58 in: D. D. Bridgewater (ed.), *Saving the Lion Marmoset*. Wild Animal Propagation Trust, Wheeling.

Epple, G. 1973. The role of pheromones in the social communication of marmoset monkeys (Callitrichidae). *Journal of Reproductive Fertility*, Supp. 19:447-454.

Epple, G. 1974a. Olfactory communication in South American primates. *Annals of the New York Academy of Science*, 237-278.

Epple, G. 1974b. *Primate pheromones. (ed). Pheromones*. Pp. 366-385 in: H. Birch, Elsevier, Amsterdam.

Epple, G. 1975a. The behavior of marmoset monkeys (Callitrichidae). Pp. 195-239 in: L. A. Rosenblum (ed.), *Primate Behavior*, vol. 4. Academic Press, New York.

Epple, G. 1975b. Paternal behavior in *Saguinus fuscicollis* spp. (Callithrichidae). *Folia Primatologica* 24:221-238.

Bibliography

Epple, G. 1977. Notes on the establishment and maintenance of the pair bond in *Saguinus fuscicollis*. Pp. 231-237 in: D. G. Kleiman (ed.), *The Biology and Conservation of the Callitrichidae*. Smithsonian Institution Press, Washington, D.C.

Epple, G. 1978a. Lack of effects of castration on scent marking, displays and aggression in a South American primate (*Saguinus fuscicollis*). *Hormones and Behaviour* 11:139-150.

Epple, G. 1978b. Overt aggression and scent marking behaviour in *Saguinus fuscicollis*: the effects of castration in adulthood: a brief report. Pp. 191-195 in: H. Rothe, J. Wolters and J. Hearn (eds.). *Biology and Behavior of Marmosets*. Eigenverlag. Gottingen.

Epple, G. 1978c. Reproductive and social behavior of marmosets with special reference to captive breeding. *Primate Medicine* 10:50-62.

Epple, G. 1979. Gonadal control of male scent in the tamarin *Saguinus fuscicollis* (Callitrichidae, Primates). *Chemical Senses Flav.* 4:15-20.

Epple, G. 1981a. Effect of pair-bonding with adults on ontogenetic manifestation of aggressive behavior in a primate, *Saguinus fuscicollis*. *Behavioral Ecology and Sociobiology* 8:117-123.

Epple, G. 1981b. Effects of prepubertal castration on the development of the scent glands, scent marking and aggression in the saddleback tamarin (*Saguinus fuscicollis*, Callitrichidae, Primates). *Hormones and Behaviour* 15:54-67.

Epple, G. 1990. Sex differences in partner preference in mated pairs of saddle-back tamarins (*Saguinus fuscicollis*). *Behavioral Ecology and Sociobiology* 27:455-459.

Epple, G. and R. Lorenz 1967. Vorkommen, Morphologie, und Funcktion der Sternaldruse bei den Platyrrhini. *Folia Primatologica* 7:98-126.

Epple, G. and D. Moulton 1978. Structural organization and communicatory functions of olfaction in nonhuman primates. Pp. 1-22 in: ed. C. R. Noback (ed.), *Sensory Systems of Primates*. Plenum Press, New York.

Epple, G. and V. A. Cerny 1979. Effects of castration and social change on scent-marking behavior of *Saguinus fuscicollis* (Callitrichidae). *Folia Primatologica* 32:252-262.

Epple, G. abd V. Katz 1980. Social influences on first reproductive success and related behaviors in the saddleback tamarin (*Saguinus fuscicollis*, Callitrichidae). *International Journal of Primatology* 1:171-183.

Epple, G. and V. Katz 1984. Social influences on estrogen excretion and ovarian cyclicity in saddle back tamarins (*Saguinus fuscicollis*). *American Journal of Primatology* 6:215-28.

Epple, G., N. F. Golob and A. B. Smith III 1979. Odor communication in the tamarin, *Saguinus fuscicollis* (Callitrichidae): Behavioral and chemical studies. Pp. 117-130 in: F. J. Ritter (ed.), *Chemical Ecology: Odour Communication in Animals*. Elsevier, Amsterdam.

Epple, G., M. C. Alveario, N. F. Golob and A. B. Smith III 1980. Stability and attractiveness related to age of scent marks by saddleback tamarins (*Saguinus fuscicollis*). *Journal of Chemical Ecology* 6:735-748.

Epple, G., N. F. Golob, M. S. Cebule and A. B. Smith III 1981. Communication by scent in some Callitrichidae: an interdisciplinary approach. *Chemical Senses* 16:377-390.

Epple, G., M. C. Alveario and Y. Katz. 1982. The role of chemical communication in aggressive behavior and its gonadal control in the tamarin (*Saguinus fuscicollis*). Pp. 279-302 in: C. T. Snowdon, C. H. Brown and M. R. Peterson (eds.). *Primate Communication*. Cambridge University Press, Cambridge.

Epple, G., A. M. Belcher, I. Kuederling, U. Zeller, I. Scolnick, K. L. Greenfield and A. B. Smith, III. 1993. Making sense out of scents: Species differences in scent glands, scent-marking behaviour, and scent-mark composition in the Callitrichidae. Pp. 123 - 151 in: A. B. Rylands (ed.), *Marmosets and Tamarins: Systematics, Behaviour, and Ecology*. Oxford University Press, Oxford.

Erikson, G. E. 1963. Brachiation in New World monkeys and in anthropoid apes. *Symposia of the Zoological Society of London* 10:135-164.

Erkert, H. G. and J. Grobert 1986. Direct modulation of activity and body temperature of owl monkeys (*Aotus lemurinus griseimembra*) by low light intensities. *Folia Primatologica* 47:171-188.

Erxleben, J. C. P. 1777. *Systema regni animalis*, classis I, Mammalia. Lipsiae.

Escobar-Paramo, P. 1989a. The development of the wild black-capped capuchin (*Cebus apella*) in La Macarena, Colombia. *Field Studies of New World Monkeys, La Macarena, Colombia* 2:45-56.

Escobar-Paramo, P. 1989b. Social relations between infants and other group members in the wild black-capped capuchin (*Cebus apella*). *Field Studies of New World Monkeys, La Macarena, Colombia* 2:57-63.

Escobar-Paramo, P. 1990. Social relations between infants and other group members in the black-capped capuchin (*Cebus apella*) at Centro de Investigaciones Primatológicos La Macarena, Colombia. [Abstract] *Bulletin Ecological Society of America* 71(2, suppl.):148.

Estrada, A. 1982. Survey and census of howler monkeys (*Alouatta palliata*) in the rain forest of Los Tuxtlas, Veracruz, Mexico. *American Journal of Primatology* 2:363-372.

Estrada, A. 1983. Primate studies at the biological reserve Los Tuxtlas, Veracruz, Mexico. *Primate Specialist Group Newsletter* no. 3:21.

Estrada, A. 1984. Resource use by howler monkeys (*Alouatta palliata*) in the rain forest of Los Tuxtlas, Verzcruz, Mexico. *International Journal of Primatology* 5:105-131.

Estrada, A. and R. Coates-Estrada 1983. Rain forest in Mexico: Research and conservation at Los Tuxtlas. *Oryx* 17(4):201-204.

Estrada, A. and R. Coates-Estrada 1984a. Fruit-eating and seed dispersal by howling monkeys (*Alouatta palliata*) in the tropical rain forest of Los Tuxtlas, Mexico. *American Journal of Primatology* 6:77-91.

Estrada, A. and R. Coates-Estrada 1984b. Some observations on the present distribution and conservation of *Alouatta* and *Ateles* in southern Mexico. *American Journal of Primatology* 7:133-137.

Estrada, A. and R. Coates-Estrada 1985. A preliminary study of resource overlap between howling monkeys (*Alouatta palliata*) and other arboreal mammals in the tropical rain forest of Los Tuxtlas, Mexico. *American Journal of Primatology* 9:27-37.

Estrada, A. and R. Coates-Estrada 1986a. Use of leaf resources by howling monkeys (*Alouatta palliata*) and leaf-cutting ants (*Atta cephalotes*) in the tropical rain forest of the Los Tuxtlas, Mexico. *American Journal of Primatology* 10(1):51-66.

Estrada, A. and R. Coates-Estrada 1986b. Frugivory by howling monkeys (*Alouatta palliata*) at Los Tuxtlas, Mexico: Dispersal and fate of seeds. Pp. 933-104 in: A. Estrada and T. H. Fleming (eds.), *Frugivores and Seed Dispersal*. Dr. W. Junk, Dordrecht, The Netherlands.

Estrada, A. and R. Coates-Estrada 1991. Howler monkeys (*Alouatta palliata*), dung beetles (Scarabaeidae) and seed dispersal: Ecological interactions in the tropical rain forest of Los Tuxtlas, Mexico. *Journal of Tropical Ecology* 7:459-474.

Estrada, A. and R. Coates-Estrada 1993a. Aspects of ecological impact of howling monkeys (*Alouatta*) on their habitat: a review. *Estudios Primatologicos en Mexico* 1:87-117.

Estrada, A. and R. Coates-Estrada 1993b. [Contraction and fragmentation of forests, and populations of forest-dwelling primates: The case of Los Tuxtlas, Veracruz.] (in Spanish). *Ciencia y el Hombre* 18:45-70.

Estrada, A. and R. Coates-Estrada 1996. Tropical rain forest fragmentation and wild populations of primates at Los Tuxtlas, Mexico. *International Journal of Primatology* 17(5):759-783.

Estrada, A. and T. H. Fleming (eds.) 1986. *Frugivores and Seed Dispersal*. Dr. W. Junk, The Netherlands.

Estrada, A. and W. Trejo 1978. Dieta y selectividad en el mono aullador (*Alouatta palliata*) en la selva alta perennifolia de Estación de Biología Tropical «Los Tuxtlas» in Veracruz. *Memorias, II Congreso Nacional de Zoología, Facultad de Ciencias Biológicas, Universidad Autonoma de Nuevo Leon, Mexico*, pp. 493-517.

Bibliography

Estrada, A., W. Trejo, E. Velarde, J. Ellefson and R. Coffin 1977. Riseña de los hábitos del mono allador (*Alouatta* sp.) e informe del estudio preliminar de este primate en la Estación de Biología Tropical Los Tuxtlas, Veracruz, Mexico. *Boletín de Estudios de Medicina Biológica* 29:401-417.

Estrada, A., R. Coates-Estrada, C. Vasquez-Yanes and A. Orozco-Segovia. 1984. Comparison of frugivory by howling monkeys (*Alouatta palliata*) and bats (*Artibeus jamaicensis*) in the tropical rain forest of Los Tuxtlas, Mexico. *American Journal of Primatology* 7:3-13.

Estrada, A., G. Halfter, R. Coates-Estrada and D. A. Meritt Jr. 1993a. Dung beetles attracted to mammalian herbivore (*Alouatta palliata*) and omnivore (*Nasua narica*) dung in the tropical rain forest of Los Tuxtlas, Mexico. *Journal of Tropical Ecology* 9(2):45-54.

Estrada, A., R. Coates-Estrada, D. Merritt Jr. and D. Curiel. 1993b. Patterns of frugivore species richness and abundance in forest islands and in agricultural habitats at Los Tuxtlas, Mexico. *Vegetatio* 107/108: 245-257. also in *Frugivory and Seed Dispersal*, eds. T. H. Fleming *et al.*, 1993.

Estrada, A. A. Anzures and R. Coates-Estrada 1999a. Tropical rain forest fragmentation, howler monkeys (*Alouatta palliata*) and dung beetles at Los Tuxtlas, Mexico. *American Journal of Primatology* 48(4):253-262.

Estrada, A., S. Juan-Solano, T. Ortiz martinez and R. Coates-Estrada 1999b. Feeding and general activity patterns of a howler monkey (*Alouatta palliata*) troop living in a forest fragment at Los Tuxtlas, Mexico. *American Journal of Primatology* 48(3):167-183.

Estrada, A., Y. Garcia, D. Muñoz and B. Franco 2001. Survey of the population of howler monkeys (*Alouatta palliate*) at Yumka Park in Tabasco, Mexico. *Neotropical Primates* 9(1):12-15.

Estrada, A., L. Lluecke, S. Van Belle, K. French, D. Muñoz, Y. Garcia, L. Castellanos and A. Mendoza 2002. The black howler monkeys (*Alouatta pigra*) and spider monkey (*Ateles geoffroyi*) in the Mayan site of Yacchilan, Chiapas, Mexico: A preliminary survey. *Neotropical Primates* 10(2):89-95.

Fairbanks, L. 1974. Analysis of subgroup structure and processes in a captive squirrel monkey (*Saimiri sciureus*) colony. *Folia Primatologica* 21:209-224.

Fajardo-P., A. and J. De la Ossa 1994. Censo preliminar de los primates de una reserva protegida en la Serranía de Coraza-Montes de María, Sucre, Colombia. *Trianea* 5(5):289-303.

Fedigan, L. M. 1983. Demographic trends in the *Alouatta palliata* and *Cebus capucinus* populations of Santa Rosa National Park, Costa Rica. Pp. 285-293 in: J. Else and P. Lee (eds.), *Primate Ecology and Conservation*, vol. 2. Cambridge University Press, Cambridge.

Fedigan, L. M. 1990. Vertebrate predation in *Cebus capucinus*: meat eating in a neotropical monkey. *Folia Primatologica* 54:196-205.

Fedigan, L. M. 1993. Sex differences and intersexual relations in adult white-faced capuchins, *Cebus capucinus*. *International Journal of Primatology* 14:853-877.

Fedigan, L. M. 2003. Restoring monkeys to tropical habitats: Lessons from a Costa Rican dry forest. [Abstract] *American Journal of Primatology* 60(Suppl 1), p. 35.

Fedigan, L. M. and M. J. Baxter 1984. Sex differences and social organization in free-ranging spider monkeys (*Ateles geoffroyi*). *Primates* 25(3):279-294.

Fedigan, L. and K. Jack 2001. Neotropical primates in a regenerating Costa Rican dry forest: A comparison of howler and capuchin population patterns. *International Journal of Primatology* 22(5):689-713.

Fedigan, L. M., L. Fedigan and C. Chapman 1985. A census of *Alouatta palliate* and *Cebus capucinus* monkeys in Santa Rosa National Park, Costa Rica. *Brenesia* 23:309-322.

Fedigan, L. M., L. M. Rose and R. M. Avila 1996. See how they grow: Tracking capuchin monkey (*Cebus capucinus*) populations in a regenerating Costa Rica dry forest. Pp. 289-307 & 543-544 in: M. A. Norconk, A. L. Rosenberger and P. A. Garber (eds.), *Adaptive Radiations of Neotropical Primates*. Plenum Press, New York.

Fedigan, L. M., L. M. Rose and R. M. Avila 1998. Growth of mantled howler groups in a regenerating Costa Rican dry forest. *International Journal of Primatology* 19(3):405-432.

Feer, F. 1999. Effects of dung beetles (Scarabaeidae) on seeds dispersed by howler monkeys (*Alouatta seniculus*) in the French Guianan rain forest. *Journal of Tropical Ecology* 15(2):129-142.

Feistner, A. T. C. and W. C. McGrew 1989. Food-sharing in primates: A critical review. Pp. 21-36 in: P. K. Seth and S. Seth (eds.), *Perspectives in Primate Biology*. Vol. 3. Today & Tomorrow's Printers and Publishers, New Delhi.

Feistner, A. T. C. and E. C. Price 1990. Food-sharing in cotton-top tamarins (*Saguinus oedipus*). *Folia Primatologica* 54:34-45.

Fernandes, M. E. B. 1991. Tool use and predation of oysters (*Crassostrea rhizophorae*) by the tufted capuchin, *Cebus apella*, in brackish water mangrove swamp. *Primates* 32(4):529-531.

Ferdandes, M. E. B. 1993a. [Evidence of the adaptation of Neotropical primates to mangrove areas with emphasis on capuchin monkeys *Cebus apella apella*.] (in Portugese), *Primatologia no Brasil* 4:67-80.

Fernandes, M. E. B. 1993b. Tail-wagging as a tension relief mechanism in pitheciines. *Folia Primatologica* 61:52-56.

Ferrari, S. F. 1999. [Primate biogeography of Amazonia.] [abstract] *Livro de Resumos IX Congresso Brasileiro de Primatologia*. S. L. Mendes (ed.), Santa Teresa.

Ferreira, R., P. Lee and P. Izar (2002). Alpha female death: Implications on social dynamics of a *Cebus apella* group. *Caring for Primates. Abstracts of the XIXth Congress. The International Primatological Society*, Mammalogical Society of China, Beijing 2002:315-316.

Ferreira, R., B. D. Resende, M. Mannu, E. B. Otón and P. Izar 2002. Bird predation prey-transfer in brown capuchin monkeys (*Cebus apella*). *Neotorpical Primates* 10(2):84-88.

Fess, K. J. 1975a. Observations on feral and captive *Cebuella pygmaea*, with comparisons to *Callithrix geoffroyi* and *Oedipomidas oedipus*. *Journal of Marmoset Breeding Farm* 1:12-21.

Fess, K. J. 1975b. Observations on a breeding pair of cotton-top pinches (*Oedipomidas oedipus*) and nine twin births and three triplet births. *Journal of Marmoset. Breed Farm.* 1:4-12.

Festa, E. 1903. Viaggio de Dr. Enrico Festa nel Darien Nell-Ecuador a regioni vicine: mammiferi pt. 1 (Primates). *Bull. Mus. Zool. Comp. Anat. Torino* 18(435):1-9.

Figueroa, R. 1989. Social interactions of a fourth month adopted infant in a wild group of *Alouatta seniculus*. *Field Studies of New World Monkeys, La Macarena, Colombia*. 2:32–39.

Fleagle, J. 1988. *Primate Adaptation and Evolution*. Academic Press, San Diego, California.

Fleagle, F. 1990. New fossil platyrrhines from the Pinturas formation, southern Argentina. *Journal of Human Evolution* 19:61-85.

Fleagle, J. 1999. *Primate Adaptatin and Evolution*. 2nd Edition. Academic Press, San Diego, xvii, 596 pp.

Fleagle, J. G. and R. A. Mittermeier 1980. Locomotor behavior, body size, and comparative ecology of seven Surinam monkeys. *American Journal of Physical Anthropology* 52:301-314.

Fleagle, J. G. and T. M. Brown 1983. New primate fossils from late Oligocene (Colhuehuapian) localities of Chubut Province, Argentina. *Folia Primatologica* 41:240-266.

Fleagle, J. G. and D. J. Meldrum 1988. Locomotor behavior and skeletal morphology of two sympatric pithecine monkeys, *Pithecia pithecia* and *Chiropotes satanas*. *American Journal of Primatology* 16:227-249.

Fleagle, J. G., R. A. Mittermeier and A. L. Skopec 1981. Differencial habitat use by *Cebus apella* and *Saimiri sciureus* in Central Surinam. *Primates* 22:361-367.

Fleagle, J. G., D. W. Powers, G. C. Conroy and J. P. Walters 1987. New fossil platyrrhines from Santa Cruz Province, Argentina. *Folia Primatologica* 48:65-77.

Fleagle, J. G., R. F. Kay and M. R. L. Anthony 1997. Fossil New World monkeys. Pp. 473-495 in: R. F. Kay, R. H. Madden, R. L. Cifelli and J. J. Flynn (eds.), *Vertebrate Paleontology in the Neotropics*. The Smithsonian Institution Press, Washington, D. C.

Bibliography

Fleagle, J. G., C. H. Janson and K. E. Reed (eds.) 1999a. *Primate Communities*. Cambridge University Press, Cambridge.

Fleagle J. G., C. H. Janson and K. E. Reed 1999b. Spatial and temporal scales in primate community structure. Pp. 284-288 in: J. G. Fleagle, C. H. Janson and K. E. Reed (eds.), *Primate Communities*. Cambridge University Press, Cambridge.

Flynn, J. J., A. R. Wyss, R. Charrier and C. C. Swisher 1995. An early Miocene anthropoid skull from the Chilean Andes. *Nature* 373(6515):603-607.

Fobes, J. S. and J. E. King 1982. Vision: the dominant primate modality. Pp. 219-243 in: J. L. Foges and J. E. King (eds.), *Primate Behavior*. Academic Press, New York.

Fontaine, R. 1980. Observations on the foraging association of double-toothed kites and white-faced capuchin monkeys. *The Auk* 97:94-98.

Fontaine, R. 1981. The uakaris, genus *Cacajao*. Pp. 443-493 in: A. F. Coimbra-Filho and R. A. Mittermeier (eds.), *Ecology and Behavior of Neotropical Primates*. Academia Brasileira de Ciências, Rio de Janeiro.

Fooden, J. 1963. A revision of the woolly monkey (genus *Lagothrix*). *Journal of Mammalogy* 44(2):213-247.

Ford, S. 1986. Subfossil platyrrhine tibia (Primates: Callitrichidae) from Hispaniola: a possible further example of island gigantism. *American Journal of Physical Anthropology* 70:47-62.

Ford, S. F. 1988. Postcranial adaptations of the earliest platyrrhine. *Journal of Human Evolution* 17(1-2):155-192.

Ford, S. F. 1990a. Platyrrhine evolution in the West Indies. *Journal of Human Evolution* 19:237-254.

Ford, S. F. 1990b. Locomotor adaptations of fossil platyrrhines. *Journal of Human Ecology* 19:141-173.

Ford, S. M. 1994. Taxonomy and distribution of the owl monkey. Pp. 1-57 in: J. F. Baer, R. E. Weller and I. Kakoma (eds.), *Aotus: The Owl Monkey*. Academic Press, New York.

Ford, S. and G. S. Morgan. 1986. A new ceboid femur from the late Pleistocene of Jamaica. *Journal of Vertebrate Paleontology* 6:461-481.

Forero, O. 1987. Contribución al conocimiento del uso espacial de *Callicebus torquatus lugens,* en el bajo Apaporis, Vaupés. Unpublished bachelor's thesis, Pontificia Universidad Javeriana, Santafe de Bogotá .

Fragaszy, D. M. 1978. Contrasts in feeding behavior in squirrel and titi monkeys. Pp. 363-367 in: D. J. Chivers and J. Herbert (eds.), *Recent Advances in Primatology*, vol. 1. Academic Press, New York.

Fragaszy, D. M. 1990. Early behavioral development in capuchins (*Cebus*). *Folia Primatology* 54:119-128.

Fragaszy, D. M., S. Schwarz and D. Shimosaka 1982. Longitudinal observations of care and development of infant titi monkeys (*Callicebus moloch*). *American Journal of Primatology* 2:191-200.

Frailey, C. D., E. L. Lavina, A. Rancy and J. Pereira de Souza Filho 1988. A proposed Pleistocene/Holocene lake in the Amazon basin and its significance to Amazonian geology and biogeography. *Acta Amazonica* 18:119-143.

Fragaszy, D. M., E. Visalberghi and J. G. Robinson 1990. Variability and adaptability in the genus *Cebus*. *Folia Primatology* 54(3-4):114-118.

Freese, C. 1975. A census of non-human primates in Peru. Pp. In: Primate *Censusing Studies in Peru and Colombia*. Report to the National Academy of Sciences on project amro-0719. Pan American Health Organization, Washington, D. C.

Freese, C. 1976a. Censusing *Alouatta palliata, Ateles geoffroyi* and *Cebus capucinus* in the Costa Rican dry forest. Pp. 4-9 in: R. W. Thorington, Jr. and P. G. Heltne (eds.), *Neotropical Primates: Field Studies and Conservation*. National Academy of Sciences, Washington, D. C.

Freese, C. 1976b. Predation on swollen-thorn acacia ants by white-faced monkeys *Cebus capucinus*. *Biotropica* 8(4):278-281.

Freese, C. 1977a. Food habits of white-faced capuchins *Cebus capucinus* l. (Primates: Cebidae) in Santa Rosa National Park, Costa Rica. *Brenesia* 10/11:43-56.

Freese, C. 1977b. Population densities and niche separation in some Amazonian monkey communities. Unpublished Ph.D. thesis, Johns Hopkins University, Maryland.

Freese, C. 1978. The behavior of white-faced capuchins (*Cebus capucinus*) at a dry-season waterhole. *Primates* 19(2):276-286.

Freese, C. 1983. *Cebus capucinus* (mono cara blanca, white-faced capuchin). Pp. 458-460 in: D. H. Janzen (ed.), *Costa Rican Natural History*. University of Chicago Press, Chicago.

Freese, C. H. and J. R. Oppenheimer 1981. The capuchin monkeys, genus *Cebus*. Pp. 331-391 in: A. F. Coimbra-Filho and R. A. Mittermeier (eds.), *Ecology and Behavior of Neotropical Primates*, vol. 1. Academica Brasileira de Ciências, Rio de Janeiro.

Freese, C. H., M. Neville and R. Castro 1976. The conservation status of Peruvian primates. *Laboratory Primate Newsletter* 15:1-9.

Freese, C. H., M. A. Freese and N. Castro 1977. The status of callitrichids in Peru. Pp. 121-130 in: D. G. Kleiman (ed.), *The Biology and Conservation of the Callitrichids*, Smithsonian Institution Press, Washington, D. C.

Freese, C. H., P. G. Heltne, R. N. Castro and H. Whitesides 1982. Patterns and determinants of monkey densities in Peru and Bolivia, with notes on distributions. *International Journal of Primatology* 3:53-90.

French, J. A. 1982. The role of scent marking in social and sexual communication in the tamarin, *Saguinus o. oedipus*. Unpublished Ph.D. thesis, University of Wisonsin, Madison.

French, J. A. and C. T. Snowdon 1981. Sexual dimorphism in response to unfamiliar intruders in the tamarin, *Saguinus oedipus*. *Animal Behaviour* 29:822.

French, J. A. and J. Cleveland 1984. Scent marking in the tamarin, *Saguinus oedipus*: Sex differences and ontogeny. *Animal Behaviour* 32:615-623.

French, J. A., D. H. Abbott, G. Scheffler, J. A. Robinson and R. W. Goy 1983. Cyclic excretion of urinary oestrogens in female tamarins (*Saguinus oedipus*). *Journal of Reproduction and Fertility* 68(1):177-184.

French, J. A., D. H. Abbott and C. T. Snowdon 1984. The effect of social environment on estrogen excretion, scent marking, and sociosexual behavior in tamarins (*Saguinus oedipus*). *American Journal Primatology* 6:155-167.

Froehlich, J. W. and P. H. Froehlich 1986. Dermaglyphics and subspecific systematics of mantled howler monkeys (*Alouatta palliata*). Pp. 107-121 in: D. M. Taub and F. A. King (eds.), *Current Perspectives in Primate Biology*. Van Nostrand Reinhold Co., New York.

Froehlich, J. W. and R. W. Thorington, Jr. 1982a. The genetic structure and socioecology of howler monkeys (*Alouatta palliata*) on Barro Colorado Island. Pp. 291-305 in: E. G. Leigh, Jr., A. S. Rand and D. M. Windsor (eds.), *The Ecology of a Tropical Forest: Seasonal Rhythms and Long-term Changes*. Smithsonian Institution Press, Washington, D. C.

Froehlich, J. W. and R. W. Thorington, Jr. 1982b. Food limitation on a small island and the regulation of population size in mantled howling monkeys (*Alouatta palliata*). *American Journal of Physical Anthropology* 57:190.

Froehlich, J. W., R. W. Thorington, Jr. and J. S. Otis 1981. The demography of howler monkeys (*Alouatta palliata*) on Barro Colorado Island, Panama. *International Journal of Primatology* 2:207-236.

Froehlich, J. W., Supriatna, J. and P. H. Froehlich 1991. Morphometric analyses of *Ateles*: systematic and biogeographic implications. *American Journal of Primatology* 25:1-22.

Galbreath, G. J. 1983. Karyotypic evolution in *Aotus*. *American Journal of Primatology* 4:245-251.

Galef, B. G., R. A. Mittermeier and R. C. Bailey 1976. Predation by the tayra (*Eira barbara*). *Journal of Mammalogy* 57:760-761.

Bibliography

Galetti, M. 1990. Predation on the squirrel, *Sciurus aestuans* by capuchin monkeys, *Cebus apella*. *Mammalia* 54(1):152-154.

Galetti, M., F. Pedroni and M. Paschoal 1994. Seasonal diet of capuchin monkeys (*Cebus apella*) in a semideciduous forest in south-east Brazil. *Journal of Tropical Ecology* 10(1):27-39.

Galindo, P. and S. Srihongse 1967. Evidence of recent jungle yellow-fever activity in eastern Panama. *Bulletin of the World Health Organization* 36:151-161.

Garber, P. A. 1980a. Locomotor behavior and feeding ecology of the Panamanian tamarin (*Saguinus oedipus geoffroyi*, Callitrichidae, primates. Unpublished Ph.D. thesis, Washington University, St. Louis.

Garber, P. A. 1980b. Locomotor behavior and feeding ecology of the Panamanian tamarin (*Saguinus oedipus geoffroyi*, Callitrichidae, Primates). *International Journal of Primatology* 1:185-201.

Garber, P. A. 1984. Proposed nutritional importance of plant exudates in the diet of the Panamanian tamarin, *Saguinus oedipus geoffroyi*. *International Journal of Primatology* 5:1-15.

Garber, P. A. 1988a. Diet, foraging patterns, and resource defense in a mixed species troop of *Saguinus mystax* and *Saguinus fuscicollis* in Amazonian Peru. *Behaviour* 105:18-34.

Garber, P. A. 1988b. Foraging decisions during nectar feeding by tamarin monkeys (*Saguinus mystax* and *Saguinus fuscicollis*, Callitrichidae, Primates) in Amazonian Peru. *Biotropica* 20:100-106.

Garber, P. A. 1990a. Role of spatial memory in primate foraging patterns: *Saguinus mystax* and *Saguinus fuscicollis*. *American Journal of Primatology* 19:203-216.

Garber, P. 1990b. A comparative study of positional behavior in three species of tamarin monkeys. [Abstract] *American Journal of Physical Anthropology* 81(2):225.

Garber, P. A. 1991a. A comparative study of positional behavior in three species of tamarin monkeys. *Primates* 32(1):219-230.

Garber, P. A. 1991b. Seasonal variation in diet and ranging patterns in two species of tamarin monkeys. *American Journal of Physical Anthropology* (suppl.) 12:75.

Garber, P. A. 1992. Vertical clinging, small body size, and the evolution of feeding adaptations in the callitrichinae. *American Journal of Physical Anthropology* 88(4):469-482.

Garber, P. A. 1993a. Seasonal patterns of diet and ranging in two species of tamarin monkeys: Stability versus variability. *International Journal of Primatology* 14(1):145-166.

Garber, P. A. 1993b. Feeding ecology and behaviour of the genus *Saguinus*. Pp. 273-295 in: A. B. Rylands (ed.), *Marmosets and Tamarins: Systematics, Behaviour, and Ecology*. Oxford University Press, Oxford,.

Garber, P. A. 1994. Aspects of fruit eating and seed dispersal in Panamanian (*Saguinus geoffroyi*) and moustached tamarins (*Saguinus mystax*). *AZA [American Zoo and Aquarium ASS] Regional Conference Proceedings* pp. 364-369.

Garber, P. A. 1995. Fruit feeding and seed dispersal in two species of tamarin monkeys (*Saguinus geoffroyi* and *Saguinus mystax*). *American Journal of Physical Anthropology* Suppl 20:95-96.

Garber, P. A. 2000a. Evidence for the use of spatial, temporal, and social information by some primate foragers. Pp. 261-298 in: S. Boinski and P. A. Garber (eds.), *On the Move: How and Why Animals Travel in Groups.* University of Chicago Press, Chicago.

Garber, P. A. 2000b. The behavioral ecology of mixed species troops of *Callimico goeldii, Saguinus labiatus* and *S. fuscicollis* in northwestern Brazil. [Abstract] *American Journal of Physical Anthropology* (Suppl 30):155.

Garber, P. A. and M. F. Teaford 1986. Body weights in mixed species troops of *Saguinus mystax mystax* and *Saguinus fuscicollis nigrifrons* in Amazonian Peru. *American Journal of Physical Anthropology* 71(3):331-336.

Garber, P. A. and L. M. Paciulli 1997. Experimental field study of spatial memory and learning in wild capuchin monkeys (*Cebus capucinus*). *Folia Primatologica* 68(3-5):236-253.

Garber, P. A. and J. A. Rehg 1998. Preliminary field study of positional behavior and habitat preference in *Callimico goeldii*. *American Journal of Physical Anthropology* (Suppl 26):85-86.

Garber, P. A. and J. A. Rehg 1999. The ecological role of the prehensile tail in white-faced capuchins (*Cebus capucinus*). *American Journal of Physical Anthropology* 110(3):325-339.

Garber, P. A. and S. R. Leigh 2001. Patterns of positional behavior in mixed-species troops of *Callimico goeldii, Saguinus labiatus,* and *Saguinus fuscicollis* in northwestern Brazil. *American Journal of Primatology* 54(1):17-31.

Garber, P. A. and J. C. Bicca-Marques 2002. Evidence of predator sensitive foraging and traveling in single- and mixed-species tamarin troops. Pp. 138-153 in: L. E. Miller (ed.), *Eat or Be Eaten: Predator Sensitive Foraging Among Primates.* Cambridge University Press, New York

Garber, P. A., J. D. Pruetz and A. Lavalle 1999a. A preliminary study of mantled howling monkey (*Alouatta palliata*) ecology and conservation on Isla de Ometepe, Nicaragua. *American Journal of Physical Anthropology Suppl.* 28:133.

Garber, P. A., J. D. Pruetz, A. C. Lavallee and S. G. Lavallee 1999b. A preliminary study of mantled howling monkey (*Alouatta palliate*) ecology and conservation on Isla de Ometepe, Nicaragua. *Neotropical Primates* 7(4):113-117.

Garcia, J. E. 1993. Comparisons of estimated densities computed for *Saguinus fuscicollis* and *Saguinus labiatus* using line-transect sampling. *Primate Report* 37:19-29.

Garcia del Valle, Y., D. Muñoz, M. Magana-Alejandro, A. Estrada and B. Franco 2001. [Use of plant foods by howler monkeys, *Alouatta palliata,* in Yumka Park, Tabasco, Mexico.] *Neotropical Primates* 9(3):112-118.

Gaulin, S. J. C. 1977. The ecology of *Alouatta seniculus* in Andean cloud forest. Unpublished Ph.D. thesis, Harvard University, Cambridge, Mass.

Gaulin, S. and C. Gaulin 1982. Behavioral ecology of *Alouatta seniculus* in Andean cloud forest. *International Journal of Primatology* 3:53-90.

Gaulin, S. J. C., D. H. Knight and C. K. Gaulin 1980. Local variance in *Alouatta* group size and food availability on Barro Colorado Island. *Biotropica* 12:137-43.

Gebo, D. L. 1988. Foot morphology and locomotor adaptation in Eocene primates. *Folia Primatologica* 50(1–2):3–41.

Gebo, D. L. 1989. New platyrrhine tali from La Venta. [Abstract] *American Journal of Physical Anthropology* 78(2):226.

Gebo, D. L. 1992. Locomotor and postural behavior in *Alouatta palliata* and *Cebus capucinus*. *American Journal of Primatology* 26:277-290.

Gebo, D. L. and E. L. Simons 1987. Morphology and locomotor adaptations of the foot in early Oligocene anthropoids. *American Journal of Physical Anthropology* 74(1):83-102

Gebo, D. L., M. Dagosto, A. L. Rosenberger and T. Setoguchi 1990. New platyrrhine tali from La Venta, Colombia. *Journal of Human Evolution* 19:737-746.

Gentry, A. H. 1982. Patterns of neotropical plant species diversity. Pp. 1-84 in: Hecht, Wallace and Prance (eds.), *Evolutionary Biology* 15:1-84.

Gentry, A. H. 1989. Diversidad florística y fitogeográfica de la Amazonia. Pp. 65-70 in: J. Hernández C., J. V. Rodríguez M., H. Chiriví G. and S. Sánchez P., *Investigación y Manejo de la Amazonia.* INDERENA, Bogotá.

Gentry, A. H. 1990. La región Amazónica. Pp. 53-90 in: B. Villegas J. and C. Uribe H. (eds.), *Selva Húmeda de Colombia.* Villegas Editores, Bogotá.

Geoffroy, I. 1843. Sur les singes Américains composant les genres nyctipithèque, *Saimiri* et callitriche. *Compt. rendus hebdom. Séanc. Acad. Sci. Paris.* 16:1150-1153.

Gil, G. and S. Heinonen 1993. [Sighting of the tufted capuchin (*Cebus apella*) in Formosa province, Argentina.] *Boletín Primatológico Latinoamericano* 4(1):15-17.

Bibliography

Gilbert, K. A. 1995. Endoparasitic infection in red howling monkeys (*Alouatta seniculus*) in the central Amazonian basin: A cost of sociality?—[publ. 1994] *Dissertation Abstracts International* A55(12):3901.

Gingerich, P. D. 1980. Eocene Adapidae, paleobiogeography and the origin of South American Platyrrhini. Pp. 123-138 in: R. L. Chiochon and A. B. Chiarelli (eds.), *Evolutionary Biology of the New World Monkeys and Continental Drift.* Plenum Press, New York.

Glander, K. E. 1974. Baby-sitting, infant sharing, and adoptive behavior in mantled howling monkeys. *American Journal of Physical Anthropology* 41:482.

Glander, K. E. 1975a. Habitat and resource utilization: An ecological view of social organization in mantled howling monkeys. Unpublished Ph.D. thesis, University of Chicago, Chicago.

Glander, K. E. 1975b. Habitat description and resource utilization: a preliminary report on mantled howling monkey ecology. Pp. 37-57 in: R. Tuttle (ed.), *Socioecology and Psychology of Primates.* Mouton, The Hague.

Glander, K. E. 1977. Poison in a monkey's garden of eden. *Natural History* 86:34-64.

Glander, K. E. 1978a. Howling monkey feeding behavior and plant secondary compounds: a study of stategies. Pp. 561-573 in: G. G. Montgomery (ed.), *The Ecology of Arboreal Folivores.* Smithsonian Institution Press, Washington, D. C.

Glander, K. E. 1978b. Drinking from arboreal water sources by mantled howling monkeys (*Alouatta palliata* Gray). *Folia Primatologica* 29:206-217.

Glander, K. E. 1979. Feeding associations between howling monkeys and basilisk lizard. *Biotropica* 11:235-236.

Glander, K. E. 1980. Reproduction and population growth in free-ranging mantled howling monkeys. *American Journal of Physical Anthropology* 53:25-36.

Glander, K. E. 1981. Feeding patterns in mantled howling monkeys. Pp. 231-257 in: A. C. Kamil and T. D. Sargent (eds.), *Foraging Behavior: Ecological, Ethological, and Psychological Approaches.* Garland Press, New York.

Glander, K. E. 1982. The impact of plant secondary compounds on primate feeding behavior. *Yearbook of Physical Anthropology* 25:1-18.

Glander, K. E. 1983. *Alouatta palliata* (congo, howling monkey, howler monkey). Pp. 448-449 in: D. H. Janzen (ed.), *Costa Rican Natural History.* The University of Chicago Press, Chicago.

Glander, K. E. 1992a. Dispersal patterns in Costa Rican mantled howling monkeys. *International Journal of Primatology* 13(4):415-436.

Glander, K. E. 1992b. Selecting and processing food. Pp. 65-68 in: S. Jones, R. Martin and D. Pilbeam (eds.), *The Cambridge Encyclopedia of Human Evolution.* Cambridge University Press, Cambridge.

Glander, L. E., L. M. Fedigan, L. Fedigan and C. Chapman 1991. Field methods for the capture and measurement of three monkey species in Costa Rica. *Folia Primatologica* 57:70-82.

Goldizen, A. W. 1986. Tamarins and marmosets: Communal care of their offspring. Pp. 34-43 in: B. B. Smuts, D. L. Cheny, R. M. Seyfarth, R. W. Wrangham and T. T. Struhsaker (eds.), *Primate Societies.* University of Chicago Press, Chicago.

Goldizen, A. W. 1987a. Facultative polyandry and the role of infant-carrying in wild saddle-back tamarins (*Saguinus fuscicollis*). *Behavioral Ecology and Sociobiology* 20:99-109.

Goldizen, A. W. 1987b. Tamarins and marmosets: communal care of offspring. Pp. 34-43 in: B. B. Smuts, D. L. Cheney, R. M. Seyfarth, R. W. Wrangham and T. T. Struhsaker (eds.), *Primate Societies.* The University of Chicago Press, Chicago.

Goldizen, A. W. 1989. Social relationships in a cooperatively polyandrous group of tamarins (*Saguinus fuscicollis*). *Behavioral Ecology and Sociobiology* 24:79-89.

Goldizen, A. W. 1990. A comparative perspective on the evolution of tamarin and marmoset social systems. *International Journal of Primatology* 11(1):63-83.

Goldizen, A. A. and J. Teborgh 1986. Cooperative polyandry and helping behavior in saddle-backed tamarins (*Saguinus fuscicollis*). Pp. 191-198 in: J. G. Else and P. C. Lee (eds.), *Primate Ecology and Conservation*. Cambridge University Press, Cambridge.

Goldizen, A. W. and J. Terborgh 1989. Demography and dispersal patterns of a tamarin population: possible causes of delayed breeding. *The American Naturalist* 134(2):208-224.

Goldizen, A. W., J. Mendelson, M. van Vlaardingen and J. Terborgh 1996. Saddle-back tamarind (*Saguinus fuscicollis*) reproductive strategies: Evidence from a thirteen-year study of a marked population. *American Journal of Primatology* 38(1):57-83.

Goldman, E. A. 1914. Descriptions of five new mammals from Panama. *Smithsonian misscelaneous Collections* 63(5):1-7.

Gómez-Posada, C. 2003. Variación en el uso del tiempo y el espacio de *Cebus apella* (Primates: Cebidae) de acuerdo a la disponibilidad de los recursos principales en la dieta. Maestría, Postgrado en Ciencias Biológicas, Universidad del Valle, Cali.

Gonzalez-Kirchner, J. P. 1999. Habitat use, population density and subgrouping pattern of the Yucatan spider monkey (*Ateles geoffroyi yucatanensis*) in Quintana Roo, Mexico. *Folia Primatologica* 70(1):55-60.

Goodman, M., C. A. Porter, J. Czelusniak, S. L. Page, H. Shneider, J. Shoshani, G. Gunnell and C. P. Groves 1998. Toward a phylogenetic classification of primates based on DNA evidence complemented by fossil evidence. *Molecular Phylogenetics and Evolution* 9(3):585-598.

Gray, J. A. 1845. On the howling monkeys (*Mycetes*, Illiger). *Ann. Mag. Nat. Hist.* 16:217-221.

Gray, J. A. 1849. On some new or little-known species of monkeys. *Proc. Zool. Soc. Lond.* 7-10.

Green, K. M. 1978. Primate censusing in northern Colombia: A comparison of two techniques. *Primates* 19:537-550.

Green, R., R. E. Whalen, B. Rutley and C. Battie 1972. Dominance hierarchy in squirrel monkeys (*Saimiri sciureus*): Role of the gonads and androgen on genital display and feeding order. *Folia Primatologica* 18:185-195.

Greenlaw, J. S. 1967. Foraging behavior of the double-toothed kite in association with white-faced monkeys. *The Auk* 84:596-597.

Gros-Louis, J. 2001. Food-associated calls in white-faced capchin monkeys (*Cebus capucinus*): Different functions from the perspective of the signaler and the recipient. Unpublished Ph.D. Dissertation Abstracts International B62(5):2463.

Gros-Louis, J. 2002. Contexts and behavioral correlates of trill vocalizations in wild white-faced capuchin monkeys (*Cebus capucinus*). *American Journal of Primatology* 57(4):189-202.

Gros-Louis, J., S. Perry and J. H. Manson 2003. Violent coalitionary attacks and intraspecific killing in wild white-faced capuchin monkeys (*Cebus capucinus*). *Primates* 44:341-346.

Groves, C. P. 1989. *A Theory of Human and Primate Evolution*. Clarendon Press, Oxford. 375 pp.

Groves, C. 2001. *Primate Taxonomy*. Smithsonian Institution Press, Washington, D. C. & London.

Guillotin, M. G. Dubost and D. Sabatier 1994. Food choice and food competition among the three major primate species of French Guiana. *Journal of Zoology* 233(3):551-579.

Gunther, A. 1876. On some new mammals from tropical America. *Processes Zoological Society Lond.* 743-751.

Haffer, J. 1974. *Avian Speciation in Tropical South America*. Nuttall Ornithological Club, Cambridge, 398 pp.

Haffer, J. 1982. General aspects of the refuge theory. Pp. 6-24 in: G. T. Prance (ed.), *Biological Diversification in the Tropics*. Columbia University Press, New York.

Haffer, J. 1987a. Quaternary history of tropical America. Pp. 1-18 in: T. C. Whitmore and G. T. Prance (eds.), *Biogeography and Quaternary History of Tropical America*. Clarendon Press, Oxford.

Haffer, J. 1987b. Biogeography of neotropical birds. Pp. 105-150 in: T. C. Whitmore and G. T. Prance (eds.), *Biogeography and Quaternary History of Tropical America*. Clarendon Press, Oxford.

Bibliography

Hall, C. L. and L. M. Fedigan 1997. Spatial benefits afforded by high rank in white-faced capuchins. *Animal Behaviour* 53(5):1069-1082.

Hampton, J. K., Jr. 1973. Diurnal heart rate and body temperature in marmosets. *American Journal of Physical Anthropology* 38:339-342.

Hampton, J. K., Jr. and S. H. Hampton 1965. Marmosets (Haplidae): Breeding seasons, twinning and sex of offspring. *Science* 150:915-917.

Hampton, S. H. and J. K. Hampton, Jr. 1977. The detection of reproductive cycles and pregnancy in tamarins (*Saguinus* spp.). Pp. 173-179 in: D. G. Kleiman (ed.), *The biology and Conservation of the Callitrichidae.* Smithsonian Institution Press, Washington, D. C.

Hampton, J. K. , S. H. Hampton and B. T. Landwehr 1966. Observations on a successful breeding colony of the marmoset *Oedipomidas oedipus. Folia Primatologica* 4:265-287.

Hanihara, T. and M. Natori 1987. Preliminary analysis of numerical taxonomy of the genus *Saguinus* based on dental measurements. *Primates* 28(4):517-523.

Hanson, A. M. and L. M. Porter 2000. Nutritional composition and distribution of fungal sporocarps consumed by Goeldi's monkey in northern Bolivia [Abstract]. *American Journal of Primatology* 51(Suppl 1):60.

Happel, R. 1981. Natural history and conservation of *Pithecia hirsuta* in Peru. Unpublished paper presented at 50th meeting ot the American Association of Physical Anthropologists.

Happel, R. 1982. Ecology of *Pithecia hirsuta* in Peru. *Journal of Human Evolution* 11:581-90.

Hardie, S. M. 1996. The behaviour of tamarin mixed-species groups (*Saguinus labiatus* and *Saguinus fuscicollis*). *Primate Eye* 58:18-19.

Hardie, S. M. 1998. Mixed-species tamarin groups (*Saguinus fuscicollis* and *Saguinus labiatus*) in northern Bolivia. *Primate Report* 50:39-62.

Hardie, S. M. and H. M. Buchanan-Smith 1997. Vigilance in single- and mixed-species groups of tamarins (*Saguinus labiatus* and *Saguinus fuscicollis*). *International Journal of Primatology* 18(2):217-234.

Hardie, S. M. and H. M. Buchanan-Smith 1998. The distribution of *Saguinus imperator* and *S. fuscicollis* in Bolivia in relation to rorest type. [Abstract] *Congress of the International Primatological Society, Abstracts* 17:3.

Hare, B., E. Addessi, J. Call, M. Tomasello and E. Visalberghi 2003. Do capuchin monkeys, *Cebus apella*, know what consepcifics do and do not see? *Animal Behaviour* 65(1):131-142.

Harris, R. A. 1996. Infant caretaking and sexual behavior in the pygmy marmoset (*Cebuella pygmaea*). [Abstract] *IPS/ASP Congress Abstracts* 1996:690.

Harrison, R. M. 1973. Ovulation in *Saimiri sciureus*: Induction, detection and inhibition. *Dissertation Abstracts International* 1973, B34:2876-2877.

Hartwig, W. C. 1995. A giant New World monkey from the Pleistocene of Brazil. *Journal of Human Evolution* 28:189-195.

Hartwig, W. C. and C. Cartelle 1996. A complete skeleton of the giant South American primate *Protopithecus. Nature,* London 381:307-311.

Hartwig, W. C., A. L. Rosenberger and T. Setoguchi 1990. New fossil Platyrrhines from the late Oligocene and middle Miocene. *American Journal of Physical Anthropologist* 81:237.

Hartwig, W. C., A. L. Rosenberger, P. W. Garber and M. A. Norconk. 1996. On atelines. Pp. 427-431 in: M. A. Norconk, A. L. Rosenberger and P. A. Garber (eds.)'. *Adaptive Radiations of Neotropical Primates.* Plenum Press, New York.

Harvey, P., R. D. Martin and T. H. Clutton-Brock 1987. Life histories in comparative aspect. Pp. 181-196 in: B. B. Smuts, D. L. Cheney, R. M. Seyfarth, R. W. Wrangham and T. T. Struhsaker (eds.), *Primate Societies.* The University of Chicago Press, Chicago.

Heinemann, H. 1970. The breeding and maintenance of captive goeldi's monkey, *Callimico goeldii. International Zoo Yearbook* 10:72-78.

Heltne, P. 1977. Census of *Aotus* in the north of Colombia. Unpublished manuscript.

Heltne, P. and C. A. Mejia 1978. *Aotus* in northern Colombia: Distribution, habitat status and possible management alternatives. unpublished paper presented at II Inter-American Conference on Conservation and Utilization of American Non-human Primates in Biomedical Research. Belem-Pará, Brazil, Oct. 24-27, 1978.

Heltne, P., D. Turner and J. Wolfhandler 1973. Maternal and paternal periods in the development of infant *Callimico goeldii. American Journal of Physical Anthropology* 38:555-560.

Heltne, P. G., D. C. Turner and N. J. Scott, Jr. 1976. Comparison of census data on *Alouatta palliata* from Costa Rica and Panama. Pp. 10-19 in: R. W. Thorington Jr. and P. G. Heltne (eds.), *Neotropical Primates: Field Studies and Conservation.* National Academy of Sciences, Washington, D. C.

Heltne, P. G., J. F. Wojcik and A. G. Pook 1981. Goeldi's monkey genus *Callimico*. Pp. 169-209 in: A. F. Coimbra-Filho and R. A. Mittermeier (eds.), *Ecology and Behavior of Neotropical Primates*, vol. 1. Academia Brasileira de Ciências, Rio de Janeiro.

Henry, R. E. and L. Winkler 2001. Foraging, feeding and defecation site selection as a parasite avoidance strategy of *Alouatta palliata* in a dry tropical forest. *American Journal of Physical Anthropology* Suppl. 32:79.

Hernández-Bacca, V. and C. I. Castillo-Ayala 2002. Cambios en el uso del espacio por una manada de *Callicebus torquatus lugens* (bajo Apaporis, Vaupés). Graduation thesis in Biology, Universidad Nacional de Colombia, Bogotá.

Hernandez Camacho, J. and H. Sanchez P. 1972. Biomas terrestres de Colombia. Pp. 153-174 in: G. Halffter (ed.), *La Diversidad Biológica de Iberoamérica I, Acta Zoológica Mexicana*, Instituto de Ecología, Xalapa, México.

Hernández Camacho, J. and R. W. Cooper 1976. The nonhuman primates of Colombia. Pp. 35-69 in: R. W. Thorington, Jr. and P. G. Heltne (eds.), *Neotropical Primates: Field Studies and Conservation.* National Academy of Sciences, Washington, D. C.

Hernández Camacho, J. and T. R. Defler 1985. Some aspects of the conservation of non-human primates in Colombia. *Primate Conservation* 6:42-50.

Hernández Camacho, J. and T. R. Defler 1989. Algunos aspectos de la conservación de primates no-humanos en Colombia. Pp. 67-100 in: C. J. Saavedra, R. A. Mittermeier and I. Bastos Santos (eds.), *La Primatología en Latinoamerica.* WWF-U.S., Washington, D. C.

Hernández Camacho, J., H. Sanchez P. and J. P. Latorre 1984. Colombia: Parques Nacionales. INDERENA, Bogotá , 263 pp.

Hernández Camacho, J., A. H. Guerra, R. Ortiz Q. and T. Walschburger 1992a. Unidades biogeográficas de Colombia. Pp. 105-152 in: G. Halffter (ed.), *La Diversidad Biológica de Iberoamerica* I, *Acta Zoológica Mexicana* (n.s.), Instituto de Ecología, Xalapa, México.

Hernández Camacho, J., T. Walschburger, R. Ortiz Q. and A. Hurtado G. 1992b. Orígen y distribución de la biota Suramericana y Colombiana. Pp. 55-104 in: G. Halffter (ed.), *La Diversidad Biológica de Iberoamérica I, Acta Zoológica* Mexicana, Instituto de Ecología, Xalapa, México.

Herrick, J. R., G. Agoramoorthy, R. Rudran and J. D. Harder 2000. Urinary progesterone in free-ranging red howler monkeys (*Alouatta seniculus*): Preliminary observations of the estrous cycle and gestation. *American Journal of Primatology* 51(4):257-263.

Hershkovitz, P. 1949. Mammals of northern Colombia. Preliminary report No. 4: Monkeys (primates), with taxonomic revisions of some forms. *Proceedings of the United States National Museum* 98:323 pp.

Hershkovitz, P. 1955. Notes on American monkeys of the genus *Cebus. Journal of Mammalogy* 36:449-452.

Hershkovitz, P. 1963. A systematic and zoogeographic account of South American titi monkeys, genus *Callicebus* (Cebidae) of the Amazonas and Orinoco river basins. *Mammalia* 27(1):1-80.

Bibliography

Hershkovitz, P. 1968. Metachromism or the principle of evolutionary change in mammalian tegumentary colors. *Evolution* 22:556-575.

Hershkovitz, P. 1969. The evolution of mammals in southern continents. VI. the recent mammals of the Neotropical region: a zoogeographic and ecological review. *Quarterly Review of Biology* 44:1-70.

Hershkovitz, P. 1970. Notes on Tertiary platyrrhine monkeys and description of a new genus from the late Miocene of Colombia. *Folia Primatologica* 12:1-37.

Hershkovitz, P. 1972a. Notes on New World monkys. *International Zoo Yearbook* 12:3-12.

Hershkovitz, P. 1972b. The recent mammals of the Neotropical region: a zoogeographic and ecological review. Pp. 311-431 in: A. Keast, F. C. Erk and B. Glass (eds.), *Evolution, Mammals, and Southern Continents.* State University of New York Press, Albany.

Hershkovitz, P. 1974. A new genus of Late Oligocene monkey (Cebidae, Platyrrhini) with notes on postorbital closure and platyrrhine evolution. *Folia Primatologica* 21:1–35.

Hershkovitz, P. 1977. *Living New World Monkeys (Platyrrhini).* Vol. 1., University of Chicago Press, Chicago.

Hershkovitz, P. 1979. The species of sakis, genus *Pithecia* (Cebidae, Primates), with notes on sexual dichromatism. *Folia Primatologica* 31:1-22.

Hershkovitz, P. 1982a. Supposed squirrel monkey affinities of the late Oligocene *Dolichocebus gaimanensis. Nature* 298:201-202.

Hershkovitz, P. 1982b. Subspecies and geographic distribution of the black-mantle tamarins (*Saguinus nigricollis* spix) (Primates: Callittrichidae). *Processes of the Biological Society of Washington* 95(4):647-656.

Hershkovitz, P. 1983. Two new species of night monkeys, genus *Aotus* (Cebidae, Platyrrhini): A preliminary report on *Aotus* taxonomy. *American Journal of Primatology* 4:209-243.

Hershkovitz, P. 1984. Taxonomy of squirrel monkey genus *Saimiri* (Cebidae, Platyrrhini): A preliminary report with description of a hitherto unnamed form. *American Journal of Primatology* 6:257-312.

Hershkovitz, P. 1987a. Uacaries, New World monkeys of the genus *Cacajao* (Cebidae, Platyrrhini): A preliminary taxonomic review with the description of a new subspecies. *American Journal of Primatology* 12:1-53.

Hershkovitz, P. 1987b. The taxonomy of South American sakis, genus *Pithecia* (Cebidae, Platyrrhini): A preliminary report and critical review with the description of a new species and a new subspecies. American Journal of Primatology 12:387-468.

Hershkovitz, P. 1988a. Origin, speciation, dispersal of South American titi monkeys, genus *Callicebus* (family Cebidae, Platyrrhini). Proceedings of the Academy of Natural Sciences of Philadelphia 140(1):240-272.

Hershkovitz, P. 1990. Titis, New World monkeys of the genus *Callicebus* (Cebidae, Platyrrhini): A preliminary taxonomic review. *Fieldiana (Zoology, New Series,* no. 55):1-109.

Hershkovitz, P. 1993. Male external genitalia of non-prehensile tailed South American monkeys: Part I. Subfamily Pitheciinae, family Cebidae. *Fieldiana Zoology* 73:1-17.

Heymann, E. W. 1990a. Reactions of wild tamarins, *Saguinus mystax* and *Saguinus fuscicollis* to avian predators. *International Journal of Primatology* 11(4):327-337.

Heymann, E. W. 1990b. Interspecific relations in a mixed-species troop of moustached tamarins, *Saguinus mystax*, and saddle-back tamarins, *Saguinus fuscicollis* (Platyrrhini: Callitrichidae), at the Río Blanco, Peruvian Amazonia. *American Journal of Primatology* 21(2):115-127.

Heymann, E. W. 1990c. Further field notes on red uacaris, *Cacajao calvus ucayalii*, from the Quebrada Blanco, Amazonian Peru. *Primate Conservation* 11:7-8.

Heymann, E. W. 1992. Associations of tamarins (*Saguinus mystax* and *Saguinus fuscicollis*) and double-toothed kites (*Harpagus bidentatus*) in Peruvian Amazonia. *Folia Primatologica* 59(1):51-55.

Heymann, E. W. 1993. Field studies on tamarins, *Saguinus mystax* and *Saguinus fuscicollis*, in northeastern Peru. *Neotropical Primates* 1(4):10-11.

Heymann, E. W. 1995. Sleeping habits of tamarins, *Saguinus mystax* and *Saguinus fuscicollis*) in north-eastern Peru. *Journal of Zoology* 237(2):211-226.

Heymann, E. W. 1996a. Ecological and evolutionary considerations of mixed-species troops in the genus *Saguinus*. [Abstract] *Folia Primatologica* 67(2):106-107.

Heymann, E. W. 1996b. Ecological and evolutionary aspects of interspecific associations (mixed-species troops) in tamarins (Genus *Saguinus*). [Abstract] *Primate Report* 44:20.

Heymann, E. W. 1997. The relationship between body size and mixed-species troops of tamarins (*Saguinus* ssp.). *Folia Primatologica* 68(3-5):287-295.

Heymann, E. W. 1998. Giant fossil New World primates: arboreal or terrestrial? *Journal of Human Evolution* 34:99-101.

Heymann, E. W. 2001. Scent marking and sexual selection in neotropical primates. [Abstract] *Primate Report* (Sp iss 60-1):24 pp.

Heymann, E. W. and U. Bartecki 1990. A young saki monkey, *Pithecia hirsuta*, feeding on ants, *Cephalotes atratus*. *Folia Primatologica* 55(3-4):181-184.

Heymann, E. W. and A. C. Smith 1999. When to feed on gums: Temporal patterns of gummivory in wild tamarins, *Saguinus mystax* and *Saguinus fuscicollis* (Callitrichinae). *Zoo Biology* 18(6):459-471.

Heymann, E. W. and P. Soini 1999. Offspring number in pygmy marmosets, *Cebuella pygmaea*, in relation to group size and the number of adult males. *Behavioral Ecology and Sociobiology* 46(6):400-404.

Heymann, E. W. and H. M. Buchanan-Smith 2000. The behavioural ecology of mixed-species troops of callitrichine primates. *Biological Reviews of the Cambridge Philosophical Society* 75(2):169-190.

Heymann, E. W., C. Knogge and E. R. Tirado Herrera 2000a. Vertebrate predation by sympatric tamarins, *Saguinus mystax* and *S. fuscicollis*. [Abstract] *Folia Primatologica* 71(4):230-231.

Heymann, E. W., C. Knogge and E. R. Tirado Herrera 2000b. Vertebrate predation by sympatric tamarins, *Saguinus mystax* and *Saguinus fuscicollis*. *American Journal of Primatology* 51(2):153-158.

Heymann, E. W., F. Encarnación-C. and J. E. Cnaquin-Y. 2002c. Primates of the Río Curaray, Northern Peruvian Amazon. *International Journal of Primatology* 23(1):191-201.

Hill, W. C. O. 1960. *Primates: Comparative Anatomy and Taxonomy, IV, Cebidae*, part a. Edinburgh University Press, Edinburgh.

Hill, W. C. O. 1962. *Primates: Comparative Anatomy and Taxonomy, V, Cebidae*, part b. Edinburgh University Press, Edinburgh.

Hilton-Taylor, C. 2000. 2000 IUCN Red List of Threatened Species. Cambridge, U.K., pp. xviii, 61 pp. & CD. http://www.redlist.org

Hilton-Taylor, C. and A. Rylands 2002. The 2002 IUCN Red List of Threatened Species. Neotropical Primtes 10(3):149-153.

Hirabuki, Y and K. Izawa 1990. Chemical properties of soils eaten by wild red howler monkeys (*Alouatta seniuclus*): A preliminary study. *Field Studies of New World Monkeys, La Macarena, Colombia*. 3:25-28.

Hirsch, B. T. 2000. Ecological and behavioral correlates of vigilance in brown capuchin monkeys (*Cebus apella*) in Iguazu, Argentina. [Abstract] *American Journal of Physical Anthropology* (Suppl 30):180.

Hirsch, B. T. 2002. Social monitoring and vigilance behavior in brown capuchin monkeys (*Cebus apella*). *Behavioral Ecology and Sociobiology* 52(6):458-464.

Hirsch, A., E. C. Landau, A. C. de Teseschi and J. O. Menegheti 1991. Estudo comparativo das esécies do genero *Alouatta* Lacèpéde, 1799 (Platyrrhini, Atelidae) e sua distribuión geográfica na américa do sul. *Primatología. no brazil* 3:239-262.

Hladik, C. M. 1967. Surface relaive du tractus digestif de quelques primates, morphologie de sillosites intestinales et correlations avec le regime alimentair. *Mammalia* 31:120-147.

Bibliography

Hladik, C. M. 1972a. Les hurleurs de Barro-Colorado. *Science et Nature* 110:29-35.

Hladik, C. M. 1972b. L'atele de geoffroy, ce singe-araignée. *Science et Nature* 111:1-11.

Hladik, C. M. 1975. Ecology, diet, and social patterns in old and New World primates. Pp. 3-35 in: R. H. Tuttle (ed.), *Socioecology and Psychology of Primates*. Mouton, The Hague.

Hladik, C. M. 1978. Adaptive strategies of primates in relation to leaf-eating. Pp. 373-395 in: G. G. Montgomery (ed.), *The Ecology of Arboreal Folivores*. Smithsonian Institution Press, Washington, D. C.

Hladik, C. M. 1990. [The feeding strategies of primates.] (in French) Pp. 35-52 in: J. -J. Roeder and J. R. Anderson (eds.), *Primates: Recherches Actuelles*. Masson, Paris.

Hladik, A. and C. M. Hladik 1969. Rapports trophiques entre végétation et primates dans la forêt de Barro Colorado (Panama). *Le Terre et la Vie* 23:25-117.

Hladik, C. M., A. Hladik, J. Bousset, P. Valdegbouze, G. Viroben and J. Delort-Laval 1971. Le régime alimentaire del primates de l'œle de Barro Colorado (Panama): résultats des analyses quantitatives. *Folia Primatologica* 16:85-122.

Hodun, A., C. T. Snowdon and P. Soini 1981. Subspecific variation in the long calls of the tamarin, *Saguinus fuscicollis*. *Zeitschrift Für Tierpsychology* 57:97-110.

Hoffmannsegg, G. von 1807. Beschreibung vier affenartiger Thiere aus Brasilian. *Magazin gesellschaft naturforschungen Freunde, Berlin* 1:83-104.

Hoffstetter, R. 1968. Un gisement de mammiferes Déséadiens (Oligocene Inférieur) en Bolivie. *Comptes-Rendus de LÁcadémie de Science de l'Académie de Paris* (d) 267(13):1095-1097.

Hoffstetter, R. 1969. Un primate de l'Oligocene Inferieur Sud-Americain: *Branisella Boliviana* gen. et sp. nov. *Compte Rendus de l''Académie des Science (Paris)* sér. D 269:434-437.

Hoffstetter, R. 1972. Relationships, origins and history of the ceboid monkeys and caviomorph rodents: a modern reinterpretation. Pp. 323-347 in: T. Bobzhansky, M. K. Hecht and W. C. Steere (eds.), *Evolutionary Biology*. Appleton-Century, New York.

Hoffstetter, R. 1974. Phylogeny and geografic deployment of the primates. *Journal of Human Evolution* 3:327-350.

Hoffstetter, R. 1980. Origin and deployment of New World monkeys emphasizing the southern continental route. Pp. 103-138 in: R. L. Ciochon and A. B. Chiarelli (eds.), *Evolutionary Biology of the New World Monkeys and Continental Drift*. Plenum Press, New York.

Hoffstetter, R. 1981. Biogeografa histórica de los mamíferos sur-americanos: problemas y aplicaciones. *Acta Geologica Hispánica* 16(1-2a):71-88.

Holdridge, L. R. 1947. Determination of world plant formations from simple climatic data. *Science* 105:267-368.

Honacki, J. H., K. E. Kinman and J. W. Koeppl 1982. *Mammal Species of the World*. Allen Press and The Association of Systematics Collections, Lawrence, Kansas.

Horovitz, I. 1999. A phylogenetic study of living and fossil platyrrhines. *American Museum Novitates* 3269:1-40.

Horovitz, I. and A. Meyer 1997. Evolutionary trends in the ecology of New World monkeys inferred from a combined phylogenetic analysis of nuclear, mitochondrial, and morphological data. Pp. 189-224, in: T. G. Givnish and K. J. Sytsma (eds.), *Molecular Evolution and Adaptive Radiation*. Cambridge University Press, Cambridge.

Horovitz, I. and R. D. E. MacPhee 1999. The Quaternary Cuban platyrrhine *Paralouatta varonai* and the origin of Antillean monkeys. *Journal of Human Evolution* 36(1):33-68.

Horovitz, I., R. Zardoya and A. Meyer 1998. Platyrrhine systematics: A simultaneous analysis of molecular and morphological data. *American Journal of Physical Anthropology* 106(3):261-281.

Houdun, H., C. T. Snowdon and P. Soini 1981. Subspecific variation in the long calls of the tamarin *Saguinus fuscicollis*. *Zeitschrift für Tierpsychology* 57:97-110.

Howe, H. F. 1980. Monkey dispersal and waste of a neotropical fruit. *Ecology* 61:944-959.

Howe, H. F. 1993. Aspects of variation in a Neotropical seed dispersal system. *Vegetatio* 107/108:149-162.

Humboldt, A. von and A. Bonpland 1812. *Recueil d'observations de zoologie et d'anatomie comparée, faites dans l'ocean atlantique dans l'interieur du nouveau continent et dans la mer de sud pendant les années 1799, 1800, 1801, 1802 et 1803.* pt. 2, vol.1, paris, viii + 368 pp., 40 pls.

Hunter, J., R. D. Martin, A. F. Dixson and B. C. C. Rudder 1979. Gestation and interbirth intervals in the owl monkey (*Aotus trivirgatus griseimembra*). *Folia Primatologica* 31:165-75.

Huston, M. A. 1994. *Biological Diversity: The Coexistence of Species on Changing Landscapes.* Cambridge University Press, Cambridge.

Hutchinson, G. E. 1967. *A Treatise on Limnology.* vol. ii, Wiley, New York.

Inaba, A. 2000. [Intergroup relationships of wild spider monkeys in Macarena, Colombia.] (in Japanese) [Abstract] *Reichorui Kenkyu/Primate Res.* 16(3):257.

Inaba, A. 2002. [Social relationships of wild spider monkeys in Macarena, Colombia.] (in Japanese)[Abstract] *Reichorui Kenkyu/ Primate Research* 18(3):416.

IUCN 1995a. A new system for classifying threatened status. *Neotropical Primates* 3(suppl.):104-112.

IUCN 1995b. Red list categories of the IUCN. *Neotropical Primates* 3(suppl.):1-14.

Iwanaga, S. and S. F. Ferrari 2001. Party size and diet of syntopic atelids (*Ateles chamek* and *Lagothrix cana*) in southwestern Brazilian Amazonia. *Folia Primatologica* 72(4):217-227.

Iwanaga, S. and S. F. Ferrari 2002. Geographic distribution of red howlers (*Alouatta seniculus*) in southwestern Brazilian Amazonia, with notes on *Alouatta caraya*. *International Journal of Primatology* 23(6):1245-1256.

Izar, P. and T. Sato 1997. [Seed dispersal by black-capped Cebus (*Cebus apella*) at Mata Atlantica.] (in Portuguese) *Programa e Resumos do VIII Congresso Brasileiro de Primatologia & V Reuniao Latino-Americano de Primatologia* Soc. Bras. Primatol., Joao Pessoa 1997:45.

Izawa, K. 1975. Foods and feeding behavior of monkeys in the upper Amazon basin. *Primates* 17:503-512.

Izawa, K. 1976. Group sizes and compositions of monkeys in the upper Amazon basin. *Primates* 17:503-512.

Izawa, K. 1978a. A field study of the ecology and behavior of the black-mantle tamarin (*Saguinus nigricollis*). *Primates* 19:241-74.

Izawa, K. 1978b. Frog-eating behavior of wild black-capped capuchin (*Cebus apella*). *Primates* 19:633-42.

Izawa, K. 1979a. Foods and feeding behavior of wild black-capuchin (*Cebus apella*). *Primates* 21:57-76.

Izawa, K. 1979b. Studies on peculiar distribution pattern of *Callimico*. *Kyoto University Oversea Research Reports of New World Monkeys* 1: 1-19.

Izawa, K. 1980. Social behavior of the wild black-capped capuchin (*Cebus apella*). *Primates* 21:443-67.

Izawa, K. 1988a. Preliminary report on social changes of red howlers (*Alouatta seniculus*). *Field Studies of New World Monkeys, La Macarena, Colombia.* 1:29-34.

Izawa, K. 1988b. Preliminary report on social changes of black-capped capuchins (*Cebus apella*). *Field Studies of New World Monkeys, La Macarena, Colombia* 1:13-18.

Izawa, K. 1989a. The adoption of an infant observed in a wild group of red howler monkeys (*Alouatta seniculus*). *Field Studies of New World Monkeys, La Macarena, Colombia* 2:33-36.

Izawa, K. 1989b. The development of the wild black-capped capuchin (*Cebus apella*) in La Macarena, Colombia. *Field Studies of New World Monkeys, La Macarena, Colombia* 1:13-18.

Izawa, K. 1990a. Social changes within a group of wild black-capped capuchins (*Cebus apella*) in Colombia (ii). *Studies of New World Monkeys, La Macarena, Colombia* 3:1-6.

Izawa, K. 1990b. Rat predation by wild capuchins (*Cebus apella*). *Field Studies of New World Monkeys, La Macarena, Colombia* 3:19-24.

Bibliography

Izawa, K. 1990c. Chemical properties of soils eaten by red howler monkeys (*Alouatta seniculus*). *Field Studies of New World Monkeys, La Macarena, Colombia* 4:27-37.

Izawa, K. 1990d. Chemical properties of special water drunk by wild spider monkeys (*Ateles belzebuth*) in La Macarena, (Colombia). *Field Studies of New World Monkeys, La Macarena, Colombia* 4:38-46.

Izawa, K. 1990e. Frequency of soil-eating by a group of wild howler monkeys (*Alouatta seniculus*) in La Macarena, (Colombia). *Field Studies of New World Monkeys, La Macarena, Colombia* 4:47-56.

Izawa, K. 1990f. The development of the wild black-capped capuchin (*Cebus apella*) in La Macarena, Colombia. *Field Studies of New World Monkeys, La Macarena, Colombia* 2:45-56.

Izawa, K. 1992. Social changes within a group of wild black-capped capuchins (*Cebus apella*) iii. *Field Studies of New World Monkeys, La Macarena, Colombia* 7:9-14.

Izawa, K. 1993. Soil-eating by *Alouatta* and *Ateles*. *International Journal of Primatology* 14(2):229-242.

Izawa, K. 1994a. Social changes within a group of wild black-capped capuchins, IV. *Field Studies of New World Monkeys, La Macarena, Colombia*. 9:15-21.

Izawa, K. 1994b. Group division of wild black-capped capuchins. *Field Studies of New World Monkeys, La Macarena, Colombia* 9:5-14.

Izawa, K. 1997a. Social changes within a group of wild black-capped capuchins, V. *Field Studies of Fauna and Flora, La Macarena, Colombia* 11:1-10.

Izawa, K. 1997b. Social changes within a group of red howler monkeys, VI. *Field Studies of Fauna and Flora, La Macarena, Colombia* 11:19-34.

Izawa, K. 1997c. Stability of the home range of red howler monkeys. *Field Studies of Fauna and Flora, La Macarena, Colombia* 11:41-46.

Izawa, K. 1999. Social changes within a group of red howler monkeys, VII. *Field Studies of Fauna and Flora, La Macarena, Colombia*. 13:15-17.

Izawa, K. 2002. The mobbing call and long call of wild spider monkeys. II. [Abstract] *Anthropological Science* 110(1):89.

Izawa, K. and A. Mizuno 1977. Palm-fruit cracking behaviour of wild black-capped capuchin (*Cebus apella*). *Primates* 18:773-792.

Izawa, K. and G. Bejarano 1981. Distribution ranges and patterns of nonhuman primates in western Pando, Bolivia. *Kyoto University Overseas Research Reports of New World Monkeys* (1981):1-12.

Izawa, K. and A. Nishimura 1988. Primate fauna at the study site. La Macarena, Colombia. *Field Studies of New World Monkeys, La Macarena, Colombia* 1:5-11.

Izawa, K. and H. Lozano M. 1989. Social changes within a group and reproduction of wild howler monkeys (*Alouatta seniculus*) in Colombia. *Field Studies of New World Monkeys, La Macarena, Colombia* 2:1-6.

Izawa, K. and H. Lozano M. 1990a. Frequency of soil-eating by a group of wild howler monkeys (*Alouatta seniculus*) in La Macarena, Colombia. *Field Studies of New World Monkeys, La Macarena, Colombia* 4:47-56.

Izawa, K. and H. Lozano M. 1990b. River crossing by a wild howler monkey (*Alouatta seniculus*). *Field Studies of New World Monkeys, La Macarena, Colombia*. 3:29-34.

Izawa, K. and A. Mizuno 1990. Chemical properties of special water drunk by wild spider monkeys (*Ateles belzebuth*) in La Macarena, Colombia. *Field Studies of New World Monkeys, La Macarena Colombia* 4:38-46.

Izawa, K. and M. H. Lozano 1991. Social changes within a group of red howler monkeys (*Alouatta seniculus*). III. *Field Studies of New World Monkeys, La Macarena, Colombia* 5:1–16.

Izawa, K. and M. H. Lozano 1992. Social changes with a group of red howler monkeys (*Alouatta seniculus*). IV. *Field Studies of New World Monkeys, La Macarena, Colombia*. 7:15–27.

Izawa, K. and M. H. Lozano 1994. Social changes within a group of red howler monkeys, V. *Field Studies of New World Monkeys, La Macarena, Colombia*. 9:33–39.

Izawa, K., K. Kumura and A. Samper 1979. Grouping of the wild spider monkey. *Primates* 20:503-512.

Izawa, K., K. Kimura and Y. Ohnishi 1990. Chemical properties of soils eatern by red howler monkeys (*Alouatta seniculus*), II. *Field Studies of New World Monkeys, La Macarena Colombia* 4:27-37.

Jack, K. M. 2001. Life history patterns of male white-faced capuchins (*Cebus capucinus*): Male-bonding and evolution of multimale groups. Unpublished Ph.D. *Dissertation Abstracts International* A62(5):1877.

Jacobs, G. H. 1977a. Brightness preference in nocturnal and diurnal South American monkeys. *Folia Primatologica* 28:231-240.

Jacobs, G. H. 1977b. Visual capacities of the owl monkey (*Aotus trivirgatus*). I. Spectral sensitivity and color vision. *Vision Research* 17:811-820.

Jacobs, G. H. 1977c. Visual capacities of the owl monkey (*Aotus trivirgatus*). II. Spatial contrast sensitivity. *Vision Research* 17:821-825.

Janson, C. H. 1982. The mating system of the brown capuchin (*Cebus apella*). *American Journal of Physical Anthropology* 57(2):198-199.

Janson, C. H. 1983a. Adaptation of fruit morphology to dispersal agents in a Neotropical rainforest. *Science* 219:187-189.

Janson, C. H. 1983b. Social divergence despite ecological similarity in two Neotropical monkey species. *American Zoologist* 23(40):933.

Janson, C. H. 1984. Female choice and mating system of the brown capuchin monkey (*Cebus apella*). *Zeitschrift für Tierpsychologie* 85:177-200.

Janson, C. H. 1985a. Ecological and social consequences of food competition in brown capuchin monkeys. Unpublished Ph.D. thesis, University of Washington, Seattle.

Janson, C. H. 1985b. The mating system as a determinant of social evolution in capuchin monkeys (*Cebus*). In: J. Else and P. Lee (eds.), *Proceedings of the XIIth International Congress of Primatology*. Cambridge University Press, Cambridge.

Janson, C. H. 1986a. Capuchin counterpoint. *Natural History* 95(2):44-53.

Janson, C. H. 1986b. The mating system as a determinant of social evolution in capuchin monkeys (*Cebus*). Pp. 169-179 in: J. G. Else and P. C. Lee (eds.), *Primate Ecology and Conservation*. Cambridge University Press, Cambridge.

Janson, C. H. 1988. Food competition in brown capuchin monkeys (*Cebus apella*): Quantitative effects of group size and tree productivity. *Behaviour* 105(1-2):53-76.

Janson, C. H. 1990a. Social correlates of individual spatial choice in foraging groups of brown capuchin monkeys, *Cebus apella. Animal Behaviour* 40(5):910-921.

Janson, C. H. 1990b. Ecological consequences of individual spatial choice in foraging groups of brown capuchin monkeys, *Cebus apella. Animal Behaviour* 40(5):922-934.

Janson, C. H. 1994. Sex roles in predator defense and vigilance. *Evolutionary Anthropology* 3(2):69.

Janson, C. H. 1998. Experimental evidence for spatial memory in foraging wild capuchin monkeys, *Cebus apella. Animal Behaviour* 55(5):1229-1243.

Janson, C. H. 1999. Predation, life histories, and the evolution of sociality. [Abstract] *Primate Report* Special Issue 54(1):21-22.

Janson, C. H. and J. W. Terborgh 1979. Age, sex, and individual specialization in foraging behavior of the brown capuchin (*Cebus apella*). *American Journal of Physical Anthropology* 50(3):452.

Janson, C. H. and P. C. Wright 1980. Parent-offspring relations in the brown capuchin (*Cebus apella*). *American Journal of Physical Anthropology* 52(2):241.

Janson, C. H. and J. W. Terborgh 1985a. Censando primates en el bosque lluvioso con referencia a la Estación Biológica de Cocha Cashu, Parque Nacional del Manu, Peru. In: *Reporte Manu*, capítulo 15. Centro de Datos para la Conservación, U. N. Agraria, La Molina, Peru.

Bibliography

Janson, C. H. and J. W. Terborgh 1985b. Censusing primates in rainforest. In: *Estudios Ecológicos en el Parque Nacional del Manu*. Ministerio de Agricultura, Lima.

Janson, C. H. and S. Boinski 1992. Morphological and behavioral adaptations for foraging in generalist primates: The case of the cebines. *American Journal of Physical Anthropology* 88(4):483-498.

Janson, C. H. and C. P. van Schaik 1993. Ecological risk aversion in juvenile primates: Slow and steady wins the race. Pp. 57-74 in: M. E. Pereira and L. A. Fairbanks (eds.), *Juvenile Primates: Life History, Development and Behaviour*. Oxford University Press, Oxford.

Janson, C. H. and C. P. van Schaik 2000. The behavioral ecology of infanticide by males. Pp. 469-494 in: C. P. van Schaik and C. H. Janson (eds.), *Infanticide by Males and Its Implications*. Cambridge University Press, Cambridge.

Janson, C. H., J. Terborgh and L. H. Emmons 1981. Non-flying mammals as pollinating agents in the Amazonian forest. *Biotropica* 13(Suppl. 2):1-6.

Johns, A. 1986. Effects of habitat disturbance on rain forest wildlife in Brazilian Amazonia. Final report to World Wildlife Fund U.S., project us-302, pp. 111.

Johns, A. D. 1991. Forest disturbance and Amazonian primates. Pp. 115-135 in: H. O. Box (ed.), *Primate Responses to Environmental Change*. Chapman & Hall, London.

Johnston, L. D., A. J. Petto and P. K. Sehgal 1991. Factors in the rejection and survival of captive cotton-top tamarins (*Saguinus oedipus*). *American Journal of Primatology* 25:91-102.

Jones, C. B. 1978. Aspects of reproductive behavior in the mantled howler monkey, *Alouatta palliata* Gray. Unpublished Ph.D. thesis, Cornell University, Ithaca, New York.

Jones, C. B. 1979. Grooming in the mantled howler monkey, *Alouatta palliata* Gray. *Primates* 20:289-292.

Jones, C. B. 1980a. The functions of status in the mantled howler monkey, *Alouatta palliata* Gray: Intraspecific competition for group membership in a folivorous neotropical primate. *Primates* 21(3):389-405.

Jones, C. B. 1980b. Seasonal parturition, mortality, and dispersal in the mantled howler monkey *Alouatta palliata* Gray. *Brenesia* 17:1-10.

Jones, C. B. 1981. The evolution and socioecology of dominance in primate groups: a theoretical formulation, classification and assessment. *Primates* 22:70-83.

Jones, C. B. 1982. A field manipulation of spatial relations among male mantled howler monkeys. *Primates* 23:130-134.

Jones, C. B. 1983. Do howler monkeys feed upon legume flowers preferentially at flower opening time? *Brenesia* 21:41-46.

Jones, C. B. 1985. Reproductive patterns in mantled howler monkeys: Estrus, mate choice and copulation. *Primates* 26(4):130-142.

Jones, C. B. 1994. Injury and disease of the mantled howler monkey in fragmented habitats. *Neotropical Primates* 2(4):4-5.

Jones, C. B. 1995. Alternative reproductive behaviors in the mantled howler monkey (*Alouatta palliata* Gray): Testing carpenter's hypothesis. *Boletín de Primatología Latinoamericana* 5(1):1-5.

Jones, C. B. 2000. *Alouatta palliate* politics: Empirical and theoretical aspects of power. *P;imate Report* 56:3-21.

Jones, C. B. 2002a. A possible example of coercive mating in mantled howling monkeys (*Alouatta palliata*) related to sperm competition. *Neotropical Primates* 10(2):95-96.

Jones, C. B. 2002b. How important are urinary signals in *Alouatta*? *Laboratory Primate Newsletter* 41(3):15-17.

Jones, C. and S. Anderson 1978. *Callicebus moloch. Mammalian Species* 112:1-5.

Julliot, C. 1994a. Diet diversity and habitat of howler monkeys. Pp. 67-71 in: B. Thierry, J. R. Anderson, J. J. Roeder and N. Herrenschmidt (eds.), *Current Primatology*, vol. I: *Ecology and Evolution*. University Louis Pasteur, Strasbourg.

Julliot, C. 1994b. Frugivory and seed dispersal by red howler monkeys: evolutionary aspect. *Revue d'Ecologie (Terre et Vie)* 49(4):331-341.

Julliot, C. 1996a. Seed dispersal by red howling monkeys (*Alouatta seniculus*) in the tropical rain forest of French Guiana. *International Journal of Primatology 17(2):239-258.*

Julliot, C. 1996b. Fruit choice by red howler monkeys (*Alouatta seniculus*) in a tropical rain forest. *American Journal of Primatology* 40(3):261-281.

Julliot, C. 1996c. Impact of seed dispersal by howler monkeys on the forest regeneration. [Abstract] *IPS/ASP Congress Abstracts* 1966:656.

Julliot, C. 1997. Impact of seed dispersal by red howler monkeys *Alouatta seniculus* on the seedling population in the understorey of tropical rain forest. *Journal of Ecology* 85(4):431-440.

Julliot, C. and D. Sabatier 1993. Diet of the red howler monkey (*Alouatta seniculus*) in French Guiana. *International Journal of Primatology* 14(4):527-550.

Julliot, C. and B. Simmen 1998. Food partitioning among a community of neotropical primates. [Abstract] *Folia Primatologica* 69(1):43-44.

Jurke, M. H. 1996. Behavioral and hormonal aspects of reproduction in captive goeldi's monkeys (*Callimico goeldii*) in a comparative and evolutionary context. *Primates* 37(1):109-119.

Jurke, M. H., C. R. Pryce and M. Doebeli 1995. An investigation into sexual motivation and behavior in female goeldi's monkey (*Callimico goeldii*): Effect of ovarian state, mate familiarity, and mate choice. *Hormones and Behavior* 29(4):531-553.

Kaplan, J. N., A. Winship-Ball and L. Sim 1978. Maternal discrimination of infant vocalizations in the squirrel monkey. *Primates* 19:187-93.

Kavanagh, M. and L. Dresdale 1975. Observations on the woolly monkey (*Lagothrix lagothricha*) in northern Colombia. *Primates* 16:285-294/

Kavanagh, M. and E. Bennett 1984. A synopsis of legislation and the primate trade in habitat and user countries. Pp. 19-48 in: D. Mack and R. A. Mittermeier (eds.), *The International Primate Trade*, Vol. 1: *Legislation, Trade and Captive Breeding*. Traffic (U.S.A.), Washington, D.C.

Kay, R. F. 1980. Platyrrhine origins: A reappraisal of the dental evidence. Pp. 159-188 in: R. L. Ciochon and A. B. Chiarelli (eds.), *Evolutionary Biology of the New World Monkeys and Continental Drift*. Plenum Press, New York.

Kay, R. F. 1989. A new small platyrrhine from the Miocene of Columbia and the phyletic position of the callitrichines [Abstract] *American Journal of Physical Anthropology* 78(2):251.

Kay, R. F. 1990. The phylogenetic relationships of extant and fossil pithecinae (Platyrrhini, Anthropoidea). *Journal of Human Evolution* J 19:175-208.

Kay, R. F. 1994. Giant tamarin from the Miocene of Colombia. *American Journal of Physical Anthropology* 95(3):333–353.

Kay, R. F. and C. D. Frailey 1993. Large fossil platyrrhines from the Río Acre local fauna, Late Miocene, western Amazonia. *Journal of Human Evolution* 25:319-327.

Kay, R. F. and R. H. Madden 1997. Paleogeography and paleoecology. Pp. 520–550 in: R. F. Kay, R. H. Madden, R. L. Cifelli and J. J. Flynn (eds.), *Vertebrate Paleontology in the Neotropics: The Miocene Fauna of La Venta, Colombia*. Smithsonian Institution Press, Washington, D. C.

Kay, R. F. and D. J. Meldrum 1997. A new small platyrrhine and the phyletic position of Callitrichinae. Pp. 435–458 in: R. F. Kay, R. H. Madden, R. L. Cifelli and J. J. Flynn (eds.), *Vertebrate Paleontology in the Neotropics: The Miocene Fauna of La Venta, Colombia*. Smithsonian Institution Press, Washington, D. C.

Kay, R. F., R. Madden, J. M. Plavcan, R. L. Cifelli and J. G. Diaz 1987. *Stirtonia victoriae*, a new species of Miocene Colombian primate. *Journal of Human Evolution* 16:173-196.

Bibliography

Keleman, G. and J. Sade 1960. The vocal organ of the howling monkey (*Alouatta palliata*). *Journal of Morphology* 107:123-140.

Kellogg, R. and E. A. Goldman 1944. Review of the spider monkeys. *Processes of U.S. National Museum* 96:1-45.

Kessler, P. 1998. Primate densities in the Natural Reserve of Nouragues, French Guiana. *Neotropical Primates* 6(2):45-46.

Kimura, K. 1992. Demographic approach to the social group of wild red howler monkeys (*Alouatta seniculus*). *Field Studies of New World Monkeys, La Macarena, Colombia* 7:29-34.

Kimura, K. 1995. [Social stability of membership of wild red howler monkey group]. (in Japanese) *Reichorui Kenkyu/Primate Res.* 11(3):292.

Kimura, K. 1997. Males life history and their social relations of wild red howler monkeys. *Field Studies of Fauna and Flora, La Macarena, Colombia* 11:45-40.

Kimura, K. 1999. Home ranges and inter-group relations among the wild red howler monkeys. *Field Studies of Fauna and Flora, La Macarena, Colombia.* 13:19-24.

Kinzey, W. G. 1975. The ecology of locomotion in *Callicebus torquatus. American Journal of Physical Anthropology* 42:312.

Kinzey, W. G. 1977a. Diet and feeding behaviour of *Callicebus torquatus.* Pp. 127-151 in: T. H. Clutton-Brock (ed.), *Primate Ecology: Studies of Feeding and Ranging Behavior in Lemurs, Monkeys and Apes.* Academic Press, London.

Kinzey, W. G. 1977b. Positional behavior and ecology in *Callicebus torquatus. Yearbook of Physical Anthropology* 20:468-480.

Kinzey, W. G. 1977c. Dietary correlates of molar morphology in *Callicebus and Aotus. American Journal of Physical Anthropology* 47:142.

Kinzey, W. G. 1978. Feeding behavior and molar features in two species of titi monkey. Pp. 373-385 in: D. J. Chivers and J. Herbert (eds.), *Recent Advances in Primatology*, vol. 1, Behaviour, Academic Press, London.

Kinzey, W. G. 1981. The titi monkeys, genus *Callicebus.* Pp. 241-276 in: A. F. Coimbra-Filho and R. A. Mittermeier (eds.), *Ecology and Behavior of Neotropical Primates.* Academia Brasileira de Ciências, Rio de Janeiro.

Kinzey, W. G. 1982. Distribution of primates and Forest Refuges. Pp. 455-482 in: G. T. Prance (ed.), *Biological Diversifications in the Tropics.* Columbia University Press, New York.

Kinzey, W. G. 1992. Dietary and dental adaptations in the Pitheciinae. *American Journal of Physical Anthropology* 88(4):499-514.

Kinzey, W. G. and A. L. Rosenberger 1975. Vertical clinging and leaping in a Neotropical anthropoid. *Nature* 255(5506):327-328.

Kinzey, W. G. and A. H. Gentry 1979. Habitat utilization in two species of *Callicebus.* Pp. 89-100 in: R. W. Sussman (ed.), *Primate Ecology: Problem-Oriented Field Studies.* John Wiley & Sons, New York.

Kinzey, W. G. and P. C. Wright 1982. Grooming behavior in the titi monkey, *Callicebus torquatus. American Journal of Primatology* 3:267-75.

Kinzey, W. G. and J. G. Robinson 1983. Intergroup loud calls, range size, and spacing in *Callicebus torquatus. American Journal of Physical Anthropology* 60:539-44.

Kinzey, W. G. and M. A. Norconk 1990. Hardness as a basis of fruit choice in two sympatric primates. *American Journal of Physical Anthropology* 81(1):5-15.

Kinzey, W. G., A. L. Rosenberger, P. S. Heisler, D. Prowse and J. Trilling 1977. A preliminary field investigation of the yellow-handed titi monkey, *Callicebus torquatus torquatus*, in northern Peru. *Primates* 18:159-181.

Kirkwood, J. K. 1983. The effects of diet on health, weight, and litter size in captive cotton-top tamarins *Saguinus oedipus oedipus. Primates* 24:515-20.

Kirkwood, J. K., M. A. Epstein and A. J. Terlecki 1983. Factors influencing population growth of a colony of cotton-top tamarins. *Laboratory Animal Newslettter* 17:35-41.

Klein, L. L. 1971. Observations on copulation and seasonal reproduction of two species of spider monkeys, *Ateles belzbuth* and *A. geoffroyi*. *Folia Primatologica* 15:233-248.

Klein, L. L. 1972. The ecology and social organization of the spider monkey, *Ateles belzebuth*. Unpublished Ph.D. thesis, University of California, Berkeley.

Klein, L. L. 1974. Agonistic behavior in neotropical primates. Pp. 77-122 in: R. L. Holloway (ed.), *Primate Aggression, Territoriality, and Xenophobia: A Comparative Perspective*. Academic Press, San Francisco.

Klein, L. L. and D. J. Klein 1971a. Aspects of social behavior in a colony of spider monkeys, *Ateles geoffroyi*, at the San Francisco Zoo. *International Zoo Yearbook* 11:175-181.

Klein, L. L. and D. J. Klein 1971b. Case study: Primates in the field. Pp. 64-75 in: *Anthropology Today*. CRM Books, Del Mar, California.

Klein, L. L. and D. J. Klein 1973a. Social and ecological contrasts between four taxa of neotropical primates. Pp. 59-85 in: T. H. Tuttle (ed.), *Socioecology and Psychology of Primates*. Mouton Publishers, The Hague.

Klein, L. L. and D. J. Klein 1973b. Observations on two taxa of neotropical primate intertaxa associations. *American Journal of Physical Anthropology* 38:649-54.

Klein, L. L. and D. J. Klein 1975. Social and ecological contrasts between four taxa of neotropical primates. Pp. 59-85 in: R. H. Tuttle (ed.), *Socioecology and Psychology of Primates*. Mouton Publishers, The Hague,.

Klein, L. L. and D. J. Klein 1976. Neotropical primates: aspects of habitat usage, population density, and regional distribution in La Macarena, Colombia. Pp. 70-78 in: R. W. Thorington, Jr. and P. G. Heltne (eds.), *Neotropical Primates: Field Studies and Conservation*. National Academy of Sciences, Washington, D. C.

Klein, L. L. and D. J. Klein 1977. Feeding behavior of the Colombian spider monkey. Pp. 153-181 in: T. H. Clutton-Brock (ed.), *Primate Ecology: Studies of Feeding and Ranging Behavior in Lemurs, Monkeys and Apes*. Academic Press, New York.

Knogge, C. and E. W. Heymann 1995. Field observations of twinning in the dusky titi monkey, *Callicebus cupreus. Folia Primatologica* 65(2):118-120.

Kobayashi, S. 1995. A phylogenetic study of titi monkeys, genus *Callicebus*, based on cranial measurementes: I. Phylogenetic groups of *Callicebus. Primates* 36(1):101-120.

Kobayashi, M. and K. Izawa 1992. Demographic approach to the social group of wild red howler monkeys (*Alouatta seniculus*). *Field Studies of New World Monkeys, La Macarena, Colombia* 7:29-34.

Koiffmann, C. P. and P. H. Saldanha 1981. The karyotype of *Cacajao melanocephalus* (Platyrrhini, Primates). *Folia Primatologica* 36:150-155.

Konstant, W., R. A. Mittermeier and S. D. Nash 1985. Spider monkeys in captivity and in the wild. *Primate Conservation* 5:82-109.

Kraglievich, J. L. 1951. Contribuciones al conocimiento de los primates fósiles de la Patagonia. Diagnosis prévia de un mono nuevo primate fósil Oligoceno Superior (Colhuehuapiano) de Gaiman, Chubut. *Comm. Inst. Nac. Cient. Nat.* 2:57-82.

Kuehlhorn, F. 1943. [Observation on the biology of Cebus apella.] (in German) *Zoologische Garten* 1943. 15:221-234

Kuhl, H. 1820. Bieträge zur Zoologie und vergleichenden Anatomie. Frankfurt am Main. *Erste Abth.*, pp. 1-152.

Kunkel, L. M., P. G. Heltne and D. S. Borgaonkar 1980. Chromosomal variation and zoogeography in *Ateles. International Journal of Primatology* 1(3):223-232.

Lambert, J. E. and P. A. Garber 1998. Evolutionary and ecological implications of primate seed dispersal. *American Journal of Primatology* 45(1):9-28.

Bibliography

Lang, J. 1967. The estrous cycle of the squirrel monkey (*Saimiri sciureus*). *Laboratory Animal Care* 17:442-45?

Lara, A. C. P. and J. P. Jorgenson 1998. Notes on the distribution and conservation status of spider an howler monkeys in the state of Quintana Roo, Mexico. *Primate Conservation* 18:25-29.

Larson, H., M. Hagelin and M. Hjern 1982. Observations on a breeding group of pygmy marmosets *Cebuella pygmaea*, at Skansen Aquarium. *International Zoo Yearbook* 22:88-93.

Latta, J., S. Hopf and D. Ploog 1967. Observation on mating behavior and sexual play in the squirrel monke (*Saimiri sciureus*). *Primates* 8:229-246.

Lavocat, R. 1969. La systématique des rangeurs histricomorphes en la derive des continents. *Compte Revu Académie Science Sér.* D. 269:1496-1497.

Leca, J. B., I. Fornasieri, and O. Petit 2002. Aggression and reconciliation in *Cebus capucinus*. *International Journal of Primatology* 23(5):979.

Le Gros Clark, W. E. 1959. *The Antecedents of Man: an Introduction to the Evolution of the Primates.* First edition Edinburgh University Press, Edinburgh.

Lehman, C. 1959. Contribuciones al estudio de la fauna de Colombia. XIV. Nuevos observaciones sobr *Oroaetus isidora* (Desmurs). *Nov. Colomb. Contribuciones Científicas del Museo de Historia ded la Universidad c Cauca, Popayan* 1(4):169-195.

Lehman, S. M. and K. L. Robertson 1994. A preliminary survey of *Cacajao melanocephalus melanocephalu. International Journal of Primatology* 15(6):927-934.

Leighton, M. and D. R. Leighton 1982. The relationship of size of feeding agregate to size of food patch: howlin monkeys (*Alouatta palliata*) feeding in *Trichilia cipo* food trees on Barro Colorado. *Biotropica* 14:81-90.

Lima, E. M. and S. F. Ferrari 2003. Diet of a free-ranging group of squirrel monkeys (*Saimiri sciureus*) i eastern Brazilian Amazonia. *Folia Primatologica* 74(3):150-158.

Lima, E. M., A. L. C. B. Pina and S. F. Ferrari 1997. Behaviour of the squirrel monkey (*Saimiri sciureu.* Platyrrhini, Cebidae) at the Fazenda Monte Verde, Peixe-Boi, Para. [Abstract] *Programa e Resumos do VI. Congresso Brasileiro de Primatologia & V Reuniao Latino-Americano de Primatología.* Soc. Brasileira de Primatologia Joao Pessoa, pp. 44.

Lima, E. M., A. L. C. B. Pina and S. F. Ferrari 2000. Behaviour of free-ranging squirrel monkeys *Saimiri sciureu.* (Platyrrhini: Cebidae) at the Fazenda Monte Verde, Peixe-Boi, Para. *A Primatologia no Brasil* 7:171-180.

Lindsay, N. B. C. 1980. A report on the field study of geoffroy's tamarin. *Saginus oedipus* geoffroyi. *Dodo Journal of the Jersey Wildlife Preservation Trust* 17:27-51.

Link, A. 2002. Costs and behavioral effects of breeding twins in free-ranging *Ateles belzebuth belzebuth* i Colombia. [Abstract] *Caring for Primates.m* [Abstract] *Abstracts of the XIXth Congress. The Internationa Primatological Society. Beijing: Mammalogical Society of China,* pp. 351-352.

Linnaeus, C. 1758. *Systema Naturae, ed. 10, vol. i, Regnum Animale.* Holmiae.

Linnaeus, C. 1766. *Systema Naturae. ed.12, vol. i, Regnum Animale.* Holmiae.

Lipald, L. K. 1988. A census of primates in Cabo Blanco Absolute Nature Reserve, Costa Rica. *Brenesi. 29:101-105.

Lipald, L. K. 1989. A wet season census of primates at Cabo Blanco Absolute Nature Reserve, Costa Rica *Brenesia* 31:93-97.

Long, J. O. and R. W. Cooper 1968. Physical growth and dental eruption in captive-bred squirrel monkeys *Saimiri sciureus* (Leticia, Colombia). Pp. 193-205 in: L. A. Rosenblum and R. W. Cooper (eds.), *The Squirre Monkey.* Academic Press, New York.

Longino, J. T. 1984. True anting by the capuchin, *Cebus capucinus. Primates* 25(2):243-245.

Lorenz, R. and H. Heinemann 1967. Beitrag zur Morphologie und korperlichen Jugendwicklung des Springtamarin (*Callimico goeldii*)(Thomas, 1904). *Folia Primatologica* 6:1-27.

Lorenz, R. 1971. Goeldi's monkey, *Callimico goeldii*, Thomas 1904 preying on snakes. *Folia Primatologica* 15:133-142.

Lorenz, R. 1972. Management and reproduction of the goeldi's monkey Callimico goeldii (Thomas, 1904), Callimiconidae, Primates. Pp. 92-109 in: D.D. Bridgewater (ed.), *Saving the Lion Marmoset*, The Wild Animal Propagation Trust, Wheeling, W. Va.

Lorenz, R. and H. Heinemann 1967. [Study of the morphology and infant growth of the spring tamarin *Callimico goeldii* (Thomas, 1904).] *Folia Primatologica* 6:1-27.

Luchterhand, K., R. F. Kay and R. H. Madden 1986. *Mohanimico hershkovitzi*, gen. et sp. nov., un primate du Miocene Moyen d'Amerique du Sud. . *Comptes-Rendus de lÁcadémie de Science de l'Académie de Paris*. sér. 3 303:753-758.

Lund, P. W. 1837. Note sur les osemens fossiles des terrains Tertiaires de Simorre de Sansan, etc., dans le departement du Gers, et sur la decouverte recente d'une machoire de singe fossile. *Annales des Science Naturelles (Paris)* 7:116-123.

Lynch, J. W. 1998. Mating behavior in wild tufted capuchins (*Cebus apella nigritus*) in Brazil's Atlantic forest. *American Journal of Physical Anthropology* (Suppl 26):153.

Lynch, J. W. 2002. Male behavior and endocrinology in wild tufted capuchin monkeys, *Cebus apella nigritus*. Unpublished Ph.D. *Dissertation Abstracts International* A62(11):3844.

Lynch, J. W. and J. Rimoli. 2000. Demography of a group of tufted capuchin monkeys (*Cebus apella nigritus*) at the Estacao Biológica de Caratinga, Minas Gerais, Brazil. *Neotropical Primates* 8(1):44-49.

Ma, N. S. F. 1981. Chromosome evolution in the owl monkey, *Aotus trivirgatus* ii. spatial contrast sensitivity. *Vision Research* 17:821-825.

Ma, N. S. F., Elliott, M. W., Morgan, L. M., Miller, A. C. and T. C. Jones 1976a. Translocation of y chromosome to an autosome in the Bolivian owl monkey, *Aotus*. *American Journal of Physical Anthropology* 45(2):191-202.

Ma, N. S. F., T. C. Jones, A. G. Miller, L. M. Morgan and E. A. Adams 1976b. Chromosome polymorphism and banding patterns in the owl monkey (*Aotus*). *Laboratory Animal Science* 26(6,ii):1022-1036.

Ma, N. S. F., T. C. Jones, M. T. Bedard, A. C. Miller, L. M. Morgan and E. A. Adams 1977. The chromosome complement of an *Aotus* hybrid *Journal of Heredity* 68:409-412.

Ma, N. S. F., R. N. Rossan, S. T. Kelley, J. S. Harper, M. T. Bedard and T. C. Jones 1978. Banding patterns of the choromosomes of two new karyotypes of the owl monkeys, *Aotus*, captured in Panama. *Journal of Medical Primatology* 7:146-155.

Ma, N. S. F., D. M. Renquist, R. Hall, P. K. Sehgal, T. Simone and T. C. Jones 1980. xx/xo sex determination system in a population of Peruvian owl monkeys, *Aotus*. *Journal of Heredity* 71(5):336-342.

Ma, N. S. F., R. Aquino and W. E. Collins 1985. The new karyotypes in the Peruvian owl monkey (*Aotus trivirgatus*). *American Journal of Primatology* 9(4):333-341.

Mack, D. 1979. Growth and development of infant red howling monkeys (*Alouatta seniculus*) in a free-ranging population. Pp. 127-136 in: J. F. Eisenberg (ed.), *Vertebrate Ecology in the Northern Neotropics*. Smithsonian Institution Press, Washington, D.C.

Mack, D. and R. A. Mittermeier (eds.) 1984. *The International Primate Trade*. Vol. 1: *Legislation, Trade and Captive Breeding*. Traffic (U.S.A.), Washington, D. C., 185 pp.

MacArthur, R. H. 1965. Patterns of species diversity. *Biological Review* 40:510-533.

MacArthur, R. H. and J. H. Connell 1966. *The Biology of Populations*. John Wiley & Sons, New York.

MacFadden, B. J. 1990. Chronology of Cenozoic primate localities in South America. *Journal of Human Evolution* 19.

MacPhee, R. D. E. and M. Rivero de la Calle 1996. Accelerator mass spectrometry 14C age determination for the alleged Cuban spider monkey, *Ateles (=Montaneia) anthropomorphus*. *Journal of Human Evolution* 30(1):89-94.

MacPhee, R. D. E. and I. Horovitz 2002. Extinct Quaternary platyrrhines of the Greater Antilles and Brazil Pp. 189–200 in: W. C. Hartwig (ed.), *The Primate Fossil Record*. Cambridge University Press, New York.

Mandujano, S. 2002. [Demografic analysis, behavior and genetics of howler moneys (*Alouatta palliata mexicana*) and spider monkeys (*Ateles geoffroyi vellerosus*) in a fragmented habitat at Veracruz, Mexico]. *Laboratory Primate Newsletter* 41(2):3.

Manson, J. H. 1999. Infant handling in wild *Cebus capucinus*: testing bonds between females? *Animal Behaviour* 57:911-921.

Manson, J. H. and S. Perry 2000. Correlates of self-directed behavior in wild white-faced capuchins. *Ethology* 106:301-317.

Manson, M. H., L. M. Rose and J. Gros-Louis 1999. Dynamics of female-female relationships in wild *Cebus capucinus*. Data from two Costa Rican sites. *International Journal of Primatology* 679-706.

Mantilla Meluk, H. and L. F. Barrios Rodríguez 1999. Ecología Básica de *Cebus apella* en la Región de Bajo Apaporis, Amazonia colombiana. Graduation Thesis in biology, Universidad Nacional de Colombia, Bogotá.

Marks, J. 1987. Social and ecological aspects of primate cytogenetics. Pp. 139-151 in: W. G. Kinzey (ed.), *Primate Models: the Evolution of Human Behavior*. State University of New York Press, Albany.

Marsh, L. K., C. A. Chapman, M. A. Norconk, S. F. Ferrari, K. A. Gilbert, J. C. Bicca-Marques and J. Wallis 2003. Fragmentation: Specter of the future or the spirit of conservation? Pp. 381-398 in L. K. Marsh (ed.), *Primates in Fragments: Ecology and Conservation*. Kluwer Academic/Plenum Publ., New York.

Martin, R. D. 1990. *Primate Origins and Evolution: A Phylogenetic Reconstruction*. Princeton University Press, Princeton.

Martinez, R. A., R. A. Moscarella, M. Aguilera and E. Marquez 2000. Update on the status of the Margarita Island capuchin, *Cebus apella margaritae*. *Neotropical Primates* 8(1):34-35.

Masataka, N. 1981a. A field study of the social behavior of Goeldi's monkeys (*Callimico goeldii*) in North Bolivia. I. Group composition, breeding cycle, and infant development. *Kyoto University Overseas Research Reports of New World Monkeys* 2:23-32.

Masataka, N. 1981b. A field study of the social behavior of Goeldi's monkeys (*Callimico goeldii*) in North Bolivia. II. Grouping pattern and intragroup relationship. *Kyoto University Overseas Research Reports of New World Monkeys* 2:33-41.

Masataka, N. 1983. Categorical responses to natural and synthesized alarm calls in Goeldi's monkeys (*Callimico goeldii*). *Primates* 24(1):40-51.

Mason, W. A. 1965. Territorial behavior in *Callicebus* monkeys. *American Zoologist* 5:675.

Mason, W. A. 1966a. Social organization of the South American monkey, *Callicebus moloch*: A preliminary report. *Tulane Studies in Zoology* 13:23-28.

Mason, W. A. 1966b. Communication in the titi monkey *Callicebus*. *Journal of the Zoological Society of London* 150:77-127.

Mason, W. A. 1968. Use of space by *Callicebus* groups. Pp. 398-419 in: P. C. Jay (ed.), *Primates: Studies in Adaptation and Variability*. Holt, Rinehart and Winston, New York.

Mason, W. A. 1971. Field and laboratory studies of social organization in *Saimiri* and *Callicebus*. Pp. 107-137 in: L. A. Rosenblum (ed.), *Primate Behavior: Developments in Field and Laboratory Research*, vol. 2. Academic Press, New York.

Mason, W. A. 1974a. Comparative studies of social behavior in *Callicebus* and *Saimiri*: behavior of male-female pairs. *Folia Primatologica* 22:1-8.

Mason, W. A. 1974b. Differential grouping patterns in two species of South American monkey. Pp. 153-169 in: N. F. White (ed.), *Ethology and Psychiatry*. University of Toronto, Toronto.

Mason, W. A. 1975. Comparative studies of social behavior in *Callicebus* and *Saimiri*: strength and specificity of attraction between male-female cagemates. *Folia Primatologica* 23:113-123.

Mason, W. A. 1976a. Primate social behavior: Pattern and process. Pp. 425-455 in: R. B. Masterton *et al.*(eds.), *Evolution of Brain and Behavior in Vertebrates*. Lawrence Erlbaum Associates, New Jersey.

Mason, W. A. 1978. Ontogeny of social systems. Pp. 5-14 in: D. J. Chivers and J. Herbert, *Recent Advances in Primatology, vol. 1, Behaviour*. Academic Press, London,.

Mason, W. A. and G. Epple 1969. Social organization in experimental groups of *Saimiri* and *Callicebus*. Pp. 59-65 in: C. R. Carpenter (ed.), *Proceedings of the Second International Congress of Primatology, Vol. 1: Behavior.* S. Karger, Basel.

Massey, J. 1987. A population survey of *Alouatta palliata, Cebus capucinus* and *Ateles geoffroyi* at Palo Verde, Costa Rica. *Revista de Biología Tropical* 35:345-347

Mast, R. B., J. V. Rodriguez and R. A. Mittermeier 1993. The Colombian cotton-top tamarin in the wild. Pp. 3-43 in: N. K. Clapp (ed.), *A Primate Model for the Study of Colitis and Colonic Carcinoma: the Cotton-Top Tamarin, Saguinus oedipus.* CRC Press, Boca Raton.

Matsushita, K. and A. Nishimura 1999. [Group and party of spider monkey, *Ateles belzebuth.*] (in Japanese) [resumen] *Reishorui Kenkyu/ Primate Res.* 15(3):412.

Mayr, E. 1969. *Animal Species and Evolution.* 2nd Edition, Belknap Press, Harvard University Press, Cambridge, Mass.

Mayr, E. 1971. *Populations, Species, and Evolution: an Abridgment of Animal Species and Evolution.* The Belknap Press of Harvard University Press, Cambridge, Mass.

Mayr, E. and P. D. Ashlock 1991. *Principles of Systematic Zoology.* 2nd Edition. McGraw-Hill, Inc., New York.

McCann, C., K. Williams-Guillen, F. Koontz, A. A. R. Espinoza, J. C. Martínez-Sanchez and C. Koontz 2003. Shade coffee plantations as wildlife refuge for mantled howler monkeys (*Alouatta palliata*) in Nicaragua. Pp. 321-341 in: L. K. Marsh (ed.), *Primates in Fragments: Ecology and Conservation.* Kluwer Academic/ Plenum Publ., New York.

McConnel, P. B. and C. T. Snowdon 1986. Vocal interactions between unfamiliar groups of captive cotton-top tamarins. *Behaviour* 97:273-296.

McFarland, M. J. M. 1986. Ecological determinants of fission-fusion sociality in *Ateles* and *Pan*. Pp. 181-190 in: J. G. Else and P. C. Lee (eds.), *Primate Ecology and Conservation.* Cambridge University Press, Cambridge, England.

McFarland, M. J. M. 1990. Fission-fusion social organization in *Ateles* and *Pan. International Journal of Primatology* 11(1):47-61.

McGrew, W. C. and E. C. McLuckie, E. C. 1986. Philopatry and dispersion in the cotton-top tamarin, *Saguinus o. oedipus*: An attempted laboratory simulation. *International Journal Primatology* 7:401-422.

Medeiros, M. A., R. M. S. Barros, J. C. Pieczark, C. Y. Nagamacho, M. Ponsa, M. Garcia, F. Garcia and J. Egozcue 1997. Radiation and speciation of spider monkeys, genus *Ateles*, from the cytogenetic viewpoint. *American Journal of Primatology* 42(3):167-178.

Meireles, C. M. M., J. Czelusniak, M. P. C. Schneider, J. A. P. C. Muñoz, M. C. Brigido, H. S. Ferreira, and M. Goodman 1999. Molecular phylogeny of ateline New World monkeys (Platyrrhini, Ateleinae) based on gamma-globin gene sequences: Evidence that *Brachyteles* is the sister group of *Lagothrix. Molecular Phylogenetics and Evolution* 12(1):10-30.

Meldrum, D. J. and J. G. Fleagle 1988. Morphological affinities of the postcranial skeleton of *Cebupithecia sarmientoi*. [Abstract] *American Journal of Physical Anthropology* 75(2):249–250.

Meldrum, D. J. and P. Lemelin 1991. Axial skeleton of *Cebupithecia sarmientoi* (Pitheciinae, Platyrrhini) from the Middle Miocene of La Venta, Colombia. *American Journal of Primatology* 25(2):69-90.

Meldrum, D. J. and R. F. Kay 1997. *Nuciruptor rubricae*, a new pitheciin seed predator from the Miocene of Colombia. *American Journal of Physical Anthropology* 102(3):407-427.

Meldrum, J., J. Fleagle and R. F. Kay 1990. Partial humeri of two Miocene Colombian primates. *American Journal of Physical Anthropology* 81:413-422.

Bibliography

Meltz, K. 2002. Effect of ecological conditions on the daily activity budget of adult male mantled howler monkeys (*Alouatta palliata*) living in a forest fragment at Bocas del Toro Province, Republic of Panama. *American Journal of Anthropology* Suppl. 34:112.

Mendel, F. C. 1975. The locomotor anatomy of *Alouatta palliata*: The utility of *Alouatta* as a model for early hominoid locomotion. Unpublished Ph.D. thesis, University of California, Davis.

Mendel, F. C. 1976. Postural and locomotor behavior of *Alouatta palliata* on various substates. *Folia Primatologica* 26:36-53.

Mendelson, J. 1994. The social system of the saddle-back tamarin (*Saguinus fuscicollis*): An examination of the adaptive basis of reproductive strategies based on a thirteen year study of demography and behavior at Coca Caschu Biological Station, Bamu National Park, Peru. Unpulished Ph.D. thesis, *Dissertation Abstracts International* 1995, B55(10):4223.

Mendes Pontes, A. R. 1997. Habitat partitioning among primates in Maraca Island, Roraima, northern Brazilian Amazonia. *International Journal of Primatology* 18(2):131-157.

Mendes Pontes, A. R. 1999. Environmental determinants of primate abundance in Maraca Island, Roraima, Brazilian Amazonia. *Journal of Zoology* 247(2):189-199.

Mendes, F. D. C., L. B. R. Martins, J. A. Pereira and R. F. Marquezan 2000. Fishing with a bait: A note on behavioural flexibility in *Cebus apella*. *Folia Primatologica* 71(5):350-352

Menzel, C. R. and E. W. Menzel, Jr. 1980. Head-cocking and visual exploration in marmosets (*Saguinus fuscicollis*). *Behaviour* 75:219-234.

Merrit, D., Jr. 1993. The owl monkey, (*Aotus trivirgatus*): reproductive parameters. *Estudios Primatológicos en México* 1:305-315.

Miller, G. S. Jr. 1916. The teeth of a monkey found in Cuba. *Smithsonian Miscelaneous Collection* 82(5):1-16.

Milton, K. 1975. Urine rubbing behavior in the mantled howler monkey, *Alouatta palliata*. *Folia Primatologica* 23:105-112.

Milton, K. 1977. The foraging strategy of the howler money in the tropical forest of Barro Colorado Island, Panama. Unpublished Ph.D. thesis, New York University, New York.

Milton, K. 1978. Behavior adaptations to leaf-eating by the mantled howler monkey (*Alouatta palliata*). Pp. 535-549 in: G. G. Montgomery (ed.), *The Ecology of Arboreal Folivores*. Smithsonian Institution Press, Washington, D. C.

Milton, K. 1980. *The Foraging Strategy of Howler Monkeys*. Columbia University Press, New York.

Milton, K. 1981a. Estimates of reproductive parameters for free-ranging *Ateles geoffroyi*. *Primates* 22:574-579.

Milton, K. 1981b. Food choice and digestive strategies of two sympatric primate species. *American Naturalist* 117:496-505.

Milton, K. 1982. Dietary quality and population regulation in a howler-monkey population. Pp. 273-289 in: E. G. Leigh, Jr., A. S. Rand and M. Windsor (eds.), *The Ecology of a Tropical Forest: Seasonal Rhythms and Long-term Changes*. Smithsonian Institution Press, Washington, D. C.

Milton, K. 1982. Dietary quality and population regulation in a howler-monkey population. Pp. 273-289 in: E. G. Leigh, Jr., A. S. Rand and M. Windsor (eds.), *The Ecology of a Tropical Forest: Seasonal Rhythms and Long-term Changes*. Smithsonian Institution Press, Washington, D. C.

Milton, K. 1993a. Factors implicated in demographic regulation of Panamanian howler monkeys. [Abstract] *American Journal of Physical Anthropology*. 1993. (Suppl. 16). Pgs: 146-147.

Milton, K. 1993b. Diet and social organization of a free-ranging spider monkey population: The development of species-typical behavior in the absence of adults. . Pp. 173-181 in: M.E. Pereira and L.A. Fairbanks (eds.), *Juvenile Primates: Life History, Development, and Behavior*. Oxford University Press, New York.

Milton, K. 1998. Physiological ecology of howlers (*Alouatta*): Energetic and digestive considerations and comparison with the Colobinae. *International Journal of Primatology* 19(3):513-548.

Milton, K. 2001. Leafeaters of the New World: Diet and energy conservation in howler monkeys. Pp. 354-355 in: D. Macdonald (ed.), *The New Encyclopedia of Mammals.* Oxford University Press, Oxford.

Milton, K. and R. A. Mittermeier 1977. A brief survey of the primates of Coiba Island, Panama. *Primates* 18:931-936.

Milton, K. and J. L. Nessimian 1984. Evidence for insectivory in two primate species (*Callicebus torquatus lugens* and *Lagothrix lagothricha lagothricha*) from northwestern Amazonia. *American Journal of Primatology* 6(4):367-371.

Milton, K., T. M. Casey and K. K. Casey 1979. The basal metabolism of mantled howler monkeys (*Alouatta palliata*). *Journal of Mammalogy* 60:373-376.

Milton, K., P. J. van Soest and J. B. Robertson 1980. Digestive efficiencies of wild howler monkeys. *Physiological Zoology* 53:402-409.

Mitchell, C. L. 1990. The ecological basis for female social dominance: A behavior study of the squirrel monkey (*Saimiri sciureus*) in the wild. Unpublished Ph.D. thesis, Princeton University.

Mitchell, C. L. 1994. Migration alliances and coalitions among adult male South American squirrel monkeys (*Saimiri sciureus*). *Behaviour* 130(3-4):169-190.

Mitchell, C. L., S. Boinski and C. P. van Schaik 1991. Competitive regimes and female bonding in two species of squirrel monkeys (*Saimiri oerstedi* and *S. sciureus*). *Behavioral. Ecology and Sociobiology* 28:55-60.

Mitchell, K., J. Floyd and L. Winkler 2002. Adaptive strategies and resource utilization of the mantled howling monkey (*Alouatta palliata*) in a small forest fragment in Nicaraugua. *American Journal of Physical Anthropology* Suppl. 34:114.

Mittermeier, R. A. 1973. Group activity and population dynamics of the howler monkey on Barro Colorado Island. *Primates* 14(1):1-19.

Mittermeier, R. A. 1978. Locomotion and posture in *Ateles geoffroyi* and *Ateles paniscus*. *Folia Primatologica* 30:161-193.

Mittermeier, R. A. 1986a. Primate conservation priorities in the Neotropical region. Pp. 221-240 in: K. Benirschke (ed.), *Primates: The Road to Self-Sustaining Populations.* Springer-Verlag, New York.

Mittermeier, R. A. 1986b. Primate prospects. *Iucn Bulletin* 17(1-3):24-25.

Mittermeier, R. A. 1986c. Strategies for the conservation of highly endangered primates. Pp. 1013-1022 in: K. Benirschke (ed.), *Primates: The Road to Self-Sustaining Populations.* Springer-Verlag, New York.

Mittermeier, R. A. 1987a. Effects of hunting on rain forest primates. Pp. 109-146 in: C. W. Marsh and R. A. Mittermeier (eds.), *Primate Conservation in the Tropical Rain Forest.* Alan R. Liss, New York.

Mittermeier, R. A. 1987b. Framework for primate conservation in the Neotropical region. Pp. 305-320 in: C. W. Marsh, R. A. Mittermeier (eds.), *Primate Conservation in the Tropical Rain Forest.* Alan R. Liss, New York.

Mittermeier, R. A. 1988. Primate diversity and the tropical forest: Case studies from Brazil and Madagascar and the importance of megadiversity countries. Pp. 145-154 in: E. O. Wilson and F. M. Peter (eds.), *Biodiversity.* National Academy Press, Washington, D. C.

Mittermeier, R. A. and J. G. Fleagle 1976. The locomotor and postural repertoires of *Ateles geoffroyi* and *Colobus guereza*, and a reevaluation of the locomotor category semibrachiation. *American Journal of Physical Anthropology* 45:235-255.

Mittermeier, R. A. and A. F. Coimbra-Filho 1977. Primate conservation in Brazilian Amazonia. Pp. 117-166 in: Prince Ranier of Monacho and G. H. Bourne (eds.), *Primate Conservation.* Academic Press, New York.

Mittermeier, R. A. and A. F. Coimbra-Filho 1981. Systematics: species and subspecies. Pp. 29-109 in: A. F. Coimbra-Filho and R. A. Mittermeier (eds.), *Ecology and Behavior of Neotropical Primates*, vol. 1. Academia Brasileira de Ciências, Río de Janeiro.

Mittermeier, R. A. and M. G. M. van Roosmalen 1981. Preliminary observations on habitat utilization and diet in eight Surinam monkeys. *Folia Primatologica* 36:1-39.

Mittermeier, R. A. and J. J. Oates 1985. Primate diversity: The world's top countries. *Primate Conservation* 5:41–48.

Mittermeier, R. A., A. B. Rylands, A. F. Coimbra-Filho and G. A. B. da Fonseca (eds.) 1988. *Ecology and Behavior of Neotropical Primates*. Vol. 2. World Wildlife Fund, Washington, D.C., 610 pp.

Mittermeier, R. A., I. Tattersall, W. R. Konstant and R. Mast 1994a. *Lemurs of Madagascar*. Conservation International, Washington, D. C.

Mittermeier, R. A., W. R. Konstant and R. B. Mast 1994b. Use of Neotropical and Malagasy primate species in biomedical research. *American Journal of Primatology* 34(1):73-80.

Mittermeier, R. A., W. R. Konstant and A. B. Rylands 2000. The world´s top 25 most endangered primates. *Neotropical Primates* 8(1):49.

Moheno, M. B. 2002. [Prevalence of gastrointestinal parasites in primates (*Alouatta pigra* and *Ateles geoffroyi yucatenensis*) located in conserved and fragmented habitats in Quintana Roo, Mexico.] *Laboratory Primate Newsletter* 41(4):10.

Mondolfi, E. and J. F. Eisenberg 1979. New records for *Ateles belzebuth hybridus*, in northern Venezuela. Pp. 93-96 in: J. F. Eisenberg (ed.), *Vertebrate Ecology in the Northern Neotropics*. Smithsonian Intitute Press, Washington, D. C.

Montoya, E., C. Málaga and E. Villavicencio 1990. Evaluación de una colonia de reproducción de *Aotus vociferans*. Pp. 616-624 in. N. Castro (ed.), *La Primatología en el Perú*. Proyecto Peruano de Primatología, Lima.

Montoya, G. E., J-P. Vernot and M. E. Patarroyo 2002. Partial characterization of the CD45 Phosphatase cDNA in the owl monkey (*Aotus vociferans*). *American Journal of Primatology* 57(1):1-11.

Moody, M. I. and E. W. Menzel 1976. Vocalizations and their behavioral contexts in the tamarin, *Saguinus fuscicollis*. *Folia Primatologica* 25:73-94.

Moore, A. J. and J. M. Cheverud 1992. Systematics of *Saguinus oedipus* group of the bare-faced tamarins: Evidence from facial morphology. *American Journal of Physical Anthropology* 89:73-84.

Moore, L., J. Cleland and W. C. McGrew 1991. Visual encounters between families of cotton-top tamarins, *Saguinus oedipus*. *Primates* 32(1):23-33.

Morales-Hernández, K. 2002. Wild populations of spider monkeys (*Ateles geoffroyi*) in El Salvador, Central America. *Neotropical Primates* 10(3):153-154.

Morales-Jimenez, A. L. 2002. [Density of howler monkeys (*Alouatta seniculus*) in premontane forest of the Andes, Risaralda, Colomiba]. *Neotropical Primates* 10(3):141-144.

Moreira, M. A. M. 2002. SRY evolution in Cebidae (Platyrrhini: Primates). *Journal of Molecular Evolution* 55(1):92-103.

Moreno, L. I., I. C. Salas and K. E. Glander 1991. Breech delivery and birth-related behaviors in wild mantled howling monkeys. *American Journal of Primatology* 23(3):197-199.

Morton, L. S. 1996. Allomaternal and maternal behaviour in squirrel monkeys (*Saimiri sciureus*) in the wild. [Abstract] IPS/ASP CONGRESS ABSTRACTS 1996:636.

Moscow, D. 1987. Troop movement and food habits of white-faced monkeys in a tropical-dry forest. *Revista de Biología Tropical* .35(2):287-297.

Moura, A. C. A. 2002. Tool use by wild groups of *Cebus apella libidinosus* living in the dry Caatinga Forest of northeastern Brazil. [Abstract] *Caring for Primates. Abstracts of the XIXth Congress. The International Primatological Soceity*, pp. 244-245

Moura, A. C. A. 2003. Ecological pressures driving tool use in capuchin monkeys. [Abstract] *Folia Primatologica* 74(4):209.

Moynihan, M. 1964. Some behavioral patterns of platyrrhine monkeys, 1: The night monkey *Aotus trivirgatus*. *Smithsonian Miscellaneous. Collection.* 146:1-84.

Moynihan, M. 1966. Communication in the titi monkey, *Callicebus. Journal of Zoology* 150:71-127.

Moynihan, M. 1967. Comparative aspects of communication in New World primates. Pp. 306-342 in: D. Morris (ed.), *Primate Ethology*. Weidenfeld and Nicolsen, London.

Moynihan, M. 1970. Some behavioral patterns of platyrrhine monkeys, 2: *Saguinus geoffroyi* and some other tamarins. *Smithsonian Contributions in Zoology* 28:1-27.

Moynihan, M. 1976a. *The New World Primates*. Princeton University Press, Princeton, N. J.

Moynihan, M. 1976b. Notes on the ecology and behavior of the pygmy marmoset (*Cebuella pygmaea*) in Amazonian Colombia. Pp. 79-84 in: R. W. Thorington and P. G. Heltne (eds.), *Neotropical Primates: Field Studies and Conservation*. National Academy of Sciences, Washington, D. C.

Muckenhirn, N. A. 1976. Addendum to the nonhuman primate trade in Colombia. . Pp. 99-100 in: R. W. Thorington, Jr. and P. G. Heltne (eds.), *Neotropical Primates: Field Studies and Conservation*. National Academy of Sciences, Washington, D.C.

Müller, A. 1994. Duettieren bei Springaffen (*Callicebus cupreus*). Graduation thesis, University of Zürich, Zürich.

Müller, A. 1995. Duetting in the titi monkey, *Callicebus cupreus. Neotropical Primates* 3(1):18-19.

Müller, A. E. and G. Anzenberger 2002. Duetting in the titi monkey *Callicebus cupreus*: Structure, pair specificity and development of duets. *Folia Primatologica* 73(2-3):104-115.

Müller, P. 1973. *The Dispersal Centres of Terrestrial Vertebrates in the Neotropical Realm. Biogeographica* 2. Junk, The Hague.

Muñoz, J. 1991. Algunos aspectos de la dispersión, estructura social y uso del espacio habitado, en un grupo de *Lagothrix lagothricha* (Humboldt, 1812) - Primates, Cebidae - en la Amazonia Colombiana. Unpublished bachelor's thesis, Universidad Nacional de Colombia, Bogotá, 196 pp.

Muñoz, D., Y. Garcia del Valle, B. Franco, A. Estrada and M. Magana-A. 2002. [Study of the activity pattern of howler monkeys (*Alouatta palliata*) in Yumka Park, Tabasco, Mexico.] *Neotropical Primates* 10(1):11-17.

Muskin, A. and A. J. Fischgrund 1981. Seed dispersal of *Stemmadenia* (Apocynaceae) and sexually dimorphic feeding strategies by *Ateles* in Tikal, Guatemala. *Biotropica Supplement Reproductive Botany* 13:78-80.

Nagy, K. A. and K. Milton 1979a. Aspects of dietary quality, nutrient assimilation and water balance in wild howler monkeys (*Alouatta palliata*). *Oecologia* 39:249-258.

Nagy, K. A. and K. Milton 1979b. Energy metabolism and food consumption by wild howler monkeys (*Alouatta palliata*). *Ecology* 60(3):475-480.

Nakatsukasa, M., N. Takai and T. Setoguchi 1997. Functional morphology of the postcranium and loco-motor behavior of *Neosaimiri fieldsi*, a *Saimiri*-like Middle Miocene platyrrhine. *American Journal of Physical Anthropology* 102(4):515-544.

Napier, J. R. and P. H. Napier 1967. *A Handbook of Living Primates*. Academic Press, London.

Nash, R. 1983. *Wilderness and the American Mind*. 3rd ed.. Yale University Press, New Haven.

NRA (National Research Council) 1981. *Techniques for the Study of Primate Population Ecology*. National Academy of Sciences, Washington, D.C.

Natori, M. 1994. Craniometrical variations among eastern Brazilian marmosets and their systematic relationships. *Primates* 35(2):167-176.

Natori, M. and T. Hanihara 1988. An analysis of interspecific relationships of *Saguinus* based on cranial measurements. *Primates* 29(2):255-262.

Natori, M. and T. Hanihara 1992. Variations in dental measurements between *Saguinus* species and their systematic relationships. *Folia Primatologica* 58:84-92.

Bibliography

Neves, A. M. S. and A. B. Rylands 1991. Diet of a group of howling monkeys, *Alouatta seniculus*, in an isolated forest patch in central Amazonia. *Primatologia no Brasil* 3:263-274.

Neville, M. K. 1972a. The population structure of red howler monkeys (*Alouatta seniculus*) in Trinidad and Venezuela. *Folia Primatologica* 18:47-77.

Neville, M. K. 1972b. Social relations within troops of red howler monkeys (*Alouatta* seniculus). *Folia Primatologica* 18:47-77.

Neville, M. K. 1976a. Census of primates in Peru. Pp. 19-29 in: *Primate Censusing Studies in Peru and Colombia*, Report to the National Academy of Sciences on Project Amro-0719, First Inter-American Conference on Conservation and Utilization of American nonhuman Primates in Biomedial REsearch. Pan American Health Organization, Washington, D. C.

Neville, M. K. 1976b. The population and conservation of howler monkeys in Venezuela and Trinidad. Pp. 1010-109 in: R. W. Thorington and P. G. Heltne (eds.), *Neotropical Primates: Field Studies & Conservation*. National Academy of Science, Washington, D. C.

Neville, M. K. 1983. Social affinities in the riverbanks howler monkeys. Pp. 1-9 in: P. K. Seth (ed.), *Perspectives in Primate Biology*. Today and Tomorrow's Printers, New Delhi.

Neville, M. K., K. E. Glander, F. Braza and A. B. Rylands 1988. The howling monkeys, genus *Alouatta*. Pp. 349-453 in: R. A. Mittermeier, A. B. Rylands, A. B. Coimbra-Filho and G. A. B. Fonseca (eds.), *Ecology and Behavior of Neotropical Primates*, vol. 2. World Wildlife Fund-U. S., Washington, D. C.

Neyman, P. F. 1977. Aspects of the ecology and social organization of free-ranging cotton-top tamarins (*Saguinus oedipus*) and the conservation status of the species. Pp. 39-71 in: D. G. kleiman (ed.), *Biology and Conservation of the Callitrichidae*. Smithsonian Institute Press, Washington, D. C.

Neyman, P. F. 1978. Ecology and social organization of the cotton- top tamarin (*Saguinus oedipus*). Unpublished Ph.D. thesis, University of California, Berkely.

Nickle, D. A. and E. W. Heymann 1996. Predation on Orthoptera and other orders of insects by tamarin monkeys, *Saguinus mystax mystax* and *Saguinus fuscicollis nigrifrons* (Primates: Callitrichidae), in north-eastern Peru. *Journal of Zoology* 239(4):799-819.

Nilsson, G. 1983. *The Endangered Species Handbook*. The Animal Welfare Institute, Washington, D. C.

Nino-Vasquez, J. J., D. Vogel, R. Rodríguez, A. Moreno, M. E. Patarroyo, G. Pluschke and C. A. Daubenberger 2000. Sequence and diversty of DRB genes of *Aotus nancymaae*, a primate model for human malaria parasites. *Immunogenetics* 51:219-230.

Nishimura, A. 1987. Sociological characteristics of woolly monkeys (*Lagothrix lagotricha*) in the upper Caquetá, Colombia. [Abstract] *International Journal of Primatology* 8(5):521.

Nishimura, A. 1988. Mating behavior of woolly monkeys, *Lagothrix lagothricha*, at La Macarena, Colombia. *Field Studies of New World Monkeys, La Macarena, Colomibia* 1:19-27.

Nishimura, A. 1990a. Mating behavior of woolly monkeys (*Lagothrix lagotricha*) at La Macarena, Colombia (II): Mating relationships. *Field Studies of New World Monkeys, La Macarena, Colombia.* 3:7-12.

Nishimura, A. 1990b. A sociological and behavioral study of woolly monkeys, *Lagothrix lagotricha* in the upper Amazon. *The Science and Engineering Review of Doshisha University* 31(2):1-121.

Nishimura, A. 1990c. Mating patterns of woolly monkeys, *Lagothrix lagotricha* at La Macarena, Colombia (II). [Abstract] *Jinruigadu Zasshi/j. Anthropological Society Nippon* 98(2):198.

Nishimura, A. 1992. [Reproductive behaviors of woolly monkeys, *Lagothrix lagothricha*.] (in Japanese) *Reichorui Kenkyu/Primate Res.* 1992 8(2):188.

Nishimura, A. 1994. Social interaction patterns of woolly monkeys (*Lagothrix lagotricha*): A comparison among the Atelines. *Science and Engineering Revies of Doshisha University* 35(2):235-254.

Nishimura, A. 1996. Co-feeding relation of woolly monkeys within the group. [Abstract] *IPS/APS Congress Abstracts* 1996:658.

Nishimura, A. 1997a. Co-feeding relation of woolly monkeys, *Lagothrix lagothricha,* within a group at La Macarena, Colombia. *Field Studies of Fauna and Flora, La Macarena, Colombia* 1997:11-18.

Nishimura, A. 1997b. [A face-whitened disease observed in a group of woolly monkeys at La Macarena, Colombia.] (in Japanese) *Reichorui Kenkyu/Primate Res.* 13(3):268.

Nishimura, A. 1998a. Feeding behavior of woolly monkeys, *Lagothrix lagothricha,* at La Macarena, Colombia. [Abstract] 1998:189.

Nishimura, A. 1998b. [A face-whitened disease observed in a group of woolly monkeys at La Macarena, Colombia.] (in Japanese) [resumen] *Reichorui Kenkyu/Primate Research* 1998:14(3):211-212.

Nishimura, A. 1998c. [Mating and reproductive patterns of female woolly monkeys at La Macarena, Colombia. [resumen] (in Japanese) *Reichorui Kenkyu/Primate Research* 14(3):235.

Nishimura, A. 1999a. Estimation of the retention times and distances of seed dispersed by two monkey species, *Alouatta seniculus* and *Lagothrix lagothricha,* in a Colombian forest. *Ecological Ressearch* 14(2):179-191.

Nishimura, A. 1999b. [An unusual mother-infant relation observed in wild woolly monkeys – a female's simultaneous care of her own and adopted infants.] (in Japanese) [resumen] *Reichorui Kenkyu/Primate Research* 15(3):440.

Nishimura, A. 1999c. Face-whitened disease observed in wild woolly monkeys, *Lagothrix lagotricha,* at La Macarena, Colombia. *Field Studies of Fauna and Flora, La Macarena, Colombia* 13:7-13.

Nishimura, A. 2001. [Home range of woolly monkeys (*Lagothrix lagothricha*) at La Macarena, Colombia: Seasonal change and a comparison with sympatric spider monkeys.] (in Japanese) [Abstract] *Reichorui Kenkyu/Primate Research* 17(2):165.

Nishimura, A. 2002a. Home range of woolly monkeys (*Lagothrix lagothricha*) at La Macarena, Colombia: Seasonal change and a comparison with sympatric spider monkeys (*Ateles belzebuth*). [Abstract] *Anthropological Science* 110(1):126.

Nishimura, A. 2002b. [Copulation pattern and mating relationship of wild woolly monkey (*Lagothrix lagothricha*) at La Macarena, Colombia.] (in Japanese) [resumen] *Reichorui Kenkyu/Primate Research* 18(3):384.

Nishimura, A. 2002b. Home range of woolly monkeys (*Lagothrix lagothricha*) at La Macarena, Colombia: Seasonal change and a comparison with sympatric spider monkeys (*Ateles belzebuth*). [Abstract] *Anthropological Science* 110(1):126.

Nishimura, A. 2003. Reproductive parameters of wild female *Lagothrix lagotricha*. *International Journal of Primatology* 24(4):707-722.

Nishimura, A. and K. Izawa 1975. The group characteristics of woolly monkeys (*Lagothrix lagothricha*) in the upper Amazonian basin. Pp. 351-357 in: S. Kondo, M. Kawai, A. Ehara and S. Kowamura (eds.), *Proceedings from the Symposia of the Fifth Congress of the International Primatological Society.* Japan Science Press, Tokyo.

Nishimura, A., A. V. Wilches and C. Estrada 1992a. Mating behaviors of woolly monkeys, *Lagothrix lagothricha,* at La Macarena Colombia (III): Reproductive parameters viewed from a longterm study. *Field Studies of New World Monkeys, La Macarena, Colombia* 7:1-7.

Nishimura, A., A. Wilches and C. Estrada 1992b. Reproductive behaviors of woolly monkeys, *Lagothrix lagothricha,* viewed from longterm studies. [Abstract] *XIVth Congress of the International Primatological Society,* Strasbourg *IPS* 1992:313-314.

Nishimura, A., K. Izawa and K. Kimura 1995. Long-term studies of primates at La Macarena, Colombia. *Primate Conservation* no. 17:7-14.

Nolte, A. and G. Dëcker 1959. Jugendwicklung eines Kapuzineraffen (*Cebus apella* l.) mit besonderer Berücksightingung des wechselseitigen Verhaltens von Mütter un Kind. *Behaviour* XIV(4):335-373.

Nugent, M., F. Bayart, F. Crozier, O. Louguet *et al.* 1999. Food, diet and feeding behavior of squirrel monkeys (*Saimiri sciureus sciureus*) in relation to food availability on a small island in French Guiana. [resumen]. *Folia Primatologica* 70(4):217.

Bibliography

Nunes, A. 1995a. Status, distribution and viability of wild populations of *Ateles belzebuth marginatus*. *Neotropical Primates* 3(1):17-18.

Nunes, A. 1995b. Foraging and ranging patterns in white-bellied spider monkeys. *Folia Primatologica* 65(2):85-99.

Nunes, A. 1998. Diet and feeding ecology of *Ateles belzebuth belzebuth* at Maraca Ecological Station, Roraima Brazil. *Folia Primatologica* 69(2):61-76.

Olivares, A. 1962. Aves de la región sur de la Sierra de la Macarena, Colombia. *Revista de la Academia Colombiana de Ciencias Exactas, Físicas y Naturales* 11(44):305-346.

Olmedes, A. and J. B. Carroll 1980. A comparative study of pair behaviors of four Callitrichid species and the Goeldi's monkey. *Dodo: Journal of Jersey Wildlife Preservation Trust* 17:51-62.

Oppenheimer, J. F. 1967a. The diet of *Cebus capucinus* and the effect of *Cebus* on the vegetation. *Bulletin of the Ecological Society of America* 48:138.

Oppenheimer, J. F. 1967b. Vocal communication in the white-faced monkey, *Cebus capucinus*. *Bulletin of the Ecological Society of America* 48:149.

Oppenheimer, J. F. 1968. Behavior and ecology of the white-faced monkey, *Cebus capucinus*, on Barro Colorado Island, Canal Zone. Unpublished Ph.D. thesis, University of Illinois.

Oppenheimer, J. F. 1969a. Changes in forehead patterns and group composition of the white-faced monkey (*Cebus capucinus*). Pp. 36-42 in: C. R. Carpenter (ed.), *Processes of the 2nd International Congress of Primatology*, Atlanta, Ga 1968, Vol. I. Karger, Basel, New York.

Oppenheimer, J. R. 1969b. *Cebus capucinus*: play and allogrooming in a monkey group. *American Zoologist* 9:1070.

Oppenheimer, J. R. 1973. Social and communicatory behavior in the *Cebus* monkey. Pp. 251-271 in: C. R. Carpenter (ed.), *Behavioral Regulators of Behavior in Primates*. Bucknell University Press, Lewisburg, Pennsylvania.

Oppenheimer, J. R. 1977a. Communication in New World monkeys. Pp. 851-889 in: T. A. Sebeok (ed.), *How Animals Communicate*. Indiana University Press, Bloomington.

Oppenheimer, J. R. 1977b. Forest structure and its relation to activity of the capuchin monkey (*Cebus*). Pp. 74-84 in: M. R. N. Prasad and T. C. A. Kumar (eds.), *Use of Non-human Primates in Biomedical Research*. Indian National Academy of Science, New Delhi.

Oppenheimer, J. R. 1982. *Cebus capucinus*: home range, population dynamics, and interspecific relationships. Pp. 253-270 in: E. G. Leigh, A. S. Rand and D. M. Windsor (eds.), *Ecology of a Tropical Forest: Seasonal Rhythms and Long-term Changes*. Smithsonian Institution Press, Washington, D. C.

Oppenheimer, J. R. 1996. Cebus capucinus: Home range, population dynamics, and interspecific relationships. Pp. 253-272 in: E. G. Leigh, Jr. and D. M. Rand (eds.), *Behavioural Regulators of Behavior in Primates*. Smithsonian Institute Press, Washington, D.C.,.

Oppenheimer, J. R. and G. E. Lang 1969. *Cebus* monkeys: Effect on branching of *Gustavia* trees. *Science* 165:187-188.

Orlosky, F. and D. R. Swindler 1975. Origins of New World monkeys. *Journal of Human Evolution* 4:77-83.

Ortiz, S. E. 1940. Lingüística colombiana, familia Choko. *Universidad Católica del Bolívar (Medellín)* 6(18):46-77.

Ortiz-Martinez, T. J., S. J. Solano, A. Estrada and R. Coates-Estrada 1999. [Activitry patterns of *Alouatta palliata* in forest fragments at Los Tuxtlas, Mexico.] *Neotropical Primates* 7(3):80-83.

Otis, J. S., J. W. Froehlich and R. W. Thorington Jr. 1981. Seasonal and age-related differential mortality by sex in the mantled howler monkey, *Alouatta palliata*. *International Journal of Primatology*. 2:197-205.

Ottoni, E. B. and M. Mannu 2001. Semifree-ranging tufted capuchins (*Cebus apella*) spontaneously use tools to crack open nuts. *International Journal of Primatology* 22(3):347-358.

Oversluijs Vasquez, M. R. and E. . W. Heymann 2001. Crested eagle (*Morphnus guianensis*) predation on infant tamarins (*Saguinus mystax* and *Saguinus fuscicollis*, Callitrichinae). *Folia Primatologica* 72(5):301-303.

Palacios, E. 2001. Infanticide following immigration of a pregnant red howler, *Alouatta seniculus*. *Neotropical Primates*.vol.8 (3).

Palacios, E. 2002. Large vertebrate densities in undisturbed Amazonian forests: implications for minimum spatial requirements of viable populations. MSc. dissertation. University of East Anglia, Norwich, UK.

Palacios Acevedo, E. 2003. Uso del espacio por *Alouatta seniculus* en el bajo río Apaporis, Amazonia Colombiana. Pp. 85-95 in: V. Pereira, F. Nassar and A. Savage (eds.), *Primatología del Nuevo Mundo*. Fundación Araguatos, Bogotá.

Palacios, E. and A. Rodríguez 1994. Caracterización de la dieta y comportamiento alimentario de *Callicebus torquatus lugens*. Unpublished bachelor's thesis, Universidad Nacional de Colombia, Bogotá.

Palacios Acevedo, E. and A. Rodríguez 2001. Ranging pattern and use of space in a group of red howler monkeys (*Alouatta seniculus*) in a Southeastern Colombian Rainforest. *American Journal of Primatology* 55(4):233-251.

Palacios A., E., A. Rodriguez R. and T. R. Defler 1997. Diet of a group of *Callicebus torquatus lugens* during the annual resource bottleneck. *International Journal of Primatology* 18(4):503-522.

Palacios, E., Rodríguez, A. and C. Castillo In press. Preliminary observations on the Mottled-faced tamarin (*Saguinus inustus*) at the lower Caquetá River, Colombian Amazonia. *Neotropical Primates*.

Palacios, E. and C. A. Peres. In press. Primate population densities at three nutrient-poor Amazonian terra firme forests of Southeastern Colombia. *Folia Primatologica*.

Panger, M. 1998. Object-use in free-ranging white-faced capuchins (*Cebus capucinus*) in Costa Rica. *American Journal of Physical Anthropology* 106:311-321.

Panger, M. 1999a. Capuchin object manipulation. Pp. 115-120 in: P. Dolhinow and A. Fuentes (eds.), *The Nonhuman Primates*. Mayfield Publications, Mountain View.

Panger, M. 1999b. Object-use in free-ranging white-faced capuchins (*Cebus capucinus*) in Costa Rica. *American Journal of Physical Anthropology* 106(3):311-321.

Panger, M. A., S. Perry, L. Rose, J. Gros-Louis, E. Vogel, K. C. MacKinnon and M. Baker 2002. Cross-site differences in foraging behavior of white-faced capuchins (*Cebus capucinus*). *American Journal of Physical Anthropology* 119(1):52-66.

Pastorini, M., M. R. J. Forstner, R. D. Martin, D. J. Melnick 1998. A reexamination of the phylogenetic position of *Callimico* (Primates) incorporating new mitochondrial DNA sequence data. *Journal of Molecular Evolution* 47(1):32-41.

Peetz, A., M. A. Norconk and W. G. Kinzey 1992. Predation by jaguar on howler monkeys (*Alouatta seniculus*) in Venezuela. *American Journal of Primatology* 28(3):223-228.

Peres, C. A. 1987. Conservation of primates in western Brazilian Amazonia. A progress report to the World Wildlife Fund U.S., unpublished manuscript.

Peres, C. A. 1988. Primate community structure in western Brazilian Amazonia. *Primate Conservation* 9:83-87.

Peres, C. A. 1990. Effects on hunting of western Amazonian primate communities. *Biological Conservation* 54(1):47-59.

Peres, C. A. 1991a. Ecology of mixed-species groups of tamarins in Amazonian primate communities. Unpublished Ph.D. thesis, University of Cambridge, Cambridge, England.

Peres, C. A. 1991b. Seed predation of *Cariniana micrantha* (Lecythidaceae) by brown capuchin monkeys in central Amazonia. *Biotropica* 23(3):262-270.

Peres, C. A. 1991c. Humboldt's woolly monkeys decimated by hunting in Amazonia. *Oryx* 25(2):89-95.

Bibliography

Peres, C. A. 1992a. Prey-capture benefits in a mixed-species group of Amazonian tamarins, *Saguinus fuscicolli* and *S. mystax*. *Behavioral Ecology and Sociobiology* 31(5):339-347.

Peres, C. A. 1992b. Consequences of joint-territoriality in a mixed-species group of tamarin monkeys. *Behaviour* 123(3-4):220-246.

Peres, C. A. 1993a. Structure and spatial organization of an Amazonian terra firme forest primate community. *Journal of Tropical Ecology* 9:259-276.

Peres, C. A. 1993b. Diet and feeding ecology of saddle-backed (*Saguinus fuscicollis*) and moustached (*S. mystax*) tamarins in an Amazonian terra firme forest. *Journal of Zoology (London)* 230:567-592.

Peres, C. A. 1993c. Anti-predation benefits in a mixed-species group of Amazonian tamarins. *Folia Primatologica* 61(2):61-76.

Peres, C. A. 1994a. Diet and feeding ecology of gray woolly monkeys (*Lagothrix lagotricha*) in central Amazonia; comparisons with other Atelines. *International Journal of Primatology* 15(3):333-372.

Peres, C. A. 1994b. Primate responses to phenological changes in Amazonian *terra firme* forest. *Biotropica* 26(1):98-112.

Peres, C. A. 1994c. Which are the largest New World monkeys? *Journal of Human Evolution* 26(3):245-249.

Peres, C. A. 1994d. Plant resource use and partitioning in two tamarin species and woolly monkeys in an Amazonian *terra firme* forest. Pp. 57-66 in: B. Thierry Jr., J. J. Anderson, J. J. Roeder and N. Herrenschmidt (eds.), *Current Primatology, vol. i: Ecology and Evolution*. University Louis Pasteur, Strasbourg.

Peres, C. A. 1995. Effects of selective hunting and forest types on the structure of Amazonian primate communities. [Abstract] *Primate Eye* 57:11-12.

Peres, C. A. 1996. Food patch structure and plant resource partitioning in interspecific associations of Amazonian tamarins. *International Journal of Primatology* 17(5):695-723.

Peres, C. A. 1998a. Effects of subsistence hunting and forest types on the structure of Amazonian primate communities. [Abstract] *Congress of the International Primatological Society, Abstracts*. 17:294.

Peres, C. A. 1998b. Assessing the impact of hunting: Lessons from standardized line-transect censusing in the neotropics. [Abstract] *Primate Eye* 65:16.

Peres, C. A. 2000. Territorial defense and the ecology of group movements in small-bodied neotropical primates. Pp. 1000-123 in: S. Boinski and P. A. Garber (eds.), *On the Move: How and Why Animals Travel in Groups*. University of Chicago Press, Chicago.

Peres, C. A., J. L. Patton and M. N. F. da Silva 1996. Riverine barriers and gene flow in Amazonian saddle-back tamarins. *Folia Primatologica* 67(3):113-124.

Peres, C. A. 1997. Primate community structure at twenty western Amazonian flooded and unflooded forests. *Journal of Tropical Ecology* 13:381-405.

Peres, C. A. and M. van Roosmalen 2002. Primate frugivory in two species-rich Neotropical forests: Implications for the demography of large-seeded plants in overhunted areas. Pp. 407-421 in: D. J. Levey, W. R. Silva and M. Galetti (eds.), *Seed Dispersal and Frugivory: Ecology, Evolution and Conservation*. CABI, New York.

Perez-Ruiz, A. L. and R. Mondragón 1998. Food availability: Effects on affiliative behavior of a spider monkey community in the Lacandona Rain Forest, Mexico. [Abstract] *Congress of the International Primatological Society* 17:88.

Perez-Ruiz, A. L. and R. Mondragón-Ceballos 2001. Diet diversity of spider monkeys (*Ateles geoffroyi*) in the Lacandona Rain Forest. [Abstract] The 19th *Congress of the Internatinal Primatological Society. Primates in the New Millennium. Abstracts and Programme*. Adelaid: IPS:60.

Perry, S. 1996. Intergroup encounters in wild white-faced capuchins (*Cebus capucinus*). *International Journal of Primatology* 17(3):309-330.

Perry, S. 1997. Male-female social relationships in wild white-faced capuchins (*Cebus capucinus*). *Behaviour* 134(7-8):477-510.

Perry, S. 1998a. Male-male social relationships in wild white-faced capuchins, *Cebus capucinus*. *Behaviour* 135:139-172.

Perry, S. 1998b. A case report of a male rank reversal in a group of wild white-faced capuchins. *Primates* 39(1):51-70.

Perry, S. 1999. Intergroup variation in social conventions in wild white-faced capuchins. [Abstract] *American Journal of Primatology* 49(1):87.

Perry, S. and L. Rose 1994. Begging and transfer of coati meat by white-faced capuchin monkeys, *Cebus capucinus*. *Primates* 35(4):409-415.

Perry, S., M. Baker, L. Fedigan, J. Gros-Louis, K. Jack, K. C. MacKinnon, J. H. Manson, M. Panger, K. Pyle and L. Rose 2003. Social conventions in wild white-faced capuchin monkeys. *Current Anthropology* 44(2):241-268.

Phillips, K. A. 1994. Resource patch use and social organization in Cebus capucinus. [Abstract] *American Journal of Primatology* 33(3):233.

Phillips, K. A. 1995. Foraging-related agonism in capuchin monkeys (*Cebus capucinus*). *Folia Primatologica* 65(3):159-162.

Phillips, K. A. 1998a. Tool use in wild capuchin monkeys (*Cebus albifrons trinitatis*). *American Journal of Primatology* 46(3):259-261.

Phillips, K. A. 1998b. Conservation of capuchin and howler monkeys in Trinidad. *ASP [American Society of Primatology] Bulletin* 22(4):5 pp.

Phillips, K. A. 1999. Functions of Hug and Arrawh vocalizations in white-fronted capuchin monkeys (*Cebus albifrons*). [Abstract] *Advances in Ethology* 34:118.

Phillips, K. A. and K. Newlon 2000. Female-female social relationships in white fronted capuchins (*Cebus albifrons*): Testing hypotheses about resource size and quality. [Abstract] *American Journal of Primatology* 51(Suppl 1):81.

Phillips, K. A. and K. Newlon 2001. Extractive foraging in white-fronted capuchin monkeys (*Cebus albifrons*) [Abstract] *Advances in Ethology* 36:238.

Phillips, K. A. and L. M. Shauver 2001. Reunion displays in tufted capuchins (*Cebus apella*). [Abstract] *American Journal of Primatology* 54(Suppl 1):83

Phillips, K. A. and C. L. Abercrombie 2003. Distribution and conservation status of the primates of Trinidad. *Primate Conservation* 19:19-22.

Phillips, K. A., I. S. Bernstein, E. L. Dettmer, H. Devermann and M. Powers 1994. Sexual behavior in brown capuchins (*Cebus apella*). *International Journal of Primatology* 15(6):907-917.

Phillips, K., C. L. Abercrombie and S. Ramsubhag 1998. Population status of capuchin (*Cebus albifrons trinitatis*) and howler (*Alouatta seniculus insulanus*) monkeys in Trinidad. [Abstract] *American Journal of Primatology* 45(2):200-201.

Phillips, K. A., B. W. Grafton and M. E. Haas 2003. Tap-scanning for invertebrates by capuchins (*Cebus apella*). *Folia Primatologica* 74(3):162-164.

Phillips, M. J. and W. A. Mason 1976. Comparative studies of social behavior in *Callicebus* and *Saimiri*: social looking in male-female pairs. *Bulletin Psychonom. Society* 7:55-56.

Pielou, E. C. 1979. *Biogeography*. John Wiley & Sons, Inc., New York.

Pinto, L. P. and E. Z. F. Setz 2000. Sympatry and new locality for *Alouatta belzebul discolor* and *Alouatta seniculus* in the southern Amazon. *Neotropical Primates* 8(4):150-151.

Ploog, D. W. and P. D. MacLean 1963 Display of penile erection in squirrels (*Saimiri sciureus*). *Animal Behavior* 11:32-39.

Bibliography

Ploog, D. W., J. Blitz and F. Ploog 1963. Studies on social and sexual behavior of the squirrel monkey (*Saimiri sciureus*). *Folia Primatologica* 1:29-66.

Ploog, D. W., S. Hopf and P. Winter 1967. Ontogenese des Verhaltens von Totenkopfaffen (*Saimiri sciureus*). *Psychologische Forschung* 31:1-41.

Pocock, R. I. 1925. Additional notes on the external characters of some platyrrhine monkeys. *Proceedings of the Zoological Society of London* 1925:27-47.

Podolsky, R. D. 1990. Effects of mixed-species association on resource use by *Saimiri sciureus* and *Cebus apella*. *American Journal of Primatology* 21:147-158.

Pola, Y. V. & C. T. Snowdon 1975. The vocalizations of pygmy marmosets (*Cebuella pygmaea*). *Animal Behavior* 23:826-842.

Polanco O., R. L. H. 1992. Aspectos etológicos y ecológicos de *Callicebus cupreus ornatus* Gray, 1970 (Primates: Cebidae) en el Parque Nacional Natural Tinigua, La Macarena, Meta, Colombia. Unpublished bachelor' thesis, Biology, Universidad Nacional de Colombia, Bogotá.

Polanco-Ochoa, R. and A. Cadena 1993. Use of space by *Callicebus cupreus ornatus* (Primates: Cebidae) in La Macarena, Colombia. *Field studies of New World monkeys, La Macarena, Colombia* 8:19-32.

Polanco Ochoa, R., J. E. Garcia and A. Cadena 1994. Utilización del tiempo y patrones de actividad de *Callicebus cupreus* (Primates: Cebidae) en La Macarena, Colombia. *Trianea* 5:305-322,

Politis, G. G. and J. Rodriguez 1994. Algunos aspectos de subsistencia de los Nukak de la Amazonia Colombiana. *Colombia Amazónica* 7(1-2):169-207.

Pook, A. G. 1977. A comparative study of the use of contact calls in *Saguinus fuscicollis* and *Callithrix jacchus*. Pp. 271-280 in: D. G. Kleiman (ed.), *The Biology and Conservation of the Callitrichidae*. Smithsonian Institute Press.

Pook, A. G. and G. Pook 1979a. The conservation status of the Goeldi's monkey *Callimico goeldii* in Bolivia. *Dodo: Journal of the Jersey Wildlife Preservation* Trust 16:40-45.

Pook, A. G. and G. Pook 1979b. A field study on the status and socioecology of the goeldi's monkey (*Callimico goeldii*) and other primates in northern Bolivia. Unpublished report to the New York Zoological Society.

Pook, A. G. and G. Pook 1981. A field study of the socio-ecology of the goeldi's monkey (*Callimico goeldii*) in northern Bolivia. *Folia Primatologica* 35:288-312.

Pook, A. G. and G. Pook 1982. Polyspecific association between *Saguinus fuscicollis, Saguinus labiatus, Callimico goeldii* and other primates in north-western Bolivia. *Folia Primatologica* 38:196-216.

Pope, T. R. 1990. The reproductive consequences of male cooperation in the red howler monkey: Paternity exclusion in multi-male and single-male troops using genetic markers. *Behavioral Ecololgy and Sociobiology* 27:439-446.

Pope, T. R. 1992. The influence of dispersal patterns and mating system on genetic differentiation within and between populations of the red howler monkey (*Alouatta seniculus*). *Evolution* 46(4):1112-1128.

Pope, T. R. 1996. Influence of social dynamics on mtDNA diversity in red howler monkey populations. [Abstract] *American Journal of Physical Anthropology* 1966 (Suppl 22):188.

Pope, T. R. 1998. Effects of demographic change on group kin structure and gene dynamics of populations of red howling monkeys. *Journal of Mammalogy* 79(3):692-712.

Porras, M. 2000. [Vocal communication and its relation to activities, social structure, and behavioral context in *Callicebus cupreus ornatus*.] *A Primatologia no Brasil* 7:265-274.

Porter, C. A., J. Czelusniak, H. Schneider, M. P. C. Schneider, I. Sampaio and M. Goodman 1997a. Sequences of the primate epsilon-globin gene: Implications for systematics of the marmosets and other New World primates. *Gene* 205(1-2):27-47.

Porter, C. A., S. L. Page, J. Czelusniak, H. Schneider, M. P. C. Schneider, I. Sampaio and M. Goodman 1997b. Phylogeny and evolution of selected primates as determined by sequences of the epsilon-globin locus and 5'flanking regions. *International Journal of Primatology* 18(2):261-295.

Porter, C. A., J. Czelusniak, H. Schneider, M. P. C. Schneider, I. Sampaio and M. Goodman 1999. Sequences from the 5' flanking region of the epsilon-globin gene support the relationship of *Callicebus* with the pitheciins. *International Journal of Primatology* 48:69-75.

Porter, L. M. 2000a. *Callimico goeldii*: Understory monkeys of northern Bolivia. [Resumen] *American Journal of Primatology* 51(Suppl 1):82.

Porter, L. M. 2000b. *Callimico goeldii* and *Saguinus*: Dietary differences between sympatric callitrichines in northern Bolivia. [Abstract] *American Journal of Physical Anthropology* (Suppl 30):252.

Porter, L. M. 2001. Benefits of polyspecific associations for the Goeldi's monkey (*Callimcio goeldii*). *American Journal of Primatology* 54(3):143-158.

Porter, L. M. 2001a. Dietary differences among sympatric Callitrichinae in Northern Bolivia: *Callimico goeldii, Saguinus fuscicollis* and *S. labiatus*. *International Journal of Primatology* 22(6):961-992.

Porter, L. M. 2001b. The behavior and ecology of the Goeldi's monkey (*Callimcio goeldii*) in northern Bolivia. Unpublished Ph.D. *Dissertations Abstracts International* A62(1):229 pp.

Porter, L. M. 2001c. Social organization, reproduction and rearing strategies of *Callimcio goeldii*: New clues from the wild. *Folia Primatologica* 72(2):69-79.

Porter, L. M. 2001d. Benefits of polyspecific associations for the Goeldi´s monkey (*Callimico goeldii*). *American Journal of Primatology* 54(3):143-158.

Porter, L. M. 2002. Habituation of wild Goeldi's monkeys (*Callimico goeldii*) at San Sebastian Departamento Pando, Bolivia. [Abstract] *American Journal of Primatology* 57(Suppl 1):80-91.

Porter, L. M. and A. Christen 2002. Fungus and *Callimico goeldii*: New insights into *Callimico goeldii* behavior and ecology. *Evolutionary Anthropology* 11(Suppl 1):87-90.

Porter, L. M., A. M. Hanson and E. N. Becerra 2001. Groups demographics and dispersal in a wild group of Goeldi's monkeys (*Callimico goeldii*). *Folia Primatologica* 72(2):108-110.

Poveda, K. 2000. Uso de hábitat de dos grupos de tití de pies blancos (*Saguinus leucopus*) en Mariquita, Colombia. Unpublished bachelors thesis, Santafé de Bogotá.

Poveda, K., A. Cadena and P. Sánchez 2001. Habitat use of two groups of white footed tamarins (*Saguinus leucopus*) in Mariiquita, Colombia. [Abstract] The 18th Congress of the International Primatological Society. Primates in the New Millennium. Abstracts and Programme. Adelaid:IPS, p. 58.

Pozo, R. 2001. Social behavior and diet of the spider monkey, *Ateles belzebuth*, in the Yasuni National Park, Ecuador. *Neotropical Primates* 9(2):74.

Prance, G. T. 1973. Phytogeographic support for the theory of Pleistocene forest refuges in the Amazon basin, based on evidence from distribution patterns in Caryocaraceae, Chrysobalanaceae, Dichapetalaceae and Lecythidaceae. *Acta Amazónica* 3(3):55-28.

Prance, G. T. 1980. A note on the probable pollination of *Combretum* by *Cebus* monkeys. *Biotropica* 12:239.

Prance, G. T. 1982. Forest refuges: Evidence from woody angiosperms. Pp. 137-181 in: G. T. Prance (ed.), *Biological Diversification in the Tropics*. Columbia University Press, New York.

Preslock, J. P., S. H. Hampton and J. K. Hampton Jr. 1973. Cyclic variations of serum progestins and immunoreactive estrogens in marmosets. *Endocrinology* 92:1096-1101.

Price, E. C. 1992a. Changes in the activity of captive cotton-top tamarins (*Saguinus oedipus*) over the breeding cycle. *Primates* 33(1):99-106.

Bibliography

Price, E. C. 1992b. Adaptation of captive-bred cotton-top tamarins (*Saguinus oedipus*) to a natural environment. *Zoo Biology* 11(2):107-120.

Price, E. C., A. McGivern-M. and L. Ashmore 1991. Vigilance in a group of free-ranging cotton-top tamarins *Saguinus oedipus*. *Dodo, Journal of the Jersey Wildlife Preservation Trust* 27:41-49.

Pruetz, J. D. and H. C. Leasor 2000. Comparison of census methods to record density and group size of *Ateles geoffroyi, Alouatta palliata* and *Cebus capusinus* in lowland tropical rainforest in Costa Rica. *American Journal of Physical Anthropology* Suppl. 30:255.

Pruetz, J. D. and H. C. Leasor 2002. Survey of three primate species in forest fragments at La Suerte Biological Field Station, Costa Rica. *Neotropical Primates* 10(1):4-9.

Pruscha, H. and M. Maurus 1976. The communicative function of some agonistic behavior patterns in squirrel monkeys: The relevance of social contact. *Behavioral Ecology Sociobiology* 1:185-214.

Pucheran, J. 1845. Description de quelques mammifères américains. *Revue Zoologique* 8:335-337.

Puertas, P. E., R. Aquino and F. Encarnación 1995. Sharing of sleeping sites between *Aotus vociferans* with other mammals in the Peruvian Amazon. *Primates* 36(2):281-287.

Pulido, M. T. 1997. Notes on the dispersal behavior of howler monkeys (*Alouatta seniculus*). *Field Studies of Fauna and Flora, La Macarena, Colombia* 10:7-11.

Queralt, A. M. and J. J. Vea 1998. Parental division of infant-care in the pygmy marmoset (*Cebuella pygmaea*) and the cotton-top tamarin (*Saguinus oedipus*). *Primate Report* 50:3-13.

Raez-Luna, E. F. 1993. Modelling hunted populations of *Alouatta seniculus, Ateles paniscus* and *Lagothrix lagothricha*, (Primates: Cebidae): Chances of persistence and lessons for conservation. Unpublished M.A. thesis, University of Florida, Gainseville.

Raez-Luna, E. F. 1995. Hunting large primates and conservation of the Neotropical rain forests. *Oryx* 29(1):43-48.

Ramirez C., J. 1985. S.O.S. for the cotton-top tamarin (*Saguinus oedipus*). *Primate Conservation* 6:17-19.

Ramirez, M. 1980. Grouping patterns of the woolly monkey, *Lagothrix lagotricha* at the Manu National Park, Peru. *American Journal of Physical Anthropology* 52:269 (abstract).

Ramirez, M. 1988. The woolly monkey, *Lagothrix*. Pp. 539-575 in: A. F. Coimbra-Filho and R. A. Mittermeier (eds.), *Ecology and Behavior of Neotropical Primates*, Vol. 2. Academia Brasileira de Ciências, Rio de Janeiro,.

Ramirez, M., C. H. Freese and J. Revilla 1977. Feeding ecology of the pygmy marmoset, *Cebuella pygmaea*, in northwestern Peru. Pp. 91-104 in: D. G. Kleiman (ed.), *The Biology and Conservation of the Callitrichidae*. Smithsonian Institution Press, Washington, D.C.

Ramírez-Orjuela, C. and I. M. Sanchez-Dueñas. 2002. Conservación del mono aullador negro (*Alouatta palliata aequatorialis*) en un bosque húmedo tropical del Chocó Biogeografico de Colombia. Informe final. Centro de Primatología Araguatos - Margot Marsh Biodiversity Foundation. Bogotá, D. C., 74 pp.

Ramírez-Orjuela, C. and A. Cadena. 2003. Dieta de *Alouatta palliata aequatorialis* en un bosque húmedo de la Costa Pacífica del Chocó colombiano. Pp. 71-84 in: V. Pereira Bengoa, F. Nassar-Montoya, A. Savage *et al.* (eds.), *Primatología del Nuevo Mundo: Biología, Medicina, Manejo y Conservación*. Araguatos, Bogotá, D.C.

Ramos-Fernandez, G. and B. Ayala-Orozco 2003. Population size and habitat use of spider monkeys at Punta Laguna, Mexico. Pp. 191-209 in: L. K. Marsh (ed.) *Primates in Fragments: Ecology and Conservation*. Kluwer Academic/Plenum Publishing, New York.

Rasmussen, D. R. 1998. Changes in range use of Geoffroy's tamarins (*Saguinus geoffroyi*) associated with habituation to observers. *Folia Primatologica* 69(3):153-159.

Re, G. E. F. and Laudato F. E. L. 1984. *Um Mergulho na Pré-história os últimos Yanomami?* C. Poit Couleur Edições, Turim, p. 245.

Redford, K. H. and J. G. Robinson 1991. Park size and the conservation of forest mammals in Latin America. Pp. 227-234 in: M. A. Mares and D. J. Schmidly (eds.), *Latin American Mammalogy: History, Biodiversity, and Conservation*. University of Oklahoma Press, Norman.

Reed, K. E. and J. G. Fleagle 1995. Geographic and climatic control of primate diversity. *Processes of the National Academy of Sciences USA* 92:7874-7876.

Rehg, J. A. 2003. Polyspecific associations of *Callimico goeldii, Saguinus labiatus,* and *Saguinus fuscicollis* in Acre, Brazil. Unpublished Ph.D. thesis, *Dissertation Abstracts International* A64(3): 973 pp.

República de Colombia 1991. *Colombia: Informe Nacional Para Cnumad (Conferencia de Naciones Unidas sobre el Medio Ambiente y el Desarrollo).* República de Colombia, Bogotá.

Resende, B. D. and E. B. Ottoni 2002. Ontogeny of nutcracking behavior in a semifree-ranging group of tufted capuchin monkeys. [Abstracts] *International Primatological Society. Beijing: Mammalogical Soceity of China,* pp. 319-320.

Rettig, N. 1978. Breeding behavior of the harpy eagle, *Harpia harpyja. Auk* 95:629-643.

Rey, M. P. 1997. [Vocal communication from the perspective of activities, social structure and behavioral context in *Callicebus cupreus ornatus*.] [Abstract] Primatologia and V. Reuniao Latino-Americano de Primatología. Joao Pessoa: Soc Bras de Primatol. 1997. Pgs: 191

Richard, A. 1970a. A comparative study of the activity patterns and behavior of *Alouatta villosa* and *Ateles geoffroyi. Folia Primatologica* 12:241-263.

Richard, A. 1970b. A comparative study of the activity patterns and behavior of *Alouatta villosa* and *Ateles geoffroyi. Folia Primatologica* 25:122-142.

Richard-Hansen, C. and J.-C. Vie 1996. Post translocation behavior of red howler monkeys (*Alouatta seniculus*) in Frenche Guiana. [Abstract] *IPS/ASP Congress Abstracts* 1996:213.

Richard-Hansen, C., N. Bello and J. C. Vie 1998. Tool use by a red howler monkey (*Alouatta seniculus*) towards a two-toed sloth (*Cholepus didactylus*). *Primates* 39(4):545-548.

Richard-Hansen, C., J. C. Vie and B. Thoisy 2000. Translocation of red howler monkeys (*Alouatta seniculus*) in French Guiana. *Biological Conservation* 93(2):247-253.

Rimoli, J. and S. F. Ferrari 1997. [Behavior and ecology of black-capped capuchins (*Cebus apella nigritus,* Goldfuss, 1809) at the Caratinga Biological Station.] [Abstract] Programa e Resumos do VIII Congresso Braqsileiro de Primatologia & V Reuniao Latino-Americano de Primatologia. Joao Pessa: Soc. Bras. de Primatol. 1997:231.

Rimoli, R. O. 1977. Una nueva especie de monos (Cebidae: Saimirinae: *Saimiri*) de la Hispaniola. *Cuadernos del Cendia* 242:1-8.

Rivero, M. and O. Arrendondo 1991. *Paralouatta veronai,* a new Quaternary platyrrhine from Cuba. *Journal of Human Evolution* 21:1-12.

Robbins, D., C. A. Chapman and R. W. Wrangham 1991. Group size and stability: Why do gibbons and spider monkeys differ? *Primates* 32(3):301-305.

Robinson, J. G. 1977. The vocal regulation of spacing in the titi monkey, *Callicebus moloch*. Unpublished Ph.D. thesis, University of North Carolina.

Robinson, J. G. 1979a. Vocal regulation of use of space by groups of titi monkeys *Callicebus moloch. Behavioral Ecology and Sociobiology* 5:1-15.

Robinson, J. G. 1979b. An analysis of the organisation of vocal communication in the titi monkey *Callicebus moloch. Zeitschrift der Tierpsychologie* 49:381-405.

Robinson, J. G. 1981. Vocal regulation of inter- and intragroup spacing during boundary encounters in the titi monkey, *Callicebus moloch. Primates* 22:161-72.

Bibliography

Robinson, J. G. 1982a. Vocal systems regulating within-group spacing. Pp. 94-116 in: C. T. Snowdon, C. H. Brown and M. Peterson (eds.), *Primate Communication*. Cambridge University Press, Cambridge.

Robinson, J. G. 1982b. Intrasexual competition and mate choice in primates. *American Journal of Primatology*. suppl. 1:131-44.

Robinson, J. G. and J. Ramírez C. 1982. Conservation biology of neotropical primates. Pp. 329-344 in: M. A. Mares and H. H. Genoways (eds.), *Mammalian Biology in South America*. V. 6, Special Publication Series, Pymatuning Laboratory of Ecology, University of Pittsburgh, Pittsburgh.

Robinson, J. G. and C. H. Janson 1987. Capuchins, squirrel monkeys, and atelines: socioecological convergence with Old World primates. Pp. 69-82 in: B. B. Smuts, D. L. Cheney, R. M. Seyfarth, R. W. Wrangham and T. T. Struhsaker (eds.), *Primate Societies*. University of Chicago Press, Chicago.

Robinson, J. G. and K. H. Redford 1991. Determinants of local rarity in Neotropical primates. *Primatologia no Brasil* 3:331-346.

Robinson, J. G. and K. H. Redford 1994. Measuring the sustainability of hunting in tropical forests. *Oryx* 28(4):249-256.

Robinson, J. G., P. C. Wright and W. G. Kinzey 1987. Monogamous cebids and their relatives: Intergroup calls and spacing. Pp. 44-53 in: B. B. Smuts, D. L. Cheney, M. Seyfarth, R. W. Wrangham and T. T. Struhsaker (eds.), *Primate Societies*. The University of Chicago Press, Chicago.

Rodríguez-Toledo, E. M., S. Mandujano and F. Garcia-Orduña 2003. Relationships between forest fragments and howler monkeys (*Alouatta palliata mexicana*) in southern Veracruz, Mexico. Pp. 79-97 in: L. K. Marsh (ed.), *Primates in Fragments: Ecology and Conservation*. Kluwer Academic/Plenum Publ., New York.

Rocha, V. J., N. R. dos Reis and M. L. Sekiama 1998. [Tool use in *Cebus apella* (Linnaeus) (Primate, Cebidae) to get Coleoptera larvae that parasitize seeds of *Syagrus romanzoffianum* (Cham.) Glassm.(Arecaceae).] *Revista Brasileira de Zoologia* 15(4):945-950.

Rockwood, L. L. and K. E. Glander 1979. Howling monkeys and leaf cutting ants: Comparative foraging in a tropical deciduous forest. *Biotropica* 11:1-10.

Rodriguez M., J. V., J. Hernandez C., T. R. Defler, M. Alberico, R. B. Mast, R. A. Mittermeier and A. Cadena 1995. Mamíferos colombianos: sus nombres comunes e indígeras. *Occasional Paper no. 3, Conservation International*, Washington, C. C., 56 pp.

Rondinelli, R. and L. L. Klein 1976. An analysis of adult social spacing tendencies and related social interactions in a colony of spider monkeys, *Ateles geoffroyi*, at the San Francisco Zoo. *Folia Primatologica* 25:122-142.

Rose, K. D. and J. G. Fleagle 1981. The fossil history of nonhuman primates in the Americas. Pp. 111-167 in: A. F. Coimbra-Filho and R. A. Mittermeier (eds.), *Ecology and Behavior of Neotropical Primates*. Academia Brasiliera de Ciências, Rio de Janeiro.

Rose, L. M. 1992. Sex differences in diet and foraging behavior and benefits and costs of resident males to females in white-faced capuchins. Unpublished M.A. thesis, University of Alberta, Edmonton.

Rose, L. M. 1994a. Sex difference in diet and foraging behavior in white-faced capuchins (*Cebus capucinus*). *International Journal of Primatology* 15(1):95-114.

Rose, L. M. 1994b. Benefits and costs of resident males to females in white-faced capuchins, *Cebus capucinus*. *American Journal of Primatology* 32(4):235-248.

Rose, L. M. 1997. Vertebrate predation and food-sharing in *Cebus* and *Pan*. *International Journal of Primatology* 18(5):727-765.

Rose, L. M. 1999. Behavioral ecology of white-faced capuchins (*Cebus capucinus*) in Costa Rica. Unpublished Ph.D. Dissertation Abstracts International A60(2):471.

Rose, L. M. and L. M. Fedigan 1995. Vigilance in white-faced capuchins, *Cebus capucinus*, in Costa Rica. *Animal Behaviour* 49(1):63-70.

Rose, K. D. and J. G. Fleagle 1981. The fossil history of nonhuman primates in the Americas. Pp. 111–167 in: A. F. Coimbra–Filho and R. A. Mittermeier (eds.), *Ecology and Behavior of Neotropical Primates*. Academia Brasileira de Ciêcias, Rio de Janeiro.

Rose, L. M., S. Perry, M. A. Panger, K. Jack, J. H. Manson, J. Gros-Louis, K. C. Mackinnon and E. Vogel 2003. Interspecific interactions between *Cebus capucinus* and other species: Data from three Costa Rican sites. *International Journal of Primatology* 24(4):759-796.

Rosenberger, A. L. 1977. *Xenothrix* and ceboid phylogeny. *Journal of Human Evolution* 6:461–481.

Rosenberger, A. L. 1979. Cranial anatomy and implications of *Colichocebus*, a late Oligocene ceboid primate. *Nature* (London) 279:416-418.

Rosenberger, A. L. 1980. Gradistic views and adaptive radiation of platyrrhine primates. *Zeitschrift für Morphologie und Anthropologie* 71:157-153. reprinted 1985, pp. 133-135 in: R. L. Ciochon and J. G. Fleagle (eds.), *Primate Evolution and Human Origins*. The Benjamin-Cummings Publishing Co., Inc., Menlo Park, Cal.

Rosenberger, A. L. 1981. Systematics: the higher taxa. Pp. 9-28 in: A. F. Coimbra-Filho and R. A. Mittermeier (ed.), *Ecology and Behavior of Neotropical Primates*. Academia Brasileira de Ciências, Rio de Janeiro.

Rosenberger, A. L. 1982. Supposed squirrel monkey affinities of the late Oligocene *Dolichocebus gaimanensis*. *Nature* (London) 298:202.

Rosenberger, A. L. 1984. A mandible of *Branisella boliviana* (Platyrrhini, Primates) from the Oligocene of South America. *International Journal of Primatology* 2(1):1-7.

Rosenberger, A. L. and A. F. Coimbra-Filho 1984. Morphology, taxonomic status and affinities of the lion tamarins, *Leontopithecus* Callitrichinae, Cebidae). *Folia Primatologica* 42(3-4):149-179.

Rosenberger, A. L., T. Setoguchi and N. Shigehara 1990. The fossil record of Callitrichine primates. *Journal of Human Evolution* 19:209-236.

Rosenberger, A. L., T. Setoguchi and W. C. Hartwig. 1991a. *Laventiana annectens*, new genus new specis: Fossil evidence for the origins of callitrichine New World monkeys. *Proceedings of the National Academy of Sciences of the United States of America* 88(6):2137-2140.

Rosenberger, A. L., W. C. Hartwig, M. Takai, T. Setoguchi and N. Shigehara. 1991b. Dental variability in *Saimiri* and the taxonomic status of *Neosaimiri fieldsi*, an early squirrel monkey from La Venta, Colombia. *International Journal of Primatology* 12(3):291-301.

Rosenberger, A. L., W. C. Hartwig and R. G. Wolff 1991c. *Szalatavus attricuspis,* an early platyrrhine primate. *Folia Primatologica* 56(4):225–233.

Rosenblum, A. L. 1968. Some aspects of female reproductive physiology in the squirrel monkey. Pp. 147-169 in: L. A. Rosenblum and R. W. Cooper (eds.), *The Squirrel Monkey*, Academic Press, New York

Rosenblum, L. A. 1969. Mother-infant relations and early behavioral development in the squirrel monkey. Pp. 207-333 in: L. A. Rosenblum and R. W. Cooper (eds.), *The Squirrel Monkey*. Academic Press, New York.

Rossan, R. N. and D. C. Baerg 1977. Laboratory and feral hybridization of *Ateles geoffroyi panamensis* Kellogg and Goldman 1944 and *A. fusciceps robustus* Allen 1914 in Panama. *Primates* 18:235-237.

Rowell, T. E. and B. J. Mitchell 1991. Comparison of seed dispersal by guenons in Kenya and capuchins in Panama. *Journal of Tropical Ecology* 7:269-274.

Rudran, R. 1979. The demography and social mobility of a red howler, (*Alouatta seniculus*) populations in Venezuela. Pp. 107-126 in: ed. J. F. Eisenberg (ed.), *Vertebrate Ecology in the Northern Neotropics*. Smithsonian Institution Press, Washington, D.C.

Rudran, R. and E. Fernandez-Duque 2003. Demographich changes over thirty years in a red howler population in Venezuela. *International Journal of Primatology* 24(5):925-947.

Ruíz-García, M. and D. Alvarez 2003. RFLP analysis of mtDNA from six platyrrhine genera: Phylogenetic inferences. *Folia Primatologica* 74:59-70.

Bibliography

Ruiz-Vidal, R., Peres-Ruiz and G. Ramos-Hernandez 1994. A study on the behavioral ecology of the spider monkey, *Ateles geoffroyi*, in the Montes Azules Biosphere Reserve, Chiapas, Mexico. *Neotropical Primates* 2(3):10-11.

Rumbaugh, D. M. 1965. Maternal care in relation to infant behavior in the squirrel monkey. *Psychological Reports* 16:171-176.

Rumbaugh, D. M. 1968. The learning and sensory capacities of the squirrel monkey in phylogenetic perspective. Pp. 255-317 in: L. A. Rosenblum and R. W. Cooper (ed.), *The Squirrel Monkey*. Academic Press, New York.

Runestadconnour, J. A. and K. E. Glander 2001. Description of a feral *Alouatta palliata* population observed during three decades. *American Journal of Physical Anthropology* Suppl. 32:128-129.

Rusconi, C. 1935. Las especies de primates del Oligoceno de Patagonia. *Revista Argentina Paleontología y Anthropología Ameghino* i: 39-68, 7-100, 103-125.

Russo, S. E., C. J. Campbell, J. L. Dew, P. R. Stevenson and M. McFarland 2003. A multi-site comparison of dietary preferences and seed dispersal by spider monkeys (*Ateles* spp.). [Abstract] *American Journal of Physical Anthyropology* (Suppl 36):181-182.

Rylands, A. B. 1985. Conservation areas protecting primates in Brazilian Amazonia. *Primate Conservation* 5:24-27.

Rylands, A. B. 1993a. *Marmosets and tamarins: Systematics, behavior and ecology*. Oxford: Oxford University Press.

Rylands, A. B. 1993b. The bare-face tamarins *Saguinus oedipus oedipus* and *Saguinus oedipus geoffroyi*: subspecies or species? *Neotropical Primates* 1(2):4-5.

Rylands, A. B. 1994. [Black headed uakari *Cacajao melanocephalus* (Humboldt, 1812).] Pp. 239-245 in: G. A. B. da Fonseca, A. B. Rylands, C. M. R. Costa, R. B. Machado, Y. L. R. Leite (eds.), *Livro Vermelho dos Mamíferos Brasileiros Ameaçados de Extincao*. Fundaçao Biodiversitas, Belo Horizonte.

Rylands, A. B. 1997. The Callitrichidae: A biological overview. Pp. 1-22 in: C. Pryce, L. Scott and C. Schnell (eds.), *Handbook:Marmosets and Tamarins in Biological and Biomedical Research*. Salisbury, UK: DSSD Imagery.

Rylands, A. B. 2001. Marmosets and tamarin species. Pp. 339-341 in: D. Macdonald (ed.), *The New Encyclopedia of Mammals*. Oxford University Press, Oxford.

Rylands, A. B., R. A. Mittermeier and E. Rodriguez Luna 1995. A species list for the New World Primates (Platyrrhini): distribution by country, endemism, and conservation status according to the Mace-Land system. *Neotropical Primates* 3 (suppl.):113-160.

Rylands, A. B., R. A. Mittermeier and E. Rodríguez-Luna 1997. Conservation of neotropical primates: Threatened species and an analysis of primate diversity by country and region. *Folia Primatologica* 68:134-160.

Rylands, A. B., H. Schneider, A. Langguth, R. A. Mittermeier, C. P. Groves and E. Rodríguez-Luna 2000. An assessment of the diversity of New World primates. *Neotropical Primates* 8(2):61-93.

Sakurai, Y and A. Nishimura 1999. [An unusual mother-infant relation observed in wild woolly monkeys – a female's simultaneous care of her own and adopted infants.] (in Japanese) [Abstract] *Reichorui Kenkyu/ Primate Research* 15(3):440.

Sampaio, I., M. P. Schneider and H. Schneider 1996. Taxonomy of the *Alouatta seniculus* group: Biochemical and chromosome data. *Primates* 37(1):165-173.

Sánchez P., H., J. Hernandez C., J. V. Rodriguez M. and C. Castaño C. 1990. *Nuevos Parques Nacionales: Colombia*. Inderena, Bogotá .

Sánchez-Villagra, M. R., T. R. Pope and V. Salas 1998. Relation of intergroup variation in allogrooming to group social structure and ectoparasite loads in red howlers (*Alouatta seniculus*). *International Journal of Primatology* 19(3):473-491.

Santamaria-Gomez, M. 2000. [Diet and seed dispersal in a group of *Alouatta seniculus* in an Amazonian terra firma forest, Central Brazil]. *Laboratory Primate Newsletter* 39(2):15.

Santos Mello, R. and M. Thiago de Mello 1986. Cariótipo de *Aotus trivirgatus* (Macaco-da-Noite) das proximidades de Manaus, Amazonas. Nota prliminar. Pp. 388 in: M. Thiago de Mello (ed.), *A Primatologia no Brasil – 2*. Sociedade Brasileira de Primatologia, Brasilia.

Sanz, V. and I. Marquez 1994. Conservación del mono capuchin de Margarita (*Cebus apella margaritae*) in la Isla de Margarita, Venezuela. *Neotropical Primates* 2(2):5-8.

Sassenrath, E. N., W. A. Mason, M. D. Fitzgerald and M. D. Kenney 1980. Comparative endocrine correlates of reproductive states in *Callicebus* (titi) and *Saimiri* (sqirrel) monkeys. *Anthropology Contemporary* 3:265.

Savage, A. 1988. Proyecto titi: the reintroduction of cotton-top tamarins to a semi-natural environment and the development of conservation education programs in Colombia. *AAAZPA (American Association of Zoological Parks and Aquariums) Annual Conference* (1988): 78-84.

Savage, A. 1989a. The ecology, biology, and conservation of the cotton-top tamarin in Colombia. Invited address, British Ecological Society, London.

Savage, A. 1989b. Proyecto titi: Conservation of the cotton-top tamarin in Colombia. Unpublished manuscript.

Savage, A. 1990. The reproductive biology of the cotton-top tamarin (*Saguinus oedipus oedipus*) in Colombia. Unpublished Ph.D. thesis, University of Wisconsin, Madison.

Savage, A. 1995a. Proyecto Tití: Saving Columbia´s endangered cotton-top tamarin (*Saguinus oedipus*). *Proceedings of the National Conference of the AAZK [American Association of Zoo Keepers, Inc.]* 22:196.

Savage, A. 1995b. Proyecto Tití: Developing global support for local conservation. *AZA [American Zoo Aquarium Association] Annual Conference Proceedings* 1995:459-461.

Savage, A. 1995c. SSP reports: Cotton-top tamarin (*Saguinus oedipus*). *AZA Communique* 1995(Oct):5 pp.

Savage, A. 1996a. SSP reports: Cotton-top tamarin SSP. *AZA Communique* 1996(Oct):44 pp.

Savage, A. 1996b. The field training program of Proyecto Tití: Collaborative efforts to conserve species and their habitat in Columbia. *AZA [American Zoo Aquarium Association] Annual Conference Proceedings* 1996:311-313.

Savage, A. 1997. Proyecto Tití: Conservation of the cotton-top tamarin in Colombia. *Conservationist Newsletter* 2(1):10-13.

Savage, A. 2002. Proyecto Tití: A multidisciplinary approach to the conservation of the cotton-top tamarin in Colombia [Abstract] *American Journal of Primatology* 57(Suppl): 32 pp.

Savage, A. and A. J. Baker 1996. Callitrichid social structure and mating system: Evdence from field studies. *American Journal of Primatology* 38(1):1-3.

Savage, A. and H. Giraldo 1990. Proyecto titi: an effective conservation education program in Colombia. *American Journal of Primatology* 20(3):229-230.

Savage, A., L. G. Giraldo, L. G. Soto and C. T. Snowdon 1986. Demography, group composition, and dispersal in wild cotton-top tamarin, (*Saguinus oedipus*) groups. *American Journal of Primatology* 38(1):85-100.

Savage, A., L. A. Dronzek and C. T. Snowdon 1987. Color discrimination by the cotton-top tamarin (*Saguinus oedipus oedipus*) and its relation to fruit coloration. *Folia Primatologica* 49(2):57-69.

Savage, A., T. E. Ziegler and C. T. Snowdon 1988. Sociosexual development, pair-bond formation and the mechanisms of fertility suppression in female cotton-top tamarins (*Saguinus o. oedipus*). *American Journal of Primatology* 14:345-359.

Savage, A., C. T. Snowdon and H. Giraldo 1990a. The ecology of the cotton-top tamarin in colombia. [Abstract] *American Journal of Primatology* 20(3):230.

Savage, A., C. T. Snowdon and H. Giraldo 1990b. Proyecto titi: A hands-on approach to conservation education in Colombia. *AAZPA (American Association Zoological Parks and Aquariums) Annual Conference Proceedings* (1989):605-606.

Bibliography

Savage, A., C. T. Snowdon, H. Giraldo and J. V. Rodriguez 1990c. The ecology of the cotton-top tamarin (*Saguinus o. oedipus*) in Colombia—Progress Report, October, 1989. In: G. D. Aquilina (ed.), *Regional Cotton-top Tamarin Studbook*. Buffalo Zoological Park, Buffalo, N. Y.

Savage, A., L. G. Giraldo, E. S. Blumer, L. H. Soto, W. Burger and C. T. Snowdon 1993. Field techniques for monitoring cotton-top tamarins (*Saguinus oedipus oedipus*) in Colombia. *American Journal of Primatology* 31(3):189-196.

Savage, A., D. S. Zirofsky, L. H Soto, L. H. Giraldo, and J. Causado 1996a. Proyecto tití: Developing alternatives to forest destruction. *Primate Conservation* 17:127-130.

Savage, A., L. G. Giraldo, L. G. Soto and C. T. Snowdon 1996b. Parental care patterns and vigilance in wild cotton-top tamarin (*Saguinus oedipus*). Pp. 187-199 & 539 in: M. Norconk, A. Rosenberger and P. Garber (eds.), *Adaptive Radiations of Neotropical Primates*. Plenum Press, New York.

Savage, A., S. E. Shideler, L. H. Soto, J. Causado, L. H. Giraldo, B. L. Lasley and C. T. Snowdon 1997. Reproductive events of wild cotton-top tamarins (*Saguinus oedipus*) in Colombia. *American Journal of Primatology* 43(4):329-337.

Schneider, H. and A. L. Rosenberger 1996. Molecules, morphology, and platyrrhine systematics. Pp. 3-19 & 533 in: M. A. Norconk, A. L. Rosenberger, and P. A. Garber (eds.), *Adaptive Radiations of Neotropical Primates*. Plenum Press, New York.

Schneider, H., M. P. C. Schneider, M. I. Sampaio, M. I. Harada, M. Stanhopes, J. Czelusniak and M. Goodman 1993. Molecular phylogeny of the New World monkeys (Platyrrhini, Primates). *Molecular Phylogenetics and Evolution* 2(3):225-242.

Schneider, H., M. P. C. Schneider, M. I. Sampaio, N. M. Carvalho-Filho, F. Encarnación, E. Montoya and F. M. Salzano 1995. Biochemical diversity and genetic distances in the Pitheciinae subfamily (Primates, Platyrrhini). *Primates* 36(1):129-134.

Schneider, H., I. Sampaio, M. L. Harada, C. M. L. Barroso, M. P. C. Schneider, J. Czelusniak and M. Goodman 1996. Molecular phylogeny of the New World monkeys (Platyrrhini, Primates) based on two unlinked nuclear genes: IRBP intron 1 and epsilon-globin sequences. American *Journal of Physical Anthropology* 100(2):153-179.

Schneider, H., F. C. Canavez, I. Sampaio, M. A. M. Moreira, C. H. Tagliaro and H. N. Seuanez 2001 Can molecular data place each Neotropical monkey in its own branch? *Chromosoma* 109(8):515-523.

Schön Ybarra, M. A. 1984. Locomotion and postures of red howlers in a deciduous forest-savanna interface. *American Journal of Physical Anthropology* 63:65-76.

Schön Ybarra, M. A. 1986. Loud calls of adult male red howling monkeys (*Alouatta seniculus*). *Folia Primatology* 47:2-4-216.

Schön Ybarra, M. A. 1987. Positional behavior and limb bone adaptations in red howling monkeys (*Alouatta seniculus*). *Folia Primatology* 49:70-89.

Schön Ybarra, M. A. 1988. Morphological adaptation for loud phonations in the vocal organ of howling monkeys. *Primate Report* 22:19-24.

Schön Ybarra, M. A. 1998. Arboreal quadrupedalism and forelimb articular anatomy of red howlers. *International Journal of Primtology*. 19(3):599-613.

Schott, D. 1975. Quantitative analysis of the vocal repertoire of squirrel monkeys (*Saimiri sciureus*). *Zeitschrift der Tierpsychologie* 38:225-250.

Schroepel, M. 1998. Multiple simultaneous breeding females in a pygmy marmoset group (*Cebuella pygmaea*). *Neotropical Primates* 6(1):1-7.

Schwarz, E. 1951. A new marmoset monkey from Brazil. *Amer. Mus. Nov. No.* 1508:1-3.

Sclater, P. L. 1872. Additional notes on rare or little-known animals now or lately living in the society's gardens. *Processes of the Zoological Society of London* 688-690.

Scollay, P. A. and P. Judge 1981. The dynamics of social organization in a population of squirrel monkeys (*Saimiri sciureus*) in a seminatural environment. *Primates* 22(1):60-69.

Scott, N. J., A. F. Scott and L. A. Malmgren 1976a. Capturing and marking howler monkeys for field behavioral studies. *Primates* 17:527-533.

Scott, N. J., Jr., T. T. Struhsaker, K. Glander and H. Chirivi 1976b. Primates and their habitats in northern Colombia with recommendations for future management and research. Pp. 30-50 in: *First Inter-American Conference on Conservation and Utilization of American Nonhuman Primates in Biomedical Research*. Pan American Health Organization.

Scott, N. J., L. A. Malmgren and K. E. Glander 1978. Grouping behavior and sex ratio in mantled howler monkeys. Pp. 183-185 in: D. J. Chivers and J. Herbert (eds.), *Recent Advances in Primatology*, vol. 1. Academic Press, New York.

Scott, W. B. 1937. *A History of the Land Mammals of the Western Hemisphere*. 2nd ed., Macmillan, New York.

Sekulic, R. 1981. The significance of howling in the red howler monkey (*Alouatta seniculus*). Unpublished Ph. D. thesis, University of Maryland, College Park.

Sekulic, R. 1982a. Behavior and ranging patterns of a solitary female red howler (*Alouatta seniculus*). *Folia Primatologica*. 39(3-4):217-232.

Sekulic, R. 1982b. Daily and seasonal patterns of roaring and spacing in four red howler (*Alouatta seniculus*) troops. *Folia Primatologica* 39:22-48.

Sekulic, R. 1982c. The function of howling in red howler monkeys (*Alouatta seniculus*). *Behaviour* 81:38-54.

Sekulic, R. 1982d. Birth in free-ranging howler monkeys *Alouatta seniculus*. *Primates* 23(4):580-582.

Sekulic, R. 1982e. The significance of howling in the red howler monkey *Alouatta seniculus*. Unpublished Ph.D. *Dissertation Abstracts International* B43(1):65.

Sekulic, R. 1983a. Male relationships and infant deaths in red howler monkeys (*Alouatta seniculus*). *Zeitschrift für Tierpsychologie* 61:185-202.

Sekulic, R. 1983b. The effect of female call on male howling in red howler monkeys (*Alouatta seniculus*). *International Journal of Primatology* 4(3):291-305.

Sekulic, R. and J. F. Eisenberg 1983. Throat-rubbing in red howler monkeys. In: D. Muller-Schwarze and R. M. Silverstein, *Chemical Signals in Vertebrates*, vol. 3. Plenum Press, New York.

Sekulic, R. and D. Chivers 1986. The significance of call duration in howler monkeys. *International Journal of Primatology* 7(2):183-190.

Serio-Silva, J. C. and V. Rico-Gray 2002. Influence of microclimate at different canopy heights on the germination of *Ficus* (*Urostigma*) seeds dispersed by Mexican howler monkeys (*Alouatta palliata mexicana*). *Interciencia* 27(4):186-190.

Serio-Silva, J. C. and V. Rico-Gray 2003. Howler monkeys (*Alouatta palliata mexicana*) as seed dispersers of strangler figs in disturbed and preserved habitat in southern Veracruz, Mexico. Pp. 267-281 in: L. K. Marsh (ed.), *Primates in Fragments: Ecology and Conservation*. Kluwer Academic/Plenum Publ., New York.

Serio-Silva, J. C., L. T. Hernández-Salazar and V. Rico-Gray 1999. Nutritional composition of the diet of *Alouatta palliata mexicana* females in different reproductive status. *Zoo Biol.* 18(6):507-513.

Serio-Silva, J. C., V. Rico-Gray, L. T. Hernandez-Salazar and R. Espinosa-Gomez 2002. The role of *Ficus* (Moracea) in the diet and nutrition of a troop of Mexican howler monkeys, *Alouatta palliata mexicana*, released on an island in southern Veracruz, Mexico. *Journal of Tropical Ecology* 18(6):913-928.

Serio-Silva, J. C., V. Rico-Gray and G. Ramos-Fernandez 2003. Distribution and conservation status of wild primates in the Yucatán Peninsula. [Abstract] *American Journal of Primatology* 60(Suppl 1):65.

Setoguchi, T. and A. L. Rosenberger 1985. Miocene marmosets: first fossil evidence. *International Journal of Primatology* 6:615-625.

Bibliography

Setoguchi, T. and A. L. Rosenberger 1987. A fossil owl monkey from La Venta, Colombia. *Nature* (London) 326:692-694.

Setoguchi, T., T. Watanabe and T. Mouri 1981. The upper dentition of *Stirtonia* (Ceboidea, Primates) from the Miocene of Colombia, South America and the origin of the postero/internal cusp of upper molars of howler monkeys (*Alouatta*). *Kyoto University Overseas Research Reports of New World Monkeys* 2:51–60.

Sherman, P. T. 1991. Harpy eagle predation on a red howler monkey. *Folia Primatologica* 56(10):53-56.

Shimooka, Y. 2000. [Grouping pattern of wild long-haired spider monkey.] (in Japanese) [resumen] *Reichorui Kenkyu/Primate Research* 16(3):282.

Shimooka, Y. 2002a. Association pattern in grouping of wild spider monkeys in La Macarena, Colombia. [Abstract] Caring for Primates. Abstracts of the XIXth Congress. The International Primatological Society. Beijing: Mammalogical Society of China, pp. 211-212.

Shimooka, Y. 2002b. Intraday grouping pattern of wild spider monkeys. [Abstract] *Anthropological Science* 110(1):89.

Shimooka, Y. 2002. [Ranging behavior of spider monkeys at La Macarena, Colombia.] (in Japanese) [Abstract] *Reichorui Kenkyu/Primate Research* 18(3):417.

Shimooka, Y. 2003. Seasonal variation in association patterns of wild spider monkeys (*Ateles belzebuth belzebuth*) at La Macarena, Colombia. *Primates* 44(2):83-90.

Siemers, B. M. 2000. Seasonal variation in food resource and forest strata use by brown capuchin monkeys (*Cebus apella*) in a disturbed fragment. *Folia Primatologica* 71(3):181-184.

Silva-Lopez, G. and J. Jimenez-Huerta 2000. A study of spider monkeys (*Ateles geoffroyi vellerosus*) in the forest of the crater of Santa Marta, Veracruz, Mexico. *Neotropical Primates* 8(4):148-150.

Silva-Lopez, G., J. Motta-Gill and A. I. Sanchez-Hernandez 1998. Distribution and status of primates of Guatemala. *Primate Conservation* 18:30-41.

Silveira, G., J. C. Bicca-Marques and C. A. Nunes 1998. On the capture of titi monkeys (*Callicebus cupreus*) using the Peruvian method. *Neotropical Primates* 6(4):114-115.

Simmen, B. 1992. Competitive utilization of *Bagassa* fruits by sympatric howler and spider monkeys. *Folia Primatologica* 58(3):155-160.

Simmen, B. and D. Sabatier 1996. Diets of some French Guianan primates: Food composition and food choices. *International Journal of Primatology* 17(5):661-693.

Simmen, B., C. Julliot, F. Bayart, E. Pages-Feuillade 2001. Diet and population densities of the primate community in relation to fruit supplies. Pp. 89-101 in: F. Bongers, P. Charles Dominque and P. M. Forget (eds.), *Nouragues: Dynamics and Plant-Animal Interactions in a Neotropical Rainforest*. Kluwer Academic Publishers, Dordrecht.

Simpson, G. G. 1941. Some Carib Indian mammal names / by George Gaylord Simpson. *American Museum Novitatis* 1119:1-10.

Skinner, C. 1985. A field study of Geoffroy's tamarin (*Saguinus geoffroyi*) in Panama. American *Journal of Primatology* 9(1):15-26.

Skinner, C. 1986a. A life history of the Geoffroy's tamarin, *Saguinus geoffroyi*, with emphasis on male-female relationships in captive animals. Unpublished Ph.D. thesis, Kent State University.

Skinner, C. 1986b. Justification for reclassifying Geoffroy's tamarin from *Saguinus oedipus geoffroyi* to *Saguinus geoffroyi*. *Primate Report* 31:77-83.

Sleeper, B. 1983. The family life of the yellow-handed titi monkey. *Wildlife* 25(9):352-257.

Smith, A. C. 1999. Potential competitors for exudates eaten by saddleback (*Saguinus fuscicollis*) and moustached (*Saguinus mystax*) tamarins. *Neotropical Primates* 7(3):73-75.

Smith, A. C. 2000a. Interspecific differences in prey captured by associating saddleback (*Saguinus fuscicollis*) and moustached (*Saguinus mystax*) tamarins. *Journal of Zoology* 251(3):315:324.

Smith, A. C. 2000b. Composition and proposed nutritional importance of exudates eaten by saddleback (*Saguinus fuscicollis*) and mustached (*Saguinus mystax*) tamarins. *International Journal of Primatology* 21(1):69-83.

Smith, H. J., J. D. Newman and D. Symmes 1982. Vocal concomitants of affiliative behavior in squirrel monkeys. Pp. 30-49 in: C. T. Snowdon, C. H. Brown and M. R. Petersen (eds.). Cambridge University Press, New York.

Smith, C. C. 1977. Feeding behavior and social organization in howling monkeys. Pp. 97-126 in: T. H. Clutton-Brock (ed.), *Primate Ecology: Studies of Feeding and Ranging Behaviour in Lemurs, Monkeys, and Apes*. Academic Press, London.

Smith, H. J., J. D. Newman and D. Symmes 1982. Vocal concomitantes of affiliative behavior in squirrel monkeys. Pp. 30-49 in: C. T. Snowdon, C. H. Brown and M. Petersen (eds.), *Primate Communication*. Cambridge University Press, New York.

Smith, H. J., J. C. Newman, D. E. Bernhards and D. Symmes 1983. Effects of reproductive state on vocalizations in squirrel monkeys (*Saimiri sciureus*). *Folia Primatologica* 40(4):233-246.

Snow, C. A. 1986. A life history study of the Geoffroy's tamarin, *Saguinus geoffroyi*, with emphasis on male-female relationships in captive animals. Unpublished Ph.D. thesis, Kent State University.

Snowdon, C. T. 1993a. A vocal taxonomy of the callitrichids. Pp. 78-94 in: A. B. Rylands (ed.), *Marmosets and Tamarins: Systematics, Behaviour, and Ecology*. Oxford University Press, Oxford.

Snowdon, C. T. 1993b. The rest of the story: Grooming, group size and vocal exchanges in Neotropical primates. *Behavioral and Brain Sciences* 16(4):718.

Snowdon, C. T. and Y. Pola 1978. Interspecific and intraspecific responses to synthesized pygmy marmoset vocalizations. *Animal Behaviour*. 26:192-206.

Snowdon, C. T. and J. Cleveland 1980. Individual recognition of contact calls in pygmy marmosets. *Animal Behaviour* 28:717-27.

Snowdon, C. T. and A. Hodun 1981. Acoustic adaptations in pygmy marmoset contact calls: Locational cues vary with distance between conspecifics. *Behavioral Ecology and Sociobiology* 9:295-300

Snowdon, C. T. and J. Cleveland 1984. Conversations among pygmy marmosets. *American Journal of Primatology* 7:15-20.

Snowdon, C. T. and P. Soini 1988. The tamarins, genus Saguinus. Pp. 223-298 in: R. A. Mittermeier, A. B. Rylands, A. Coimbra-Filho and G. A. B. Fonseca, *Ecology and Behavior of Neotropical Primates*, vol. 2. WWF-U.S., Washington, D. C.

Snowdon, C. T. and S. De la Torre 2002. Multiple environmental contexts and communication in pygmy marmosets (*Cebuella pygmaea*). *Journal of Comparative Psychology* 116(2):182-188.

Snowdon, C. T., J. Cleveland and J. A. French 1983. Responses to context- and individual-specific cues in cotton-top long calls. *Animal Behaviour* 31(1):92-101.

Snowdon, C. T., A. Savage and P. B. McConnell 1985. A breeding colony of cotton-top tamarins (*Saguinus oedipus*). *Laboratory Animal Science* 35(5):477-480.

Soini, P. 1972. The capture and commerce of live monkeys in the Amazonian region of Peru. *International Zoo Yearbook* 12:26-36.

Soini, P. 1981. Informe de Pacaya no. 4: Ecología y dinámica poblacional del pichico *Saguinus fuscicollis* (Primates, Callitrichidae). Unpublished manuscript.

Soini, P. 1982. Ecology and population dynamics of the pygmy marmoset, *Cebuella pygmaea*. *Folia Primatologica* 39:1-21.

Soini, P. 1986a. A synecological study of a primate community in the Pacaya-Samiria National Reserve, Peru. *Primate Conservation* 7:63-71.

Bibliography

Soini, P. 1986b. Informe de Pacaya no. 21: Resumen comparativo de la ecología y dinámica poblacional de la familia Callitrichidae (Primates). Unpublished manuscript.

Soini, P. 1986c. Informe preliminar de la ecología y dinámica poblacional del choro *Lagothrix lagothricha* (Primates). Unpublished manuscript.

Soini, P. 1987a. Sociosexual behavior of a free-ranging *Cebuella pygmaea* (Callitrichidae, platyrrhini) troop during postpartum estrus of its reproductive female. *American Journal of Primatology* 13:223-230.

Soini, P. 1987b. Informe de pacaya no. 24. desarrollo dentario y la estimación de la edad en *Cebuella pygmaea, Saguinus fuscicollis* y *Saguinus mystax* (Callitrichidae, Primates). Unpublished manuscript.

Soini, P. 1987c. Ecology of the saddle-back tamarin *Saguinus fuscicollis illigeri* on the río Pacaya, northeastern Peru. *Folia Primatologica* 49:11-32.

Soini, P. 1987d. Informe de Pacaya no. 25. La dieta del mono huapo (*Pithecia monachus*). Unpublished report.

Soini, P. 1988a. The pygmy marmoset, genus *Cebuella*. Pp. 79-130 in: R. A. Mittermeier, A. B. Rylands, A. Coimbra-Filho, and G. A. B. Fonseca (eds.), *Ecology and Behavior of neotropical primates*, WWF-U.S., Washington, D. C.

Soini, P. 1988b. El huapo (*Pithecia monachus*): Dinámica poblacional y organización social. Unpublished report.

Soini, P. 1990a. Ecología y dinámica poblacional del "choro" (*Lagothrix lagothricha*, Primates) en Río Pacaya, Perú. Pp. 382-395 in: E. Castro R. (ed.), *La Primatología en Perú: Investigaciones Primatológicas (1973-1985)*. Proyecto Peruano de Primatología, Manuel Moro Sommo, Lima, Peru.

Soini, P. 1990b. Ecología dinámica poblacional del pichico común *Saguinus fuscicollis* (Callitrichidae, Primates). Pp. 202-253 in: E. Castro R. (ed.), *La Primatología en el Perú: Investigaciones Primatológicas (1973-1985)*. Proyecto Peruano de Primatología, Lima, Perú.

Soini, P. 1993. The ecology of the pygmy marmoset, *Cebuella pygmaea*: some comparisons with two sympatric tamarins. Pp. 257-261 in: A. B. Rylands (ed.), *Marmosets and Tamarins: Systematics, Behavior and Ecology*. Oxford University Press, Oxford.

Soini, P. 1995a. La dieta del mono huapo (*Pithecia monachus*). Pp. 273-278 in: P. Soini, A. Tovar and U. Valdez (eds.), Reporte Pacaya-Samiria: Investigaciones en la Estación Biológica Cahuana 1979-1994. Fundación Peruana para la Conservación de la Naturaleza and Centro de Datos para la Conservación, Universidad Nacional Agraria La Molina, Lima.

Soini, P. 1995b. El huapo (*Pithecia monachus*): Dinámica poblacional y organización social. Pp. 289-302 in: P. Soini, A. Tovar and U. Valdéz (eds.), *Reporte Pacaya-Samiria: Investigaciones en la Estación Biológica Cahuana 1979-1994*. Fundación Peruana para la Conservación, Universidad Nacional Agraria La Molina, Lima.

Soini, P. 1995c. Desarrollo dentario y la estimación de la edad en *Cebuella pygmaea, Saguinus fuscicollis* y *Saguinus mystax* (Callitrichidae, Primates). Pp. 257-271 in: P. Soini N., A. Tovar N. and U. Valdez O. (eds.), *Reporte Pacaya-Samiria: Investigaciones en La Estación Biológica Cahuana 1979-1994*. Fundación Peruana para la Conservación, Universidad Nacional Agraria La Molina, Lima.

Soini, P. 1995c. *Reporte Pacaya-Samiria: Investigaciones en la Estación Biológica Cahuana 1979-1994*. Fundación Peruana para la Conservación de la Naturaleza, Lima.

Soini, P. and M. Soini 1982. Distribución geográfica y ecología poblacional del *Saguinus mystax* (Primates, Callitrichidae). Pp. 65-93 in: P. Soini, A. Tovar N. and U. Valdez O. (eds.), Reporte Pacaya-Samairia: Investigaciones en la Estación Biológica Cahuana 1979-1994.

Solano, C. 1995. Patrón de actividad y area de acción del mico nocturno *Aotus brumbacki* Hershkovitz, 1983 (Primates: Cebidae) Parque Nacional Natural Tinigua, Meta, Colombia. Unpublished bachelor's thesis, Pontificia Universidad Javeriana, Bogotá.

Solano, S. J., T. J. O. Martinez, A. Estrada and R. Coates-Estrada 1999. [Use of plants for food by *Alouatta palliata* in a forest fragmenta t Los Tuxtlas, Mexico.] *Neotropical Primates* 7(1):8-11.

Soltis, J., D. Bernhards, H. Donkin and J. D. Newman 2002. Squirrel monkey chuck call: Vocal response to playback chucks based on acoustic structure and affiliative relationship with the caller. *American Journal of Primatology* 57(3):119-130.

Sorensen, T. C. 1997. Influence of a regeneration gradient on three species of monkeys within tropical dry forest. [Abstract] *Bulletin of the Ecological Society of America* 78(4, suppl):189.

Sorensen, T. C. 1999. Tropical dry forest regeneration and its influence on three species of Costa Rican monkeys (*Cebus capucinus, Aloutta palliata, Ateles geoffroyi*). Masters Abstracts 36(6):1540.

Sorensen, T. C. and L. M. Fedigan 2000. Distribution of three monkey species along a gradient of regenerating tropical dry forest. *Biological Conservation* 92(2):227-240.

Southwick, C. H. 1955. The black howlers of Barro Colorado. *Animal Kingdom* 58:104-109.

Southwick, C. H. 1962. Patterns of intergroup social behavior in primates, with special reference to rhesus and howling monkeys. *Annals of the New York Academy of Sciences* 102:436-454.

Southwick, C. H. 1969. Social behavior of nonhuman primates. Pp. 299-300 in: B. K. Sladen & F. B. Bang (eds.), *Biology of Populations*. Elsevier, New York.

Spironelo, W. R. 1987. Range size of a group of *Cebus a. paella* in central Amazonia. [Abstract] *International Journal of Primatology* 8(5):522.

Spironelo, W. R. 1991. Importancia dos frutos de palmeiras (Palmae) na dieta de *Cebus apella* (Cebidae, Primates) na Amazónia central. *Primatología no Brasil* 3:285-296.

Spix, H. de 1823. *Simiarum et vespertiliarum Brailienses Species Novae; ou histoire naturelle des espêces nouvelles de singes et de chauvesouris observées t recuelli,s pendant le voyage dans l'intérieur du brásil*. Monaco, viii + 72 pp.

Sponsel, L. E., D. S. Brown, R. C. Bailey and R. A. Mittermeier 1974. Evaluation of squirrel monkey ranching on Santa Sofía Island, Amazonas, Colombia. *International Zoo Yearbook* 14:233-240.

Spurlock, L. B. 2002a. Reproductive suppression in the pygmy marmoset *Cebuella pygmaea*. Unpublished Ph.D. Dissertation Abstracts International B62(9):3872.

Spurlock, L. B. 2002b. Behavior and reproductive physiology of captive pygmy marmoset daughters. [Abstraqct] *American Journal of Primatology* 57 (Suppl 1):37-38.

Starin, P. R. 1992. Food transfer by wild titi monkeys (*Callicebus torquatus torquatus*). *Folia Primatologica* 30:145-51.

Stevenson, P. R. 1992. Diet of woolly monkeys (*Lagothrix lagotricha*), at La Macarena, Colombia. *Field Studies of New World Monkeys, Colombia*. 6:3-14.

Stevenson, P. R. 1997a. Vocal behavior of woolly monkeys (*Lagothrix lagothricha*) at Tinigua National Park, Colombia. *Field Studies of Fauna and Flora, La Macarena, Colombia*. 10:17-28.

Stevenson, P. R. 1997b. Notes on the mating behavior of woolly monkeys (*Lagothrix lagotricha*) at Tinigua National Park, Colombia. *Field Studies of Fauna and Flora, La Macarena, Colombia*. 10:13-15.

Stevenson, P. R. 1998a. Proximal spacing between individuals in a group of woolly monkeys (*Lagothrix lagotricha*) in Tinigua National Park, Colombia. *International Journal of Primatology* 19(2):299-311

Stevenson, P. R. 1998b. Seed shadows generated by woolly monkeys at Tinigua N. P., Colombia. [Abstract] *Congress of the International Primatological Society*, Abstracts 17:093.

Stevenson, P. R. 2000. Seed dispersal by woolly monkeys (*Lagothrix lagothricha*) at Tinigua National Park, Colombia: Dispersal distance, germination rates, and dispersal quantity. *American Journal of Primatology* 50:275-289.

Stevenson, P. R. 2001. The relationship between fruit production and primate abundance in Neotropical communities. *Biological Journal of the Linnean Society* 72(1):161-178.

Stevenson, P. R. 2002a. Frugivory and seed dispersal by woolly monkeys at Tinigua National Park, Colombia. Unpublished Ph. D. Thesis, State University of New York at Stony Brook, 411 pp.

Bibliography

Stevenson, P. R. 2002b. Weak relationships between dominance and foraging efficiency in Colombian woolly monkeys (*Lagothrix lagothricha*) at Tinigua Park. [Abstract] *American Journal of Primatology* 57(Suppl 1):68.

Stevenson, P. R. 2003. Como medir la dieta natural de un primate: variaciones interanuales en *Lagothrix lagothricha lugens*. Pp. 3–22 in: V. Pereira–Bengoa V., F. Nassar-Montoya and A. Savage (eds.), *Primatología del Nuevo Mundo: Biología, Medicina, Manejo y Conservación*. Centro de Primatología Araguatos, Bogotá.

Stevenson, P. R. 2004. Fruit choice by woolly monkeys in Tinigua National Park, Colombia. *International Journal of Primatology* 25(2):367-381.

Stevenson D., P. and M. Quiñones 1993. Vertical stratification of four new world primates at Tinigua National Park, Colombia. *Field Studies of New World Monkeys, La Macarena, Colombia* 8:11-18.

Stevenson, P. R. and J. Ahumada 1994. Ecological strategies of woolly monkeys (*Lagothrix lagotricha*) at La Macarena, Colombia. *American Journal of Primatology* 32:123-140.

Stevenson, P. and M. C. Castellanos 2000. Feeding rates and daily path range of the Colombian woolly monkeys as evidence for between- and within-group competition. *Folia Primatologica* 71:399-408.

Stevenson, P. R. and M. C. Castellanos 2001. New evidence for large variations in daily path length related to differences in habitat quality in troops of Colombian woolly monkeys, *Lagothrix lagothricha*. [Abstract] *The 18th Congress of the International Primatological Society. Primates in the New Millennium. Abstracts and Programme*. Adelaid:IPS, p. 446.

Stevenson, P. R. and A. del Pilar Medina 2003. Dispersión de semillas por micos churucos (*Lagothrix lagothricha*) en el Parque Nacional Tinigua, Colombia. Pp. 122–135 in: V. Pereira–Bengoa, F. Nassar–Montoya and A. Savage (eds.), *Primatología del Nuevo Mundo: Biología, Medicina, Manejo y Conservación*. Centro de Primatología Araguatos, Bogotá,

Stevenson D., P. R., M. J. Quiñones F. and J. A. Ahumada P. 1992. Relación entre la abundancia de frutos y las estrategias alimenticias de cuatro especies de primates en La Macarena, Colombia. Unpublished manuscript, report to Banco de la República.

Stevenson, P. R., M. Quiñones and J. Ahumada 1994. Ecological strategies of woolly monkeys (*Lagothrix lagothricha*) at Tinigua National Park, Colombia. *American Journal of Primatology* 32:123-140.

Stevenson, P. R., M. Quiñones and J. Ahumada 1998. Effects of fruit patch availability on feeding subgroup size and spacing patterns in four primate species at Tinigua National Park, Colombia. *International Journal of Primatology* 19(2):313-324.

Stevenson, P. R., M. J. Quinones and J. A. Ahumada 2000. Influence of fruit availability on ecological overlap among four neotropical primates at Tinigua National Park, Colombias. *Biotropica* 32(3):533-544.

Stevenson, P. R., M. C. Castellanos, J. C. Pizarro and M. Garabilo 2002. Effects of seed dispersal by three ateline monkey species on seed germination at Tinigua National Park, Colombia. *International Journal of Primatology* 23(6):1155-1168.

Stirton, R. A. 1951. Ceboid monkeys from the Miocene of Colombia. Bulletin of the University of California *Publications in Geological Science* 28(11):315-356.

Stirton, R. A. and D. E. Savage 1951. A new monkey from the La Venta Miocene of Colombia. *Compilacion de los Estudios Geológicos Oficiales en Colombia* (Bogotá) 8:345–356.

Stoner, K. E. 1994. Population density of the mantled howler monkey (*Alouatta palliata*) at La Selva Biological Reserve, Costa Rica: a new technique to analyze census data. *Biotropica* 26:332-340.

Stoner, K. E. 1996a. Habitat selection and seasonal patterns of activity and foraging of mantled howling monkeys (*Alouatta palliata*) in northeastern Costa Rica. *International Journal of Primatology* 17(1):1-30.

Stoner, K. E. 1996b. Prevalence and intensity of intestinal parasites in mantled howling monkeys (*Alouatta palliata*) in northeastern Costa Rica: Implications for conservation biology. *Conservation Biology* 10(2):539-546.

Strushaker, T. T. and L. Leland 1977. Palm-nut smashing by *Cebus a. apella* in Colombia. *Biotropica* 9:124-126.

Struhsaker, T. T., K. Glander, H. Chirivi and N. J. Scott 1975. A survey of primates and their habitats in northern Colombia. Pp. 43-79 in: *Primate Censusing Studies in Peru and Colombia: Report to the National Academy of Sciences on the Activities of Project AMRO-0719.* Pan American Health Organization, Washington, D. C.

Suárez, C. E., E. M. Gamboa, P. Claver and F. Nassar-Montoya 2001. Survival and adaptation of released group of confiscated capuchin monkeys. *Animal Welfare* 10(2):191-203.

Suárez, S. 2001a. Feeding patch choice in free-ranging *Ateles belzebuth belzebuth*: Implications for cognitive foraging skills. [Abstract] *American Journal of Primatology* 54(Suppl 1):41.

Suárez, S. 2001b. Quantifying fission-fusion behavior and social dynamics in free-ranging spider monkeys (*Ateles belzebuth*). [resumen] *American Journal of Physical Anthropology* (Suppl 32):145-146.

Symington, M. N. 1987a. Ecological and social correlates of party size in the black spider monkey, (*Ateles paniscus chamek*). Unpublished Ph.D. thesis, Princeton University, Princeton, N.J.

Symington, M. M. 1987b. Sex ratio and maternal rank in wild spider monkeys: when daughters disperse. *Behavioral Ecology and Sociobiology* 20:421-425.

Symington, M. M. 1988a. Food competition and foraging party size in the black spider monkey (*Ateles paniscus chamek*). *Behaviour* 105(1-2):117-134.

Symington, M. M. 1988b. Demography, ranging patterns and activity budgets of black spider monkeys (*Ateles paniscus chamek*) in the Manu National Park, Peru. *American Journal of Primatology* 15:45-67.

Symington, M. M. 1988c. Environmental determinants of population densities in *Ateles*. *Primate Conservation* 9:74-78.

Symington, M. M. 1990. Fission-fusion social organization in *Ateles* and *Pan*. *International Journal of Primatology* 11(1):47-61.

Szalay, F. S. and E. Delson 1979. *Evolutionary History of the Primates.* Academic Press, New York.

Tagliaro, C. H., M. P. C. Schneider, H. Schneider, I. C. Sampaio and M. J. Stanhope 1997. Marmoset phylogenetics, conservations perspectives, and evolution of the mtDNA control region. *Molecular Biology and Evolution* 14(6):674-684.

Tagliaro, C. H., M. P. C. Schneider, H. Schneider, I. C. Sampaio, and M. J. Stanhope 2000. Molecular studies of *Callithrix pygmaea* (Primates, Platyrrhini) based on Transferrin intronic and ND1 regions: Implications for taxonomy and conservation. *Gen. Mol. Biol.* (cited in Rylands *et al.*, 2000)

Takai, M. 1994. New specimens of *Neosaimiri fieldsii* from La Venta, Colombia: a middle Miocene ancestor of the living squirrel monkey. *Journal of Human Evolution* 27:329-360.

Takai, M. and F. Anaya 1994. New specimens of the oldest fossil Platyrrhine, *Branisella boliviana*, from Salla, Bolivia. *American Journal of Physical Anthropology* 99(2):301-317.

Takai, M. and T. Setoguchi 1992. [Dental variability of *Neosaimiri* (Middle Miocene platyrrhine fossil) and its similarity to extant squirrel monkeys.] (in Japanese) *Reichorui Kendyu/Primate Research* 8(2):226.

Takehara, A. and P. R. Stevenson 1997. Stem size and date of some mono/species dominated forests in La Macarena, Colombia. *Field Studies of Fauna and Flora, La Macarena, Colombia* 11:57–71.

Takai, M., K. Takemura, A. Takemura, A. C. Villarroel, A. Hayashida, T. DAnhara, T. Ohno, N. R. Franco, T. Setoguchi and Y. Nogami 1992a. Geology of La Venta, Colombia, South America. *Kyoto University Overseas Research Reports of New World Monkeys* 8:1–17.

Takemura, A., M. Takai, T. Danhara and T. Setoguchi 1992b. Fission/track ages of the Villavieja Formation of the Miocene Honda Group in La Venta, Department of Huila, Colombia. *Kyoto University Overseas Research Reports of New World Monkeys* 8:19–27.

Tardif, S. D. 1984. Social influences on sexual maturation of female *Saguinus oedipus oedipus*. *American Journal of Primatology* 6:199-209.

Tardif, S. D. and C. B. Richter 1981. Competition for a desired food in family groups of the common marmoset (*Callithrix jacchus*) and the cotton-top tamarin (*Saguinus oedipus*). *Laboratory Animal Science* 31:52-55.

Tardif, S. D. and R. Colley 1988. *International Cotton-top Studbook*. Oak Ridge Associated Universities, Oak Ridge, Tennessee, 58 pp.

Tardif, S. D., C. B. Richter and R. L. Carson 1984. Reproductive performance of three species of Callitrichidae. *Laboratory Animal Science* 272-275.

Tardif, S. D., R. L. Carson and B. L. Gangaware 1990. Infant-care behavior of mothers and fathers in a communal-care primate, the cotton-top tamarin (*Saguinus oedipus*). *American Journal of Primatology* 22(2):73-85.

Tardif, S. D., M. L. Harrison and M. A. Simek 1993. Communal infant care in marmosets and tamarins: Relation to energetics, ecology, and social organization. Pp. 220-234 in: A. B. Rylands (ed.), *Marmosets and Tamarins: Systematics, Behaviour, and Ecology*. Oxford University Press, Oxford.

Teaford, M. F. and K. E. Glander 1991. Dental microwear in live, wild-trapped *Alouatta palliata* from Costa Rica. *American Journal of Physical Anthropology* 85(3):313-319.

Terborgh, J. 1983. Five *New World Primates: A Study in comparative ecology*. Princeton: Princeton University Press.

Terborgh, J. 1986a. The social systems of New World primates: an adaptationist views. Pp. 199-211 in: J. G. Else and P. C. Lee (eds.), *Primate Ecology and Conservation*. Cambridge University Press, New York.

Terborgh, J. 1986b. Conserving new world primates: present problems and future solutions. Pp. 355-366 in: J. G. Else and P. C. Lee (eds.), *Primate Ecology and Conservation*. Cambridge University Press, Cambridge.

Terborgh, J. 1990. Mixed flocks and polyspecific associations: Costs and benefits of mixed groups to birds and monkeys. *American Journal of Primatology* 21:87-100.

Terborgh, J. and W. A. Goldizen 1985. On the mating system of the cooperatively breeding saddle-backed tamarin (*Saguinus fuscicollis*). *Behavioral Ecology and Sociobiology* 16:293-299.

Terborgh, J. and M. Stern 1987. The surreptitious life of the saddle-backed tamarin. *American Scientist* 75(3):260-269.

Thierry, B., D. Wunderligh and C. Gueth 1989. Possession and transfer of objects in a group of brown capuchins (*Cebus apella*). *American Journal of Primatology* 15:349-360.

Thomas, O. 1904. New forms of *Callithrix*, *Midas*, *Felis*, *Rhipidomys* and *Proechimys* from Brazil and Ecuador. *Annales of the. Magazine of Natural History* (7)14:188-196.

Thomas, O. 1927a. A remarkable new monkey from Peru. *Ann. Magazine of Natural History* 19(9):156-157.

Thomas, O. 1927b. The Godman-Thomas Expedition to Peru. VI. On mammals from the upper Huallaga and neighbouring highlands. Ann. Magazine Natural History 19(20):594-608.

Thorington, R. W. Jr. 1967. Feeding and activity of *Cebus* and *Saimiri* in a Colombian forest. Pp. 180-184 in: D. Starck, R. Schnieder & H. Kuhn (eds.), *Progress in Primatology*, Gustav Fischer, Stuttgart.

Thorington, R. W. Jr. 1968. Observations of squirrel monkeys in a Colombian forest. Pp. 69-85 in: L. A. Rosenblum and R. W. Cooper (eds.), *The Squirrel Monkey*. Academic Press, New York.

Thorington, R. W. Jr. 1969. The study and conservation of New World monkeys. *Anais da Academia Brasileira de Ciencias* 41(Suppl.): 253-260.

Thorington, R. W. Jr. 1985. The taxonomy and distribution of squirrel monkeys (*Saimiri*). Pp. 1-33 in: L. A. Rosenblum and C. L. Coe (eds.), *Handbook of Squirrel Monkey Research*. Plenum Press, New York.

Thorington, R. W., Jr., N. A. Muckenhirn and G. G. Montgomery 1976. Movements of a wild night monkey (*Aotus trivirgatus*). Pp. 32-34 in: T. W. Thorington, Jr. and P. G. Heltne (eds.), *Neotropical Primates: Field Studies and Conservation*. National Academy of Sciences, Washington, D. C.

Thorington, R. W. Jr., R. Rudran and D. Mack 1979. Sexual dimorphism of *Alouatta seniculus* and observations on capture techniques. Pp. 97-106 in: J. F. Eisenberg (ed.), *Vertebrate Ecology in the Northern Neotropics*. Smithsonian Institution Press, Washington, D. C.

Tirado Herrera, E. R. and E. W. Heymann 2000. Mom needs more protein – Sex differences in the diet compositor of red titi monkeys, *Callicebus cupreus*. [Abstract] *Folia Primatologica* 71(4):244.

Tokuda, K. 1968. Group size and vertical distribution of New World monkeys in the basin of the Río Putumayo, the upper Amazon. Pp. 260-261 in: Processes of the 8th Congress of Anthropological Science, vol. I, Anthropology.

Tokuda, K. 1988. Some social traits of howling monkeys (*Alouatta seniculus*) in La Macarena, Colombia. *Field Studies of New World Monkeys, La Macarena, Colombia* 1:35-38.

Tomblin, D. C. and J. A. Cranford 1994. Ecological niche differences between *Alouatta palliata* and *Cebus capucinus* comparing feeding modes, branch use, and diet. *Primates* 35(3):265-274.

Torres, O. M., S. Enciso, F. Ruiz, E. Silva and I. Yunis 1998. Chromosome diversity of the genus *Aotus* from Colombia. *American Journal of Primatology* 44(4):255-275.

Torres de Assumçsao, V. 1981. *Cebus apella* and *Brachyteles arachnoides* (Cebidae) as potential pollinators of *Mabea fistulifera* (Euphorbiaceae). *Journal of Mammalogy* 62:386-388.

Townsend W. R. 2001. *Callithrix pygmaea*. *Mammalian Species* 665:1-6.

UICN (Unión Mundial para la Naturaleza) 1984. Categorias de las listas rojas de la UICN. UICN, Gland, Suiza.

Ulloa, V. R. 1986. Primer registro de *Callicebus torquatus* (Cebidae, Platyrrhini) en Ecuador.] *Pulbicaciones del Museu Ecuatoriano de Ciencias Naturales Serie Revista* 7(5):123-136.

Umaña, J. A., J. Ramirez C., C. A. Espinal T. and E. Saboghal M. 1984. Establishment of a colony of nonhuman primates (*Aotus lemurinus griseimembra*) in Colombia. *Bulletin of the Pan American Health Organization*. 18(3):221-229.

Vaitl, E. 1977a. Experimental analysis of the nature of social context in captive groups of squirrel monkeys (*Saimiri sciureus*). *Primates* 18:849-59.

Vaitl, E. 1977b. Social context as a structuring mechanism in captive groups of squirrel monkeys *(Saimiri sciureus)*. *Primates* 18:861-74.

Vaitl, E. 1978. Nature and implications of the complexly organized social system in non-human primates. Pp 17-30 in: D. J. Chivers and J. Herbert (eds.), *Recent Advances in Primatology*, vol. 1, Academic Press, New York.

Valenzuela, N. 1992. Early development of three wild infant *Cebus apella* at La Macarena, Colombia. *Field Studies of New World Monkeys, La Macarena, Colombia* 6:15-24.

Valenzuela, N. 1993. Social contacts between infants and other group members in a wild group of *Cebus apella* in La Macarena, Colombia. *Field Studies of New World Monkeys, La Macarena, Colombia* 8:1-9.

Valenzuela, N. 1994. Early behavioral development of three wild infant *Cebus apella* in Colombia. Pp. 297-302 in: J. J. Roeder, B. Thierry, J. R. Anderson and N. Herrenschmidt (eds.), *Current Primatology, vol. II: Social Development, Learning and Behaviour*. Universidad Louis Pasteur, Strasbourg.

Vallejo-E., J. 1929. Vocabulario Baudó. Revista de Colombia, Bogotá , p. 134. Reproduction from *Idearium*, vol. I, Pasto, 1937, p. 259.

Van der Hammen, T. 1972. Changes in vegetation and climate in the Amazon basin and surrounding areas during the Pleistocene. *Geol. En Mijnbouw* 51:641-643.

Van der Hammen, T. 1974. The Pleistocene changes of vegetation and climate in tropical South America. *Journal Biography* 1:3-26.

Van der Hammen, T. 1978. Stratigraphy and environments of the Upper Quaternary of the El Abra corridor and rock shelters (Colombia). *Palaeogeogr., Palaeoclimat., Palaeoecol.* 25:111-162.

Bibliography

Van der Hammen, T. 1982. Paleoecology of tropical South America. Pp. 60-66 in: G. T. Prance (ed.), *Biological Diversification in the Tropics*. Columbia University Press, New York.

Van Geel, B. and T. van der Hammen 1973. Upper Quaternary vegetational and climatic sequence of the Fúquene area. *Palaeogeogr., Palaeoclimat., Palaeoecol.* 14:9-92.

Van Roosmalen, M. G. M. 1985. Habitat preferences, diet, feeding strategy and social organization of the black spider monkey (*Ateles paniscus paniscus* Linnaeus 1758) in Surinam. *Acta Amazonica* 15(3/4, suppl):1-238.

Van Roosmalen, M. G. M. and L. L. Klein 1988. The spider monkeys, genus *Ateles*. Pp. 455-537 in: R. A. Mittermeier, A. B. Rylands, A. Coimbra-Filho and G. A. B. Fonseca. *Ecology and Behavior of Neotropical Primates*. Vol. II, WWF-U.S., Washington, D. C.

Van Roosmalen, M. G. M. and T. van Roosmalen 1997. An eastern extension of the geographical range of the pygmy marmoset, *Cebuella pygmaea*. *Neotropical Primates* 5(1):3-6.

Van Roosmalen, M. G. M., T. Van Roosmalen and R. A. Mittermeier 2002. A taxonomic review of the titi monkeys, genus *Callicebus* Thomas, 1903, with the description of two new species, *Callicebus bernhardi* and *Callicebus stephennashi*, from Brazilian Amazonia. *Neotropical Primates* 10(Suppl.):1-52.

Van Schaik, C. P. and J. A. R. A. M. van Hooff 1983. On the ultimate causes of primate social systems. *Behaviour* 85(1-2):91-117.

Van Schaik, C. P. and M. A. van Noordwijk 1989. The special role of male *Cebus* monkeys in predation avoidance and its effect on group composition. *Behavioral Ecology and Sociobiology* 24:265-276.

Van Schaik, C. P. and M. Horstermann 1994. Predation risk and the number of adult males in a primate group: A comparative test. *Behavioral Ecology and Sociobiology* 35(4):261-272.

Vanzolini, P. E. 1970. Zoologa sistemática, geografia e a origem das espécies. *Inst. Geogr.. Univ. S. Paulo*, Sér. Teses e Monogr. 3.

Vanzolini, P. E. 1973. Paleoclimate, relief, and species multiplication in equatorial forests. Pp. 255-258 in: B. J. Meggers, E. S. Ayensu and W. D. Duckworth (eds.), *Tropical Forest Ecosystems in Africa and South America*. Smithsonian Inst. Press, Washington, D.C.

Vanzolini, P. E. and E. E. Williams 1970. South American anoles: The geographic differentiation and evolution of the *Anolis chrysolepis* species group (Sauria, Iguanidae). *Arq. Zool. São Paulo* 19:1-298.

Vargas T., N. 1992. Patrones de actividad diaria de *Saguinus nigricollis hernandezi* Hershkovitz, 1982 (Primates: Callitrichidae), Parque Nacional Natural Tinigua, Departamento del Meta. Colombia. Unpublished bachlor's thesis, Pontificia Universidad Javeriana, Bogotá.

Vargas T., N. 1994a. Evaluación de las poblaciones de primates en dos sectores del Parque Nacional Natural «Las Orquideas», Departamento de Antioquia. Unpublished manuscript.

Vargas T., N. 1994b. Activity patterns of *Saguinus nigricollis hernandezi* at the Tinigua National Park, Colombia. *Field Studies of New World Monkeys La Macarena Colombia* 9:23-32.

Vargas T., N. and C. Solano 1996a. Evaluation of the condition of two populations of *Saguinus leucopus* Gunther 1817, in order to determine potential conservation areas in Middle Magdalena, Colombia. [Abstract] *IPS/ASP Congress Abstracts* 1996:373.

Vargas T., N. and C. Solano 1996b. Evaluación de los problemas en determinar potenciales areas de conservación para *Saguinus leucopus* en una sección del medio valle del Magdalena. *Neotropical Primates* 4(1):13-15.

Vassart, M., A. Guedant, J. C. Vie, J. Keravec, J. A. Seguela and V. T. Volobouev 1996. Chromosomes of *Alouatta seniculus* (Platyrrhini, Primates) from French Guiana. *Journal of Heredity* 87(4):331-334.

Vercauteren Drubbel, R. and J. Gautier-P. 1993. On the occurrence of nocturnal and diurnal loud calls, differing in structure and duration, in red howlers (*Alouatta seniculus*) of French Guyana. *Folia Primatologica* 60(4):195-209.

Vick, L. G. and D. M. Taub 1996a. Feeding opportunism, social organization, and breeding peaks in spider monkeys (*Ateles geoffroyi*) at Punta Laguna, Mexico. [Abstract] *IPS/ASP Congress Abstracts* 1996:#039.

Vick, L. B. and D. M. Taub 1996b. Poco a poco: Steps toward spider monkey conservation in Mexico's Yucatan peninsula. [Abstract] *Chimpanzoo Converence Proceedings* 1995:91.

Viegas Pequignot, E., C. P. Koiffmann and B. Dutrillaux 1985. Chromosomal phylogeny of *Lagothrix, Brachyteles,* and *Cacajao. Cytogenetics and Cell Genetics* 39:99-104.

Visalberghi, E. 1990. Tool use in *Cebus. Folia Primatologica* 54(3-4):146-154.

Visalberghi, E. 1993. Tool use in a South American monkey species: an overview of the characteristics and limits of tool use in *Cebus apella*. Pp. 118-131 in: A. Berthelet and J. Chavaillon (ed.), *The Use of Tools by Human and Non-human Primates*. Claredon Press, Oxford.

Visalberghi, E. 2002a. Insight from capuchin monkeys studies: Ingredientes de, recipes for, and flaws in capuchins' success. Pp. 405-411 in: M. Bekoff, C. Allen and G. M. Burghardt (eds.), *The Cognitive Animal: Empirical and Theoretical Perspectives on Animal Cognition*. MIT Press, Cambridge.

Visalberghi, E. 2002b. Food for thoughts: Experiments on social biases on feeding behavior in tufted capuchins. [Abstract] *Caring for Primates. Abstracts of the XIXth Congress. The International Primatological Society & Mammalogical Soceity of China* 2002:154.

Visalberghi, E. and F. Antinucci 1986. Tool use in the exploitation of food resources in *Cebus apella*. Pp. 57-62 in: J. G. Else and P. C. Lee, *Primate Ecology and Conservation*, Cambridge University Press, Cambridge.

Visalberghi, E. and E. Addessi 2003. Food for thought: Social learning about food in capuchin monkeys. Pp. 187-212 in: D. M. Fragaszy and S. Perry (eds.), *The Biology of Traditions: Models and Evidence*. Cambridge University Press, New York.

Visalberghi, E., C. H. Janson and I. Agostini 2003. Response toward novel foods and novel objects in wild *Cebus apella. International Journal of Primatology* 24(3):653-675.

Vogt, J. L. 1978a. The social behavior of a marmoset social (*Saguinus fuscicollis*) group. II. Behavioral patterns and interaction. *Primates* 19:287-300.

Vogt, J. L. 1978b. The social behavior of a marmoset social (*Saguinus fuscicollis*) group. III. Spatial analysis of structure. *Folia Primatologica* 29:250-267.

Vogt, J. L., H. Carlson and E. Menzel 1978. Social behavior of a marmoset (*Saguinus fuscicollis*) group, 1: Parental care and infant development. *Primates* 19:715-26.

Von Dornum, M. and M. Ruvolo 1999. Phylogenetic relationships of the New World monkeys (Primates, Platyrrhini) based on nuclear G6PD DNA sequences. *Molecular Phylogenetics and Evolution* 11 (3):459-476.

Wallace, R. B. and R. L. E. Painter 1999. A new primate record for Bolivia: An apparently isolated population of common woolly monkeys representing a southern range extension for the genus *Lagothrix. Neotropical Primates* 7(4):111-112.

Wang, E. and K. Milton 2003. Intragroups social relationships of male *Alouatta palliata* on Barro Colorado Island, Republica of Panama. International Journal of Primatology 24(6):1227-1243.

Warkentin, I. G. 1993. Presumptive foraging association between sharp-shinned hawks (*Accipiter striatus*) and white-faced capuchin monkeys (*Cebus capucinus*). *Journal of Raptor Research* 27(1):46-47.

Webb, S. D. and A. Rancy 1996. Late Cenezoic evolution of the neotropical mammal fauna. Pp. 335-358 in: J. B. C. Jackson, A. F. Budd and A. G. Coates (eds.), *Evolution and Environment in Tropical America*. The University of Chicago Press, Chicago.

Weghorst, J. A. 2001. Behavioral ecology of the Central American spider monkey (*Ateles geoffroyi panamensis*) in Costa Rican wet forest: Pilot study results. [Abstract] *American Journal of Primatology* 54(Suppl 1):97.

Wehncke, E. V., S. P. Hubbell, R. B. Foster and J. W. Dalling 2003. Seed dispersal patterns produced by white-faced monkeys: Implications for the dispersal limitation of neotropical tree species. *Journal of Ecology* 91:677-685.

Weigel, R. M. 1974. The facial expressions of the brown capuchin monkey (*Cebus apella*). Unpublished M.S. thesis, University of Illinois, Urbana, 66 pp.

Bibliography

Welker, C. 1979a. Zum Sozialverhalten des Kapuzineraffen *Cebus apella cay* Illiger, 1815, in Gefangenschaft. *Philippia* iv/2:154-168.

Welker, C. 1979b. Zur Sozialstructur des Kapuzineraffen, *Cebus apella cay*, (Cebidae, Platyrrhina; Primates) in *Gefangenschaft. Verhaltniss Deutscher Zoologischen. Geschaft* 240.

Welker, C. and B. Lührman 1978. Social behavior in a family group of *Saguinus oedipus oedipus*. Pp. 281-288 in: H. Rothe, -J. Wolters and J. P. Hearn (eds.), *Bioloby and Behaviour of Marmostes. Proceedings of the Marmoset Workshop.* Mercke-Druck, Goettingen.

Welker, C., W. Meinel, M. Grebian and B. Lährmann 1980. Zur lokomotorischen Aktivität des Lisztäffchens *Saguinus oedipus oedipus* (Linnaeus, 1758) in Gefangenschaft. *Zeitschrift für Säugertierkunde* 45:39-44.

Welker, C., C. Brinkmann and C. Schöfer 1981. Zum Sozialverhalten des Kapuzineraffen *Cebus apella cay* Illiger, 1815, in Gefangenschaft. *Philippia* iv/4:331-342.

Welker, C., P. Becker, H. Hoehmann *et al.* 1987. Social relations in groups of the black-capped capuchin *Cebus apella* in captivity. Interactions of group-born infants during their first 6 months of life. *Folia Primatologica* 49:33-47.

Welker, C., P. Becker, H. Hoehmann and C. Schaefer-Witt 1990. Social relations in groups of black-capped capuchins (*Cebus apella*) in captivity. Interactions of group-born infants during their second half-year of life. *Folia Primatologica* 54:16-33.

Welker, C., H. Hoehmann-Kroeger and G. A. Doyle 1992a. Social realtions in groups of black-capped capuchin monkeys, (*Cebus apella*) in captivity: Sibling relations from the second to the fifth year of life. *Zeitschrift für Säugetierkunde* 57(5):269-274.

Welker, C., H. Hoehmann-Kroeger and G. A. Doyle 1992b. Social relations of black-capped capuchin monkeys, (*Cebus apella*) in captivity: Mother-juvenile relations from the second to the fifth year of life. *Zeitschrift für Säugetierkunde* 57(2):70-76.

Welker, C., B. Jantschke and A. Klaiber-Shuh 1998a. Behavioural data on the titi monkey *Callicebus cupreus* and the owl monkey *Aotus azarae boliviensis*. A contribution to the discussion on the correct systematic classification os these species. Part IV: Breeding biology. *Primate Report* 51:43-53.

Welker, C., B. Jantschke and A. Klaiber-Schuh 1998b. Behavioural data on the titi monkey *Callicebus cupreus* and the owl monkey *Aotus azarae boliviensis*. A contribution to the discussion on the correct systematic classification of these species. Part I: Introduction and behavioural differences. *Primate Report* 51:3-18.

Welker, C., B. Jantschke and A. Klaiber-Schuh 1998c. Behavioural data on the titi monkey *Callicebus cupreus* and the owl monkey *Aotus azarae boliviensis*. A contribution to the discussion on the correct systematic classification of these species. Part II: Pair-formation and relatins between mates. *Primate Report* 51:19-27.

Welker, C., B. Jantschke and A. Klaiber-Schug 1998d. Behavioural data on the titi monkey *Callicebus cupreus* and the owl monkey *Aotus azarae boliviensis*. A contribution to the discussion on the correct systematic classification of these species. Part II: Living in family groups. *Primate Report* 51:29-42.

Welker, C., B. Jantschke and A. Klaiber-Schug 1998e. Behavioural data on the titi monkey *Callicebus cupreus* and the owl monkey *Aotus azarae boliviensis*. A contribution to the discussion on the correct systematic classification of these species. Part V: Miscellaneous notes and final discussion. *Primate Report* 51:55-71.

Westergaard, G. C. 1994. The subsistence technology of capuchins. *International Journal of Primatology* 15(6):899-906.

Westergaard, G. C. and D. M. Fragaszy 1987. The manufacture and use of tools by capuchin monkeys (*Cebus apella*). *Journal of Comparative Psychology* 2:159-168.

Westergaard, G. C. and S. J. Suomi 1994a. Aimed throwing of stones by tufted capuchin monkeys (*Cebus apella*). *Human Evolution* 9(4):323-329.

Westergaard, G. C. and S. J. Suomi 1994b. Hierarchichal complexity of combinatorial manipulation in capuchin monkeys (*Cebus apella*). *American Journal of Primatology* 32(3):171-176.

Westergaard, G. C., A. L. Lundquist, H. E. Kuhn and S. J. Suomi 1997. Ant-gathering with tools by captive tufted capuchins (*Cebus apella*). *International Journal of Primatology* 18(1):95-103.

Westergaard, G. C. and G. Byrne and S. J. Suomi 1998. Early lateral bias in turfted capuchins (*Cebus apella*). *Developmental Psychobiology* 32(1):45-50.

Westergaard, G. C., M. K. Waynie and A. L. Lundquist 1999. Carrying, sharing and hand preference in tufted capuchins (*Cebus apella*). *International Journal of Primatology* 20(10:153-162.

Westergaard, G. C., A. Cleveland, A. M. Rocca, E. L. Wendt and M. J. Brown 2003. Throwing behavior and the mass distribution of rock selection in tufted capuchin monkeys (*Cebus apella*). [Abstract] *American Journal of Physical Anthropology* (Suppl 36): 228.

White, B. C., S. E. Dew, J. R. Prather, M. J. Stearns, E. Schneider and S. Taylor 2000. Chest-rubbing in captive woolly monkeys (*Lagothrix lagotricha*). *Primates* 41(2):185-188.

Whitehead, J. M. 1985. Long-distance vocalizations and spacing in mantled howling monkeys, *Alouatta palliata*. Unpublished Ph.D. dissertation, University of North Carolina, Chapel Hill.

Whitehead, J. M. 1986a. Development of feeding selectivity in mantled howling monkeys, *Alouatta palliata*. Pp. 105-117 in: J. G. Else and P. C. Lee (eds.), *Primate Ontogeny, Cognition and Social Behaviour*. Cambridge University Press, New York.

Whitehead, J. M. 1986b. Long-distance vocalizations and spacing in mantled howling monkeys, *Alouatta palliata*. Unpublished Ph.d. thesis, Dissertation Abstracts International. 1986. B46(11). Pgs: 3753

Whitehead, J. M. 1987. Vocally mediated reciprocity between neighboring groups of mantled howling monkey, *Alouatta palliata palliata*. *Animal Behaviour* 35:1615-1627.

Whitehead, J. M. 1989. The effect of the location of simulated intruder on responses to long-distance vocalizations of mantled howling monkeys, *Alouatta palliata palliata*. *Behaviour* 108:73-103.

Whitehead, J. M. 1992. Acoustic correlates of social contexts and inferred internal states in howling monkeys (*Alouatta palliata*). *Journal of Acoustics Society of America* 91:246.

Whithead, J. M. 1994. Acoustic correlates of internal states in free-ranging primates: the example of the mantled howling monkey *Alouatta palliata*. Pp. 221-226 in: J. J. Roeder, B. Thierry, J. F. Anderson and N. Herrenschmidt (eds.), *Current Primatology, vol. II: Social Development, Learning and Behaviour*. University Louis Pasteur, Strasbourg.

Whitehead, J. M. 1995. Vox Alouattinae: A preliminary survey of the acoustic characteristics of long-distance calls of howling monkeys. *International Journal of Primatology* 16(1); 121-144.

Whittemaze, T. 1970. Observations of *Callimico goeldii* and *Saguinus mystax*. Unpublished report.

Williams, J. 1932. *Bib. Ling. Anthropos*. Viii (quoted by Simpson, 1941).

Williams, L. 1967. *Man and Monkey*. Andre Deutsch, London, 203 pp.

Williams-Guillen, K. and C. M. McCann 2002. Ranging behavior of Nicaraguan howling monkey (*Alouatta palliata*) as evidence for within-group competition. *American Journal of Physical Anthropology* Suppl. 34:166.

Wilson, A. C., G. L. Bush, S. M. Case and M. –C. King 1975. Social structuring of mammalian populations and rate of chromosomal evolution. *Processes of the National Academy of Sciences* U.S.A. 72:5061-5065.

Wilson, D. E., F. R. Cole, J. D. Nichols, R. Rudran and M. S. Foster (eds.) 1996. *Measuring and monitoring biological diversity: standard method for mammals*. Smithsonian Institute Press, Washington, D. C.

Winkler, L. A. 2000. Patterns of fission-fusion social organization in the mantled howling monkey (*Alouatta palliata*) in Nicaragua. *American Journal of Physical Anthropology* Suppl. 30:324-325.

Winter, P. 1972. Observations on the vocal behaviour of free-ranging squirrel monkeys (*Saimiri sciureus*). *Zeitschrift für Tierpsychologie* 31:1-7.

Bibliography

Winter, P. D. Ploog and J. Latta 1966. Vocal repertoire of the squirrel monkey (*Saimiri sciureus*), its analysis and significance. *Experimental Brain Research* 1:359-384.

Wiswall, O. B. 1965. Gestation and maturation in the *Saimiri sciureus* (squirrel monkey). *Excerpta Med. Int. Congr.* Seri 99, e, 52.

Wolf, R. C. 1984. New specimens of the primate *Branisella bolivariana* from the early Oligocene of Salle. Bolivia. *Journal of Vertebrate Paleontology* 4(4):570-574.

Wolfheim, J. H. 1983. *Primates of the World: Distribution, Abundance, and Conservation*. University of Washington Press, Seattle.

Wolters, H., -J. 1980. Artificial rearing of cotton-top tamarins. Unpubl. Report. Callitrichid Station, Department of Ethology, University of Bielefeld.

Wright, E. M., Jr. and D. E. Bush 1977. The reproductive cycle of the capuchin (*Cebus apella*). *Laboratory Animal Science* 27:651-54.

Wright, K. A. 2001. A comparison of the locomotor behavior and habitat use of *Cebus olivaceus* and *Cebus apella* in Guyana. [Abstract] *American Journal of Physical Anthropology* Suppl. 32:167.

Wright, K. A. 2003. Differences in patterns of locomotor behavior and habitat use in adult and juvenile *Cebus apella* and *Cebus olivaceus*. [Abstract] *American Journal of Physical Anthropology* Supp 36:228.

Wright, P. C. 1978. Home range, activity pattern, and agonistic encounters of a group of night monkeys (*Aotus trivirgatus*) in Peru. *Folia Primatologica* 29:43-55.

Wright, P. C. 1981. The night monkeys, genus *Aotus*. Pp. 211-240 in: A. F. Coimbra-Filho and R. A. Mittermeier (eds.), *Ecology and Behavior of Neotropical Primates*, vol. 1, Academia Brasileira de Ciências, Rio de Janeiro.

Wright, P. C. 1984. Biparental care in *Aotus trivirgatus* and *Callicebus moloch*. Pp. 59-75 in: M. F. Small (ed.), *Female Primates: Studies by Women Primatologists*. Alan R. Liss, Inc., New York.

Wright, P. C. 1985. What do monogamous primates have in common. [Abstract] *American Journal of Physical Anthropology* 66(2):244.

Wright, P. C. 1986a. The costs and benefits of nocturnality for *Aotus trivirgatus* (the night monkey). Unpublished Ph.D. thesis, City University of New York.

Wright, P. C. 1986b. Ecological correlates of monogamy in *Aotus* and *Callicebus*. Pp. 159-167 in: J. G. Else and P. C. Lee (eds.), *Primate Ecology and Conservation*. Cambridge University Press, New York.

Wright, P. C. 1990. Patterns of paternal care in primates. *International Journal of Primatology* 11(2):89-102.

Wright, P. C. 1994a. Night watch on the Amazon. *Natural History* 103(5):44-51.

Wright, P. C. 1994b. The behavior and ecology of the owl monkey. Pp. 97-112 in: J. F. Baer, R. E. Weller and I. Kakoma (eds.), *The Owl Monkey*. Academic Press, San Diego.

Yarger, R. G., A. B. Smith, III, G. Preti and G. Smith 1977. The major volatile constituents of the scent mark of a South American primate, *Saguinus fuscicollis*, Callitrichidae. *Journal of Chemical Ecology* 3:45-56.

Yoneda, M. 1981. Ecological studies of *Saguinus fuscicollis* and *Saguinus labiatus* with reference to habitat segregation and height preference. *Kyoto University Overseas research reports of New World Monkeys* 2:43-50.

Yoneda, M. 1984a. Ecological study of the saddle backed tamarin (*Saguinus fuscicollis*) in northern Bolivia. *Primates* 25(1):1-12.

Yoneda, M. 1984b. Comparative studies on vertical separation, foraging behavior and travelling of saddle-backed tamarins (*Saguinus fuscicollis*) and red-chested moustached tamarins (*Saguinus labiatus*) in northern Bolivia. *Primates* 25:414-422.

Yoneda, M. 1988. Habitat utilization of six species of monkeys in río Duda, Colombia. *Field Studies of New World Monkeys, La Macarena, Colombia* 1:39-45.

Yoneda, M. 1990. The difference of tree size used by five cebid monkeys in Macarena Colombia. *Field Studies of New World Moneys, La Macarena, Colombia*. 3:13–18.

Yoshihiko, H. and K. Izawa 1990. Chemical properties of soils eaten by wild red howler monkeys (*Alouatta seniculus*). *Field Studies of New World Monkeys, La Macarena, Colombia* 3:25-28.

Youlatos, D. 1993. Passages within a discontinuous canopy: bridging in the red howler monkey (*Alouatta seniculus*). *Folia Primatologica* 61(3):144-147.

Youlatos, D. 1998. Seasonal variation in the positional behavior of red howling monkeys (*Alouatta seniculs*). *Primates* 39(4):449-457.

Youlatos, D. 1999a. Positional behavior of *Cebuella pygmaea* in Yasuni National Park, Ecuador. *Primates* 40(4):543-550.

Youlatos, D. 1999b. Tail-use in capuchin monkeys. *Neotropical Primates* 7(1):16-20.

Youlatos, D. 2002. Positional behavior of black spider monkeys (*Ateles paniscus*) in French Guiana. *International Journal of Primatology* 23(5):1071-1093.

Youlatos, D. and J. P. Gasc 1994. A preliminary study of head-first descent of lianas in the red howler monkey, *Alouatta seniculus*, in a primary rain forest of French Guiana. Pp. 203-210 in: B. Thierry, J. F. Anderson, J. J. Roeder and N. Herrenschmidt (eds.), *Current Primatology, vol. I: Ecology and Evolution*. University Louis Pasteur, Strasbourg.

Youlatos, D. and W. P. Rivera 1999. Preliminary observations on the songo songo (dusky titi monkey, *Callicebus moloch*) of northeastern Ecuador. *Neotropical Primates* 7(2):45-46.

Young, O. P. 1981a. Copulation-interrupting behavior between females within a howler monkey troop. *Primates* 22:135-136.

Young, O. P. 1981b. Chasing behavior between males within a howler monkey troop. *Primates* 22:424-426.

Young, O. P. 1982a. Tree-rubbing behavior of a solitary male howler monkey. *Primates* 23:303-306.

Young, O. P. 1982b. Aggressive interaction between howler monkeys and turkey vultures: The need to thermoregulate behaviorally. *Biotropica* 14:228-231.

Young, O. P. 1983. An example of apparent dominance-submission behavior between adult male howler monkeys (*Alouatta palliata*). *Primates* 24(2):283-287.

Young-Owl, M., R. H. Johnson, A. L. Jones, R. Namiki et al. 1992. A preliminary survey of primates at the Los Cedros Biological Reserve, Ecuador [Abstract] *American Journal of Primatology* 27(1):65.

Yumoto, T., K. Kimura and A. Nishimura 1999. Estimation of the retention times and distances of seed dispersed by two monkey species, *Alouatta seniculus* and *Lagothrix lagotriocha*, in a Colombian forest. *Ecological Research* 14(2):179-191.

Zhang, S. Y. 1995. Activity and ranging patterns in relation to fruit utilization by brown capuchins (*Cebus apella*) in French Guiana. *International Journal of Primatology* 16(3):489-507.

Zhang, S. and L. Wang 1995. Fruit consumption and seed dispersal of *Ziziphus cinnamomum* (Rhamnaceae) by two sympatric primates (*Cebus apella* and *Ateles paniscus*) in French Guiana. *Biotropica* 27(3):397-401.

Zingg, J. and R. D. Martin 2001. Temporal pattern of exudates feeding in pygmy marmosets (*Cebuella pygmaea*) in Ecuador. [abstract] *Folia Primatologica* 72(3):193.

Zucker, E. L. and M. R. Clarke 1992. Developmental and comparative aspects of social play of mantled howling monkeys in Costa Rica. *Behaviour* 123(1-2):144-171.

Zucker, E. L. and M. R. Clarke 1998. Agonistic and affiliative relationships of adult female howlers (*Alouatta palliata*) in Costa Rica over a 4-year period. *International Journal of Primatology* 19(3):433-449.

Zucker, E. L. and M. R. Clarke 2003. Longitudinal assessment of immature to adult ratios in two groups of Costa Rican *Alouatta palliata*. *International Journal of Primatology* 24(1):87-101.

Index

This edition was printed
at Panamericana Formas e Impresos, S.A.
August, 2004
Bogotá D.C, Colombia.